Chemistry

Providing a holistic overview of general chemistry and its foundational principles, this textbook is an essential accompaniment to students entering the field. It is designed with the reader in mind, presenting the historical development of ideas to frame and center new concepts as well as providing primary and summative sources for all topics covered. These sources help to provide definitive information for the reader, ensuring that all information is peer-reviewed and thoroughly tested.

Features:

- The development of key ideas is presented in their historical context.

- All information presented is supported through citations to chemical literature.

- Problems are incorporated throughout the text and full, worked-out solutions are presented for every problem.

- International Union of Pure and Applied Chemistry style and technical guidelines are followed throughout the text.

- The problems, text, and presentation are based on years of classroom refinement of teaching pedagogy.

This textbook is aimed at an advanced high school or general college audience, aiming to engage students more directly in the work of chemistry.

William B. Tucker's passion for chemistry was inspired by his high school teacher Gary Osborn. He left Maine to pursue Chemistry at Middlebury College, and after graduating in 2010 he decided to pursue a Ph.D. in Organic Chemistry at the University of Wisconsin-Madison. At the University of Wisconsin-Madison, he worked in the laboratory of Dr. Sandro Mecozzi, where he developed semifluorinated triphilic surfactants for hydrophobic drug delivery. After earning his Ph.D. in 2015, he took a fellowship at Boston University as a Postdoctoral Faculty Fellow. There he co-taught organic chemistry while working in the laboratory of Dr. John Caradonna. In the Caradonna laboratory, he worked on developing a surface-immobilized iron-oxidation catalyst for the oxidation of C–H bonds using dioxygen from the air as the terminal oxidant. Throughout all of this work, his passion has always been for teaching and working with students both in and out of the classroom. He has been lucky for the past six years to work at Concord Academy, where his students have, through their questions, pushed him to think deeper and more critically about chemistry. Their curiosity inspires him, and their inquisitiveness inspired his writing.

Chemistry

Energy, Matter, and Change

William B. Tucker

CRC Press
Taylor & Francis Group
Boca Raton London New York

CRC Press is an imprint of the
Taylor & Francis Group, an **informa** business

Open-access funding provided by the Concord Academy Class of 1972 Green Seed Fund.

First edition published 2024
by CRC Press
2385 NW Executive Center Drive, Suite 320, Boca Raton FL 33431

and by CRC Press
4 Park Square, Milton Park, Abingdon, Oxon, OX14 4RN

CRC Press is an imprint of Taylor & Francis Group, LLC

ISBN: 9781032766324 (hbk)
ISBN: 9781032766287 (pbk)
ISBN: 9781003479338 (ebk)

DOI: 10.1201/9781003479338

Typeset in Palatino
by Deanta Global Publishing Services, Chennai, India

This book is dedicated to

Gary Osborn

Beloved Winslow High School (Winslow, ME) chemistry teacher

and inspiration for my love of and interest in chemistry.

Contents

Introduction viii

Attribution. ix

Acknowledgments x

Creative Commons License xi

1 Introduction: Inclusion, Sustainability, Conventions, and Writing. 1

2 Energy 9

3 Atoms. 15

4 Binding Energy, Nuclear Stability, and Decay 29

5 Nuclear Reactions 44

6 Light and Electrons 56

7 Electrons and Quantum Numbers 63

8 Periodic Trends. 76

9 Bonding. 90

10 Molecular Shapes. 113

11 Gases 119

12 The van der Waals Equation and Intermolecular Forces. 128

13 States, Phases, and Physical Changes. 141

14 Chemical Kinetics 146

15 Enthalpy 159

16 Entropy 167

17 Gibbs Energy 170

18 Equilibrium. 177

19 Electron Transfer 192

20 Electron Sharing. 206

Appendices. 223

Answers 273

Index. 359

Introduction

This project is the culmination of my educational learning over the past decade. In this, I am immensely indebted to my students, particularly those of Concord Academy, for always asking great questions and pushing me to think more deeply about chemistry. This book is intended for any student who has had a general introduction to some of the basics of chemistry, specifically dimensional analysis, the basics of atomic structure and bonding, and an introductory coverage of reactions. Within this text, those concepts will be cursorily reintroduced before a deeper treatment is presented. Note that while the objective is to provide students with an answer to the questions of how and why, each of the chapters in this text could and often do constitute advanced courses in and of themselves. As such, for a student who is interested in pursuing chemistry, each of these topics will be more fully addressed in future courses you will take.

This book is designed to be more engaging than a typical textbook with all problems embedded within the document itself rather than coming at the end of each chapter. These problems come from discussion worksheets, homework problems, and exam problems that I have used. The solutions are worked out in the "Answers" section at the back of the book. Together, these active pieces of the book constitute roughly one-third of the text's total page count.

The traditional chemistry content is supplemented with a discussion of diversity, equity, inclusion, and belonging (DEIB) and the incorporation of both computational chemistry and green chemistry. Finally, to provide the most definitive information for the reader, all the original references (or summative references when an original reference could not be found) for the chemistry and ideas presented are cited with footnotes. These references are also meant to provide an interested reader with a point of entry into understanding the origins of these ideas, though I should caution that a working knowledge of German, French, and English is necessary to read all the references presented. Formatting, style, vocabulary, and symbology are all presented as recommended by the International Union of Pure and Applied Chemistry (IUPAC). I hope you find this supportive and helpful in your efforts to learn chemistry.

Attribution

All chemical structures were drawn using ChemDraw 22.2 (RRID:SCR_016768). Registered Trademark of PerkinElmer Informatics. https://perkinelmerinformatics.com/products/research/chemdraw (Accessed 04/30/23).

All computationally derived models and energy values were created and calculated using the platform WebMO: Schmidt, J.R. and Polik, W.F. WebMO, version 20.0, WebMO LLC: Holland, MI, 20; https://www.webmo.net (Accessed 04/30/23).

Calculations at the B3LYP/6-31G(d) and CCSD(T)/cc-pVTZ level of theory and basis set were performed using Gaussian 09: Gaussian 09 (RRID:SCR_014897), Revision A.02, M.J. Frisch, G.W. Trucks, H.B. Schlegel, G.E. Scuseria, M.A. Robb, J.R. Cheeseman, G. Scalmani, V. Barone, G.A. Petersson, H. Nakatsuji, X. Li, M. Caricato, A. Marenich, J. Bloino, B.G. Janesko, R. Gomperts, B. Mennucci, H.P. Hratchian, J.V. Ortiz, A.F. Izmaylov, J.L. Sonnenberg, D. Williams-Young, F. Ding, F. Lipparini, F. Egidi, J. Goings, B. Peng, A. Petrone, T. Henderson, D. Ranasinghe, V.G. Zakrzewski, J. Gao, N. Rega, G. Zheng, W. Liang, M. Hada, M. Ehara, K. Toyota, R. Fukuda, J. Hasegawa, M. Ishida, T. Nakajima, Y. Honda, O. Kitao, H. Nakai, T. Vreven, K. Throssell, J.A. Montgomery, Jr., J.E. Peralta, F. Ogliaro, M. Bearpark, J.J. Heyd, E. Brothers, K.N. Kudin, V.N. Staroverov, T. Keith, R. Kobayashi, J. Normand, K. Raghavachari, A. Rendell, J.C. Burant, S.S. Iyengar, J. Tomasi, M. Cossi, J.M. Millam, M. Klene, C. Adamo, R. Cammi, J.W. Ochterski, R.L. Martin, K. Morokuma, O. Farkas, J.B. Foresman, and D.J. Fox, Gaussian, Inc., Wallingford, CT, 2016.

Visualizations of orbitals and electrostatic potential maps were produced using NBO 7: E.D. Glendening, J.K. Badenhoop, A.E. Reed, J.E. Carpenter, J.A. Bohmann, C.M. Morales, P. Karafiloglou, C.R. Landis, and F. Weinhold, Theoretical Chemistry Institute, University of Wisconsin, Madison, WI, 2018.

Acknowledgments

I want to thank the following people for their thoughtful review of this work, their feedback, their criticism, and their assistance that has helped to make this work better.

Max Hall, *Concord Academy, Editor*
Mary Katherine McElroy, *Middlebury College '10, Editor*
Susan Flink, *Concord Academy, Editor*
Jake Klineman, *Concord Academy, Editor*
Chris Labosier, *Concord Academy*
James McNeely, *Boston University*
Brian Esselman, *University of Wisconsin-Madison*
Jeff Byers, *Middlebury College*
Edoardo Takacs, *Concord Academy*
Aidan Quealy, *Concord Academy*
Isabella Ginsburg, *Concord Academy*
Anghelo Chavira, *Concord Academy*
Chiara Wanandi, *Concord Academy*
Mei Reed, *Concord Academy*
Bobby Skrivanek, *Concord Academy*
Nick Brady, *Concord Academy*
Elspeth Yeh, *Concord Academy*
Axel Bostrom, *Concord Academy*

Creative Commons License

1 Introduction: Inclusion, Sustainability, Conventions, and Writing

Chemistry is a branch of science that focuses on the study of matter and change. This is the traditional definition of chemistry, but it is also not necessarily how an average chemist thinks about their field. When a group of working chemists provided their own definitions of chemistry, the answers included: "everything between physics and biology,"[1] "the central science that studies atoms and everything involving them,"[2] and "the science of the transformation of matter."[3] As you read this text, the author encourages you to develop your own working definition of chemistry.

INCLUSION

Within this book attempts have been made to provide original references for key ideas and concepts. If you were to research the cited individuals further, there would be a stark trend: they are almost all white, they are almost all European or American, and they are almost all men. Almost all came from wealthy backgrounds, and all were, so far as we know, cisgender and heterosexual. Chemistry, as we know it today and as presented in this text, was established by and for these people. The artifacts of this distortion, a human endeavor that was developed by only a small subsection of humanity, are seen through to the present day: racism and sexism in chemistry and science broadly, the underrepresentation of Black, Pacific Islander, and Latine chemists as degree recipients and in chemistry jobs (researchers, professors, and teachers), and the lower likelihood of non-white, non-male applicants receiving grants. What created this homogeneous history of chemistry? In the 19th and early 20th centuries, only people who were wealthy, white, and male were afforded the privilege of pursuing science. And so, the system has historically been built for and by these people. With the rise of funding from the federal government, part of the Cold War, the study of chemistry started to become accessible to those who were from lower socioeconomic classes. And with the work of countless activists in the Civil Rights, Women's Liberation, and LGBTQIA Liberation movements, the study of chemistry started to become accessible to those who were non-white, non-male, non-cisgender, and non-heterosexual. With these changes, chemistry has become more diverse, but it is still harder for some people than it is for others to progress through the system. What is important as you read this text is to consider the folks – non-white, non-male, non-cisgender, and non-heterosexual – who are not cited or recognized in chemical history. These people have always been a part of chemistry and they have done significant research in the field. As chemistry slowly becomes more inclusive, it is incumbent upon us to both recognize those folks omitted by chemistry's past and to help expand and make space for those folks who can and should be part of chemistry's present and future,[4] because to not do so is actively harmful.[5] As recently underscored by Dr. H. Holden Thorp, editor-in-chief of *Science*:

> A monolithic group of scientists will bring many of the same preconceived notions to their work. But a group of many backgrounds will bring different points of view that decrease the chance that one prevailing set of views will bias the outcome. This means that the scientific consensus can be reached faster and with greater reliability. It also means that the applications and implications will be more just for all.[6]

SUSTAINABILITY AND GREEN CHEMISTRY

An important aspect of modern chemistry to highlight here is chemical sustainability and green chemistry. To think about sustainability in chemistry, let us first consider chemistry in its fullest context. Chemistry uses materials and energy to produce new materials, all of which ultimately need to be disposed of and, ideally, recycled. Production and consumption are in connection to the Earth's land, air, and water. In addition, production and consumption are in service of human society, which is sustained by and is responsible for the stewardship of Earth. Sustainability is an emergent property of these production–consumption, biosphere, and human systems. And so, making chemistry sustainable should not be thought of as a single change to a single process, but rather sustainable chemistry is a systemic and holistic approach to the work and processes in chemistry, in society, and in our interactions with the planet.[7]

While sustainability is a systemic concern, green chemistry is a focus on making individual aspects of chemical processes more sustainable. A guiding framework for helping to analyze individual processes is the 12 Principles of Green Chemistry.[8] The 12 principles aim to reduce or eliminate the use or generation of hazardous substances.

DOI: 10.1201/9781003479338-1

1. Prevention: Not making waste is a better approach than cleaning it up.

2. Atom economy: In a chemical process, what atoms are incorporated into the final material, and what atoms are wasted?

3. Less hazardous chemical syntheses: If possible, and practical, the (by)products of chemical reactions should have low toxicity to humans and the environment.

4. Designing safer chemicals: Chemicals should be developed that accomplish desired goals and minimize hazards to human and ecological health.

5. Safer solvents and auxiliaries: If they cannot be avoided altogether, then chemicals that have no to minimal impact on human and environmental health should be used for running reactions and purifying products.

6. Design for energy efficiency: When practicable, reactions should be carried out without heat or added pressure, and if either heat or pressure is needed, the energy should be considered in terms of its environmental and economic impacts.

7. Use of renewable feedstocks: If possible, and practical, starting materials should come from renewable sources.

8. Reduce derivatives: Nonproductive synthetic steps should be avoided or minimized to reduce waste.

9. Catalysis: Chemicals that increase the rate of reaction without modifying the overall change in reaction energy should be used to increase yield and to maximize energy efficiency.

10. Design for degradation: Products should be designed so that when disposed of they do not persist in the environment.

11. Real-time analysis for pollution prevention: Tools should be developed and used to monitor processes so that the formation of hazardous or toxic substances can be prevented.

12. Inherently safer chemistry for accident prevention: Improvements in safety through the reduction of hazards for researchers, technicians, and workers should be prioritized.

Through to the present day, chemistry is a materially intensive science, which has often overlooked its impact on the Earth. This has led to environmental degradation and pollution in the form of contamination of soils and waterways by perfluorinated chemicals (so-called forever chemicals), lead as a component of pipes and paints and fuels, depletion of the ozone layer with chlorofluorocarbons (CFCs), and plastic packaging and waste. While chemistry has created these problems, chemistry is also the solution to its own problems and those of human society writ large.

In this book, green chemistry will be highlighted when it intersects with the material of a chapter. In addition, you are encouraged to keep the 12 Principles of Green Chemistry in mind as you read through the book. Sections corresponding to the 12 Principles of Green Chemistry will be identified with the symbol ⤳. Advancements in green chemistry and improvements to chemistry's sustainability will require not only applying existing knowledge to current practices, but it will also require new innovations. As learners of chemistry, you are uniquely positioned to offer outside perspectives to those who "know" what can and cannot be done.

CLARITY

Clarity, by which we mean unambiguousness, is key in most areas of life; however, it is of paramount importance in science. Whether discussing feet or meters, 0.1 g or 0.01 g, or weight or mass, it is vital that the speaker (or writer) can convey the correct information to the audience.[9] In the remainder of this chapter, we will consider measurement, significant figures, and writing. These are hopefully a review for the reader but are presented as the first chapter to ensure a common and strong foundation that is necessary both for success in working through this book and in the course this book might accompany.

MEASUREMENT

Chemistry is unique (among the traditional triad of biology, chemistry, and physics) in that the objects of our investigation cannot be directly observed. While there are now tools that allow us to visualize atoms and molecules, their direct observation is obscured by the significant difference

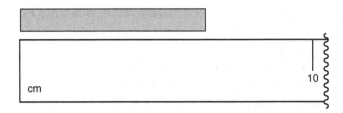

Figure 1.1 Rectangular object with a ruler that denotes every 10 cm.

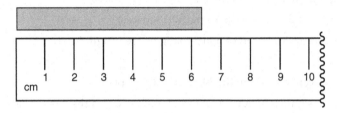

Figure 1.2 Rectangular object with a ruler that denotes every 1 cm.

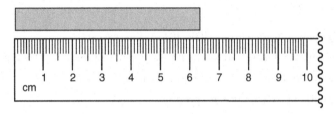

Figure 1.3 Rectangular object with a ruler that denotes every 0.1 cm.

in scale between ourselves and atoms and molecules. As such, chemistry is intensely dependent upon proxies, that is, those things that we can observe and measure, from which we can infer an understanding of the atoms and molecules that we are interested in.

Measurement involves the use of a tool or device, from which we can obtain a numerical value that somehow quantifies our object. While digital tools are now common, they are not universal, and they obscure some of how measurement works. For our discussion of measurement, consider the object and ruler in Figure 1.1. The object is less than 10 cm – how much less is up to our estimation – and one could argue that it is 6 cm. That is, we can estimate one more place (here the 1s) than our nondigital tool shows (here the 10s).

Now consider the same object but with a better ruler in Figure 1.2. Our ruler now denotes every 1 cm. We can see that the object is more than 6 cm but less than 7 cm, and so a reasonable measurement would be 6.3 cm (any value between 6.1 cm and 6.4 cm would all be reasonable).

Finally, consider the object and an even better ruler in Figure 1.3. The object is between 6.3 cm and 6.4 cm, but again where in between is up to the reader. Any value between 6.31 cm and 6.35 cm is plausible. To the author it looks like 6.33 cm.

In each case, our measurement includes some amount of estimation in the final digit, which means there is some amount of uncertainty in our measurements and the values we report. For a digital tool, it feels as though we avoid the whole problem of estimation and uncertainty, but the tool is still reporting a value that involves uncertainty and this is stated by the manufacturer. The next time you are in a laboratory, you should ask the instructor for the manufacturer specifications on any digital instruments you are using to see what is the reported precision of the device.

SIGNIFICANT FIGURES

Since any measurement produces an inexact number (this contrasts with counted objects where there is no ambiguity when you say "seven people are in the room"), we must take that into account when we do math with measured values.[10] First, when we consider a measured value, we must determine how many of the digits are significant. It should be noted that the term "significant" should be thought of as the reasonableness of the number based on the tool used. For example, 6 cm is a reasonable measurement of the length of the unknown in Figure 1.1 based on the ruler provided.

While we must be careful with our own measurements, when we consider other people's measurements, we need a method of identifying which digits are significant and which are not. There are several conventions (rules) established to determine this.

Rules for determining the number and places of significance:

1) All nonzero digits are significant.

   Examples:

   632 has three significant figures (in the 100s, 10s, and 1s places)

   2.5 has two significant figures (in the 1s and 10ths places)

2) All leading zeros (zeros to the left of nonzero digits) are insignificant.

   Examples:

   0.54 has two significant figures (in the 10ths and 100ths places)

   0.000 6636 has four significant figures (in the 10 000ths, 100 000ths, 1 000 000ths, and 10 000 000ths places)

3) All sandwiched zeros (zeros between nonzero digits) are significant.

   Examples:

   101 has three significant figures (in the 100s, 10s, and 1s places)

   1.0005 has five significant figures (in the 1s, 10ths, 100ths, 1000ths, and 10 000ths places)

4) Trailing zeros (zeros to the right of nonzero digits) are:

   a. Insignificant without a decimal point present.

      Examples:

      10 has one significant figure (in the 10s place)

      23 000 has two significant figures (in the 10 000s and 1000s places)

   b. Significant with a decimal point present.

      Examples:

      10. has two significant figures (in the 10s and 1s places)

      10.0 has three significant figures (in the 10s, 1s, and 10ths places)

      0.050 50 has four significant figures (in the 100ths, 1000ths, 10 000ths, and 100 000ths places)

Given the limitations of inexact, measured values, it is important that when we manipulate these numbers mathematically, we report results that are consistent with the precision of the tools used. The following conventions ensure that our results maintain the significance of any measured values and that we are not reporting insignificant digits (which would be comparable to reporting a value of 6.5 cm with the ruler in Figure 1.1).

In multiplication and division (including square root functions), the final number of significant figures is determined by the number that has the fewest significant figures.
$3.2 \times 2.12 \times 4.422 = 30.$
(3.2 has the fewest number of significant figures, two, and so the final answer is rounded to two significant figures.)
$1595/22.4 = 71.2$
(22.4 has the fewest number of significant figures, three, and so the final answer is rounded to three significant figures.)

In addition and subtraction, the final answer is determined by rounding to the same place (tens, ones, tenths, hundredths, etc.) as the least precise number, e.g.:

3.20
2.12
+ 4.425

9.75 (rounded to the 100ths because the least precise numbers 3.20 and 2.12 have no significant figures in the 1000ths place)

23 000
20
− 42.22

23 000 (the least precise number (23 000) has no places of significance less than the 1000s place and so the final answer is rounded to the 1000s place)

In logarithmic calculations, the final answer is rounded to have the same number of decimal places as the number of significant figures in the initial value.

$\ln(0.000\ 045\ 32) = -10.\underline{0018}$
$\log(\underline{4.1}) = 0.\underline{61}$
$\log(\underline{3.532} \times 10^6) = 6.\underline{5480}$

In exponential calculations, the final answer is rounded to have the same number of significant figures as the number of decimal places in the initial value.

$e^{4.132} = 62.3$
$10^{1.5} = 30$
$10^{1.50} = 32$

UNITS AND PREFIXES

On the topic of measurements, one of the single most important requirements for clarity is that *every* number has an appropriate unit label (Table 1.1). In the absence of a unit label, there can be no clarity and significant confusion can result. There is also, always, a space between the number and the label. This is for two purposes. First, when we write 6 m, that is a translation of the words "six meters." Writing 6m would imply the word "sixmeters." Second, 6 m is a measurement with an appropriate unit label, while $6m$ is a coefficient and a variable (variables are always italicized). One of the most frequent sources of confusion among students is whether something is a measurement and unit label or a coefficient and variable. The space between the number and label (or lack thereof) and the typeface (roman or italicized) are key identifiers as to the type of number you are looking at.

Chemistry, and science generally, uses the International System of Units (SI) for measurement. This system originated during the French Revolution (1789) as a more rational system of units than previous customary measurement systems. Today the system is used as the fundamental system of measurement in science, and all other measurement systems, including the U.S. customary

Table 1.1 List of SI Standard Units, Unit Labels, and Related Quantity and Symbol

Unit (Unit Label)	Unit of (Unit Symbol)
meter (m)	length (l)
cubic meter (m³)	volume (V)
kilogram (kg)	mass (m)
mole (mol)	amount (n)
second (s)	time (t)
kelvin (K)	temperature (T)
joule (J)	energy (E)
pascal (Pa)	pressure (p)
coulomb (C)	charge (Q)
volt (V)	electric potential (φ)
newton (N)	force (F)

Table 1.2   SI Prefixes, Orders of Magnitude, and Symbols

Prefix	Order of Magnitude	Symbol
giga	$\times 10^9$	G
mega	$\times 10^6$	M
kilo	$\times 10^3$	k
hecto	$\times 10^2$	h
deca	$\times 10^1$	da
—	$\times 10^0$	—
deci	$\times 10^{-1}$	d
centi	$\times 10^{-2}$	c
milli	$\times 10^{-3}$	m
micro	$\times 10^{-6}$	μ
nano	$\times 10^{-9}$	n

system, are based on the SI base units. The standard SI units, unit labels, and their associated quantities are shown in Table 1.1.

Finally, in terms of measurement, there are the SI prefixes. The most used prefixes in chemistry are those shown in Table 1.2. While one may be familiar with these as a reader, what is important to stress here is that the SI prefixes imply a scientific notation. That is, 500 nm corresponds to 500 × 10^{-9} m. The SI prefixes, their symbols, and implied scientific notation are assumed knowledge for readers of this book.

WRITING

In this chapter on clarity, it is also important to address the issue of writing. Good analytical writing necessitates several things. First, analytical, or argumentative, writing should always include three pieces: claim, evidence, and reasoning (CER). A claim is any inference that you draw based on your analysis of the information or data at hand. The evidence is the information from the problem or data from empirical observation or an experiment that you are basing that claim on (you need to clearly highlight this evidence to the audience because they may not see or notice the same evidence). Finally, the audience needs to see your reasoning, which is the logical connection that you have made to move from the evidence to the claim you are stating.

Second, good writing, of any kind, needs to follow the rules of prescriptive English grammar. While prescriptive English grammar is not how people usually talk, it is an established convention for ensuring that the grammatical construction of your writing does not stand in the way of your audience's understanding.[11] Some of the more salient points of prescriptive English grammar are:[12]

- Sentences end in periods, and punctuation is appropriately used to avoid run-on sentences, comma splices, or sentence fragments.

- Verbs end in -ed and -s when necessary, and appropriate helping verbs are employed.

- Verb tense shifts are avoided.

- The correct pronoun case (I/me/my or who/whom/whose, for example) is used.

- The pronoun agrees with the antecedent. The antecedent is the word the pronoun refers to.

- The plural -s/-es and possessive -'s endings are used appropriately.

- Quotation marks are only employed when citing others' work or ideas. In scientific writing, we rarely use quotation marks to cite other's ideas. But to ensure appropriate credit to the other, ideas are adequately referenced and noted with footnotes or endnotes.

- Material taken from another source is always referenced following a standard format, for example, APA, MLA, Chicago, Turabian, or ACS. ACS reference formatting is used in this book.

- Typographical errors – including misspellings – are avoided.

- There is proper sentence sense. That is, no words are omitted and ideas are not scrambled nor incomprehensible.

As a reader of this first chapter and the thoughts above on writing, you may have a question that is often received by the author. It is not a new question from folks studying chemistry, and a famous example of this question comes from Allen Hoos, an American high school student, who wrote to R.B. Woodward, one of the most famous chemists of the 20th century:[13]

```
I am a sophomore at Fairview High School and I have been study-
ing English since about the third grade. I plan to become a chemist
such as you are and I wondered why I have to take so much English. Is
English that important in the field of chemistry? It seems as though
I would be better off taking subjects that would have closer relation-
ships to chemistry. I would appreciate your opinions about English and
its usefulness, if any, in chemistry.
```

Woodward's response to Allen Hoos's query is a great summary of the importance of writing and knowledge of language to chemistry:[14]

```
It is a valuable asset to a chemist to be able to formulate [their]
ideas, describe [their] experiments, and express [their] conclusions
in clear, forceful English. Further, since thought necessarily involves
the use of words, thinking is more powerful, and its conclusions are
more valid, in the degree to which the thinker has a command of the
language.
```

NOTES

1. Dr. Joseph Moore.

2. Dr. Malika Jeffries-EL.

3. Dr. Stephen Matlin at the 2022 Biennial Conference on Chemistry Education.

4. (i) Reisman, S.E.; Sarpong, R.; Sigman, M.S.; Yoon, T.P. Organic Chemistry: A Call to Action for Diversity and Inclusion. *J. Org. Chem.*, **2020**, *85* (16), 10287–10292. DOI: 10.1021/acs.joc .0c01607.

   (ii) Sanford, M.S. Equity and Inclusion in the Chemical Sciences Requires Actions Not Just Words. *J. Am. Chem. Soc.*, **2020**, *142* (26), 11317–11318. DOI: 10.1021/jacs.0c06482.

   (iii) Ruck, R.T. and Faul, M.M. Update to Editorial "Gender Diversity in Process Chemistry." *Org. Process Res. Dev.*, **2021**, *25* (3), 349–353. DOI: 10.1021/acs.oprd.0c00471.

5. Dunn, A.L.; Decker, D.M.; Cartaya-Marin, C.P.; Cooley, J.; Finster, D.C.; Hunter, K.P.; Jacques, D.R.N.; Kimble-Hill, A.; Maclachlan, J.L.; Redden, P.; Sigmann, S.B.; Situma, C. Reducing Risk: Strategies to Advanced Laboratory Safety through Diversity, Equity, Inclusion, and Respect. *J. Am. Chem. Soc.*, **2023**, *145*, 21, 11468–11471. DOI: 10.1021/jacs.3c03627.

6. Dr. H. Holden Thorp, "It Matters Who Does Science." https://www.science.org/content/blog -post/it-matters-who-does-science (Accessed 05/13/23).

7. Matlin, S.; Mehta, G.; Cornell, S.E.; Krief, A.; Hopf, H. Chemistry and Pathways to Net Zero for Sustainability. *RSC Sustain.*, **2023**, *1*, 1704–1721. DOI: 10.1039/D3SU00125C.

8. Anastas, P.T.; Warner, J.C. *Green Chemistry: Theory and Practice.* Oxford University Press: Oxford, 1998.

9. For a clear example of what can happen when ambiguity comes into play, you should consider the debacle of the 1999 Mars Climate Orbiter.

10. (i) The Oxford English Dictionary (*Oxford English Dictionary*, s.v. "significant, adj., sense 2.b," July 2023. DOI: 10.1093/OED/1149173983) cites William Bedwell's 1614 work as the earliest English source to use the term. Bedwell, W. *De Numeris Geometricis*. 1614.

    (ii) Gauss (1809) gives an extensive treatment of the importance of precise and imprecise numbers and the impact on error in calculations: Gauss, C.F. *Theory of the Motion of the Heavenly Bodies Moving About the Sun in Conic Sections, a Translation of Gauss's "Theoria Motus."* Trans. Charles Henry Davis. Little, Brown and Company: Boston, 1857.

    (iii) The earliest complete outline of the rules and conventions presented here that the author could find come from: Holman, S.W. *Discussion of the Precision of Measurements with Examples Taken Mainly from Physics and Electrical Engineering*. Kegan Paul, Trench, Trübner, and Co., Ltd: London, 1892.

11. https://osuwritingcenter.okstate.edu/blog/2020/10/30/prescriptive-and-descriptive-grammar (Accessed 08/24/21).

12. https://www.ben.edu/college-of-liberal-arts/writing-program/upload/Policy-for-Use-of-Edited-Standard-Written-English.pdf (Accessed 08/24/21).

13. A. Hoos, letter to R. B. Woodward, Boulder, CO, October 23, 1961 (Harvard University. Records of Robert B. Woodward. HUGFP 68.8 Early subject files, 1931–1960 (bulk), 1931–1979 (inclusive), Box 8, in folder Correspondence—Requests for autographs, etc. [1960–1971]. Harvard University Archives). From: Seeman, J.I. Woodward's Words: Elegant and Commanding. *Angew. Chem. Int. Ed.*, **2016**, *55* (41), 12898–12912. DOI: 10.1002/anie.201600811.

14. R. B. Woodward, letter to A. Hoos, Boulder, CO, December 1, 1961 (Harvard University. Records of Robert B. Woodward. HUGFP 68.8 Early subject files, 1931–1960 (bulk), 1931–1979 (inclusive), Box 8, in folder Correspondence—Requests for autographs, etc. [1960–1971]. Harvard University Archives). From: Seeman, J.I. Woodward's Words: Elegant and Commanding. *Angew. Chem. Int. Ed.*, **2016**, *55* (41), 12898–12912. DOI: 10.1002/anie.201600811.

2 Energy

Energy is a central, unifying concept in science, and it will be a theme that will thread its way throughout the topics covered in chemistry and throughout this book. Given the importance of the topic, it is vital that we have a common language for and understanding of energy. Energy is defined as the ability to move an object (do work) or to produce heat,[1] and it comes in two forms: kinetic energy and potential energy. The SI unit for energy is joule (J), where one joule is one newton meter (N m) and defined, in SI base units, as m^2 kg s^{-2}. Other units, with which you may be familiar, are kilowatt-hour (kWh) or calorie (cal). For some atomic-centered topics, we will also see electronvolt (eV), where the energy is joule per elementary electric charge (J/e); the elemental electric charge (e) is 1.602 176 634 × 10^{-19} C.

KINETIC ENERGY (E_k)

Kinetic energy (E_k) is the energy of motion.[2] The amount of kinetic energy associated with a moving object can be calculated using Equation 2.1.

$$E_k = \frac{1}{2}mv^2 \tag{2.1}$$

E_k is the kinetic energy (J).
m is the mass of the object (kg).
v is the speed of the object (m/s).

PROBLEM 2.1
Calculate the kinetic energy (E_k) in each of the following examples.

a. A car (1628 kg) travels at 105 km/h.

b. A human (89 kg) walks at 4.8 km/h.

c. A proton (1.62 × 10^{-27} kg) moves at 3.00 × 10^6 m/s.

TEMPERATURE

We shall see, in a Chapter 11, that the kinetic energy of particles is an important quantity. For example, the average kinetic energy of particles in a gas is proportional to the temperature. Given this relationship, thermal energy is a type of kinetic energy. We measure thermal energy and changes in thermal energy (heat) with thermometers (Figure 2.1), which provide a macroscopic measure of average particle motion. The preferred thermodynamic temperature scale is the kelvin (K) scale,[3] which starts at absolute zero (0 K) where molecular motion, except vibrations, ceases. The most common laboratory thermometers record values in Celsius (°C), where the zero point (0 °C) is the freezing point of water. To convert between the two scales there is a straightforward relationship: K = °C + 273.15. Frequently, however, this conversion is unnecessary because what we will care about is the change in temperature (ΔT), and $\Delta T = 1$ °C is equivalent to $\Delta T = 1$ K because the magnitude of each °C is equal to the magnitude of each K.

POTENTIAL ENERGY (E_p)

Potential energy (E_p or U) is the energy of configuration.[4] Consider, for example, a rock on the ground at the bottom of a valley compared to a rock on the ground at the top of a cliff.

The rock at the bottom of a valley has low potential energy and if pushed nothing very interesting will happen. In contrast, the rock on the cliff has a lot of potential energy and if pushed (off the cliff) it will fall (turning the potential energy into kinetic energy). In this example, the difference is the configuration (or position) of the rock experiencing gravitational attraction to the Earth. Given our lived experience on Earth, we are familiar with the force due to gravity and how different configurations lead to different outcomes. The height of an object on Earth is only one type of potential energy (gravitational potential energy). Other types of potential energy, more relevant to chemistry, include the internal energy of a system (U), energy stored in a battery (electric potential energy, measured in volts [V], which is the energy [J] per charge [C]), energy stored in the arrangement of and interactions among atoms and molecules (chemical potential energy [μ] is the energy [J] per particle number [N]), and the energy stored in the arrangement of and interactions among nucleons (nuclear potential energy).

DOI: 10.1201/9781003479338-2

9

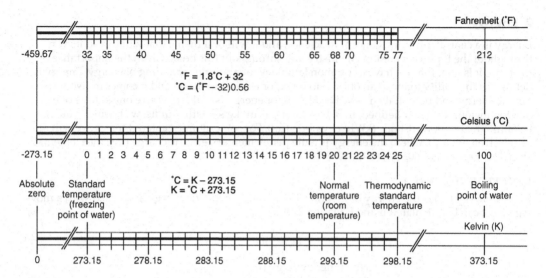

Figure 2.1 Comparison of Fahrenheit, Celsius, and Kelvin temperature scales and their inter-conversions. Standard temperature (ST) is defined by the International Union of Pure and Applied Chemistry (IUPAC) as 0 °C. The National Institutes of Standards and Technology (NIST) defines normal temperature (NT) as 20 °C and thermodynamic standard temperature as 25 °C.

SYSTEMS AND SURROUNDINGS

Potential energy is often represented with energy diagrams (Figure 2.2). Continuing with the analogy of a rock on a cliff versus a rock on a valley floor, one can see that the diagram has higher potential energy relative to the rock on a valley floor. Let's now consider in our analogy that the rock is moving from the cliff to the valley floor. For the rock on a cliff to reach the potential energy of the rock on a valley floor it would have to give off energy. In the language of thermodynamics, the rock is our system (our object of interest). Everything that is not the rock is the surroundings. And so, the rock (the system) is giving off potential energy (to the surroundings). That would mean that $\Delta U < 0$ and this is defined as exergonic (energy going out). A more familiar term might be exothermic, which is an exergonic change where the energy is lost only as heat.

For the rock on the valley floor, it has lower potential energy than the rock on the cliff. If we again consider the rock moving from the valley floor to the cliff, it will have to take in energy (someone carrying it up to the top of the cliff, for example). In the language of thermodynamics, the system is taking in energy from the surroundings. That would mean that $\Delta U > 0$ and this is

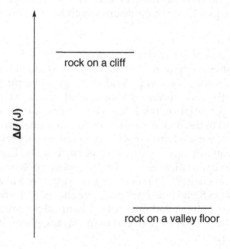

Figure 2.2 Potential energy diagram of a rock on a cliff versus a rock on a valley floor.

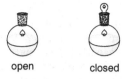

open   closed

Figure 2.3 An open system (left) and a closed system (right).

defined as endergonic (energy coming in). A more familiar term is endothermic, which is an endergonic change where the energy gained is only in the form of heat.

In the laboratory, the system is almost invariably a container (usually a flask) with chemicals inside it. There are several types of systems: open, closed, and isolated. In an open system, matter and energy can leave or enter the system, whereas in a closed system, matter cannot enter or leave the system but energy can leave or enter. If we think of a flask (Figure 2.3), the difference is whether it is an open container or a closed container. An isolated system is one where neither matter nor energy can enter or leave. A truly isolated system is hard to achieve, but we can approximate an isolated system by using insulation.

HEAT (q)

Heat (q) is the transfer of thermal energy between the system and its surroundings. Mathematically, we follow the same conventions as we saw above for ΔU. That is, for thermal energy transfer from the system to the surroundings (heat loss or exothermic processes), we define this as $q < 0$. Whereas for a thermal energy transfer from the surroundings to the system (heat gain or endothermic processes), we define this as $q > 0$. Heat can be quantified (Equation 2.2) using specific heat capacity (c), which is defined as the amount of energy (J) needed to raise 1 gram (g) of a substance by 1 kelvin (K).

$$q = mc_p\Delta T \tag{2.2}$$

q is heat (J).
m is mass (g).
c_p is specific heat capacity (J/(g K)).
ΔT is change (final less initial) in temperature (K).

Heat can also be calculated (Equation 2.3) using enthalpy (ΔH), which is the amount of heat (J) per mol for a given process. We will discuss enthalpy in detail in a future chapter. While Equation 2.2 is very useful for a single chemical increasing or decreasing in temperature, for the heat of a phase change, of mixing, or of a reaction, it becomes cumbersome because we can only indirectly measure the temperature change of these processes. Enthalpy provides a way of calculating the heat associated with these changes.

$$q = n\Delta H \tag{2.3}$$

q is heat (J).
n is the amount of substance (mol).
ΔH is the enthalpy (kJ/mol).

PROBLEM 2.2
For each of the following, calculate the thermal energy transfer (J) and identify whether each is endothermic or exothermic.

a. 155 kg water (the average mass of water in a bathtub) is heated from 25.0 °C to 38.5 °C, where $c_p(H_2O(l)) = 4.184$ J/(g K).

b. 18.0 g of water evaporates ($\Delta H = 43.9$ kJ/mol).

c. 3660 g aluminium is heated from 298.2 K to 311.7 K, where $c_p(Al(s)) = 0.89$ J/(g K).

d. 1.0 g hydrogen gas ($H_2(g)$) is cooled from 100.0 °C to –77.2 °C, where $c_p(H_2(g)) = 1.030$ J/(g K).

e. 1.5 g methane gas ($CH_4(g)$) combusts ($\Delta H = -890.4$ kJ/mol).

WORK (w)

If someone were to carry the rock (Figure 2.2) from the valley floor to the top of the cliff, the rock would gain potential energy, but its temperature would not change. To effect this change, a different type of energy is added to the system: work (w). Formally, work is the product of force (in newtons) and distance (in meters). In chemistry, we typically do not move our reactions all over the lab to try to input energy. We will be most concerned with pressure-volume (pV) work (Equation 2.4), which occurs when the volume of the system changes under some external pressure. Note that the magnitude of work is typically only significant when a gas is consumed or produced by a reaction.

$$w = -p\Delta V \tag{2.4}$$

w is work (J).
p is the external pressure (Pa).
ΔV is the change (final less initial) in volume (m^3).

The SI units for pressure and volume are pascal (Pa) and cubic meter (m^3), where 1 Pa m^3 = 1 J. In a laboratory setting, it is much more common to measure pressure in atmosphere (atm) and volume in liters (L). To convert the resulting L atm to J, one would need to use the conversion factor: 1 L atm = 101.325 J. The sign convention follows what we saw above. If $w < 0$, then the system is doing work on the surroundings (what can be called exoworkic). If $w > 0$, then the surroundings are doing work on the system (what can be called endoworkic).[5]

PROBLEM 2.3
For each of the following, calculate the work (J) and identify whether each is endoworkic or exoworkic.

a. A piston compresses from 1.5 L to 0.5 L with an external pressure of 1.05 atm.

b. A sample of 18.00 mL liquid water evaporates and becomes 31 L of steam with an external pressure of 0.985 atm.

c. If $UF_6(s)$ has a density of 5.09 g/mL and $UF_6(g)$ at 51.7 °C has a molar volume of 26.7 L/mol, determine the work (at 1 atm) associated with 200 g of $UF_6(s)$ converting to $UF_6(g)$.

⤳ENERGY SOURCES IN CHEMISTRY

Heat and work are two important means of manipulating the energy of a system. When thinking about a reaction, it is important to design for energy efficiency and think about the sources of energy that generate the heat or cause the pressure changes. Ideally, a reaction would utilize neither added heat nor pressure. To avoid using added external energy (heat and pressure), chemical energy (Chapter 17) can be used by coupling reactions together.

Increasingly, chemists are also finding ways to directly use electric potential (Chapter 19) to manipulate chemical syntheses and using light (Chapter 6) as a means of controlling reactions. In situations where added heat cannot be avoided – whether to speed up the formation of product (Chapter 14) or to ensure meaningful amounts of product are produced (Chapter 18) – the amount of heat required can be reduced by using a catalyst (Chapter 14), which is a chemical that increases the rate of reaction without modifying the overall change in reaction energy.

Whether the energy used to control and effect chemical reactions is in the form of heat, work, electric potential, or light, it has historically come from the combustion of hydrocarbons (fossil fuels), which has contributed to global warming through the production of carbon dioxide. As the electric grid is decarbonized, this energy is increasingly coming from renewable energy sources (e.g., solar [Chapter 6] and wind), from nuclear fission, and potentially from nuclear fusion (Chapter 5).

THE FIRST LAW OF THERMODYNAMICS

The change in potential energy of a closed system (ΔU) can only be affected through heat (q) or work (w). This relationship is the first law of thermodynamics (Equation 2.5):[6]

$$\Delta U = q + w \tag{2.5}$$

ΔU is the change in internal/potential energy (J).
q is heat (J).
w is work (J).

PROBLEM 2.4

Answer the following problems where a sample of gas is contained in a cylinder-and-piston arrangement. It undergoes the change in state shown in the diagram below:

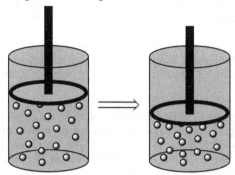

a. Assume that the cylinder and piston are perfect insulators and they do not allow thermal energy to enter or leave the enclosed volume. What is the value for q (<0, 0, or >0) for this change? What is the value for w (<0, 0, or >0) for this change? What is the value for ΔU (<0, 0, or >0) for this change?

b. Assume that the cylinder and piston are conductors and during this state change the cylinder becomes warm to the touch. What is the value for q (<0, 0, or >0) for this change? What is the value for w (<0, 0, or >0) for this change? What is the value for ΔU (<0, 0, or >0) for this change?

When carrying out calculations involving the first law of thermodynamics we can expand Equation 2.5 to show explicitly how we determine the values of heat with specific heat capacity (Equation 2.6) or enthalpy (Equation 2.7) and work.

$$\Delta U = mc_p\Delta T - p\Delta V \tag{2.6}$$

ΔU is the change in internal/potential/system energy (J).
m is mass (g).
c_p is the specific heat capacity (J/(g K)).
ΔT is the change (final less initial) in temperature (K).
p is pressure (Pa).
ΔV is the change (final less initial) in volume (m³).

$$\Delta U = n\Delta H - p\Delta V \tag{2.7}$$

ΔU is the change in internal/potential energy (J).
n is the amount of substance (mol).
ΔH is enthalpy (J/mol).
p is pressure (Pa).
ΔV is the change (final less initial) in volume (m³).

PROBLEM 2.5

For each of the following scenarios, calculate the change in internal energy (ΔU). Specify whether the system is undergoing an endergonic or an exergonic process.

a. 18.0 g of water evaporates and expands from 18.0 mL to 31 L with an external pressure of 0.985 atm.

b. 10.0 g of sodium chloride dissolves in water ($\Delta H = 3.9$ kJ/mol) with negligible volume change at 1.00 atm.

c. 14.0 g nitrogen gas is cooled from 100.0 °C to –77.2 °C, where $c_p(N_2(g)) = 1.040$ J/(g K), and changes from 15.3 L to 8.1 L at a pressure of 1.0 atm.

d. 500.0 g of copper metal is heated from room temperature (25.0 °C) to being red hot (460 °C), where $c_p(Cu(s)) = 0.386$ J/(g K), with negligible change in volume at 1.10 atm.

NOTES

1. (i) William Rankin original proposed the definition of energy, which is only slightly modified in the IUPAC definition given in this text: Rankin, W.J.M. *Miscellaneous Scientific Papers Volume 2*. W.J. Millar (Ed.), Charles Griffin and Company: London, 1881.

   (ii) "Energy." IUPAC. *Compendium of Chemical Terminology*, 2nd ed. Compiled by A.D. McNaught and A. Wilkinson, Blackwell Scientific Publications, Oxford, 1997. Online version (2019–) created by S.J. Chalk. DOI: 10.1351/goldbook.E02101.

2. The idea that subsequently became our modern kinetic energy was first termed *vis viva* (Latin for "living force") by Gottfried Leibnitz (Leibnitz, G.G. Specimen dynamicum pro admirandis naturae legibus circa corporum vires & mutuas actiones detegendis & adsuas causas revocandis. *Nova Acta Erud.*, **1695**, 145–157.). The term kinetic energy was formally introduced by Lord Kelvin (William Thomson): Thomson, W.; Tait, P.G. *Treatise on Natural Philosophy Volume 1*, Clarendon Press: Oxford, 1867.

3. Thomson, W. XXXIX. On an Absolute Thermometric Scale Founded on Carnot's Theory of the Motive Power of Heat, and Calculated from Regnault's Observations. *Proc. Camb. Philos. Soc.*, **1848**, 100–106. DOI: 10.1017/CBO9780511996009.040.

4. Originally defined by Rankin as having the capacity to do work: Rankin, W.J.M. *Miscellaneous Scientific Papers Volume 2*. W.J. Millar (Ed.), Charles Griffin and Company: London, 1881.

5. Although there is no official term for the case of $w > 0$ or $w < 0$, this text will follow the convention proposed by Bhairav D. Joshi in: Joshi, B.D. The Sign of Work: Endoworkic and Exoworkic Processes. *J. Chem. Educ.*, **1983**, *60* (10), 895. DOI: 10.10.1021/ed060p895.

6. Clausius, R. Ueber die bewegende Kraft der Wärme und die Gesetze, welche sich daraus für die Wärmelehre selbst arbeiten lassen. *Ann. Phys.* (Leipzig), **1850**, *155* (4), 500–524. DOI: 10.1002/andp.18501550306.

3 Atoms

In Chapter 2, we discussed energy and its importance as a central concept in science. In this chapter, we will turn our attention to one of the core concepts of chemistry: matter. Matter is defined, generally, as anything that has mass and occupies space. In chemistry, we define matter according to atomic theory: matter is anything that is made up of atoms. It is important to note that an atomistic philosophy is present across cultures in the ancient world and is well documented in both Greek (following the philosophical tradition of Leucippus) and Indian (following the Nyāya-Vaiśeṣika philosophical tradition) history. The modern chemistry term "atom" comes from the Greek word *atomos* meaning "indivisible." In these atomic philosophies, the common tenet is that there are small indivisible and eternal particles that make up the material world. In this chapter, we will consider the development of empirical atomic theory, the discovery of subatomic particles, and the basic structure of atoms. To help frame the narrative and give appropriate credit, the chemists who were involved in this work are named in this chapter. The interested reader is invited to peruse the cited references and biographical information, but specific knowledge of these historical figures is not required for a practicing chemist.

DEVELOPMENT OF ATOMIC THEORY

One of the foundational concepts in modern chemistry is the law of conservation of mass.[1] As originally stated by Lavoisier, "in every operation an equal quantity of matter exists both before and after the operation."[2] At its foundation, the development and study of chemistry depends upon the conservation of mass because any increase or decrease in mass over the course of a reaction would make it nearly impossible to understand how or why the change being studied occurs. In addition, (⤳) the conservation of mass makes it incumbent upon chemists to economize their use of chemical matter and to prevent waste, because mass lost as waste can persist indefinitely without remediation or recycling. While the law of conservation of mass is of central importance, two equally important philosophical questions arise with the development of this law. First, why is mass conserved? Second, what does this law tell us about the nature of matter?

PROBLEM 3.1
Early experimentation (1894)[3] to determine the density of nitrogen gas found that nitrogen gas produced from chemical reactions was reproducibly 0.4% lighter than "atmospheric" nitrogen gas. Considering this difference, answer the following questions:

a. Why is the law of conservation of mass important in this example?

b. Why is precision (the number of significant figures in a measurement) important in this example?

c. Provide a testable hypothesis for what you think might be able to explain this difference in the density of chemically produced nitrogen gas and "atmospheric" nitrogen gas.

Another foundational concept in the development of chemistry and atomic theory is the law of constant composition, also known as the law of definite proportions.[4] According to the law of constant composition, when broken down into its constituent elements a pure substance will always give the same fraction of each element. For example, analysis of water will always show that it is 10% hydrogen and 90% oxygen by mass. This law subsequently became the basis for the definition of a compound; in contrast, the composition of mixtures can show continuous variation in the relative amounts of each component. It is important to note, then, that there are limited, fixed ways in which elements can combine to produce a compound.

Finally, let us consider the law of multiple proportions.[5] This law is best thought of in the context of the specific example of carbon and oxygen combining. Dalton reported, using the terminology of the time, that carbonic oxide consisted of 1.27 parts oxygen for every 1.00 parts carbon, and carbonic acid consisted of 2.57 parts oxygen for every 1.00 parts carbon. The law of multiple proportions states that if we compare those parts-of-oxygen values (1.27 and 2.57) they give a whole number ratio (1:2). The inference that Dalton draws in this example is that there is one oxygen atom for every carbon atom in carbonic oxide (now called carbon monoxide) and two oxygen atoms for every carbon atom in carbonic acid (now called carbon dioxide). These results are summarized in the law of multiple proportions: two elements (A and B) may combine in multiple different ways to create different compounds, and if the amount of A is fixed, then the ratio of the

DOI: 10.1201/9781003479338-3

mass of B in each compound will be a small, whole number ratio. This law was presented as part of Dalton's larger development of his empirical atomic theory.

Dalton's atomic theory provides a conceptual framework that can explain the law of conservation of mass, the law of constant proportions, and the law of multiple proportions. Key postulates of Dalton's atomic theory are:

1) Each element is composed of extremely small particles called atoms.

2) All atoms of a given element are identical to one another in mass and other properties, but the atoms of one element are different from the atoms of all other elements.

3) Atoms of an element are not changed into atoms of a different element by chemical reactions; atoms are neither created nor destroyed in chemical reactions.

4) Compounds are formed when atoms of more than one element combine; a given compound always has the same kind and relative number of atoms.

PROBLEM 3.2
Provide an explanation for how atomic theory supports and explains:

a. The law of conservation of mass

b. The law of constant proportions

c. The law of multiple proportions

THE PERIODIC TABLE

In addition to developing an empirically derived theory of atoms, Dalton also proposed the first table of relative atomic mass (A_r) values. These values were and are called atomic weight, though relative atomic mass is the preferred term. In Dalton's work, the atomic mass values were all relative to the atomic mass of hydrogen. Later, chemists determined atomic mass values by defining 1/16th the atomic mass of oxygen as one atomic mass unit (1 amu). Today the relative atomic mass values are all determined by defining $A_r(^{12}C)$ as *exactly* 12 unified atomic mass units (12 u) or, equivalently, as *exactly* 12 daltons (12 Da).

Relative atomic mass values provided chemists with a means of ordering the elements from hydrogen, the smallest, to uranium, the largest naturally occurring. Lists of elements, however, do not provide a framework for understanding why elements are, or are not, like one another. Early in the 19th century, scientists noticed patterns among the elements. Döbereiner noticed that there were triads of similar elements, with the middle element showing a weight that was the average of the two end elements, e.g., chlorine (35.470), bromine (78.383), and iodine (126.470).[6] And Newlands noticed that every eighth element showed similar properties. Newlands called this periodic repetition of properties the law of octaves.[7] This and other work showed that there was an underlying structure to how elements might be arranged, but these early works were hampered by imprecise and inaccurate relative atomic mass values.

The first table that gained widespread attention was put together by Dmitri Mendeleev (Figure 3.1).[8] Mendeleev constructed his periodic table to help organize the elements for a textbook he was writing. Working from a list of elements ordered by relative atomic mass, Mendeleev then sorted the elements based on similar properties and the similarities among the compounds they formed. For example, he grouped (horizontally) Li, Na, K, Rb, Cs, and Tl because they all formed oxides with the formula R_2O.

As his textbook was finalized and published, Mendeleev revised his periodic table (Figure 3.2) to show the common chemical formulae for each element in a group when it combines with oxygen or with hydrogen. The power of Mendeleev's periodic table was twofold. First, it successfully classified elements into groups with similar properties. Second, Mendeleev's insight was the possibility that there may have been yet-undiscovered elements (shown as question marks [?] in Figure 3.1 and as long dashes [—] in Figure 3.2).

A successful theory is one that can do two things. The first is to explain current information and the second is to provide testable hypotheses. Mendeleev made predictions for four of the blanks in his table: eka-boron (with relative atomic mass 44 and oxide formula R_2O_3), eka-aluminium (with relative atomic mass 68 and oxide formula R_2O_3), eka-silicon (with relative atomic mass 72 and oxide formula RO_2), and eka-manganese (with relative atomic mass 100 and oxide formula R_2O_7). Gallium was discovered four years later, which has a relative atomic mass of 69.723 and formed

```
                                        Ti=50      Zr=90      ?=180
                                        V=51       Nb=94      Ta=182
                                        Cr=52      Mo=96      W=186
                                        Mn=55      Rh=104.4   Pt=197.4
                                        Fe=56      Ru=104.4   Ir=198
                                        Ni=Co=59   Pd=106.6   Os=199
        H=1                             Cu=63.4    Ag=108     Hg=200
              Be=9.4    Mg=24           Zn=65.2    Cd=112
              B=11      Al=27.4         ?=68       Ur=116     Au=197?
              C=12      Si=28           ?=70       Sn=118
              N=14      P=31            As=75      Sb=122     Bi=210?
              O=16      S=32            Se=79.4    Te=128?
              F=19      Cl=35.5         Br=80      I=127
        Li=7  Na=23     K=39            Rb=85.4    Cs=133     Tl=204
                        Ca=40           Sr=87.6    Ba=137     Pb=207
                        ?=45            Ce=92
                        ?Er=56          La=94
                        ?Yt=60          Di=95
                        ?In=75.6        Th=118?
```

Figure 3.1 Mendeleev's first periodic table (1869).

Row	Group I — R_2O	Group II — RO	Group III — R_2O_3	Group IV RH$_4$ RO$_2$	Group V RH$_3$ R$_2$O$_5$	Group VI RH$_2$ RO$_3$	Group VII RH R$_2$O$_7$	Group VIII — RO$_4$
1	H=1							
2	Li=7	Be=9.4	B=11	C=12	N=14	O=16	F=19	
3	Na=23	Mg=24	Al=27.3	Si=28	P=31	S=32	Cl=35.5	
4	K=39	Ca=40	—=44	Ti=48	V=51	Cr=52	Mn=55	Fe=56, Co=59,
5	(Cu=63)	Zn=65	—=68	—=72	As=75	So=73	Br=80	Ni=59, Cu=63
6	Rb=85	Sr=87	?Yt=88	Zr=90	Nb=94	Mo=96	—=100	Ru=104, Rh=104,
7	(Ag=108)	Cd=112	In=113	Sn=118	Sb=122	Te=128	I=127	Pd=106, Ag=108
8	Cs=133	Ba=137	?Di=136	?Ce=140	—	—	—	
9	(—)	—	—	—	—	—	—	
10	—	—	?Er=178	?La=180	Ta=182	W=184	—	Os=195, Ir=197,
11	(Au=199)	Hg=200	Tl=204	Pb=207	Bi=208	—	—	Pt=198, Au=199
12	—	—	—	Th=231	—	U=240	—	— — — —

Figure 3.2 Mendeleev's revised periodic table (1871). (From Mendeleev, D. Osnovy Khimii [The Principles of Chemistry]. St. Petersburg, 1869–71.)

an oxide with a chemical formula of Ga_2O_3, which corresponds to the predicted eka-aluminium. The subsequent discoveries of scandium in 1879 (relative atomic mass 44.956, oxide formula Sc_2O_3), germanium in 1886 (relative atomic mass 72.630, oxide formula GeO_2), and technetium in 1937 (most stable isotopic mass 97.907, oxide formula Tc_2O_7) conformed with the expected properties of eka-boron, eka-silicon, and eka-manganese, respectively. Together, these discoveries helped to underscore and support Mendeleev's periodic law. It should be noted that based on relative atomic mass there were some uncertainties of order (nickel and cobalt) and some anomalies like tellurium and iodine. Their properties followed the periodic law, but tellurium (128) has a relative atomic mass greater than the relative atomic mass of iodine (127).

Following the work by van den Broek and Moseley, see below, it was discovered that the ordering principle of the periodic table was not relative atomic mass but nuclear charge (Z), which is now called atomic number. This change to atomic-number order resolved the ambiguity (nickel and cobalt) and apparent discrepancies (tellurium and iodine) in Mendeleev's periodic table. The arrangement of elements by atomic-number order also leads to the periodic table more familiar to us today, which was first presented by Deming in his 1923 textbook *General Chemistry* (Figure 3.3).[9]

Two sets of elements challenged Mendeleev's periodic system, which are the elements that look out of place in Deming's periodic table (Figure 3.3): the noble or inert gases and the rare earth elements. The noble gases were discovered by Strutt (Baron Rayleigh) and William Ramsay.[10]

17

0	IA R_2O	IIA RO	IIIA R_2O_3	IVB RO_2	VB R_2O_5	VIB RO_3	VIIB R_2O_7		IB R_2O	IIB RO	IIIB R_2O_3	IVA RO_2	VA R_2O_5	VIA RO_3	VIIA R_2O_7
								1 H 1.01							
2 He 4.00	3 Li 6.94	4 Gl 9.1	5 B 10.9									6 C 12.0	7 N 14.0	8 O 16.0	9 F 19.0
10 Ne 20.2	11 Na 23.0	12 Mg 24.3	13 Al 27.0									14 Si 28.1	15 P 31.0	16 S 32.1	17 Cl 35.5
Inert Gases	Light Metals			Transition Group Valence Variable						Heavy Metals			Non-Metals		
18 A 39.9	19 K 39.1	20 Ca 40.1	21 Sc 45.1	22 Ti 48.1	23 V 51.0	24 Cr 52.0	25 Mn 54.9	26 Fe 55.8	27 Co 59.0	28 Ni 58.7	29 Cu 63.6	30 Zn 65.4	31 Ga 70.1	32 Ge 72.5	33 As 75.0	34 Se 79.2	35 Br 79.9
36 Kr 82.9	37 Rb 85.5	38 Sr 87.6	39 Yt 89.3	40 Zr 90.6	41 Cb 93.1	42 Mo 96.0	43 ? 99	44 Ru 102	45 Rh 103	46 Pd 107	47 Ag 108	48 Cd 112	49 In 115	50 Sn 119	51 Sb 120.	52 Te 128	53 I 127
54 Xe 130.	55 Cs 133	56 Ba 137	57–72 Rare Earths 139–179	73 Ta 182	74 W 184	75 ? 188	76 Os 191	77 Ir 193	78 Pt 196	79 Au 197	80 Hg 201	81 Tl 204	82 Pb 207	83 Bi 209	84 Po 210.	85 ? 219
86 Nt 222	87 ? 225	88 Ra 226	89 Ac 230.	90 Th 232	91 Pa 234	92 U 238	The rare earth elements are: 57 La 58 Ce 59 Pr 60 Nd 61 Pm 62 Sm 63 Eu 64 Gd 65 Tb 66 Dy 67 Ho 68 Er 69 Tm 70 Yb 71 Lu 72 Hf								

Figure 3.3 A more modern periodic table made by Deming in 1923. Note several differences: First, the noble gases and rare earth elements are out of place relative to a modern table. Second, several elements have different symbols (and names) compared to today: Gl for glucinium (now beryllium, Be), A for argon (now Ar), Yt for yttrium (now Y), Cb for columbium (now niobium, Nb), Nt for niton (now radon, Rn).

Following Ramsay's suggestion, Mendeleev placed the noble gases into a new group, group zero, which did not disturb the basic structure of the table (Figure 3.3).[11]

The rare earth elements proved significantly harder to fit into Mendeleev's periodic table. While the number of rare earth elements was clear, they did not show a regular trend in their properties, and when combined with oxygen all formed compounds of the R_2O_3 type, so they all seemed to belong in one group (IIIA). To explain this, an asteroid hypothesis was invoked.[12] In this hypothesis, it is useful to think of the asteroid belt between Mars and Jupiter: There are many roughly equivalent bodies in the asteroid belt that are all in the same relative position compared to the planets. In the periodic table, barium and tantalum are the planets and the rare earth elements are the asteroid belt between the two. In this way, the rare earth elements could, as shown in Figure 3.3, be shown as part of and yet separate from the rest of the periodic table. It is worth highlighting that despite their divergent behavior from other elements in their group, actinium (Z = 89), thorium (Z = 90), protactinium (Z = 91), and uranium (Z = 92) are all shown as regular transition metal elements in pre-World War II periodic tables.

In 1945, Seaborg recognized that the actinoids (Z = 89–103) showed similar behavior to the lanthanoids (Z = 57–71) and that neither behaved like the transition metals. This led him to reorganize the periodic table to show the rare earth elements, or inner transition metals, as a new set of groups in the periodic table.[13] Most periodic tables follow Seaborg's template as shown in Figure 3.4. The current IUPAC Commission on Isotopic Abundances and Atomic Weights (CIAWW) recommended relative atomic mass value for each element, $A_r°(E)$, which can be found in Appendix 2.

While most periodic tables are formatted like Figure 3.4, a more comprehensive picture, showing the placement of the rare earth elements, would format the periodic table as in Figure 3.5. Due to the size constraints of formatting a 32-column periodic table, the 18-column periodic table (Figure 3.4) is often preferred for legibility.

It should be noted that different groups and blocks of the periodic table have different collective names (Figure 3.6). The main group elements are those in groups 1, 2, 13, 14, 15, 16, 17, and 18 (group 12 elements are sometimes included in the set of main group elements). The transition metals are those elements in groups 3–12 (group 12 elements are not always included in the transition metal elements). The lanthanoids and actinoids are collectively called inner transition metals and, along with the elements in group 3, rare earth metals.

1	2	3	4	5	6	7	8	9	10	11	12	13	14	15	16	17	18
1 H 1.01																	2 He 4.00
3 Li 6.94	4 Be 9.01											5 B 10.8	6 C 12.0	7 N 14.0	8 O 16.0	9 F 19.0	10 Ne 20.2
11 Na 23.0	12 Mg 24.3											13 Al 27.0	14 Si 28.1	15 P 31.0	16 S 32.1	17 Cl 35.5	18 Ar 40.0
19 K 39.1	20 Ca 40.1	21 Sc 45.0	22 Ti 47.9	23 V 50.9	24 Cr 52.0	25 Mn 54.9	26 Fe 55.8	27 Co 58.9	28 Ni 58.7	29 Cu 63.5	30 Zn 65.4	31 Ga 69.7	32 Ge 72.6	33 As 74.9	34 Se 79.0	35 Br 79.9	36 Kr 83.8
37 Rb 85.5	38 Sr 87.6	39 Y 88.9	40 Zr 91.2	41 Nb 92.9	42 Mo 96.0	43 Tc –	44 Ru 101	45 Rh 103	46 Pd 106	47 Ag 108	48 Cd 112	49 In 115	50 Sn 119	51 Sb 122	52 Te 128	53 I 127	54 Xe 131
55 Cs 133	56 Ba 137	*	72 Hf 178	73 Ta 181	74 W 184	75 Re 186	76 Os 190.	77 Ir 192	78 Pt 195	79 Au 197	80 Hg 201	81 Tl 204	82 Pb 207	83 Bi 209	84 Po –	85 At –	86 Rn –
87 Fr –	88 Ra –	**	104 Rf –	105 Db –	106 Sg –	107 Bh –	108 Hs –	109 Mt –	110 Ds –	111 Rg –	112 Cn –	113 Nh –	114 Fl –	115 Mc –	116 Lv –	117 Ts –	118 Og –

	57 La 139	58 Ce 140	59 Pr 141	60 Nd 144	61 Pm –	62 Sm 150.	63 Eu 152	64 Gd 157	65 Tb 159	66 Dy 163	67 Ho 165	68 Er 167	69 Tm 169	70 Yb 173	71 Lu 175
*															
**	89 Ac –	90 Th 232	91 Pa 231	92 U 238	93 Np –	94 Pu –	95 Am –	96 Cm –	97 Bk –	98 Cf –	99 Es –	100 Fm –	101 Md –	102 No –	103 Lr –

Figure 3.4 Modern periodic table with the recommended relative atomic mass value of each element (Appendix 2) rounded to three significant figures.

1	2	3	4	5	6	7	8	9	10	11	12	13	14	15	16	17	18
H																	He
Li	Be											B	C	N	O	F	Ne
Na	Mg											Al	Si	P	S	Cl	Ar
K	Ca	Sc	Ti	V	Cr	Mn	Fe	Co	Ni	Cu	Zn	Ga	Ge	As	Se	Br	Kr
Rb	Sr	Y	Zr	Nb	Mo	Tc	Ru	Rh	Pd	Ag	Cd	In	Sn	Sb	Te	I	Xe
Cs	Ba	La Ce Pr Nd Pm Sm Eu Gd Tb Dy Ho Er Tm Yb Lu	Hf	Ta	W	Re	Os	Ir	Pt	Au	Hg	Tl	Pb	Bi	Po	At	Rn
Fr	Ra	Ac Th Pa U Np Pu Am Cm Bk Cf Es Fm Md No Lr	Rf	Db	Sg	Bh	Hs	Mt	Ds	Rg	Cn	Nh	Fl	Mc	Lv	Ts	Og

Figure 3.5 A 32-column periodic table showing the correct placement of the rare earth elements within the larger periodic framework. Note that the exact composition of group 3 is currently debated as it could be Sc, Y, Lu, Lr (as shown) or Sc, Y, La, Ac.

THE MOLE

With the development of atomic theory, an important conceptual division arose in the practice of chemistry, which has persisted until the recent development of techniques that allow us to study single atoms and molecules. That is, chemists measure and observe macroscopic quantities of atoms and molecules, from which they must make inferences about the nature of the elementary entities (atoms and molecules) under investigation. This necessitates a clearly defined way of linking the macroscopic to the nanoscopic, which is called the amount of substance (n) and is given the unit mole.[14] The mole and its interconnection to the macroscopic and nanoscopic scales are shown in Figure 3.7. Note that the mole was defined as the number of atoms in 0.012 kg of carbon-12, but since 2019 it has been defined as exactly $6.022\,140\,76 \times 10^{23}$ elementary entities. The molar mass (M) of carbon is 0.012 011 kg/mol. The molar mass of each element is equal to the product of its

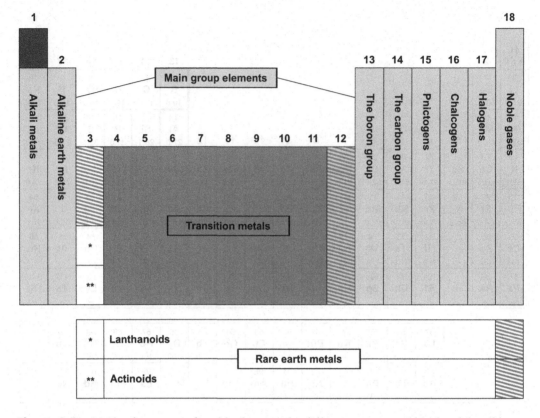

Figure 3.6 An 18-column periodic table showing the different groups and blocks of the table. Light gray groups are the main group elements, dark gray groups are transition metals, and white elements are rare earth metals.

recommended relative atomic mass (Appendix 2) and 0.999 999 999 65 × 10^{-3} kg/mol, the molar mass constant (M_u). For a chemical species, the molar mass is the sum of the recommended relative atomic mass value of each element ($A_r°(E)$) in the chemical formula.

PROBLEM 3.3
Use the mole map (Figure 3.7) to answer the following questions.

a. What is the mass (in kg) of 5.01×10^{25} carbon atoms?

b. What volume (L at STP) does 1.23 g helium gas occupy?

c. If a solution is 0.900 mol/L NaCl, how many grams of NaCl are in 500.0 mL?

DISCOVERY OF ATOMIC STRUCTURE

Dalton proposed that atoms were the smallest, most fundamental unit of matter. The idea that atoms could be composed of smaller, subatomic particles was first proposed by Prout in 1815. Prout hypothesized that hydrogen was the fundamental particle – he called it a protyle – and other atoms were groupings of hydrogen atoms.[15] Prout's hypothesis inspired later research, especially Rutherford, but it took a further 120 years of scientific research to elucidate the structure of the atom.

The Electron

The first experimental evidence for subatomic particles came in 1897 when J.J. Thomson published his results[16] studying the rays of a Crookes tube,[17] also called a cathode ray tube. The rays of a cathode ray tube can be deflected using an electric field (Figure 3.8) or using a magnetic field

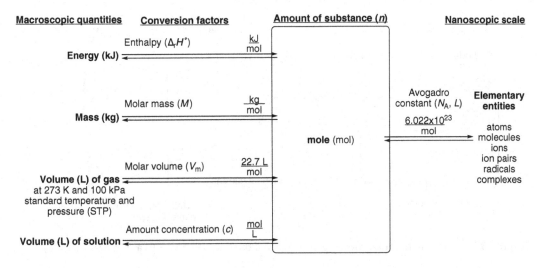

Figure 3.7 A mole map showing the centrality of the mole between the macroscopic and the nanoscopic (elementary entity) scales. Enthalpy, molar volume, and amount concentration, while all serving as connections to the mole, are all intrinsically related to the molar mass (M) of a substance. (Note: Amount concentration is commonly called molarity and mol/L is given the unit label M, e.g., 0.900 mol/L NaCl would be represented as 0.900 M NaCl. IUPAC has deprecated both the term molarity and the unit label M for amount concentration. While molarity and the unit label M are commonly used, they tend to confuse learners and IUPAC's recommendation is followed closely in this case.)

(Figure 3.9). When the cathode rays pass through an electric field (E), they experience an electric force of QE, where Q is the charge (in coulombs, C) of the cathode ray and E is the strength of the electric field (in volts, V). Similarly, when the cathode rays pass through a magnetic field (B), they experience a force of QvB, where Q is the electric charge (in coulombs, C), v is the velocity of the particle (in meters per second, m/s), and B is the strength of the magnetic field (in teslas, T). By exactly matching the magnetic and electric deflections – such that the rays strike the same spot on the cathode ray tube – the forces are equal and the charge-to-mass quotient (Q/m) of the cathode rays can be determined. Thomson was the first to measure this value for cathode rays.[18] The current CODATA-recommended value is $-1.758\,820\,010\,76 \times 10^{11}$ C/kg for cathode rays.[19] The negative

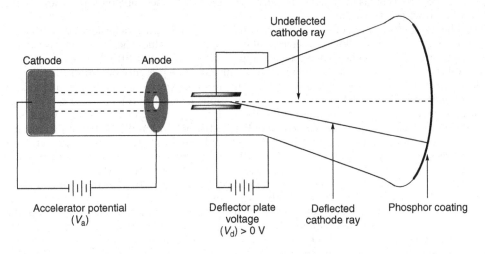

Figure 3.8 Cathode ray tube showing deflection of the cathode ray with an electric field (E).

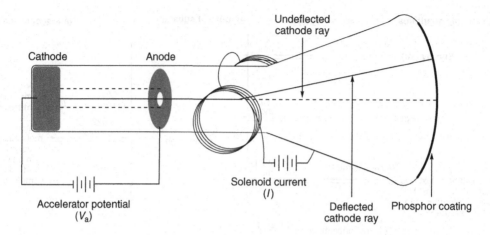

Figure 3.9 Cathode ray tube showing deflection of the cathode ray with a magnetic field (B).

charge of cathode ray particles was determined by the direction of their deflection in response to electric and magnetic fields.

PROBLEM 3.4
If the CODATA-recommended value for the charge-to-mass quotient for H⁺ is 9.578 833 1560 × 10⁷ C/kg, what are the two possible inferences one could make about cathode rays?

Thomson concluded that these negatively charged particles were a new subatomic (smaller than an atom) particle based upon his results and those of results from Philipp Lenard,[20] who found that cathode rays penetrated farther through various media than atoms typically do. This suggested that the particles were less massive than an atom. It should be noted Lenard contributed meaningfully to the field of early 20th-century physics, for which he was awarded the 1905 Nobel Prize. Lenard is not as well known today because he was an early and vociferous supporter of the National Socialist German Workers (Nazi) Party and a promoter of Deutsche Physik (German Physics). Despite science's foundation on open inquiry and objective engagement with the world, prejudice can and does negatively impact science, scientists, and society.

Thomson termed the new particles negative corpuscles. The term negative corpuscle was replaced with electron, which had been proposed for the theoretical concept three years before Thomson's work.[21] With the discovery of electrons, several questions then arise: What is the charge of the electron ($-e$)? What is the mass of the electron (m_e)? What other subatomic particles exist? What is the structure of an atom?

Since the charge-to-mass quotient of the electron ($-e/m_e$) was known, determination of the charge ($-e$) or the mass (m_e) would allow for determination of the other value. The first high-precision determination of the elementary charge (e) was reported by Millikan in 1910.[22] In this experiment, a mist of oil is dispersed within a chamber that was ionized by the presence of radium, and the time it takes a drop to fall was determined, from which the terminal velocity and the mass and radii of the drops were calculated. Then an electric field was applied, and the drops would rise upward, against the force due to gravity, because of the electric force (F_E) acting on the ionized droplets. Since the electric force is equal to the strength of the electric field (E) multiplied by the charge (Q), the value of the charge could be determined. It was found that the charge on the drops were multiples of one elementary charge (e) value 1.592×10^{-19} C. The modern CODATA value for the charge of an electron ($-e$) is $-1.602\ 176\ 634 \times 10^{-19}$ C, a less than 1% difference from Millikan's value.

With the value of the electron charge ($-e$), the mass of the electron could be determined. The modern CODATA value is $9.109\ 383\ 7015 \times 10^{-31}$ kg (a relative mass of 0.000 55), which is smaller than the hydrogen atom by a factor of 1836. While these data on the electron answer some of the questions above, we are still left with multiple questions: What other subatomic particles exist? What is the structure of an atom? What was clear is that most of an atom's mass comes from something other than the electrons and that there must be a positive charge carrier to offset the negatively charged electrons.

PROBLEM 3.5
Following Avogadro's initial hypothesis, that the volume of gas is proportional to the number of atoms,[23] the Avogadro constant was determined predominantly through gas studies. Millikan's value for the elementary charge (1.592×10^{-19} C) provided a way of determining the Avogadro constant from the Faraday constant (96 485.332 12 C/mol). What is the value of the Avogadro constant using Millikan's 1.592×10^{-19} C value for the elementary charge?

The Proton

After discovering electrons, Thomson proposed a model atomic structure to account for his new corpuscles.[24] In Thomson's model, the electrons are dispersed in a uniformly charged positive mass. This proposed model was tested experimentally by Rutherford, one of Thomson's proteges, and his team, Marsden and Geiger. They fired α particles (helium nuclei) at metallic foil.[25] The expectation, if Thomson's model was correct, is that α particles would move through the foil with minimal deflection (Figure 3.10).

While most particles did proceed through the foil without substantial deflection, about 1 in 8000 was deflected 90° or more by gold foil (Figure 3.10). These results led Rutherford to propose a model of the atomic structure with a small nucleus of positive charge that constitutes most of the mass and electrons dispersed around the nucleus in a large cloud (Figure 3.10).[26] A similar model was proposed in 1904 by Nagaoka.[27]

Based on Rutherford's model of a nuclear atom, van den Broek proposed that in a neutral atom, the number of electrons equaled the charge of the nucleus. The hypothesis, then, is that this atomic number – the nuclear charge – is the organizing principle of the periodic table rather than relative atomic mass.[28] This proposal was confirmed by Moseley using X-ray spectroscopy, which showed that the elements demonstrated a regular stepwise increase in nuclear charge.[29] The reorganization of the periodic table based on atomic number (nuclear charge, Z) served to clarify the positions of elements, like tellurium and iodine. More importantly, it also clarified the number of elements that existed. Based on relative atomic mass, many radioactive elements were found to exist, for example, radium A, radium B, radium C, radium C_1, radium C_2, and radium D. With determination of their nuclear charge, however, these were all shown to be known elements: radium A is polonium, radium B is lead, radium C is bismuth, radium C_1 is polonium, radium C_2 is thallium, and radium D is lead. The fact that so many of these *different* elements had the same nuclear charge as known elements led Soddy to identify them as isotopes; *isos* means "same" and *topos* means "place" in Greek.[30] The existence of isotopes was one of the pieces of evidence that helped lead to the discovery of neutrons (see below).

While Moseley's work with X-ray spectroscopy established that nuclear charge defined an element, the constitution and nature of the nucleus were unclear. The origin of nuclear charge came from Rutherford's laboratory. After firing α particles into nitrogen gas, hydrogen was produced.[31] After repeated experimentation to rule out other sources, Rutherford deduced that hydrogen was produced from the impact of α particles with a nitrogen nucleus, which dislodged a hydrogen

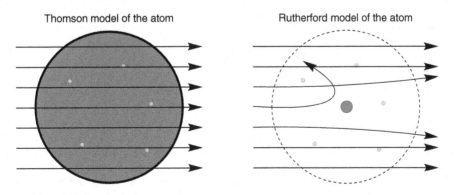

Thomson model of the atom   Rutherford model of the atom

Figure 3.10 Thomson model of the atom (left) with electrons (light gray small circles) dispersed in a positively charged mass (dark gray). Rutherford model of the atom (right) with electrons (light gray small circles) dispersed around a small, positively charged nucleus (dark gray bigger circle). The arrows represent the path of α particles through the atom.

nucleus from the nitrogen nucleus. Rutherford called this new particle a proton, partly in honor of Prout's protyle.[32] The charge is exactly opposite that of an electron ($1.602\ 176\ 634 \times 10^{-19}$ C) and the CODATA value of the proton mass (m_p) is $1.672\ 621\ 923\ 69 \times 10^{-27}$ kg. The proton resides in the nucleus, with a relative mass of 1.01, and the number of protons corresponds to the nuclear charge and atomic number (Z).

The Neutron

Electrons and protons alone could not satisfactorily explain the structure of atoms. For example, helium has a relative atomic mass of 4.00 but a nuclear charge number of 2. The nuclear charge number corresponds to two protons in the nucleus, but two protons would constitute a relative mass of only 2.02. In addition, the existence of isotopes could also not be clearly explained with electrons and protons alone. Rutherford proposed the existence of another particle with a mass equivalent to the proton and no charge to explain these inconsistencies.[33] James Chadwick discovered neutrons when investigating an unusually penetrating radiation that resulted from α particles colliding with beryllium.[34] This unusually penetrating radiation was found to be a neutral particle, called a neutron. The neutron has no charge and the CODATA value of the neutron mass (m_n) is $1.674\ 927\ 498\ 04 \times 10^{-27}$ kg. The neutron resides in the nucleus, has a relative mass of 1.01, and the number of neutrons defines the isotope.

OVERVIEW OF ATOMIC STRUCTURE AND ATOMIC SYMBOLOGY

In summary (Figure 3.11), an atom consists of a small, dense (2×10^{17} kg/m^3) nucleus that is made up of protons (p) and neutrons (n), collectively called nucleons. The number of protons is called the atomic number (Z) and corresponds to the nuclear charge, which defines the element. The number of neutrons identifies the specific isotope of each element, and the sum of the proton number (Z) and the neutron number (N) equals the mass number (A). Around the nucleus is a cloud of electrons and the number of electrons relative to protons determines the charge of an atom. The charge number (z) is the difference between the number of protons and the number of electrons.

To indicate the specific makeup of an atom, there are two systems. The first is an annotation of the element name. For example, if there were an element with eight protons, ten neutrons, and ten electrons, we would name that element oxide-18(2–). Oxygen – here it is given the -*ide* suffix because it is an anion – is the name of any atom with eight protons, and its mass number is 18. The charge number is 2– because there are two more electrons than there are protons. We could also represent the specific isotope using the atomic symbol $^{18}_{8}O^{2-}$. The general scheme for atomic symbols is:

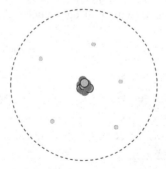

Figure 3.11 A cartoon representation of a boron-10 atom showing a proton (dark gray bigger circles) and neutron (bold outlined gray bigger circles) in the nucleus and electrons (light gray small circles) in a surrounding cloud. The dotted circle represents the atomic radius. Note this is not to scale, as the atomic radius should be about 30 000 to 60 000 times larger than the nucleus.

$$_Z^A E^z$$

E is the atomic symbol.

z is the charge number.

Z is the atomic number, proton number.

A is the mass number, nucleon number.

Note that the atomic number (Z) is often omitted from atomic symbols because it is redundant given a periodic table and the atomic symbol.

PROBLEM 3.6
Complete the following table.

Atomic Symbol	Isotope Designation	Atomic Number (Z)	Mass Number (A)	Charge Number (z)	Proton Number (Z)	Neutron Number (N)	Electron Number
		12	24	0			
$^{39}K^+$							
	iron-60(2+)		60				24
					17	18	17
$^{35}Cl^-$							
		26				30	23
			15	3–	7		
		8		2–		8	
	sodium-24(1+)						
			27		13		10
^{62}Ni	nickel-62						28

NOTES

1. Whitaker, R.D. An Historical Note on the Conservation of Mass. *J. Chem. Educ.*, **1975**, *52* (10), 658–659. DOI: 10.1021/ed052p658.

2. Lavoisier. *Traité Élémentaire de Chimie, Présenté dans un Ordre Noveau et d'Après les Découvertes Modernes.* Chez Cuchet: Paris, 1789.

3. Strutt, J.W. I. On an Anomaly Encountered in Determinations of the Density of Nitrogen Gas. *Proc. R. Soc. Lond.*, **1894**, *55* (331–335), 340–344. DOI: 10.1098/rspl.1894.0048.

4. Proust. Recherches sur le Blue de Prusse. *Journal de Physique, de Chimie, d'Histoire Naturelle et des Arts*, **1794**, 334–341.

5. Dalton, J. A New System of Chemical Philosophy, Part II. S. Russell: Manchester, 1810.

6. Döbereiner, J.W. Versuch zu einer Gruppirung der elementaren Stoffe nach ihrer Analogie. *Ann. Phys.*, **1829**, *91*, 301–307. DOI: 10.1002/andp.18290910217.

7. (i) Newlands, J.A.R. On Relations Among the Equivalents. *Chemical News*, **1864**, *10*, 94–95.

   (ii) Newlands, J.A.R. On the Law of Octaves. *Chemical News*, **1865**, *12*, 83.

8. Mendelejeff, D. Ueber die Beziehungen der Eigenschaften zu den Atomgewichten der Elemente. *Zeitschrift für Chemie*, **1869**, *12*, 405–406.

9. Deming, H.G. *General Chemistry: An Elementary Survey, emphasizing Industrial Applications of Fundamental Principles.* J. Wiley & Sons: New York, 1923.

10. (i) Strutt, J.W.; Ramsay, W. VI. Argon, a New Constituent of the Atmosphere. *Philosophical Transactions of the Royal Society of London. (A.)*, **1895**, *186*, 187–241. DOI: 10.1098/rsta.1895.0006.

    (ii) Ramsay, W. I. Helium, a Gaseous Constituent of Certain Minerals. Part I. *Proc. R. Soc. Lond.*, **1895**, *58* (347–352), 80–89. DOI: 10.1098/rspl.1895.0010.

    (iii) Ramsay, W.; Travers, M.W. On the Companions of Argon. *Proc. R. Soc. Lond.*, **1898**, *63* (389-400), 437–440. DOI: 10.1098/rspl.1898.0057.

    (iv) Ramsay, W.; Travers, M.W. On a New Constituent of Atmospheric Air. *Proc. R. Soc. Lond.*, **1898**, *63* (389–400), 405–408. DOI: 10.1098/rspl.1898.0051.

    (v) Ramsay, W.; Travers, M.W. On the Extraction from Air of the Companions of Argon and on Neon. *Report of the British Association for the Advancement of Science*, **1899**, 828–830. https://www.biodiversitylibrary.org/page/30221699 (Date Accessed: 04/15/23).

11. Mendeleeff, D. *An Attempt Towards a Chemical Conception of the Ether.* Transl. George Kamensky. Longmans, Green, and Co., 1904.

12. Brauner, B. Über die Stellung der Elemente der seltenen Erden im periodischen System von Mendelejeff. *Z. Anorg. Chem.*, **1902**, *32*, 1–30. DOI: 10.1002/zaac.19020320102.

13. (i) Seaborg, G.T. The Chemical and Radioactive Properties of the Heavy Elements. *Chem. Eng. News*, **1945**, *23* (23), 2190–2193. DOI: 10.1021/cen-v023n023.p2190.

    (ii) Seaborg, G.T. The Transuranium Elements. *Science*, **1946**, *104* (2704), 379–386. DOI: 10.1126/science.104.2704.379.

14. Ostwald, W. *Hand- und Hilfsbuch zur Ausführung Physiko-Chemischer Messungen.* Wilhelm Engelmann: Leipzig, 1893.

15. (i) Prout, W. On the Relation between the Specific Gravities of Bodies in their Gaseous State and the Weights of their Atoms. *Ann. Philos.*, **1815**, *6*, 321–330.

    (ii) Prout, W. Correction of a Mistake in the Essay on the Relation between the Specific Gravities of Bodies in their Gaseous State and the Weights of their Atoms. *Ann. Philos.*, **1816**, *7*, 111–113.

16. Thomson, J.J. Cathode-Rays. *Electrician*, **1897**, *39*, 104–109.

17. Crookes, W. V. The Bakerian Lecture. –On the Illumination of Lines of Molecular Pressure, and the Trajectory of Molecules. *Phil. Trans. R. Soc.*, **1879**, *170*, 135–164. DOI: 10.1098/rstl.1879.0065.

18. Thomson, J.J. XL. Cathode Rays. *Lond. Edinb. Dublin Philos. Mag. J. Sci. Series 5*, **1897**, *44* (269), 293–316. DOI: 10.1080/14786449708621070.

19. Committee on Data of the International Science Council (CODATA): Tiesinga, E.; Mohr, P.J.; Newell, D.B.; Taylor, B.N. CODATA Recommended Values of the Fundamental Physical Constants: 2018, *J. Phys. Chem. Ref. Data.* **2021**, *50* (3), 033105-1–033105-61. DOI: 10.1063/5.0064853.

20. Lenard, P. Ueber Kathodenstrahlen in Gasen von atmosphärischem Druck und im äussersten Vacuum. *Ann. Phys.*, **1894**, *287* (2), 225–267. DOI: 10.1002/andp.18942870202.

21. Stoney, G.J. XLIX. Of the "Electron," or Atom of Electricity. *Lond. Edinb. Dublin Philos. Mag. J. Sci. Series 5*, **1894**, *38* (233), 418–420. DOI: 10.1080/14786449408620653.

22. Millikan, R.A. XXII. A New Modification of the Cloud Method of Determining the Elementary Electrical Charge and the Most Probable Value of That Charge. *Lond. Edinb. Dublin Philos. Mag. J. Sci. Series 6*, **1910**, *19* (110), 209–228. DOI: 10.1080/14786440208636795.

23. Avogadro, A. Essai d'une manière de determiner les masses relatives des molecules élémentaires des corps. *Journal de Physique*, **1811**, *73*, 58–76.

24. Thomson, J.J. XXIV. On the Structure of the Atom: An Investigation of the Stability and Periods of Oscillation of a Number of Corpuscles Arranged at Equal Intervals Around the Circumference of a Circle; with Application of the Results to the Theory of Atomic Structure. *Lond. Edinb. Dublin Philos. Mag. J. Sci. Series 6*, **1904**, *7* (39), 237–265. DOI: 10.1080/14786440409463107.

25. (i) Geiger, H. On the Scattering of the α Particles by Matter. *Proc. R. Soc. Land. A*, **1908**, *81*, 174–177. DOI: 10.1098/rspa.1908.0067.

    (ii) Geiger, H.; Ernest, M. On a Diffusion Reflection of the α Particles. *Proc. R. Soc. Land. A*, **1909**, *82*, 495–500. DOI: 10.1098/rspa.1909.0054.

    (iii) Geiger, H. The Scattering of α Particles by Matter. *Proc. R. Soc. Land. A*, **1910**, *83*, 492–504. DOI: 10.1098/rspa.1910.0038.

    (iv) Geiger, H.; Marsden, E. LXI. The Laws of Deflexion of α Particles through Large Angles. *Lond. Edinb. Dublin Philos. Mag. J. Sci. Series 6*, **1913**, *25* (148), 604–623. DOI: 10.1080/14786440408634197.

26. Rutherford, E. LXXIX. The Scattering of α and β Particles by Matter and the Structure of the Atom. *Lond. Edinb. Dublin Philos. Mag. J. Sci. Series 6*, **1911**, *21* (125), 669–688. DOI: 10.1080/14786440508637080.

27. Nagaoka, H. LV. Kinetics of a System of Particles illustrating the Line and the Band Spectrum and the Phenomena of Radioactivity. *Lond. Edinb. Dublin Philos. Mag. J. Sci. Series 6*, **1904**, *7* (41), 445–455. DOI: 10.1080/14786440409463141.

28. (i) van den Broek, A. Das Mendelejeffsche „kubische" periodische System der Elemente und die Einordnung der Radioelemente in dieses System. *Phys. Z.*, **1911**, *12*. 490–497.

    (ii) van den Broek, A. Die Radioelemente, das periodische System und die Konstitution der Atome. *Phys. Z.*, **1913**, *14*. 32–41.

29. (i) Moseley, H.G.J. XCIII. The High-Frequency Spectra of the Elements. *Lond. Edinb. Dublin Philos. Mag. J. Sci. Series 6*. **1913**, *26* (156), 1024–1034. DOI: 10.1080/14786441308635052.

    (ii) Moseley, H.G.J. LXXX. The High-Frequency Spectra of the Elements. Part II, *Lond. Edinb. Dublin Philos. Mag. J. Sci. Series 6*. **1914**, *27* (160), 703–713. DOI: 10.1080/14786440408635141.

30. Soddy, F. Intra-Atomic Charge. *Nature*, **1913**, *92*, 399–400. DOI: 10.1038/092399c0.

31. Rutherford, E. LIV. Collision of α Particles with Light Atoms. IV. An Anomalous Effect in Nitrogen. *Lond. Edinb. Dublin Philos. Mag. J. Sci. Series 6*. **1919**, *37* (222), 581–587. DOI: 10.1080/14786440608635919.

32. Physics at the British Association. *Nature*, **1920**, *106*, 357–358. DOI: 10.1038/106357a0.

33. Rutherford, E. Bakerian Lecture: Nuclear Constitution of Atoms. *Proc. R. Soc. Lond. A*, **1920**, *97* (686), 374–400. DOI: 10.1098/rspa.1920.0040.

34. (i) Chadwick, J. Possible Existence of a Neutron. *Nature*, **1932**, *129*, 312. DOI: 10.1038/129312a0.

(ii) Chadwick, J. The Existence of a Neutron. *Proc. R. Soc. Lond. A*, **1932**, *136* (830), 692–708. DOI: 10.1098/rspa.1932.0112.

4 Binding Energy, Nuclear Stability, and Decay

In this chapter, the nucleus will be considered in greater detail. Specifically, we will focus on how the atomic nucleus is held together, a topic that has intrigued scientists since the discovery of the neutron.[1] We will then discuss what factors make the atomic nucleus stable and what factors make it unstable, and conclude by considering how unstable nuclides decay to achieve greater stability.[2]

BINDING ENERGY (E_b)

In the nucleus, there are both protons and neutrons. Together, protons and neutrons are referred to as nucleons. Consider the existence of a nucleus itself. Nuclei, dense collections of nucleons, exist. The fact that nuclei exist means there must be a benefit for nucleons to assemble into a nucleus: They must be more stable (lower in energy) than the same number of free or unbound nucleons. In this thought experiment, consider the converse. If a nucleus was less stable (higher in energy) than the same number of free nucleons, then nuclei would fall apart into free nucleons. In a universe where nuclei were less stable than unbound nucleons, it would be far less interesting, as no element aside from hydrogen-1 would be possible.

Having conceptually established that nuclei are more stable than free protons and neutrons, let us now consider the idea quantitatively. Nucleons are held together by what is called the nuclear force,[3] itself a residual effect of one of the four fundamental forces: the strong force.[4] The nuclear force is a powerfully attractive force at femtometer (1×10^{-15} m) distances. It is strong enough to overcome, for example, the nearly 2 MeV (2×10^8 kJ/mol) of repulsive energy from having two protons that close together in the helium-4 nucleus.[5] The energy released upon forming a nucleus is called the binding energy (E_b) (Figure 4.1). That is the amount of energy that is holding (binding) the nucleons together into a nucleus. Binding energy increases as the number of nucleons increases, due to the increasing number of favorable nucleon–nucleon interactions.

As you can see, binding energy (E_b) is incredibly large and so it takes a lot of energy to completely pull all the nucleons in an atom apart into free protons and neutrons. Comparing the binding energy values for different nuclides can help us understand the first factor that influences the stability of a nucleus: the number of nucleons. To be able to compare nuclides, the binding energy is normalized as binding energy per nucleon, where the absolute binding energy is divided by the number of nucleons. Figure 4.2 shows the first ten stable nuclides in detail.

As we consider Figure 4.2, we still see the general trend that bigger nuclides are more stable (helium-4 is an obvious outlier that will be discussed in more detail later). If we consider a fuller picture of all naturally occurring nuclides (Figure 4.3), we can see there is a rapid increase in the magnitude of the binding energy per nucleon until it reaches a maximum magnitude (the most negative binding energy). The nucleus with the highest binding energy per nucleon is nickel-62 (28 protons and 34 neutrons).[6] Beyond nickel-62, the binding energy gradually begins to diminish.[7] How can we understand the shape of this binding energy per nucleon curve (Figure 4.3)? The increase in the magnitude of binding energy per nucleon is due to the greater nuclear force as the number of nucleons increases. This increase in binding energy is counterbalanced by two factors. The first destabilizing factor is the rapid increase in coulombic, electrostatic, repulsion as the number of protons increases: coulombic repulsion increases in proportion to Z^2. The second destabilizing factor is the increase in the size of the nucleus as the number of nucleons increases. The nuclear force only affects nucleons over very short (femtometer) distances and so outermost nucleons are less stabilized by the nuclear force. The increasing size of the nucleus leads to more nucleons on the surface, where they are less stable than nucleons in the core of the nucleus. The balance of these stabilizing and destabilizing factors leads to nickel-62 being the most stable nuclide, and all larger nuclides show a decrease in the magnitude of binding energy per nucleon.

To calculate the binding energy, we first need to calculate the change in mass, Δm, also called the mass defect. The mass defect is the difference between the nuclide's atomic mass (m_a) in unified atomic mass units (u) and the sum of the particle masses. Let us consider the carbon-12 nuclide with an atomic mass of 12.000 00 u. Carbon-12 is made up of six protons, six neutrons, and six electrons. We will use the atomic mass of hydrogen-1 (1.007 83 u) to account for one proton and one electron and the neutron mass (1.008 66 u) to account for each neutron. The mass defect for carbon-12, then, is:

$$\Delta m = 12.000\ 00\ \text{u} - [6(1.007\ 83\ \text{u}) + 6(1.008\ 66\ \text{u})] = -0.098\ 94\ \text{u}$$

This difference, or defect, is the mass that is lost as the nucleons combine into a single nucleus. That mass is lost in the form of energy, which is what we call the nuclear binding energy (E_b). We can determine the energy value represented by this mass defect by using Equation 4.1.[8]

DOI: 10.1201/9781003479338-4

29

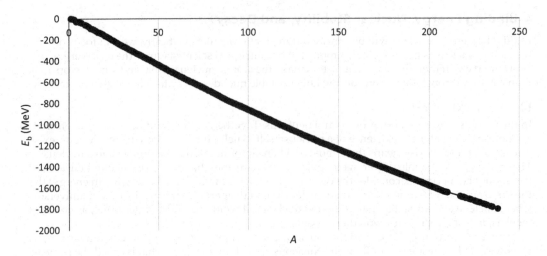

Figure 4.1 Binding energy (E_b) as a function of the nucleon number (A) for all naturally occurring nuclides.

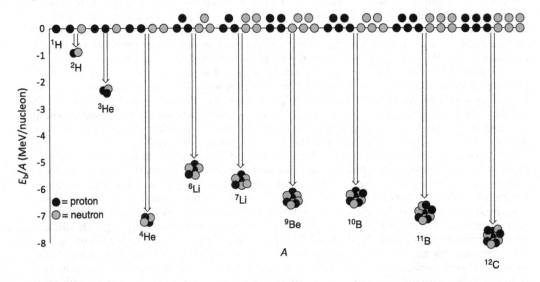

Figure 4.2 Binding energy per nucleon (E_b/A) as a function of the nucleon number (A) for the first ten stable nuclides.

$$E_b = \Delta mc^2 \tag{4.1}$$

In Equation 4.1, which provides the equivalence between energy and mass, the mass defect is multiplied by the speed of light squared (c^2). Using this equation, we find that 1 u c^2 is equal to 931.494 102 MeV. Using this mass–energy relationship, we find that the binding energy of a carbon-12 nuclide is –92.16 MeV:

$$E_b = -0.09894 \text{ u } c^2 \left(\frac{931.494\ 102 \text{ MeV}}{\text{u } c^2} \right) = -92.16 \text{ MeV}$$

To be able to compare the binding energy values of different nuclides, with different numbers of nucleons, we normalize the binding energy to the number of nucleons, which gives carbon-12 a binding energy of –7.6801 MeV/nucleon:

$$\frac{-92.16 \text{ MeV}}{12 \text{ nucleon}} = -7.680 \text{ MeV / nucleon}$$

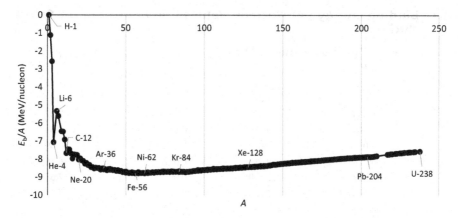

Figure 4.3 Plot of binding energy per nucleon for all naturally occurring nuclides.

By calculating the binding energy per nucleon of different nuclides, we can identify which nuclides are more stable (release more energy per mole nucleon) and which are less stable (release less energy per mole nucleon).

PROBLEM 4.1
Boron has two stable nuclides, ^{10}B (m_a = 10.012 94 u) and ^{11}B (m_a = 11.009 31 u). Calculate the binding energies per nucleon (MeV/nucleon) of these two nuclei and compare their stabilities.

Given the multitude of possible nuclides, it is important at this juncture to cement some definitional terms. First, isotopes are nuclides that possess the same number of protons (isoto_p_es) but different numbers of neutrons. Second, isotones are nuclides that possess the same number of neutrons (isoto_n_es) but different numbers of protons. Finally, isobars are nuclides that have the same mass number ("bar" comes from Greek *baros* meaning "weight"). These terms will help us to differentiate and converse about different types of nuclides in further discussions of this chapter.

PROBLEM 4.2
Consider the first seven isotopes of hydrogen and their atomic masses in Table 4.1 and the five isotones (equal number of neutrons) of N = 7 and their atomic masses in Table 4.2.

Table 4.1   Isotopes (Z = 1) of Hydrogen and Their Atomic Masses (u)

Isotope	Atomic Symbol	Atomic Mass (u)
Protium	^{1}H	1.0078
Deuterium	^{2}H	2.0141
Tritium	^{3}H	3.0161
Hydrogen-4	^{4}H	4.0264
Hydrogen-5	^{5}H	5.0353
Hydrogen-6	^{6}H	6.0450
Hydrogen-7	^{7}H	7.0527

Table 4.2   Isotones (N = 7) and Their Atomic Masses (u)

Isotope	Atomic Symbol	Atomic Mass (u)
Beryllium-11	^{11}Be	11.0217
Boron-12	^{12}B	12.0144
Carbon-13	^{13}C	13.0034
Nitrogen-14	^{14}N	14.0031
Oxygen-15	^{15}O	15.0031

Table 4.3 Binding Energy per Nucleon Values (MeV/nucleon) for Select Nuclides

Nuclide	Mass of Particles (u)	Atomic Mass (u)	Δm (u)	Binding Energy (MeV/nucleon)
^2H	2.016	2.014	–0.002	–1.11
^3He	3.024	3.016	–0.008	–2.57
^4He	4.033	4.003	–0.030	–7.07
^6Li	6.049	6.015	–0.034	–5.33
^7Li	7.058	7.016	–0.042	–5.60
^9Be	9.075	9.012	–0.062	–6.46
^{12}C	12.099	12.000	–0.099	–7.68
^{13}C	13.108	13.003	–0.104	–7.47
^{14}C	14.116	14.003	–0.113	–7.52
^{14}N	14.115	14.003	–0.112	–7.47
^{15}N	15.124	15.000	–0.124	–7.70

Explain, using Table 4.1 and Table 4.2, why it is not possible to determine the mass of a neutron or the mass of a proton from these data.

PROBLEM 4.3
Consider the graph of binding energy per nucleon (Figure 4.3).

a. What does the fact that binding energy is negative mean? Could a nucleus have positive binding energy? Explain.

b. Why does binding energy increase in magnitude from $A = 1$ to $A = 62$?

c. Why does binding energy decrease in magnitude from $A = 62$ to $A = 238$?

PROBLEM 4.4
Consider Table 4.3 of nuclear binding energy values (MeV/nucleon).

a. In general, what happens to binding energy per nucleon as atomic mass increases?

b. What produces greater stability (higher binding energy per nucleon): having an even or an odd number of neutrons? Provide specific evidence to support your claim.

c. What produces greater stability (higher binding energy per nucleon): having an even or an odd number of protons? Provide specific evidence to support your claim.

d. Consider the periodic table in Figure 4.4. Can you use your answer in part c to explain the pattern observed?

PROBLEM 4.5
Look at the periodic table in Figure 4.4. Do you notice anything significant about the stability of heavy elements? Is there a clear cut-off that you may want to hold in mind as we move forward?

RADIOACTIVE DECAY
When we talk about unstable nuclides and stable nuclides, what we are discussing are nuclides that are radioactive (unstable) and nuclides that are not radioactive (stable). When a nuclide is radioactive, or unstable, it will emit energy or particles to become more stable. The emitted energy and particles from radioactive decay are invisible but can be seen in terms of how they affect other substances: photographic film will develop without light,[9] materials will weaken as chemical bonds are broken, and living organisms can become sick and/or develop mutations. There are three types of radioactive decay: α (alpha), β (beta), and γ (gamma). The first two types of decay, α and β, involve the emission of a particle from the nucleus along with the concomitant transmutation of the initial element into a new element.[10] The last type of decay, γ, involves the emission

Periodic table (values in parentheses indicate the number of stable, nonradioactive nuclides per element):

1	2	3	4	5	6	7	8	9	10	11	12	13	14	15	16	17	18
H (2)																	He (2)
Li (2)	Be (1)											B (2)	C (2)	N (2)	O (3)	F (1)	Ne (3)
Na (1)	Mg (3)											Al (1)	Si (3)	P (1)	S (4)	Cl (2)	Ar (3)
K (2)	Ca (5)	Sc (1)	Ti (5)	V (1)	Cr (4)	Mn (1)	Fe (4)	Co (1)	Ni (5)	Cu (2)	Zn (5)	Ga (2)	Ge (5)	As (1)	Se (5)	Br (2)	Kr (6)
Rb (1)	Sr (3)	Y (1)	Zr (4)	Nb (1)	Mo (6)	Tc (0)	Ru (7)	Rh (1)	Pd (6)	Ag (2)	Cd (6)	In (1)	Sn (10)	Sb (2)	Te (6)	I (1)	Xe (9)
Cs (1)	Ba (6)	*	Hf (5)	Ta (2)	W (4)	Re (1)	Os (6)	Ir (2)	Pt (5)	Au (1)	Hg (7)	Tl (2)	Pb (4)	Bi (0)	Po (0)	At (0)	Rn (0)
Fr (0)	Ra (0)	**	Rf (0)	Db (0)	Sg (0)	Bh (0)	Hs (0)	Mt (0)	Ds (0)	Rg (0)	Cn (0)	Nh (0)	Fl (0)	Mc (0)	Lv (0)	Ts (0)	Og (0)

*	La (1)	Ce (4)	Pr (1)	Nd (5)	Pm (0)	Sm (5)	Eu (2)	Gd (6)	Tb (1)	Dy (7)	Ho (1)	Er (6)	Tm (1)	Yb (7)	Lu (1)
**	Ac (0)	Th (0)	Pa (0)	U (0)	Np (0)	Pu (0)	Am (0)	Cm (0)	Bk (0)	Cf (0)	Es (0)	Fm (0)	Md (0)	No (0)	Lr (0)

Figure 4.4 Periodic table showing the number of stable, nonradioactive nuclides per element. Dark gray corresponds to elements with three or more stable nuclides and light gray corresponds to elements with two or fewer stable nuclides. Elements left unshaded have no stable nuclides.

of electromagnetic radiation (Chapter 6) as a nuclide goes from a higher energy state to a lower energy state. In γ decay, there is no transmutation of the nuclide. We will consider each type of decay in turn.

α Decay

The first type of radioactive decay that we will consider is α decay.[11] In α decay, a nuclide emits a helium-4(2+) nuclide, an α particle, which results in a new nuclide that is four mass numbers and two atomic numbers smaller. Let's consider the α decay of uranium-238, which produces thorium-234 (called a decay product nuclide):

$$^{238}U \rightarrow {}^{234}Th + {}^{4}He$$

It should be noted that because we are focused on the nuclei and not the whole atom, the charge number is often omitted. If we were to include the charge numbers, the balanced equation would be:

$$^{238}U \rightarrow {}^{234}Th^{2-} + {}^{4}He^{2+}$$

The danger of α decay is not that it produces a helium-4 nuclide but that it produces a helium nucleus without electrons. The helium atom strongly pulls on electrons (Chapter 8) and will ionize all chemicals it encounters to fill its electron shell. This makes α decay especially dangerous in living tissue. Notably, α decay is the only source of helium gas on Earth, which comes from the radioactive decay of heavy-metal ores in the Earth's crust. Despite its constant production, helium is lost to space through both its high gas particle speed (Chapter 11) and other atmospheric effects. As such, helium gas, as of 2023, is a byproduct of drilling for oil and natural gas, which makes helium a nonrenewable resource.

Finally, note that it is also common to show α decay equations by writing α to represent the helium-4(2+) particle:

$$^{238}U \rightarrow {}^{234}Th + \alpha$$

Now the question arises, why and how does α decay occur? Nuclides above a certain size (see Figure 4.4) are highly destabilized by proton–proton repulsion in the nucleus. Hypothetically, any nucleus larger than nickel-62 could become more stable through α decay. Currently, the largest nuclide considered stable is lead-208. Bismuth-209, the primordial nuclide of bismuth, and all larger nuclides undergo α decay. In these nuclides, the proton–proton repulsion provides enough destabilization to tip the nucleus toward radioactive decay. It is worth noting that α decay happens spontaneously through tunneling. That is, the energetic barrier preventing the emission of an α particle is quite high, but through a quantum mechanical mechanism, the α particle can be ejected by going through the energy barrier (which is why it is called tunneling) rather than over it.[12] This is relevant because, unlike chemical reactions that typically proceed over energetic barriers (Chapter 14) and are therefore dependent on temperature, radioactive decay is independent of the temperature.

PROBLEM 4.6
Beryllium-8 (Z = 4) is a unique example of a nuclide that undergoes α decay even though it is significantly smaller in size than lead (Z = 82). Provide an explanation for this anomalous α decay.

PROBLEM 4.7
For each of the following, predict the α decay product nuclide or starting nuclide.

a. $^{210}Po \rightarrow$

b. $^{240}Pu \rightarrow$

c. $\rightarrow\ ^{216}Po + \alpha$

d. $\rightarrow\ ^{233}Pa + \alpha$

β Decay

The second type of radioactive decay that we will consider is β decay. There are two types of β decay: β⁻ decay[13,14] and β⁺ decay.[15,16] In β⁻ decay, a nuclide emits an electron (a β⁻ particle) and an electron antineutrino ($\bar{v}_e$) as a neutron transmutes into a proton, which results in a new nuclide that has the same mass number (A) but an atomic number one larger than the initial nuclide. Let's consider the β⁻ decay of aluminium-28, which produces silicon-28:

$$^{28}Al \rightarrow\ ^{28}Si^+ + e^- + \bar{v}_e$$

It is also common to show β⁻ decay equations by writing β⁻ to represent the electron particle and to omit the charge on the product nuclide:

$$^{28}Al \rightarrow\ ^{28}Si + \beta^- + \bar{v}_e$$

In β⁺ decay, a nuclide emits a positron (a β⁺ particle) and an electron neutrino (v_e) as a proton transmutes into a neutron, which results in a new nuclide that has the same mass number (A) but an atomic number one smaller than the initial nuclide. Let's consider the β⁺ decay of aluminium-26, which produces magnesium-26:

$$^{26}Al \rightarrow\ ^{26}Mg^- + e^+ + v_e$$

It is also common to show β⁺ decay equations by writing β⁺ to represent the positron particle and to omit the charge on the product nuclide:

$$^{26}Al \rightarrow\ ^{26}Mg + \beta^+ + v_e$$

Nuclides that undergo β⁺ decay can achieve the same transmutation through electron capture (EC),[17] where a core electron (from the innermost K or L electron shells) is captured by the nucleus as the proton transmutes into a neutron.

$$^{26}Al + e^- \rightarrow\ ^{26}Mg + v_e$$

Now the question arises, why and how does β decay occur? The simplest explanation is that β decay is the result of neutron–proton imbalance. All stable nuclides, except ^1H and ^3He, require that the nucleus has at least as many neutrons as there are protons. Nuclides that have fewer neutrons than protons are unstable and will decay through β^+ decay or electron capture. In contrast, nuclides that have too many neutrons, relative to the number of protons, will decay through β^- decay.

To give a clearer sense of when β decay occurs, we need to discuss another aspect of nuclear structure: the nuclear shell model.[18,19,20,21] To start, reconsider Figure 4.2. We can generally explain and calculate the binding energy of a nuclide as a composite of the stabilizing nuclear force and the destabilizing electrostatic repulsion between protons by treating the nucleus as a liquid drop with specific effects stemming from nucleus size, surface effects, nucleon pairing, and proton–neutron (a)symmetry.[22] This approach works quite well to semi-empirically calculate the binding energy of most nuclides. This approach fails, however, to model the binding energy of nuclides with specific numbers of nucleons. If the proton number, Z, and/or the neutron number, N, is 2, 8, 20, 28, 50, 82, or 126, then the nuclide will be exceptionally stable. Consider helium-4 (Figure 4.2), where both Z and N are the magic number 2. The helium-4 nuclide is uniquely stable among the lightest nuclides (see Problem 4.6).

Because these magic numbers of nucleons exist, scientists proposed a nuclear shell model. In this model, there are specific energy levels (Figure 4.5) that nucleons can occupy. Each level can be occupied by, at most, two nucleons.[23] The energy levels are filled from the bottom, lowest energy,

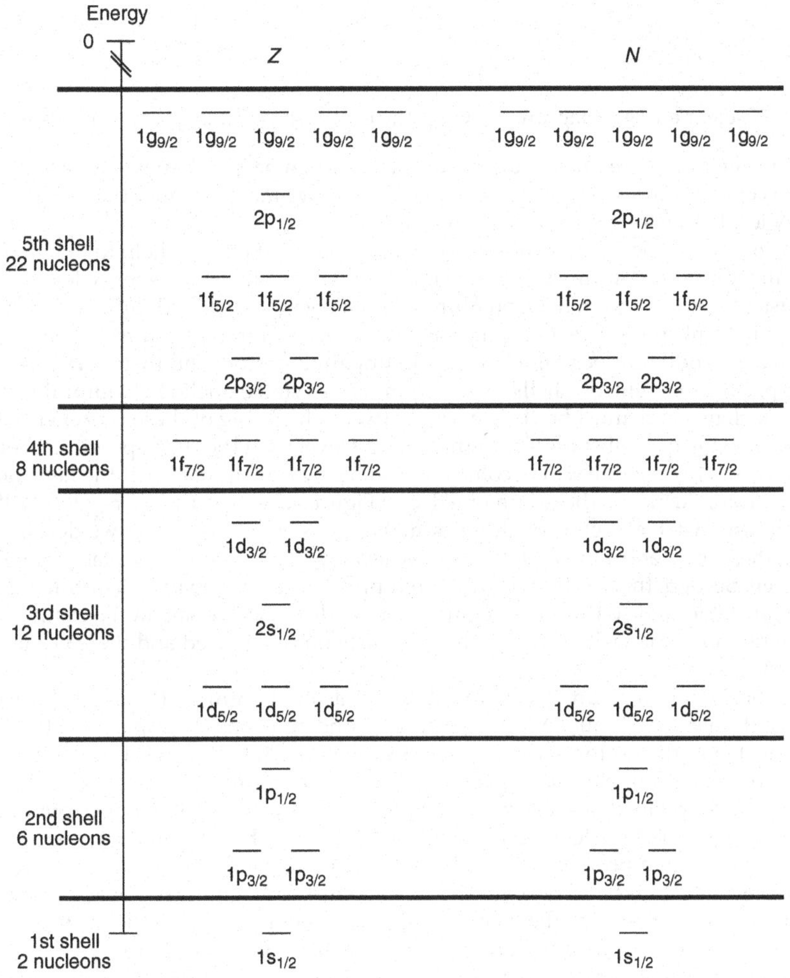

Figure 4.5 Energy diagram showing the first five shells of the nuclear shell model. The number and letter designations are quantum mechanical labels. The origin of the specific pattern of energy levels and the labels will not be considered further for nuclear energy levels, but we will look at the quantum mechanical origin of electron energy levels in Chapter 7. The subscript number indicates the total angular momentum associated with the energy level.

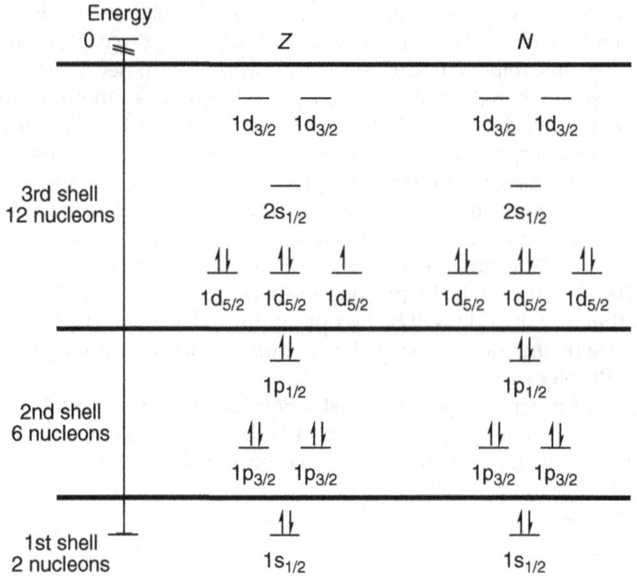

Figure 4.6 Nuclear energy diagram showing the energy-level filling for aluminium-27.

to the highest energy and the nucleons are more stable when paired than when unpaired.[24] Since there are two types of nucleons – protons and neutrons – we must consider the filling of two different energy levels, one for each species of nucleon.

Let us consider a specific example to see how these nuclear shells can help to understand β decay. We will look at aluminium. Consider Figure 4.6, which shows the energy-level filling – arrows are used to show proton/neutron filling of each energy level, with up and down arrows representing different spin states for each proton or neutron – for aluminium-27. We can see that the protons have an odd number, which leaves an unpaired proton, and that there is relative parity between the proton and neutron shells. Aluminium-27 is a stable nuclide of aluminium.

In a stable nucleus, there must be close parity between the filling of these two energy levels. That is, if a neutron is in a substantially higher energy level than the corresponding proton energy level (or vice versa), the nuclide will become more stable by transmutation of the neutron into a lower energy proton, which is the origin of β decay. Figure 4.7 shows the energy-level filling for aluminium-28, an unstable nuclide of aluminium. We can see that there are two destabilizing factors. First, there is a neutron in a higher energy level with a lower energy place available in the proton shell. Second, there is both an unpaired proton and an unpaired neutron. Unpaired nucleons are unstable, and so this nuclide undergoes β⁻ decay, as we saw in the example above, to produce a more stable silicon-28 nuclide, where all nucleons are paired and there is energetic parity between the two shells.

Now if we consider aluminium-26, an unstable nuclide of aluminium (Figure 4.8), we can see that even though there is an equal number of protons and neutrons and a parity in the energy levels of the nucleons, the proton shells and neutron shells each have an odd number of nucleons. This leaves two unpaired nucleons, which is unstable. Aluminium-26 undergoes β⁺ decay (or electron capture), which produces a stable magnesium-26 nuclide, where all the nucleons are paired. The instability of unpaired nucleons can explain the pattern in Figure 4.4. Odd-atomic-number nuclides have an unpaired proton and so there are fewer allowable neutron numbers that will produce a stable nuclide. To be a stable nuclide, then, odd-atomic-number elements generally must have an even number of neutrons that is near to the number of protons in the nucleus (nucleon parity). In contrast, even-atomic-number nuclides have all their protons paired, so stable nuclides can consist of even or odd numbers of neutrons, though even numbers will be more stable. This means that even-atomic-number nuclides have more allowable neutron numbers that will produce a stable nuclide.

The nuclear shell model can help explain the observation that above calcium, stable nuclides must have more neutrons than protons: As more protons are added to the nucleus, the proton shell is progressively destabilized (shifts to higher energy) by proton–proton repulsion, which

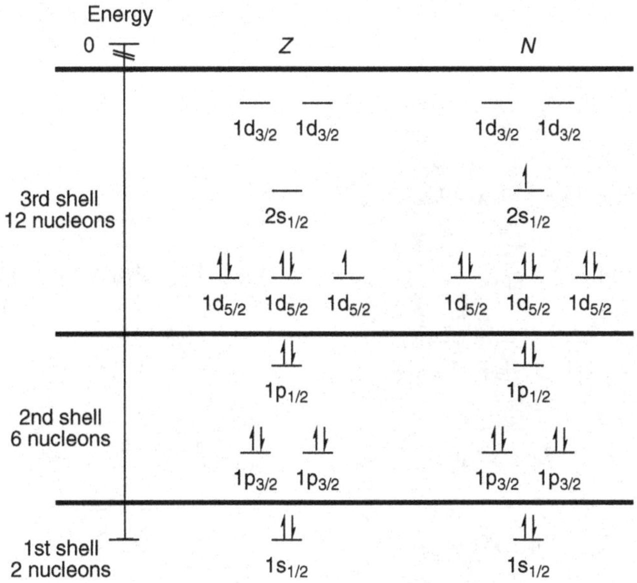

Figure 4.7 Nuclear energy levels showing the nucleon filling for aluminium-28.

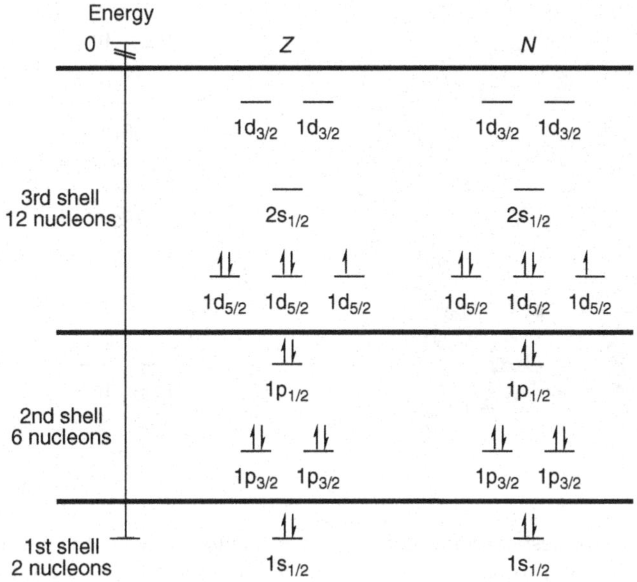

Figure 4.8 Nuclear energy levels showing the nucleon filling for aluminium-26.

means that more neutrons are required just to maintain energy parity with the protons. Consider Figure 4.9, which shows the relative proton and neutron shell energies for yttrium (no nucleons are shown).

PROBLEM 4.8
Predict the β decay product nuclide or starting nuclide.

a. $^{18}F \rightarrow$

b. $^{3}H \rightarrow$

c. $\rightarrow {}^{11}B + \beta^+ + \nu_e$

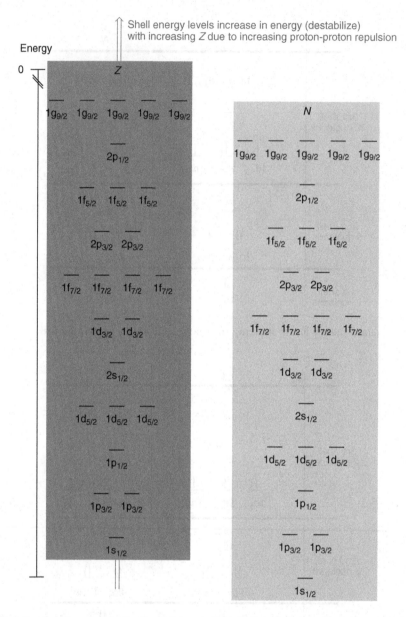

Figure 4.9 Progressive destabilization of the proton shells relative to the neutron shells as protons are added and proton–proton repulsion increases. Here the relative energies for the two shells are shown for yttrium.

d. $\rightarrow\ ^{233}\text{Pa} + \beta^- + \bar{\nu}_e$

e. $^{49}\text{V} + e^- \rightarrow$

f. $^{14}\text{C} \rightarrow$

g. $\rightarrow\ ^{133}\text{Cs} + \beta^- + \bar{\nu}_e$

h. $\rightarrow\ ^{99}\text{Ru} + \beta^- + \bar{\nu}_e$

PROBLEM 4.9
Yttrium is a monoisotopic element, that is, it possesses only one stable nuclide.

a. Explain why yttrium-89 is the only stable nuclide of yttrium.

b. What type of decay is yttrium-88 most likely to undergo? Write a nuclear equation to show this decay.

c. What type of decay is yttrium-90 most likely to undergo? Write a nuclear equation to show this decay.

γ Decay

The last type of decay that will be considered is γ decay.[25] In γ decay, a nuclide emits γ rays, a form of electromagnetic radiation (Chapter 6) as the nucleus goes from an excited state (indicated with an asterisk [*]) to a ground state, for example:

$$^{60}Ni^* \rightarrow {}^{60}Ni + \gamma$$

After α decay and β decay, the resulting product nuclide can be left in an excited state. To return to the ground state, γ rays are emitted. Whereas most of the topics discussed here are at the intersection of chemistry and physics, γ decay is almost entirely a physics concept and interested readers are encouraged to look more deeply into nuclear physics.

NUCLIDE TRANSMUTATION SUMMARY AND DECAY CHAINS

The transmutation of nuclides through radioactive decay was first summarized by Fajans and Soddy as the radioactive displacement law (Figure 4.10).[26] Understanding these changes can help us to understand not only naturally occurring radioactivity but also the results of nuclide synthesis (Chapter 5).

We will use the radioactive displacement law to conclude this chapter and consider radioactive decay chains. Figure 4.11 shows the radioactive decay chain from uranium-238 to lead-206, which is often called the uranium series or the radium series.

Decay chains, like those in Figure 4.10, represent the known pathways for naturally occurring radionuclides to decay into stable nuclides (typically lead nuclides). In addition, knowledge of this

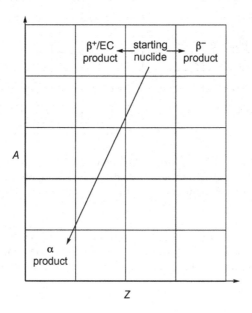

Figure 4.10 Nuclide transmutation summary and the effect on mass number (*A*) and atomic number (Z).

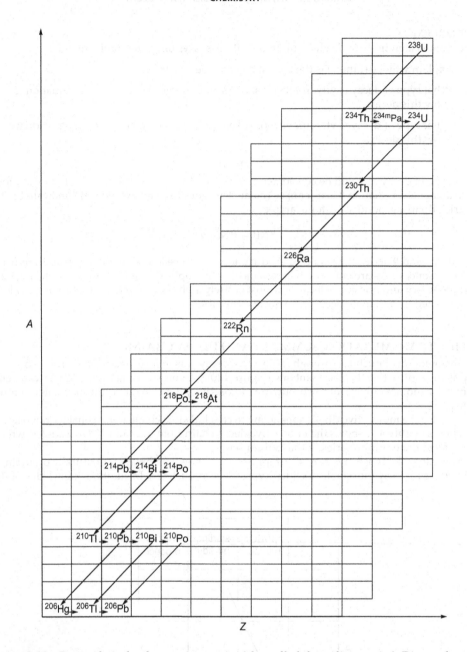

Figure 4.11 Decay chain for the uranium series (also called the radium series). Diagonal arrows represent α decay and horizontal arrows represent β⁻ decay. The stable end nuclide is lead-206. Empty cells represent other known nuclides, but they are omitted for clarity as they are not part of this decay series.

decay pathway provides us with a means of determining the age of rocks (radiometric dating). By comparing the ratio of lead-206 to uranium-238 and knowing how long it takes this process to occur (see below), we can determine the age of rocks that crystallized from 1 million years ago to over 4.5 billion years ago.

PROBLEM 4.10
Identify the decay products for each step of the thorium series. The thorium decay series starts with thorium-232 and undergoes the following decay steps to produce lead-208: α, β⁻, β⁻, α, α, α, α, β⁻, β⁻, α.

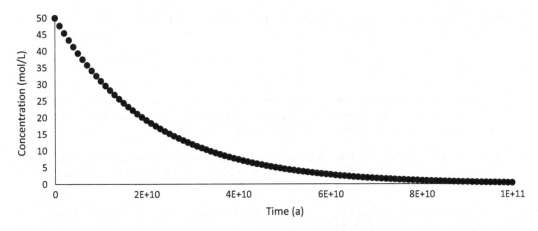

Figure 4.12 Plot of thorium-232 amount concentration over time (a = years) showing first-order, exponential decay.

Finally, you may be familiar with the fact that radioactive materials, especially radioactive waste from reactors, present a storage problem. The problem is that the original radioactive material and many of the byproducts are highly radioactive and will remain so for hundreds or thousands of years. To understand the longevity of radioactive materials, scientists study the rate of decay and quantify the rate constant for the first-order decay (Chapter 14) of each radionuclide. While the rate constant (the number of times a decay happens per second) can be useful in many contexts, in nuclear chemistry we tend to think in half-lives ($t_{1/2}$). One half-life corresponds to the amount of time it takes half of a radioactive nuclide to transmute,[27] through either α decay or β decay. After one half-life, 50% of the original radionuclide will have transmuted (into a new nuclide). After two half-lives, 75% of the original radionuclide will have transmuted. After three half-lives, 87.5%, and 95.75% after four half-lives, and so forth. In the case of thorium-232, the half-life is 1.45×10^{10} years, so it takes 1.45×10^{11} years (ten half-lives, the amount recommended by nuclear safety guidelines) to decay to a safe level (Figure 4.12). The half-life of the uranium-238 to lead-206 decay chain, useful in lead–uranium radiometric dating, is 4.47 billion years. Kinetics and the rate of change in a process will be considered in greater detail in Chapter 14.

NOTES

1. (i) Heisenberg, W. Über den Bau der Atomkerne. I. Z. *Phys.*, **1932**, *77* (1–2), 1–11. DOI: 10.1007/BF01337585.

   (ii) Heisenberg, W. Über den Bau der Atomkerne. II. Z. *Phys.*, **1932**, *78* (3–4), 156–164. DOI: 10.1007/BF01342433.

   (iii) Heisenberg, W. Über den Bau der Atomkerne. III. Z. *Phys.*, **1933**, *80* (9–10), 587–596. DOI: 10.1007/BF01335696.

2. A note on terminology: A nuclide is an atom with a specific atomic number, mass number, and nuclear energy state: Kohman, T.P. Proposed New Word: *Nuclide*. Am. *J. Phys.*, **1947**, *15* (4), 356–357. DOI: 10.1119/1.1990965.

3. (i) Yukawa, H. On the Interaction of Elementary Particles. *Proc. Phys.-Math Soc. Jpn.*, **1935**, *17*, 48–57. DOI: 10.11429/ppmsj1919.17.0_48.

   (ii) Kellogg, J.M.B.; Rabi, I.I.; Ramsey Jr, N.F.; Zacharias, J.R. An Electrical Quadrupole Moment of the Deuteron. *Phys. Rev.*, **1939**, *55*, 318–319. DOI: 10.1103/PhysRev.55.318.

   (iii) Reid Jr, R.V. Local Phenomenological Nucleon-Nucleon Potentials. *Ann. Phys.*, **1968**, *50* (3), 411–448. DOI: 10.1016/0003-4916(68)90126-7.

4. (i) Gell-Mann, M. A Schematic Model of Baryons and Mesons. *Phys. Lett.*, **1964**, *8* (3), 214–215. DOI: 10.1016/S0031-9163(64)92001-3.

   (ii) Zweig, G. An SU_3 Model for Strong Interaction Symmetry and Its Break. *CERN Report "CERN-TH-412,"* **1964**. DOI: 10.17181/CERN-TH-412.

5. Liquid drop model of the nucleus considers the electrostatic repulsion as a destabilizing factor in the overall cohesiveness of a nucleus. von Weizsäcker, C.F. Zur Theorie der Kernmassen. *Z. Phys.*, **1935**, *96* (7-8), 431–458. DOI: 10.1007/BF01337700.

6. It should be noted that even though nickel-62 is the nucleus with the greatest binding energy per nucleon, iron-56 is more abundant as it is an end-product of helium fusion in stars (Chapter 5).

7. (i) Fewell, M.P. The Atomic Nuclide with the Highest Mean Binding Energy. *Am. J. Phys.*, **1995**, *63* (7), 653–658. DOI: 10.1119/1.17828.

   (ii) Harsha, N.R.S. The Tightly Bound Nuclei in the Liquid Drop Model. *Eur. J. Phys.*, **2018**, *39* (3), 035802. DOI: 10.1088/1361-6404/aaa345.

8. Einstein, A. Ist die Trägheit eines Körpers von seinem Energieinhalt abhängig? *Ann. Phys.*, **1905**, *323* (13), 639–641. DOI: 10.1002/andp.19053231314.

9. (i) Becquerel, H. Sur les radiations émises par phosphorescence. *C. R. Acad. Sci.*, **1896**, *122*, 420–421.

   (ii) Rutherford, E. VIII. Uranium Radiation and the Electrical Conduction Produced by It. *Lond. Edinb. Dublin Philos. Mag. J. Sci. Series 5*, **1899**, *47* (284), 109–163. DOI: 10.1080/14786449908621245.

10. (i) Rutherford, E.; Soddy, F. XLI. The Cause and Nature of Radioactivity.–Part I. *Lond. Edinb. Dublin Philos. Mag. J. Sci. Series 6.* **1902**, *4* (21), 370–396. DOI: 10.1080/14786440209462856.

    (ii) Rutherford, E.; Soddy, F. LXIV. The Cause and Nature of Radioactivity.–Part II. *Lond. Edinb. Dublin Philos. Mag. J. Sci. Series 6.* **1902**, *4* (23), 569–585. DOI: 10.1080/14786440209462881.

11. Rutherford, E.; Royds, T. Spectrum of the Radium Emanation. *Nature*, **1908**, *78*, 220–221. DOI: 10.1038/078220c0.

12. Gamow, G. Zur Quantentheorie des Atomkernes. *Z. Phys.*, **1928**, *51* (3-4), 204–212. DOI: 10.1007/BF01343196.

13. Becquerel, H. The Radio-Activity of Matter. *Nature*, **1901**, *63* (1634), 396–398. DOI: 10.1038/063396d0.

14. (i) Fermi, E. Versuch einer Theorie der β-Strahlen. I *Z. Phys.* **1934**, *88* (3-4), 161–177. DOI: 10.1007/BF01351864.

    (ii) Salam, A.; Ward, J.C. Electromagnetic and Weak Interactions. *Phys. Lett.*, **1964**, *13* (2), 168–171. DOI: 10.1016/0031-9163(64)90711-5.

    (iii) Glashow, S.L. Partial-Symmetries of Weak Interactions. *Nucl. Phys.*, **1961**, *22* (4), 579–588. DOI: 10.1016/0029-5582(61)90469-2.

    (iv) Weinberg, S. A Model of Leptons. *Phys. Rev. Lett.*, **1967**, *19* (21), 1264–1266. DOI: 10.1103/PhysRevLett.19.1264.

15. When saying these aloud, β⁻ decay is read as "beta minus decay" and β⁺ decay is read as "beta plus decay."

16. Curie, I; Joliot, M.F. Un noveau type de radioactivité. *C. R. Acad. Sci.*, **1934**, *198*, 254–256

17. Alvarez, L.W. Nuclear *K* Electron Capture. *Phys. Rev.*, **1937**, *52* (2), 134–135. DOI: 10.1103/PhysRev.52.134.

18. Gapon, E.; Iwanenko, D. Zur Bestimmung der Isotopenzahl. *Naturwissenschaften*, **1932**, *20*, 792–793.

19. Wigner, E. On the Consequences of the Symmetry of the Nuclear Hamiltonian on the Spectroscopy of Nuclei. *Phys. Rev.*, **1937**, *51* (2), 106–119. DOI: 10.1103/PhysRev.51.106.

20. (i) Goeppert Mayer, M. Nuclear Configurations in the Spin-Orbit Coupling Model. I. Empirical Evidence. *Phys. Rev.*, **1950**, *78* (1), 16–21. DOI: 10.1103/PhysRev.78.16.

    (ii) Goeppert Mayer, M. Nuclear Configurations in the Spin-Orbit Coupling Model. II. Theoretical Considerations. *Phys. Rev.*, **1950**, *78* (1), 22–23. DOI: 10.1103/PhysRev.78.22.

21. Haxel, O.; Jensen, J.H.D.; Suess, H.E. On the "Magic Numbers" in Nuclear Structure. *Phys. Rev.*, **1949**, *75* (11), 1766. DOI: 10.1103/PhysRev.75.1766.2.

22. Liquid drop model of the nucleus considers the electrostatic repulsion as a destabilizing factor in the overall cohesiveness of a nucleus. von Weizsäcker, C.F. Zur Theorie der Kernmassen. *Z. Phys.*, **1935**, *96* (7–8), 431–458. DOI: 10.1007/BF01337700.

23. (i) Pauli, W. Über die Gesetzmässigkeiten des anomalen Zeemaneffektes. *Z. Phys.*, **1923**, *16*, 155-164. DOI: 10.1007/BF01327386.

    (ii) Pauli, W. Über den Einfluss der Geschwindigkeitsabhängigkeit der Elektronenmasse auf den Zeemaneffekt. *Z Phys.*, **1925**, *31*, 373-385. DOI: 10.1007/BF02980592.

    (iii)Pauli, W. Über den Zusammenhang des Abschlusses der Elektronengruppen im Atom mit der Komplexstruktur der Spektren. *Z. Phys.*, **1925**, *31*, 765-783. DOI: 10.1007/BF02980631.

24. Cottingham, W.N. *An introduction to Nuclear Physics.* Cambridge University Press: New York, 1937.

25. Rutherford, E. XV. The Magnetic and Electric Deviation of the Easily Absorbed Rays from Radium. *Lond. Edinb. Dublin Philos. Mag. J. Sci. Series 6*, **1903**, *5* (26), 177–187. DOI: 10.1080/14786440309462912.

26. (i) Fajans, K. Die radioaktiven Umwandlungen und das periodische System der Elemente. *Ber. Dtsch. Chem. Ges.*, **1913**, *46* (1), 422–439. DOI: 10.1002/cber.19130460162.

    (ii) Soddy, F. The Radio-Elements and the Periodic Law. *Chem. News.*, **1913**, *107*, 97–99.

27. Rutherford, E.; Soddy, F. LX. Radioactive Change. *Lond. Edinb. Dublin Philos. Mag. J. Sci. Series 6.*, **1903**, *5* (29), 576–591. DOI: 10.1080/14786440309462960.

5 Nuclear Reactions

In Chapter 4 we discussed the factors that contribute to nuclear stability and nuclear instability. In this chapter, we will discuss how nuclei can be altered through fusion (combining nuclei) and fission (breaking down nuclei). Since the change in the number of nucleons affects the binding energy, this means that fusion and fission reactions can produce significant amounts of energy: 1 to 30 MeV, which corresponds to 1×10^8 kJ/mol to 3×10^9 kJ/mol (0.1 TJ/mol to 3 TJ/mol). These impressively large amounts of energy help to explain how stars, nuclear reactors, and nuclear weapons can produce so much energy. We will discuss fusion first – it is the first type of nuclear reaction that scientists were able to accomplish, and it led to our understanding of atomic structure – before turning to fission. Special attention will be given to the importance of fusion and fission in a broader context.

FUSION

Fusion is a nuclear reaction where smaller particles combine to produce larger nuclides. If we reconsider the binding energy per nucleon (Figure 5.1), binding energy increases from hydrogen-1 to nickel-62. For nuclides with a larger mass than nickel-62, binding energy per nucleon decreases. This means that fusion is exergonic, i.e., releases energy, until nuclides approach a mass number of 62, and the process is endergonic, i.e., takes in energy, as mass numbers increase past 62. The stupendous release of energy when small nuclides undergo fusion is the source of energy that powers stars (see below), and ⤳ the reason why fusion reactors are of interest as sources of energy for a sustainable, decarbonized energy future.

The first nuclear fusion reaction conducted in a laboratory was the reaction that led to Rutherford's discovery of the proton.[1] In this work, α particles were fired into the air, and the fusion of an α particle with nitrogen-14 produced oxygen-17 and a proton. This fusion reaction can be shown according to the conventions of a chemical equation:

$$^{14}\text{N} + \alpha \rightarrow {}^{17}\text{O} + \text{p}$$

The recommended representation of this reaction, however, looks somewhat different:

$$^{14}\text{N}(\alpha, \text{p})^{17}\text{O}$$

In this notation, the initial nuclide is listed first, and the product nuclide is listed after the parentheses. Within the parentheses, the first term is any incoming particle(s) or quanta of light (typically a γ ray), and the second term is any outgoing particle(s) or quanta of light. If we consider another example, from James Chadwick's discovery of the neutron,[2] we can represent a collision of an α particle with beryllium-9 to produce carbon-12 and a neutron as:

$$^{9}\text{Be}(\alpha, \text{n})^{12}\text{C}$$

These two examples are exemplars of early fusion work that looked at bombarding nuclei with hydrogen nuclides (hydrons),[3] with α particles,[4] and with neutrons.[5] The use of α particles and neutrons also led to the synthesis of new nuclides that are not found in nature. Consider the synthesis, from the Joliot-Curie laboratory, of phosphorus-30 from aluminium-27:

$$^{27}\text{Al}(\alpha, \text{n})^{30}\text{P}$$

Also consider the synthesis, from the Fermi laboratory, of sodium-24 from aluminium-27:

$$^{27}\text{Al}(\text{n}, \alpha)^{24}\text{Na}$$

In both examples, the reaction involves the fusion of an incoming particle, the ejection of a particle, and the formation of a new nuclide. In addition to being examples of fusion, the new nuclides are radioactive. These syntheses of radionuclides were the first instances of artificial, or synthetic, radioactivity.

PROBLEM 5.1
Fusion requires the combination of nuclear particles. Why is it harder to fuse a proton or α particle with a nucleus than it is to fuse a neutron with a nucleus?

PROBLEM 5.2
Consider the fusion of hydrogen-2 (2.014 10 u) and hydrogen-3 (3.016 05 u) to produce helium-4 (4.002 60 u) and a neutron (1.008 66 u). Note that hydrogen-2 is called deuterium (atomic symbol D) and hydrogen-3 is called tritium (atomic symbol T):[6]

DOI: 10.1201/9781003479338-5

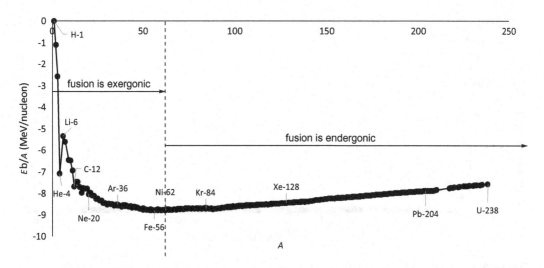

Figure 5.1 Binding energy per nucleon and division between where fusion is exergonic ($1 < A \leq 62$) and where fusion is endergonic ($A > 62$).

$$^3T(^2D, n)^4He$$

a. Determine the change in mass for this process (Δm) and the energy released by this fusion process (in MeV and in kJ/mol).

b. If burning methane (CH_4) produces 890 kJ/mol, what mass (kg) of methane needs to be combusted to produce the same energy as the fusion of 1.0 kg tritium with deuterium?

PROBLEM 5.3
While the process for heavy elements is endergonic, fusion has allowed scientists to synthesize the transuranic ($Z > 92$) elements and push the boundaries of the periodic table.

a. Curium-242 (242.058 83 u) is synthesized through the fusion of plutonium-239 (239.052 16 u) with an α particle (4.002 60 u).

$$^{239}Pu(\alpha, n)^{242}Cm$$

Determine the change in mass for this process (Δm) and the energy required to make this fusion process occur (in MeV and in kJ/mol).

b. What must be the velocity (m/s) of α particles ($M = 0.004\ 002\ 60$ kg/mol) to impart enough kinetic energy to make plutonium–α particle fusion occur (1 J = 1 kg m^2/s^2)? What percentage of the speed of light is this?

CHEMISTRY IN CONTEXT: Nucleosynthesis

The creation of nuclides is called nucleosynthesis. In this section, we will consider the origin of the nuclides that exist in our universe. At the beginning of our universe, there was a rapid transition from hot and dense to cold and dilute. The synthesis of the first stable nuclides – ^1H, ^2H, ^3He, ^4He, and ^7Li – began a few seconds after the universe started expanding, after the Big Bang,[7] and stopped after only 20 minutes, as the universe was then cool and dilute enough that further fusion was impossible.[8,9] The most abundant nuclide in the universe, hydrogen-1, consists of only a proton and is the primordial nuclide. The other nuclides are the result of proton–neutron fusion and nuclide–nuclide fusion. It should be noted that early nucleosynthesis was hampered by the fact that there are no stable nuclides with a mass number of five, which meant that helium-4 was the key end point for the majority of Big Bang nucleosynthesis, and only a tiny fraction of heavier lithium-7 was produced. At the end of Big Bang nucleosynthesis, the universe consisted of

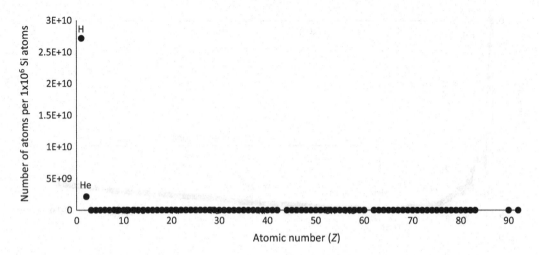

Figure 5.2 Solar system abundances of the elements. The data shows the number of atoms for every 1 × 10⁶ silicon atoms. (From (i) Suess, H.E.; Urey, H.C. Abundances of the Elements. *Rev. Mod. Phys.*, **1956**, *28* (1), 53–74. DOI: 10.1103/RevModPhys.28.53. (ii) Anders, E.; Ebihara, M. Solar-System Abundances of the Elements. *Geochim. Cosmochim. Acta.*, **1982**, *46* (11), 2363–2380. DOI: 10.1016/0016-7037(82)90208-3.)

predominantly hydrogen-1 (92%) and helium-4 (7%) nuclides (Figure 5.2). While free protons are stable, which means that all protons in the universe have existed since the Big Bang, free neutrons are unstable and undergo β⁻ decay, which means that almost all primordial neutrons exist within helium-4 nuclides.[10] Free neutrons that exist now are the result of nuclear reactions (see James Chadwick's reaction of beryllium and an α particle above).

After the initial cooling period and the end of Big Bang nucleosynthesis, the universe was a dark place that consisted of clouds of primordial nuclides, predominantly hydrogen-1 (Figure 5.2). About 155 million years after the Big Bang, the first star formed from the gravitational coalescence of hydrogen gas clouds and we call this era, the current era of the universe, the Stelliferous Era. The formation of fusion-powered stars has led to the development of nuclides larger than lithium-7.[11] Note that if the abundances in Figure 5.2 are shown on a logarithmic scale, we can see the distribution of more massive nuclides (Figure 5.3). We will consider the fusion processes that led to these more massive nuclides.

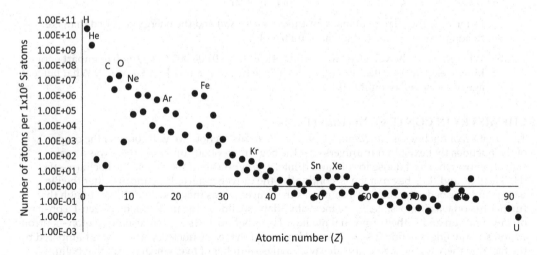

Figure 5.3 Solar system abundances of the elements. The data shows the number of atoms for every 1 × 10⁶ silicon atoms on a logarithmic scale.

The initial phase of stellar nucleosynthesis, the formation of nuclei in stars, is the fusion of hydrogen-1 into helium-4 in the sun's core. In this main-sequence phase, there are two main methods for hydrogen fusion: the proton–proton chain (p-p chain)[12] and the carbon–nitrogen–oxygen (CNO) cycle.[13] Regardless of the process, the overall transformation is that four protons (hydrogen-1 nuclides) are converted into a helium-4 nuclide, positrons, electron neutrinos, and γ rays (the exact number of γ rays depends on the cycle).

$$4\,^1\text{H} \rightarrow {}^4\text{He} + 2\,e^+ + 2\,\nu_e + 2\,\gamma$$

As helium-4 accumulates in the star's core and hydrogen is depleted, the star evolves into a new phase, which depends on the mass of the star. For small stars, fusion ceases and they become white dwarfs that slowly emit their residual heat and light. For larger stars, helium fusion occurs. One process is the triple-alpha process, where the helium-4 nuclides are converted to carbon-12.[14] This requires the fusion of two α particles to form beryllium-8, which is highly unstable and rapidly undergoes α decay. In the triple-alpha process, this short-lived nuclide is rapidly fused with another α particle to form the stable carbon-12 nuclide.

$$\alpha + \alpha \rightleftharpoons {}^8\text{Be} + \alpha \rightarrow {}^{12}\text{C}$$

Another form of helium fusion is the alpha process, also called the alpha ladder.[15] In this process, α particles are added to carbon-12 and all subsequent nuclides, which produces a series of nuclides that sequentially have a mass number difference of four and an atomic number difference of two.

$$^{12}\text{C}(\alpha, \gamma)^{16}\text{O}$$

$$^{16}\text{O}(\alpha, \gamma)^{20}\text{Ne}$$

$$^{20}\text{Ne}(\alpha, \gamma)^{24}\text{Mg}$$

$$^{24}\text{Mg}(\alpha, \gamma)^{28}\text{Si}$$

$$^{28}\text{Si}(\alpha, \gamma)^{32}\text{S}$$

$$^{32}\text{S}(\alpha, \gamma)^{36}\text{Ar}$$

$$^{36}\text{Ar}(\alpha, \gamma)^{40}\text{Ca}$$

$$^{40}\text{Ca}(\alpha, \gamma)^{44}\text{Ti}$$

$$^{44}\text{Ti}(\alpha, \gamma)^{48}\text{Cr}$$

$$^{48}\text{Cr}(\alpha, \gamma)^{52}\text{Fe}$$

$$^{52}\text{Fe}(\alpha, \gamma)^{56}\text{Ni}$$

The last four nuclides produced by the alpha process are unstable as there are too few neutrons in the nucleus. And so, titanium-44, chromium-48, iron-52, and nickel-56 all undergo two successive β^+ decays to produce calcium-44, titanium-48, chromium-52, and iron-56. The alpha process stops at iron-56 because successive fusion events are hampered by photodisintegration, where high-energy γ rays lead to the breakdown of nickel-62 and other heavier nuclides. This process produces ^{16}O, ^{20}Ne, ^{24}Mg, ^{28}Si, ^{32}S, ^{36}Ar, ^{40}Ca, ^{44}Ca, ^{48}Ti, ^{52}Cr, and ^{56}Fe. These nuclides are some of the most abundant nuclides in the solar system (Figure 5.4) because they are the product of this common helium fusion process.

After helium fusion, stars can undergo a variety of fusion processes, depending on their mass and core composition and density: carbon burning (which produces sodium-23 and

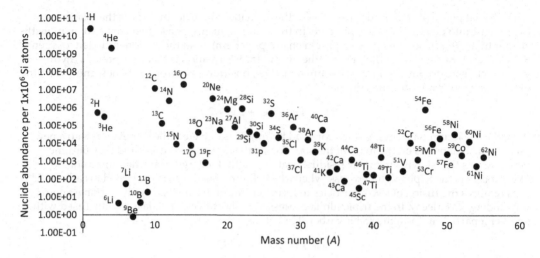

Figure 5.4 Solar system abundances of selected nuclides with mass numbers (A) between 1 and 62. The data shows the number of atoms for every 1×10^6 silicon atoms on a logarithmic scale.

magnesium-24),[16] neon burning (which produces oxygen-16 and magnesium-24),[17] oxygen burning (which produces predominantly silicon-28 and phosphorus-31),[18] and, finally, silicon burning (which produces alpha process products).[19] These processes are ultimately followed by a supernova if a star is massive enough to go through all of these different stages. In sum, stellar nucleosynthesis produces many of the common elements with which we are familiar, but a large portion of the periodic table, especially those elements larger than iron and nickel, are missing from our discussion so far. Those elements are formed through other mechanisms.

One process that produces other nuclides is spallation. In spallation, a nuclide ejects a single nucleon after being bombarded with high-energy particles or γ rays in solar wind and cosmic rays, or in supernovae. Spallation is one of the pathways of fluorine-19 nucleosynthesis.[20] The bombardment of a neon-20 nuclide with high-energy tau neutrino (ν_τ) displaces a proton and creates fluorine-19.

$$^{20}\text{Ne}(\nu_\tau, \text{p})^{19}\text{F}$$

To produce nuclides more massive than iron or nickel, however, relies upon a combination of neutron fusion and β⁻ decay events that either take place slowly in red-giant stars, the s-process,[21] or rapidly in a supernova or neutron star merger, the r-process.[22] In either of these situations, iron-56, the seed nuclide, is bombarded with neutrons produced by stellar nucleosynthesis. This can lead to a new stable nuclide, like iron-57. If enough neutrons fuse with the nucleus, however, an unstable nuclide, iron-59, can form that will then undergo β⁻ decay to produce a new nuclide; iron-59 produces cobalt-59.

$$^{59}\text{Fe} \rightarrow {}^{59}\text{Co} + \beta^- + \bar{\nu}_e$$

The difference between these processes is whether the process is slow, the s-process, or rapid, the r-process. In the case of slow neutron capture, β⁻ decay happens between neutron capture steps; in rapid neutron capture, the nuclide captures many neutrons before undergoing a series of β⁻ decay events. The r-process requires very high densities of neutrons and happens only in cosmologically catastrophic events: supernovas and neutron star mergers. The observation of a neutron star merger in 2017, by gravity-wave and electromagnetic frequency detectors, provided firm evidence for the r-process, including both the production of strontium, and the formation of an estimated ten Earth masses of gold and platinum combined.[23]

FISSION

Fission is a nuclear reaction where larger particles are split to produce smaller nuclides. Considering the binding energy per nucleon (Figure 5.5), fission is exergonic (releases energy)

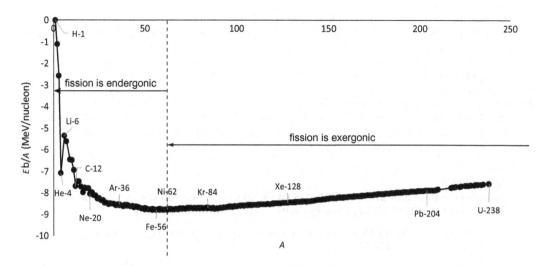

Figure 5.5 Binding energy per nucleon and division between where fission is exergonic ($A > 62$) and where fission is endergonic ($1 < A \leq 62$).

for nuclides larger than nickel-62 and is endergonic (takes in energy) as mass numbers decrease below 62.

In the early discovery of nuclear reactions, there are numerous examples of the ejection of small particles, spallation, following the bombardment of nuclides. This spallation process is how Rutherford discovered protons and Chadwick discovered neutrons. The first evidence of nuclide fragmentation rather than spallation came from bombarding lithium-7 nuclides with very fast protons, ^{7}Li(p, f), where f represents the formation of smaller nuclide fragments.[24] In this case the fragments were two α particles.

Fission, the fragmentation of a nucleus following neutron absorption, was first observed when uranium was bombarded with fast neutrons. Initially, the radioactive products were erroneously classified as newly formed transuranic ($Z > 92$) elements, the product of a laboratory-scale s-process.[25] After further work, including careful isolation of the products, it was demonstrated that uranium had fragmented into barium and other nuclides smaller than uranium.[26] This fragmentation occurs because the incoming neutron excites and deforms the nucleus, and if deformed enough the nucleus can break down.[27,28] While the range of possible nuclide products is large – over 800 possible nuclide fragments[29] – the mass ratio of each pair of products produced by a single nuclide fission is always in a roughly three-to-two atomic mass (A) ratio (Figure 5.6).[30] The

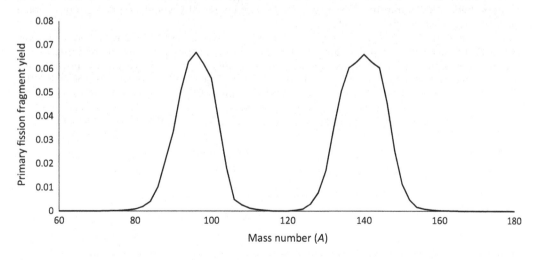

Figure 5.6 Uranium-235 fission fragment yield showing the distribution of atomic mass of the products and the fraction of their yield from fission.

three-to-two atomic mass ratio of fission fragments – rather than equal-mass products – can be understood through the lens of nuclear shell stability: The energy pathway that leads to fragments that have stable nuclear shell configurations is the lower energy pathway.[31]

Nuclides that can break down when bombarded with neutrons are called fissionable. There are a small handful of fissionable nuclides: ^{232}U, ^{233}U, ^{234}U, ^{235}U, ^{237}Np, ^{238}Pu, ^{239}Pu, ^{240}Pu, ^{241}Pu, ^{242}Pu, ^{241}Am, ^{242m}Am, ^{243}Am, ^{242}Cm, ^{243}Cm, ^{244}Cm, ^{245}Cm, ^{246}Cm, ^{247}Cm, ^{249}Cf, and ^{251}Cf.[32] Of these, the ones that are most important in both world history and contemporary international politics are fissile nuclides. Fissile nuclides are those fissionable nuclides that are fissionable with slow neutrons and capable of sustaining a chain reaction: ^{233}U, ^{235}U, and ^{239}Pu.[33] A nuclear chain reaction is the crux of how nuclear technology – reactors and weapons – works.

PROBLEM 5.4

Consider the fission of uranium-235 (235.043 93 u) with a neutron (1.008 66 u) to produce barium-144 (143.922 95 u), krypton-89 (88.917 63 u), and three neutrons.

a. Determine the change in mass for this process (Δm) and the energy released by this fusion process (in MeV and in kJ/mol).

b. If burning methane (CH_4) produces 890 kJ/mol, what mass (kg) of methane needs to be combusted to produce the same energy as the fission of 1.0 kg uranium-235?

CHEMISTRY IN CONTEXT: Nuclear Technology

A chain reaction is a self-sustaining process where the products of one event initiate another event. The hypothesis of a nuclear chain reaction was proposed by Leo Szilard shortly after the discovery of the neutron, six years before the discovery of a fission. Szilard was born in Hungary but fled to Germany and then the United Kingdom[34] as the tide of antisemitism surged across Europe. A prolific scientist, Szilard saw the potential impact of a nuclear chain reaction and its importance for the 20th century: atomic energy and, of greater concern in interwar Europe, nuclear weapons.[35]

First, let us consider what a nuclear chain reaction is. When bombarded with a neutron, a uranium-235 nucleus splits into two smaller fission fragments (Figure 5.6). In the process, there are also free neutrons that are released. It was found that there are about three neutrons produced for every fission event.[36] This process is shown pictorially in Figure 5.7.

Since uranium-235 is a larger nuclide than nickel-62, this process produces energy (Problem 5.4). Naturally occurring uranium is 0.720% uranium-235 and 99.28% uranium-238. Uranium-238 is not fissile, and so if naturally occurring uranium has a uranium-235 nuclide undergo fission, then the process will not continue as uranium-238 will absorb any produced neutrons (and through β^- decay produces plutonium-239). If uranium is enriched so that the percent of uranium-235 is greater ($\geq 20\%$), then a chain reaction can be achieved. Figure 5.8 shows the start of a chain reaction in a hypothetical 100% uranium-235 sample. Each fragmentation by a neutron produces on average three new neutrons, which initiates further fission.

The sustained fission of a sample of enriched uranium is what Leo Szilard envisioned. Its importance is that such a process could potentially be used to generate power or new, more destructive weapons: atomic bombs.[37] At this moment, the development of atomic science, nuclear power's awesome potential, and the events of world history collide. The discovery of the neutron (1932), the vision of a chain reaction (1932), and the discovery of fission (1938) all occurred in minds and in laboratories. During this time, the wider world was embroiled in the Great Depression (1929–1939), rising right-wing authoritarianism and fascism, and ultimately World War II (1937–1945). In this

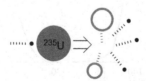

Figure 5.7 Diagrammatic representation of uranium-235 (filled circle) fission from neutron bombardment (small black circle) to produce two fission fragments (open circles) in a 3:2 mass ratio (Figure 5.6) and three neutrons.

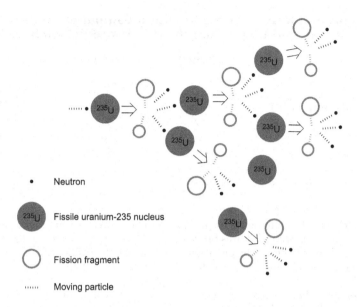

Figure 5.8 Diagrammatic representation of uranium-235 chain reaction in a 100% enriched sample.

climate, the knowledge of fission-chain-reaction research led to a secretive race to develop atomic weapons among Japan, Germany, and the United Kingdom.

The United States developed an atomic weapons committee and later a development program – the Manhattan Project – following a letter from Einstein, and Szilard, to President Franklin Delano Roosevelt.[38] In the letter, the importance, and danger, of this work was forcefully expressed:

> Some recent work by E. Fermi and L. Szilard, which has been communicated to me in manuscript, leads me to expect that the element uranium may be turned into a new and important source of energy in the immediate future. Certain aspects of the situation which has arisen seem to call for watchfulness and, if necessary, quick action on the part of the [Roosevelt] Administration. I believe therefore it is my duty to bring to your attention the following facts and recommendations:
>
> In the course of the last four months it has been made probable – through the work of Joliot in France as well as Fermi and Szilard in America – that it may become possible to set up a nuclear chain reaction in a large mass of uranium, by which vast amounts of power and large quantities of new radium-like elements would be generated. Now it appears almost certain that this could be achieved in the immediate future.
>
> This new phenomenon would also lead to the construction of bombs, and it is conceivable – though much less certain – that extremely powerful bombs of a new type may thus be constructed. A single bomb of this type, carried by boat and exploded in port, may very well destroy the whole port together with some of the surrounding territory. However, such bombs might very well prove to be too heavy[39] for transportation by air.

The Manhattan Project would ultimately succeed in producing several nuclear bombs. Two of these were dropped on Hiroshima and Nagasaki, Japan. The devastation in each of these two cities was shocking. In Hiroshima, it was found that roughly a third of the population was killed and a third were severely injured in the blast along with extensive destruction of the city. In Nagasaki, it was noted that "within a radius of one kilometer from ground zero, men and animals died almost

Table 5.1 Effort and Results of the Bombing Campaigns in Hiroshima, Nagasaki, and Tokyo from the U.S. Strategic Bombing Survey

	Hiroshima	Nagasaki	Tokyo
Planes	1	1	279
Bomb load	1 atomic	1 atomic	1667 tons incendiary
Population density (per square mile)	46 000	65 000	130 000
Square miles destroyed	4.7	1.8	15.8
Injured	70 000	40 000	102 000
Mortality rate per square mile destroyed	15 000	20 000	5300
Casualty rate per square mile destroyed	32 000	43 000	11 800

instantaneously from the tremendous blast pressure and heat; houses and other structures were smashed, crushed and scattered; and fires broke out."[40] While a high casualty rate and significant destruction occurred in both cities, the grim statistics bear out the full weight of the impact of the dropping of a single atomic bomb compared to the traditional carpet bombing that occurred in Tokyo (Table 5.1).

The terrible results of these two bombings, along with the threat of continued nuclear assault and Soviet invasion, ultimately led to Japanese capitulation to the United States and brought World War II to a close.

Important debate has and continues to occur about whether the bombing of Hiroshima and Nagasaki was the right choice, or even if it was justified. The advent of nuclear weapons has also led to a general challenge for humanity: What role should nuclear technology play in modern life? In warfare? In international relations and world politics? These are important questions, to which there are no easy answers. Consideration of these questions is important for citizens, nations, states, and the international community.

NOTES

1. Rutherford, E. LIV. Collision of α Particles with Light Atoms. IV. An Anomalous Effect in Nitrogen. *Lond. Edinb. Dublin Philos. Mag. J. Sci. Series 6.* **1919**, *37* (222), 581–587. DOI: 10.1080/14786440608635919.

2. Chadwick, J. The Existence of a Neutron. *Proc. R. Soc. Lond. A,* **1932**, *136* (830), 692–708. DOI: 10.1098/rspa.1932.0112.

3. Oliphant, M.L.E.; Harteck, P. and Rutherford, E. Transmutation Effects Observed with Heavy Hydrogen. *Proc. R. Soc. Lond. A,* **1934**, *144* (853), 692–703. DOI: 10.1098/rspa.1934.0077.

4. (i) Curie, I.; Joliot, F. Un nouveau type de radioactivité *C. R. Acad. Sci.,* **1934**, *198*, 254–256.

   (ii) Joliot, F.; Curie, I. Artificial Production of a New Kind of Radio-Element. *Nature,* **1934**, *133*, 201–202. DOI: 10.1038/133201a0.

5. (i) Fermi, E. Radioattività indotta da bombardamento di neutroni.–I. *La Ricerca Scientifica,* **1934**, 5, 282.

   (ii) Fermi, E. Radioactivity Induced by Neutron Bombardment. *Nature,* **1934**, *133*, 757. DOI: 10.1038/133757a0.

6. Urey, H.C.; Murphy, G.M.; Brickwedde, F.G. A Name and Symbol for H^2. *J. Chem. Phys.,* **1933**, *1* (7), 512–513. DOI: 10.1063/1.1749326.

7. Lemaître, G. Un Universe Homogène de Masse Constante et de Rayon Croissant, Rendant Compte de la Vitesse Radiale des Nébuleuses Extra-Galactiques. *Annales de la Société Scientifique de Burxelles,* **1927**, *A47*, 49–59.

8. Alpher, R.A.; Bethe, H.; Gamow, G. The Origin of the Chemical Elements. *Phys. Rev.*, **1948**, *73* (7), 803–804. DOI: 10.1103/PhysRev.73.803.

9. Steigman, G. Primordial Nucleosynthesis in the Precision Cosmology Era. *Annu. Rev. Nucl. Part. Sci.*, **2007**, *57* (1), 463–491. DOI: 10.1146/annurev.nucl.56.080805.140437.

10. All neutrons (other than the primordial ones in helium-4) are the result of stellar nucleosynthesis and β^+ decay.

11. (i) Eddington, A.S. The Internal Constitution of the Stars, *Science*, **1920**, *52* (1341), 233–240. DOI: 10.1126/science.52.1341.233.

    (ii) Eddington, A.S. The Internal Constitution of the Stars, *Nature*, **1920**, *106*, 14–20. DOI: 10.1038/106014a0.

12. Bethe, H.A.; Critchfield, C.L. The Formation of Deuterons by Proton Combination, *Phys. Rev.*, **1938**, *54* (4), 248–254. DOI: 10.1103/PhysRev.54.248.

13. Note that the carbon, nitrogen, and oxygen in the CNO cycle are small fractions of most stars and are present as the result of previous stars' production of heavy elements and their distribution among the interstellar medium through supernovae.

    (i) Bethe, H.A. Energy Production in Stars, *Phys. Rev.*, **1939**, *55* (1), 103–456. DOI: 10.1103/PhysRev.55.103.

    (ii) Bethe, H.A. Energy Production in Stars, *Phys. Rev.*, **1939**, *55* (5), 434–456. DOI: 10.1103/PhysRev.55.434.

    (iii) von Weizsäcker, C.F. Über Elementumwandlungen in inner der Sterne I, *Phys. Z.*, **1937**, *38*, 176–191.

    (iv) von Weizsäcker, C.F. Über Elementumwandlungen in inner der Sterne II, *Phys. Z.*, **1938**, *39*, 633–646.

14. (i) Hoyle, F. On Nuclear Reactions Occurring in Very Hot Stars. I. The Synthesis of Elements from Carbon to Nickel. *Astrophys. J. Supplement*, **1954**, *1*, 121–146. DOI: 10.1086/190005.

    (ii) Bethe, H.A. Energy Production in Stars, *Phys. Rev.*, **1939**, *55* (5), 434–456. DOI: 10.1103/PhysRev.55.434.

15. Burbidge, E.M.; Burbidge, G.R.; Fowler, W.A.; Hoyle, F. Synthesis of the Elements in Stars, *Rev. Mod. Phys.*, **1957**, *29* (4), 547–650. DOI: 10.1103/RevMOdPhys.29.547.

16. Arnett, W.D. Advanced Evolution of Massive Stars. II. Carbon Burning, *Astrophys. J.*, **1972**, *176*, 699–710. DOI: 10.1086/151672.

17. Arnett, W.D. Advanced Evolution of Massive Stars. V. Neon Burning, *Astrophys. J.*, **1974**, *193*, 169–176. DOI: 10.1086/163143.

18. Arnett, W.D. Advanced Evolution of Massive Stars. VI. Oxygen Burning, *Astrophys. J.*, **1974**, *194*, 373–383. DOI: 10.1086/163254.

19. Woosley, S.; Janka, T. The Physics of Core-Collapse Supernovae, *Nature Phys.*, **2005**, *1*, 147–154. DOI: 10.1038/nphys172.

20. Renda, A.; Fenner, Y.; Gibson, B.K.; Karakas, A.I.; Lattanzio, J.C.; Campbell, S.; Chieffi, A.; Cunha, K.; Smith, V.V. On the origin of fluorine in the Milky Way. *Mon. Not. R. Astron. Soc.*, **2004**, *354* (2), 575–580. DOI: 10.1111/j.1365-2966.2004.08215.x.

21. Clayton, D.D.; Fowler, W.A.; Hull, T.E.; Zimmerman, B.A. Neutron Capture Chains in Heavy Element Synthesis. *Ann. Phys.*, **1961**, *12* (3), 331–408. DOI: 10.1016/0003-4916(61)90067-7.

22. (i) Cameron, A.G.W. Nuclear Reactions in Stars and Nucleogenesis. *Publ. Astron. Soc. Pac.*, **1957**, *69* (408), 201–222. DOI: 10.1086/127051.

    (ii) Seeger, P.A.; Fowler, W.A.; Clayton, D.D. Nucleosynthesis of Heavy Elements by Neutron Capture, *Astrophys. J.*, **1972**, *176*, 699–710. DOI: 10.1086/151672.

    (iii) Woosley, S.; Trimble, V.; Thielemann, K.-F. The Origin of the Elements, *Phys. Today*, **2019**, *72* (2), 36–37. DOI: 10.1063/PT.3.4134.

23. (i) Arcavi, I.; Hosseinzadeh, G.; Howell, D.A.; McCully, C.; Poznanski, D.; Kasen, D.; Barnes, J.; Zaltzman, M.; Vasylyev, S.; Maoz, D.; Valenti, S. Optical Emission from a Kilonova Following a Gravitational-Wave Detected Neutron-Star Merger. *Nature*, **2017**, *551*, 64–66. DOI: 10.1038/nature24291.

    (ii) Pian, E.; D'Avanzo, P.; Benetti, S. *et al.* Spectroscopic Identification of r-Process Nucleosynthesis in a Double Neutron-Star Merger, *Nature*, **2017**, *551*, 67–70. DOI: 10.1038/nature24298.

    (iii) Smartt, S.; Chen, T.W.; Jerkstrand, A. *et al.* A Kilonova as the Electromagnetic Counterpart to a Gravitational-Wave Source. *Nature*, **2017**, *551*, 75–79. DOI: 10.1038/nature24303.

24. Cockcroft, J.D. and Walton, E.T.S. Experiments with High Velocity Positive Ions. II. -The Disintegration of Elements by High Velocity Protons. *Proc. R. Soc. Lond. A*, **1932**, *137* (831), 229–242. DOI: 10.1098/rspa.1932.0133.

25. (i) Fermi, E. Possible Production of Elements of Atomic Number Higher than 92. *Nature*, **1934**, *133*, 898–899. DOI: 10.1038/133898a0.

    (ii) Hahn, O.; Meitner, L.; Strassmann, F. Über die Trans-Urane und ihr chemisches Verhalten. *Ber. Dtsch. Chem. Ges.*, **1937**, *70* (6), 1374–1392. DOI: 10.1002/cber.19370700634.

    (iii) Meitner, L.; Hahn, O.; Strassmann, F. Über die Umwandlungsreihen des Urans, die durch Neutronenbestrahlung erzeugt warden. *Z. Phys.*, **1937**, *106* (3-4), 249–270. DOI: 10.1007/BF01340321.

26. (i) Hahn, O.; Strassmann, F. Über die Entstehung von Radiumisotopen aus Uran durch Bestrahlen mit schnellen and verlangsamten Neutronen. *Naturwissenschaften*, **1938**, *26* (46), 755–756. DOI: 10.1007/BF01774197.

    (ii) Hahn, O.; Strassmann, F. Über den Nachweis und das Verhalten der bei der Bestrahlung des Urans mittels Neutronen entstehenden Erdalkalimetalle. *Naturwissenschaften*, **1939**, *27* (1), 11–15. DOI: 10.1007/BF01488241.

27. (i) Meitner, L.; Frisch, O.R. Disintegration of Uranium by Neutrons: A New Type of Nuclear Reaction. *Nature*, **1939**, *143*, 239–240. DOI: 10.1038/143239a0.

    (ii) Hahn, O.; Strassmann, F.; Nachweis der Entsehung aktiver Bariumisotope aus Uran und Thorium durch Neutronenbestrahlung; Nachweis weiterer aktiver Bruchstücke bei der Uranspaltung. *Naturwissenschaften*, **1939**, *27* (6), 89–95. DOI: 10.1007/BF01488988.

28. Bohr, N.; Wheeler, J.A. The Mechanism of Nuclear Fission. *Phys. Rev.*, **1939**, *56* (5), 426–450. DOI: 10.1103/PhysRev.56.426.

29. England, T.R. and Rider, B.F. Evaluation and Compilation of Fission Product Yields 1993. Los Alamos National Laboratory Technical Report. 1995, 1–173. DOI: 10.2172/10103145.

30. Okumura, S.; Kawano, T.; Jaffke, P.; Talou, P. and Chiba, S. ^{235}U(n, f) Independent Fission Product Yield and Isomeric Ratio Calculated with the Statistical Hauser-Feshbach Theory, *J. Nucl. Sci. Technol.*, **2018**, *55* (9), 1009–1023. DOI: 10.1080/00223131.2018.1467288.

31. Meitner, L. Fission and Nuclear Shell Model, *Nature*, **1950**, *165*, 561. DOI: 10.1038/165561a0.

32. American Nuclear Society Standards Committee Working Group ANS-8.1. Nuclear Criticality Safety Control of Selected Actinide Nuclides. ANSI/ANS-8.15-2014 (R2019). American Nuclear Society: Le Grange Park, IL. 2014 (reaffirmed 2019). 1–19.

33. American Nuclear Society Standards Committee Working Group ANS-8.1.Nuclear Criticality Safety in Operations with Fissionable Material Outside Reactors. ANSI/ANS-8.1-2014 (R2018). American Nuclear Society: Le Grange Park, IL, 2014 (reaffirmed 2018). 1–15.

34. Szilard later moved to the US to pursue atomic energy research.

35. Rhodes, R. *The Making of the Atomic Bomb.* Simon & Schuster: New York, 1986.

36. (i)   von Halban, H.; Joliot, F.; Kowarski, L. Liberation of Neutrons in the Nuclear Explosion of Uranium. *Nature*, **1939**, *143*, 470–471. DOI: 10.1038/143470a0.

    (ii) von Halban, H.; Joliot, F.; Kowarski, L. Number of Neutrons Liberated in the Nuclear Fission of Uranium. *Nature*, **1939**, *143*, 680. DOI: 10.1038/143680a0.

    (iii) Anderson, H.L; Fermi, E.; Szilard, L. Neutron Production and Absorption in Uranium. *Phys. Rev.*, **1939**, *56* (3), 284–286. DOI: 10.1103/physrev.56.284.

37. H.G. Wells coined the term in one of his books: Wells, H.G. *The World Set Free: A Story of Mankind.* E.P. Dutton: New York, 1914.

38. A. Einstein, letter to F. D. Roosevelt, Washington, D.C., August 2nd, 1939 (FDR Library. Series 1: Franklin D. Roosevelt Significant Documents. FDR-24: Letter, Albert Einstein to FDR, August 2, 1939. President's Secretary's Files; Safe File; Sachs, Alexander (Box 5)).

39. Later work would demonstrate that the mass of uranium required for an explosive chain reaction, a critical mass, was only 52 kg rather than several tons that was estimated at the time of Einstein's letter.

40. U.S. Strategic Bombing Survey. The Effects of the Atomic Bombings of Hiroshima and Nagasaki. 1946. (Harry S. Truman Library. President's Secretary's File, The Decision to Drop the Atomic Bomb.) https://www.trumanlibrary.gov/library/research-files/u-s-strategic -bombing-survey-effects-atomic-bombings-hiroshima-and-nagasaki?documentid=NA&page- number=6 (Accessed 05/28/23).

6 Light and Electrons

In this chapter, we will move our attention out from the nucleus to the electrons. This will expand our focus from the femtometer (1 × 10⁻¹⁵ m) scale to the picometer (1 × 10⁻¹² m) scale. The development of our understanding of electronic structure begins by first working to understand light (electromagnetic radiation) and its interaction with matter, a field of physical chemistry that is called spectroscopy. Spectroscopy provided the first insights into electronic structure, which established the foundations for what is now referred to as old quantum theory, the focus of this chapter. We will continue with new quantum theory and a fully modern perspective on electron configurations in Chapter 7.

ELECTROMAGNETIC RADIATION

Visible light is a form of electromagnetic radiation that can be perceived by the human eye.[1] Light, electromagnetic radiation, more generally, is a coupled electric and magnetic (electromagnetic) wave that propagates through space. In a vacuum, light travels at 299 792 458 m/s, which is called the speed of light (c). An electromagnetic wave consists of oscillating, perpendicular electric (E) and magnetic (B) fields (Figure 6.1).[2]

Although Figure 6.1 shows the oscillating electric and magnetic fields of an electromagnetic wave, the complexity of the diagram can make it challenging to analyze and think about. Let's trace the ends of the electric field vectors (Figure 6.2).

Now if we plot that trace and show only the electric field component of light – most graphs typically only plot the electric component of an electromagnetic wave – we can see the sinusoidal behavior of the wave (Figure 6.3).

Different types of light vary in terms of their wavelength (λ), or what in mathematics is called the period, and their frequency (ν). Since all electromagnetic radiation travels at the same speed (c), the frequency and wavelength of light is given by Equation 6.1:

$$\nu = \frac{c}{\lambda}$$

(6.1)

ν is the frequency (Hz).
c is the speed of light (299 792 458 m/s).
λ is the wavelength (m).

PROBLEM 6.1
Convert each wavelength value into frequency.

a. 590 nm

b. 100.0 mm

PROBLEM 6.2
Convert each frequency value into wavelength.

a. 89.7 MHz (the frequency of WGBH in Boston)

b. 10.0 Hz

BLACKBODY RADIATION

Whereas the wave nature of light was clear in the 19th century, other aspects of light remained unsolved until the early 20th century: blackbody radiation, the photoelectric effect, and atomic emission spectra. We will first consider blackbody radiation.[3] Any object with thermal energy emits electromagnetic radiation, which is called blackbody radiation. As the temperature increases, the amount of light emitted – spectral radiance – increases, and the peak wavelength of emitted light moves to shorter wavelengths (Figure 6.4). For example, interstellar space at 2.7 K emits electromagnetic radiation with a peak wavelength of 1.07 mm, humans at 310 K emit electromagnetic radiation with a peak wavelength of 9.3 µm, and the sun with a surface temperature of 5772 K emits electromagnetic radiation with a peak wavelength of 530 nm.

The challenge for 19th-century scientists was explaining the sudden decrease in blackbody spectral radiance at wavelengths below 100 nm, a phenomenon that was retrospectively termed the ultraviolet catastrophe.[4] The solution, proposed by Max Planck, was to introduce the idea that

DOI: 10.1201/9781003479338-6

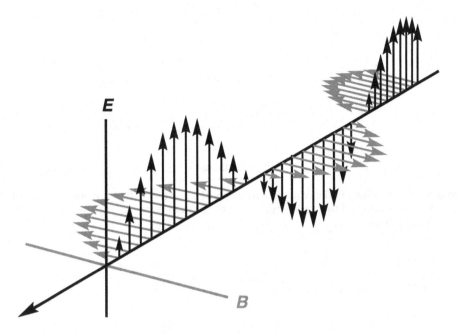

Figure 6.1   Oscillating electric and magnetic fields of an electromagnetic wave.

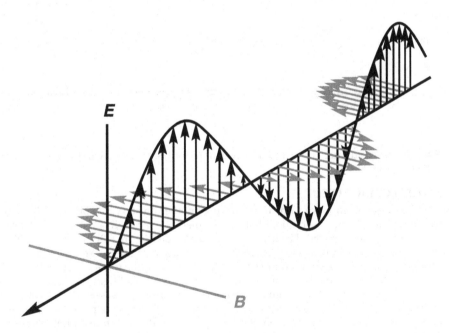

Figure 6.2   Oscillating electric and magnetic fields of an electromagnetic wave with a trace of the electric field oscillations.

an object can only have and emit energy in discrete units, or quanta. In the mathematical solution devised by Planck, these quanta of energy are always a multiple of $h\nu$.[5] As we saw above, ν is the symbol for frequency, here the frequency of some sort of resonator within the heated object, and h was a new constant that made the math work out, which we now call the Planck constant: 6.626 070 15 × 10^{-34} J s. While the introduction of this quantum idea allowed Planck to provide a mathematical fit for the blackbody radiation data (Figure 6.4), the quantum assumption was assumed to be a mere mathematical formalism, not the revolutionary discovery it would prove to be.[6]

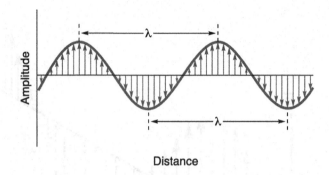

Figure 6.3 A simplified model of an electromagnetic wave (only showing one component, the electric field oscillation) and the wavelength (λ) as the peak-to-peak distance or, equivalently, the trough-to-trough distance.

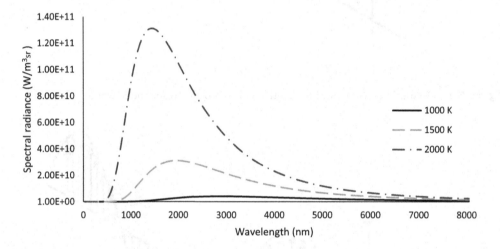

Figure 6.4 Spectral radiance compared to wavelength for the blackbody radiation from objects at 1000 K, 1500 K, and 2000 K.

THE PHOTOELECTRIC EFFECT

Support for Planck's quantum idea came in answering another mystery: the photoelectric effect. The photoelectric effect is the observation that when a metal surface is irradiated with the right type of light, electrons are emitted. The brightness – amplitude (Figure 6.3) – of the light does not matter when it comes to producing electrons, but each metal shows the ejection of electrons when the frequency of light exceeds a particular threshold, which is unique to each material. The idea that a particle, an electron, could be displaced by a continuous wave of light was still paradoxical. To explain this phenomenon, Einstein proposed that light itself possessed packets of energy, quanta.[7] That is light, while it is a wave, also could be thought of as having packet-like, quantum behavior and the energy of each packet, photon,[8] of light is specific. When that photon's energy exceeds a certain threshold, then an electron is ejected by the photon.

Planck's and Einstein's work provided new insight into light. Each type of light – with a unique frequency and wavelength – also has its own energy (Equation 6.2).

$$E = h\nu = \frac{hc}{\lambda} \tag{6.2}$$

E is the energy of a photon (J).
h is the Planck constant ($6.626\,070\,15 \times 10^{-34}$ J s).
ν is the frequency (Hz).

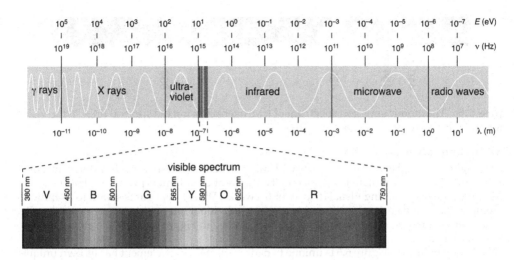

Figure 6.5 The electromagnetic spectrum showing the different regions – γ rays, X rays, ultraviolet (UV), visible, infrared (IR), microwave (μW), and radio waves – and their corresponding wavelength (λ) in meters, frequency (ν) in hertz, and photon energy (E) in electronvolts. (Based on Philip Ronan, https://commons.wikimedia.org/wiki/File:EM_spectrum.svg (Accessed on 05/22/23).)

c is the speed of light (299 792 458 m/s).
λ is the wavelength (m).

PROBLEM 6.3
Convert each wavelength value into energy (J).

a. 590 nm

b. 100.0 mm

PROBLEM 6.4
Convert each frequency value into energy (J).

a. 89.7 MHz (the frequency of WGBH in Boston)

b. 10.0 Hz

PROBLEM 6.5
On average, one square meter receives 225 kJ of solar energy every three minutes. Assuming all the light has a wavelength of 530 nm, how many photons does this correspond to?

In summary, light is an electromagnetic wave that exhibits quantum behavior. We can characterize light based on its wavelength, frequency, and the energy of a photon, a light quantum. The full spectrum (Figure 6.5) of light, electromagnetic radiation, is continuous from the incredibly small wavelengths (γ rays) to the macroscopically long (radio waves).[9]

↝ THE PHOTOVOLTAIC EFFECT
It is worth highlighting that while the photoelectric effect was useful in the development of quantum theory, a similar process is of increasing importance to modern society: the photovoltaic effect. Unlike the photoelectric effect, where electrons are ejected from a material by exposure to light, the photovoltaic effect involves only the promotion of an electron to a higher energy state, which creates an electric potential (Chapter 19).[10] With the proper engineering of the materials involved, this can be used to create electric current and is the basis for solar cells,[11] one important component of clean energy.

Figure 6.6 The atomic emission spectrum of hydrogen showing the spectral lines in the visible range and their corresponding wavelengths.

ATOMIC EMISSION SPECTRA

The last light-related phenomenon that we will study is atomic emission spectra. When an atom is heated in a flame or stimulated with electricity, different color light is produced. This fact is utilized to produce the stunning visual display of fireworks. When this light is split into its individual components, with a prism or diffraction grating, individual lines can be seen in what is called an atomic emission spectrum. An example of this is shown in Figure 6.6, with the atomic emission spectrum for hydrogen.

This pattern of lines in Figure 6.6 is unique to hydrogen, and every element has its own unique pattern of lines in its atomic emissions spectrum. As the periodic table expanded during the 19th century, atomic emission spectra were used to identify new elements.[12] Despite discovering the atomic emission spectra nearly 200 years ago, the physical basis for the origin of these spectra was mysterious; however, late 19th-century scientists identified mathematical formulae that could model these lines. The most general formula developed was that proposed by Rydberg, which stated that the reciprocal wavelength (λ) of a line (in meters) could be calculated using Equation 6.2:[13]

$$\frac{1}{\lambda} = R_H \left(\frac{1}{n_i^2} - \frac{1}{n_f^2} \right) \tag{6.2}$$

λ is the wavelength of light (m).
R_H is the Rydberg constant for hydrogen ($1.096\,775\,83 \times 10^7$ 1/m).
n_i is the initial state (any natural number, 1, 2, 3, 4, 5, 6, 7, ...).
n_f is the final state (any natural number, 1, 2, 3, 4, 5, 6, 7, ...).

Using this equation, if we were to choose n_i equal to three and n_f equal to two, then we would calculate the following:

$$\frac{1}{\lambda} = 1.096\,775\,83 \times 10^7 \, \frac{1}{m} \left(\frac{1}{3^2} - \frac{1}{2^2} \right) = -1523\,299.76 \frac{1}{m}$$

Taking the reciprocal of this value gives -6.5647×10^{-7} m or -656.47 nm (the value is negative because light is being emitted), which is in close agreement with the red line in Figure 6.6.

PROBLEM 6.6
What are n_i and n_f for the other three lines in Figure 6.6?

PROBLEM 6.7
Calculate the wavelength in nanometers for n_i equals two and n_f equals one. Can humans see this light (humans see roughly 400 nm to 800 nm light)?

BOHR MODEL

Niels Bohr was the individual who was able to take and unify several disparate threads – the Rutherford model of the atom (Chapter 3), the quantum behavior of light and atomic emission spectra, and the Rydberg equation – into the first coherent framework for electron configurations.[14] Bohr proposed, (Figure 6.7) that the electrons in an atom occupy specific, discrete energy levels (n = 1, 2, 3, 4, ...). The lines in atomic emission spectra, then, are the result of electrons moving between these discrete energy levels. The emission of light is the result of an electron going from

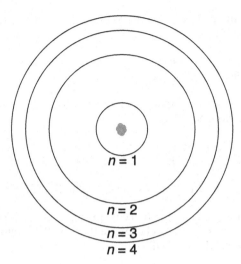

Figure 6.7 Bohr model, refining the Rutherford model, of the atom with a nucleus (neutrons were not discovered yet and so Bohr's nucleus only contains protons) and shells (*n*) from 1 to 4.

a higher energy level to a lower energy level, and the absorption of light is the result of an electron going from a lower energy level to a higher energy level.

Bohr's model worked well to explain the observed behavior of hydrogen and hydrogen-like (one-electron) atoms; it failed to fully explain the behavior of multielectron atoms and could not explain the intensity (brightness) of the lines in atomic emission spectra. This would require the further development and refinement of quantum mechanics, which we will consider in Chapter 7.

NOTES

1. (i) Discovery that an electric field creates a magnetic field: Oersted, J.C. Article IV. Experiments on the Effect of Electricity on the Magnetic Needle. *Ann. Philos.*, **1820**, *16* (4), 273–276. URL: https://www.biodiversitylibrary.org/page/15882011#page/291/mode/1up (Accessed 04/21/23).

   (ii) Discovery that magnetic fields can rotate light (1845, paragraphs in diary #7504 and #7718): Faraday, Michael. *Faraday's Diary: Being the Various Philosophical Notes of Experimental Investigation. Vol. IV, Nov. 12, 1839–June 26, 1847*. G. Bell and Sons, LTD: London, 1932.

   (iii) Maxwell's work (in the 1860s) on what are now known as Maxwell's equations formulated the concept that light traveled as an electromagnetic wave. This work is summarized by him in: Maxwell, J.C. *A Treatise on Electricity and Magnetism Vol. 1*, Clarendon Press: Oxford, 1873.

2. Magnetic field can refer to both magnetic field strength (H) and to magnetic flux density (B). The different quantities H and B are related to each other by the vacuum permeability constant (μ_0): $H = B/\mu_0$.

3. Wien, W. Ueber die Energievertheilung im Emissionsspectrum eines schwarzen Körpers. *Ann. Phys.*, **1896**, *294* (8), 662–669. DOI: 10.1002/andp.18962940803.

4. Ehrenfest, P. Welche Züge der Lichtquantenhypothese spielen in der Theorie der Wärmestrahlung eine wesentliche Rolle. *Ann. Phys.*, **1911**, *341* (11), 91–118. DOI: 10.1002/andp.19113411106.

5. (i) Planck, M. Ueber eine Verbesserung der Wien'schen Spectralgleichung. *Verhandl. Dtsch. Phys. Ges.*, **1900**, *2* (1), 202–204.

(ii) Planck, M. Zur Theorie des Gesetzes der Energieverteilung im Normalspectrum. *Verhandl. Dtsch. Phys. Ges.*, **1900**, *2* (1), 237–245.

6. Kragh, H. *Quantum Generations: A History of Physics in the Twentieth Century*. Princeton University Press: Princeton, NJ, 2002.

7. Einstein, A. Über einen die Erzeugung und Verwandlung des Lichtes betreffenden heuristischen Gesichtspunkt. *Ann. Phys.*, **1905**, *322* (6), 132–148. DOI: 10.1002/andp.19053220607.

8. Lewis, G.N. The Conservation of Photons. *Nature*, **1926**, *118*, 874–875. DOI: 10.1038/118874a0.

9. (i) Infrared (IR) light was discovered by William Herschel: Herschel, W. XIV. Experiments on the Refrangibility of the Invisible Rays of the Sun. *Phil. Trans. R. Soc.*, **1800**, *90*, 284–292. DOI: 10.1098/rstl.1800.0015.

(ii) Ultraviolet (UV) light was discovered by J.W. Ritter. The original paper in 1801 could not be found, but he references his first paper in this paper: Ritter, J.W. Versuche über das Sonnenlicht. *Ann. Phys.*, **1803**, *12* (12), 409–415. DOI: 10.1002/andp. 18030121205.

(iii) Radio waves and microwaves (μW) were discovered by Hertz in the 1880s and that work is summarized by him (translated into English by D.E. Jones) in: Hertz, H. *Electric Waves: Being Research on the Propagation of Electric Action With Finite Velocity Through Space*. Macmillan and Co.: London and New York, 1893.

(iv) X-rays were discovered by Wilhelm Röntgen: Röntgen, W.C. Ueber eine neue Art von Strahlen. *Verhandlungen der Physik.-Medizin.-Gesellschaft zu Würzburg*, 1895.

(v) γ rays were discovered by Paul Villard: Villard, P. Sur la rayonnement du radium. *C. R. Acad. Sci.*, **1900**, *130*, 1178–1179.

10. Becquerel, E. Mémoire sur les effets électriques produits sous l'influence des rayons solaires. *C. R. Acad. Sci.*, **1839**, *9*, 561–567.

11. The first solid-state solar cell: Fritts, C.E. On a New Form of Selenium Cell, and Some Electrical Discoveries Made by Its Use. *Am. J. Sci.*, **1883**, *s3-26*(156), 465–472. DOI: 10.2475/ajs. s3-26.156.465.

12. The use of emission spectra to identify elements was first reported by Charles Wheatstone (Wheatstone, C. On the Prismatic Decomposition of Electrical Light. *Br. Assoc. Adv. Sci., Report*, **1835**, 11–12) and Angstrom first demonstrated the spectra lines of gases (Angstrom, A.J. Otiska undersökningar. *Kungl. Svenska Vetenskapsakad. Handl.* **1852**, *40*, 339–366).

13. Rydberg, J.R. Recherches sur la Constitution des Spectres D'Émission des Éléments Chimiques. *Kungl. Svenska Vetenskapsakad. Handl.* **1888–1889**, *23* (11), 1–155.

14. (i) Bohr, N. I. On the Constitution of Atoms and Molecules. *Lond. Edinb. Dublin Philos. Mag. J. Sci. Series 6.*, **1913**, *26* (151), 1–25. DOI: 10.1080/1478644130864955.

(ii) Bohr, N. XXXVII. On the Constitution of Atoms and Molecules. *Lond. Edinb. Dublin Philos. Mag. J. Sci. Series 6.*, **1913**, *26* (153), 476–502. DOI: 10.1080/14786441308634993.

(iii) Bohr, N. LXXIII. On the Constitution of Atoms and Molecules. *Lond. Edinb. Dublin Philos. Mag. J. Sci. Series 6.*, **1913**, *26* (155), 857–875. DOI: 10.1080/14786441308635031.

7 Electrons and Quantum Numbers

Planck, Einstein, and Bohr were progenitors of a new, quantum interpretation of the atomic world (Chapter 6). The foundation of this new quantum understanding is the idea that at the smallest scales, energy is not a continuous quantity, but rather it comes in small, discrete packets that we call quanta. This approach allowed scientists to understand and explain what had been mystifying phenomena: blackbody radiation, the photoelectric effect, and atomic emission spectra. The quantum revolution, however, remained incomplete in the early 1920s, and the shortcomings of Bohr's model of the atom – including its inability to correctly model multielectron atoms – were evident. The development of modern quantum mechanics required further insights and developments.

WAVE–PARTICLE DUALITY

In seeking to understand the photoelectric effect (Chapter 6), Einstein hypothesized that light is made up of individual packets of energy (photons). This hypothesis invokes the idea that light, an electromagnetic wave, has particle-like behavior. The key development in further refining quantum mechanics extended this idea further: particles (matter) have wave behavior.[1] That is to say that there is a wave–particle duality, where waves exhibit particle phenomena and particles exhibit wave phenomena. Mathematically this is expressed in Equation 7.1:

$$\lambda = \frac{h}{p} \tag{7.1}$$

λ is the wavelength (m).

h is the Planck constant ($6.626\ 070\ 15 \times 10^{-34}$ J s).

p is the momentum ($\frac{\text{kg m}}{\text{s}}$), the product of mass (kg) and velocity (m/s).

PROBLEM 7.1
An electron has a velocity of 6.06×10^6 m/s and a mass (m_e) of $9.109\ 383\ 7015 \times 10^{-31}$ kg. What is the wavelength of an electron with this velocity? How does this wavelength compare to the van der Waals radius of a hydrogen atom (1.20×10^{-10} m)?

PROBLEM 7.2
An average human (89 kg) has a velocity of 1.3 m/s when walking. What is the wavelength of a human with this velocity? How does this wavelength compare to the height of the average human (1.7 m)?

PROBLEM 7.3
A neutron is found to have a de Broglie wavelength of 181 pm. What is the velocity (m/s) given that the mass of a neutron (m_n) is $1.674\ 927\ 498\ 04 \times 10^{-27}$ kg?

The wave–particle nature of matter has several important ramifications. One effect is that our approach to modeling and understanding electrons' energy and arrangement, the electron configuration, requires a substantial revision from that proposed in the Bohr model (see below). The other consequence of wave–particle duality is that there is a definitive limit on our ability to measure complementary properties of a particle, like position and momentum, which is known as the uncertainty principle.[2] Let's consider the wave shown in Figure 7.1. This particle wave has a well-defined, clearly measurable wavelength (λ), but if asked to identify the position of the particle, we would be unable to say where in this box the particle is, definitively.

Now, if we contrast the particle wave in Figure 7.2 with that in Figure 7.1, we can see that the particle in Figure 7.2 has a clear position (in the center of the box). The wavelength, however, is not clear and it would be hard to measure.

Position (x) and momentum (p) are an example of a complimentary pair. That is, measuring one necessarily limits the ability to measure the other. This is expressed mathematically in Equation 7.2:

$$\sigma_x \sigma_p \geq \frac{h}{4\pi} \tag{7.2}$$

DOI: 10.1201/9781003479338-7

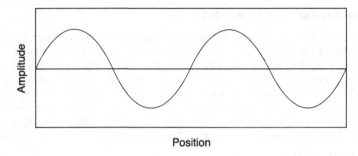

Position

Figure 7.1 A particle wave with a well-defined wavelength, from which energy and momentum can be calculated, but no clear position.

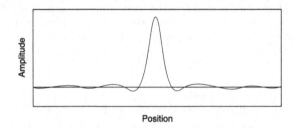

Position

Figure 7.2 A wave with a poorly defined wavelength, meaning that energy and momentum cannot be definitively determined, but a clear and identifiable position.

σ_x is the standard deviation (uncertainty) in the position of a particle (m).

σ_p is the standard deviation (uncertainty) in the momentum of a particle ($\frac{\text{kg m}}{\text{s}}$).

h is the Planck constant ($6.626\,070\,15 \times 10^{-34}$ J s).

PROBLEM 7.4
Consider the following scenarios.

a. The energy of an electron around an atom is known with a high level of precision (a very low uncertainty or low standard deviation). Can we say where the electron is around the atom?

b. The position of an electron around an atom is known precisely (almost no uncertainty). Can we say what the energy of that electron is?

PROBLEM 7.5
Consider the two scenarios in Problem 7.4. Given that there are inherent trade-offs in what we can know about the electron, which do you think is more important to know to a high level of precision: the position or the momentum/energy?

QUANTUM MECHANICAL ENERGY LEVELS

Given that electrons (as particles) are both particles and waves, the shortcomings of the Bohr model become clear. Bohr assumed that electrons were purely particles, for which we could know their precise pathway (their speed and their position). Given that electrons also have wave behavior, a new approach was necessary. Two different quantum mechanical approaches were independently developed. The first used matrices to model the observable phenomena such as atomic emission spectra.[3] The second used wave functions to model the particle waves of electrons.[4] Both approaches were shown to be mathematically equivalent.[5] The study and explanation of matrix mechanics or wave functions is beyond the scope of this chapter. What is relevant to know is that the specific energy level (known as an orbital) of an electron can be described by three quantum numbers: the principal quantum number (n), the angular momentum quantum number (ℓ), and the magnetic quantum number (m_ℓ).[6] The energy of an electron is described by n, ℓ, and m_ℓ along with the spin quantum number (m_s).

The principal quantum number (n) is any natural number (1, 2, 3, 4, 5, 6, 7, …). This number corresponds to the shell number and describes an electron's distance from the nucleus (smaller n value means closer to the nucleus) and its orbital energy level (smaller n value means lower energy level). In some fields of chemistry, particularly X-ray spectroscopy, you may see that the shells are given letter designations: the first shell is the K shell, the second shell is the L shell, the third shell is the M shell, and so on.

The angular momentum quantum number (ℓ) is a whole number (0, 1, 2, 3, 4, 5, 6, …) and for a given n value, ℓ is all of the values from 0 to $n–1$. For example, if $n = 1$, then $\ell = 0$; if $n = 2$, then $\ell = 0, 1$; if $n = 3$, then $\ell = 0, 1, 2$; if $n = 4$, then $\ell = 0, 1, 2, 3$. The angular momentum quantum number corresponds to the subshell ($\ell = 0$ is the s subshell, $\ell = 1$ is the p subshell, $\ell = 2$ is the d subshell, and $\ell = 3$ is the f subshell) and describes the shape of the orbital.

The magnetic quantum number (m_ℓ) is an integer (0, ±1, ±2, ±3, ±4, ±5, ±6, …), and for a given ℓ value, m_ℓ is all the integers from $-\ell$ to $+\ell$. For example, if $\ell = 0$, then $m_\ell = 0$; if $\ell = 1$, then $m_\ell = -1$, 0, +1; if $\ell = 2$, then $m_\ell = -2, -1, 0, +1, +2$; if $\ell = 3$, then $m_\ell = -3, -2, -1, 0, +1, +2, +3$. The magnetic quantum number identifies the direction of the orbital and indicates how many orbitals there are per subshell (e.g., if $\ell = 0$, then $m_\ell = 0$ corresponds to the s subshell and there is one s orbital per s subshell; if $\ell = 1$, then $m_\ell = -1, 0, +1$ corresponds to the p subshell and there are three p orbitals per p subshell).

The spin quantum number (m_s) describes the spin of the electron (an intrinsic property of the electron not related to our macroscopic idea of spin) and has values of $-\frac{1}{2}$ and $+\frac{1}{2}$.

Putting this together into a coherent energy diagram gives Figure 7.3, which is only a partial diagram as it shows only up to $n = 4$, $\ell = 1$.

While directly tied to and showing the quantum numbers of each level, the diagram in Figure 7.3 loses some utility and functionality as a model because of the depth of information

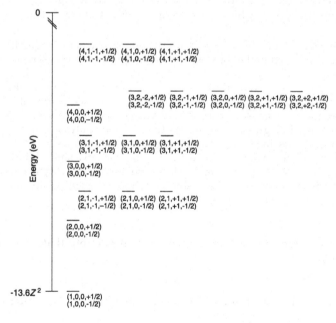

Figure 7.3 Partial orbital energy diagram with corresponding quantum number designations (Z = the atomic number of the element). The energy ordering shown here comes from the orbital energy ordering for fluorine; actual orbital energy ordering and filling may vary. Note that $-13.6Z^2$ is the lowest possible value for the $n = 1$ and $\ell = 1$ energy level, which is only valid for 1-electron atoms and ions. With more electrons, the energy levels shift upward due to electron–electron repulsion, just as we saw with the proton energy levels (Chapter 4, Figure 4.9). (Data from Kramida, A., Ralchenko, Yu., Reader, J., and NIST ASD Team (2022). NIST Atomic Spectra Database (ver. 5.10), [Online]. Available: https://physics.nist.gov/asd [2023, July 7]. National Institute of Standards and Technology, Gaithersburg, MD. DOI: 10.18434/T4W30F.)

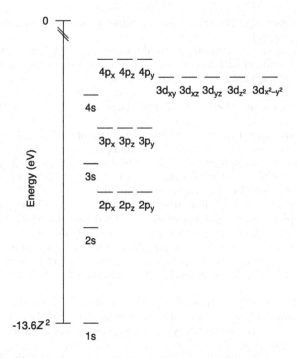

Figure 7.4 Partial orbital energy diagram with shell, orbital designations (Z = the atomic number of the element). Note that $-13.6Z^2$ is the lowest possible value for the 1s orbital, which is only valid for 1-electron atoms and ions. With more electrons, the energy levels shift upward due to electron–electron repulsion, like the proton energy levels in Chapter 4 (Figure 4.9).

presented. As such, Figure 7.4 is a more common representation where the quantum number information has been encoded in the shell and orbital designations that are more commonly used in chemistry.

Electrons occupy these atomic orbitals according to several rules and principles. According to the Aufbau principle, orbitals are filled starting from the lowest energy, typically in order of $n + \ell$ value.[7] The Pauli exclusion principle states that no electron can have the same four quantum numbers.[8] As can be seen in Figure 7.3, this means that there are only two electrons allowed in any one orbital. To show this, in Figure 7.3 and Figure 7.4 we use arrows pointed up (↑) and down (↓) to show the $+\frac{1}{2}$ and $-\frac{1}{2}$ spin values. If there are two electrons in a single orbital, they must be spin-paired (one up and one down: ↑↓). Finally, Hund's rule states that degenerate, equal energy orbitals (p, d, or f orbitals) must each be filled singly and spin-aligned (all up or all down) before two electrons can be paired in the same orbital.[9]

PROBLEM 7.6
Using Figure 7.4, determine the maximum number of electrons that can fit in the s, p, and d subshells.

PROBLEM 7.7
The angular momentum quantum number (ℓ) for the f subshell is 3. Determine:

a.   How many f orbitals there are.

b.   The maximum number of electrons that can go into the f subshell.

ELECTRON CONFIGURATIONS

Following all the rules and principles above will give the ground-state electron configuration of an atom. The one electron of a hydrogen atom could have the quantum numbers $(1,0,0,+\frac{1}{2})$ or $(1,0,0,-\frac{1}{2})$, which would be equivalent energetically. On an orbital energy diagram that would be represented as shown in Figure 7.5. To represent the hydrogen electron configuration in standard notation, we write $1s^1$.

Figure 7.5 Energetically equivalent, ground-state electron energy diagrams for hydrogen with spin quantum number (m_s) +½ (left) and –½ (right).

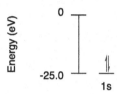

Figure 7.6 Ground-state electron energy diagrams for helium. 1s orbital energy value calculated at the CCSD(T)/cc-pVTZ level of theory and basis set.

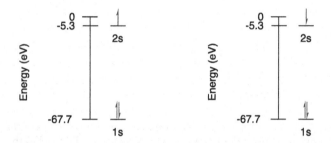

Figure 7.7 Energetically equivalent, ground-state electron energy diagrams for lithium with spin quantum number (m_s) +½ (left) and –½ (right) for the highest energy, valence, electron. Orbital energy values calculated at the CCSD(T)/cc-pVTZ level of theory and basis set.

For helium, there are two electrons, which would have the quantum numbers (1,0,0,+½) and (1,0,0,–½). The electrons would be spin-paired, as we can see in the orbital energy diagram in Figure 7.6. The standard notation would be $1s^2$ for helium. With the second electron in the 1s orbital, the orbital, subshell, and shell are filled, which is why helium is a chemically (electronically) inert noble gas.

Now as we move onto lithium, with three electrons, the electrons would fill the 1s orbital and the last electron would occupy the 2s orbital. As with hydrogen, there are two, energetically equivalent ways to show the +½ or –½ spin the electron can have (Figure 7.7).

For lithium, the standard notation to represent the electron configuration would be $1s^2 2s^1$. It is worth looking at the details of lithium's electron configuration a bit more deeply. There are two electrons in the filled first shell, which are significantly lower in energy (–67.7 eV) than the electron in the unfilled second shell (–5.3 eV). The electrons in the filled first shell are not significantly involved in bonding or in chemistry, and so we refer to those electrons as core electrons. Core electrons are any electrons that are in filled, low energy shells and not involved in bonding or reactions. Electrons that are in unfilled shells or subshells, and are highest in energy, are called valence electrons. Valence electrons are involved in both bonding and in reactions and will be the focus of our study of chemistry. Lithium has two core electrons ($1s^2$) and one valence electron ($2s^1$). Because the core electrons have the same configuration as helium, we can rewrite the configuration of lithium in noble gas notation: [He]$2s^1$. The noble gas notation highlights the valence electrons, and the core electrons are replaced with the equivalent noble gas notation.

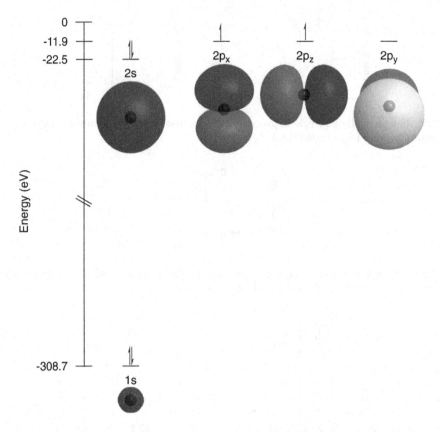

Figure 7.8 Ground-state electron energy diagram of carbon with images of the s and p orbitals. Orbital images and energy values calculated at the CCSD(T)/cc-pVTZ level of theory and basis set.

PROBLEM 7.8
Write the electron configuration for Be and B. Identify the number of core electrons and the number of valence electrons.

Let's use carbon ($1s^22s^22p^2$) as an example (Figure 7.8) to understand orbitals further: their shape. Given the well-defined energy value for each orbital, the orbital represents (due to the uncertainty principle) a region of space where an electron is likely to be found with 90% probability. Carbon has two types of orbitals: s and p. The s orbitals have a spherical shape and are radially symmetric (that is, every point at a defined distance r from the nucleus is the same). The 2s orbital is larger than the 1s orbital. As orbitals increase in energy, they increase in size, due to the greater distance between the nucleus and the electron (Chapter 8). The p orbitals have two symmetric lobes along the x, z, or y axes, and the lobes are reflected across a nodal plane that is centered on the atom.

We will continue to use carbon and Figure 7.8 to understand the rules of orbital filling: the Aufbau principle, Hund's rule, and the Pauli exclusion principle. Figure 7.8 shows the ground-state configuration for carbon ($1s^22s^22p_x^12p_z^12p_y^0$), which is the lowest energy (ground state) because it follows the Aufbau principle and Hund's rule. If either the Aufbau principle or Hund's rule is violated, the resulting configuration is an excited (higher energy) state. For example, if the Aufbau principle is violated for carbon the resulting configuration ($1s^22s^12p_x^12p_z^12p_y^1$) is 4.1 eV higher in energy than the ground-state configuration. And if Hund's rule is violated for carbon the resulting configuration ($1s^22s^22p_x^22p_z^02p_y^0$) is 1.5 eV higher in energy than the ground-state configuration. Note that while violations of the Aufbau principle and Hund's rule are possible, violation of the Pauli exclusion principle is not possible. If violation of the Pauli exclusion principle is written or shown for an electron configuration, it is impossible.

So far, we have considered only elements in the first and second periods. Elements in period 4 and beyond complicate the pictures by including d orbitals ($\ell = 2$) and then f orbitals ($\ell = 3$). Figure 7.9 shows the orbital energy diagram for iron. Iron has eight valence electrons

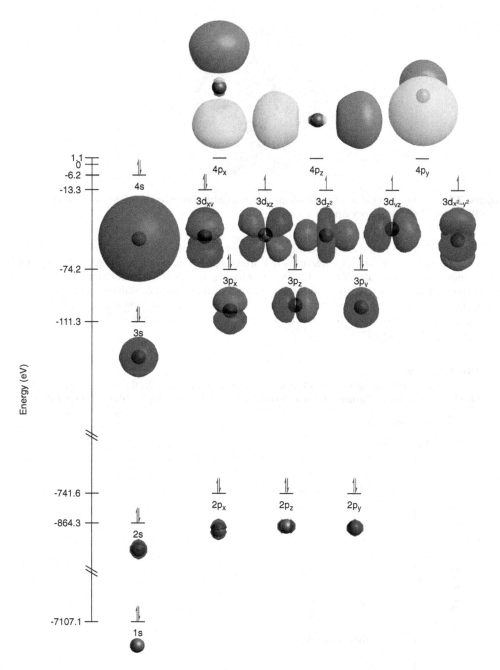

Figure 7.9 Ground-state electron energy diagram of iron with images of the s, p, and d orbitals. Orbital images and energy values calculated at the CCSD(T)/cc-pVTZ level of theory and basis set. Note, the 1s orbital is so small in volume that it is smaller than the sphere that represents the atom. Also note that the interaction of the electron spin and the orbital angular momentum leads to slightly (<1 eV) different energy values for filled or partially filled orbitals that are degenerate when unfilled: p, d, and f. The importance and effects of spin–orbit coupling are beyond the scope of this chapter, and the values in the diagrams shown here do not include the effects of spin–orbit coupling.

Figure 7.10 Example f orbital for the atom cerium. Orbital image calculated with WebMO at the B3LYP/CEP-4G level of theory and basis set.

$(1s^22s^22p^63s^23p^63d^64s^2)$ in the 3d and 4s subshells. Electrons in the d orbitals count toward the valence for transition metals, and electrons in f orbitals count toward the valence for lanthanoids and actinoids. The d orbitals all look like four-leaf clovers except the d_{z^2} orbital, which looks like a p orbital with a ring around the middle. We will not consider the f orbitals in detail here, but it is worth seeing that they are more complicated in appearance than d orbitals. Figure 7.10 shows an example f orbital for cerium. Cerium has four valence electrons in the 4f, 5d, and 6s subshells: $1s^22s^22p^63s^23p^63d^{10}4s^24p^64d^{10}5s^25p^64f^15d^16s^2$.

PROBLEM 7.9
For each electron configuration identify it as ground state, excited state, or impossible. Identify what neutral atom each configuration (that is not impossible) corresponds to.

a. $1s^22s^22p^63s^23p^1$

b. $1s^22s^22p_x^22p_z^12p_y^0$

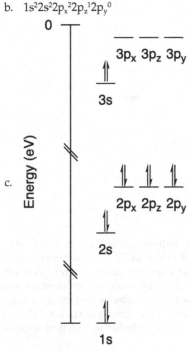

c.

d. $1s^22s^22p^63s^23p^63d^{12}4s^2$

e. $1s^12s^12p^6$

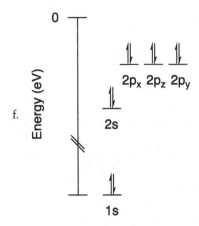

f.

ELECTRON CONFIGURATIONS AND THE PERIODIC TABLE

Until now, we have considered elements individually, but now we will turn our attention to trends in electron configuration. Appendix 3 has the complete electron configuration of every element and orbital energy diagrams for hydrogen through krypton, but we are most interested in valence electrons – those involved in bonding and reactions. The valence electron configuration (core electrons are omitted for clarity) of each element is shown in Figure 7.11.[10]

If you look at Figure 7.11, you can see that elements in the same group have, with some exceptions for transition metals, lanthanoids, and actinoids, the same valence electron configurations. Their common valence configurations explain why elements in the same group exhibit similar chemical properties. What should also become apparent is that the separate blocks of the periodic table are different based on where the valence electrons are (Figure 7.12). The main group elements are made up of the s block (groups 1 and 2) and the p block (groups 13–18). Groups 3–12, the transition metals, are called the d block. Finally, the lanthanoids and actinoids are called the f block.

1	2											13	14	15	16	17	18
1 H $1s^1$	2 He $1s^2$																
3 Li $2s^1$	4 Be $2s^2$											5 B $2s^22p^1$	6 C $2s^22p^2$	7 N $2s^22p^3$	8 O $2s^22p^4$	9 F $2s^22p^5$	10 Ne $2s^22p^6$
11 Na $3s^1$	12 Mg $3s^2$	3	4	5	6	7	8	9	10	11	12	13 Al $3s^23p^1$	14 Si $3s^23p^2$	15 P $3s^23p^3$	16 S $3s^23p^4$	17 Cl $3s^23p^5$	18 Ar $3s^23p^6$
19 K $4s^1$	20 Ca $4s^2$	21 Sc $3d^14s^2$	22 Ti $3d^24s^2$	23 V $3d^34s^2$	24 Cr $3d^54s^1$	25 Mn $3d^54s^2$	26 Fe $3d^64s^2$	27 Co $3d^74s^2$	28 Ni $3d^84s^2$	29 Cu $3d^{10}4s^1$	30 Zn $3d^{10}4s^2$	31 Ga $4s^24p^1$	32 Ge $4s^24p^2$	33 As $4s^24p^3$	34 Se $4s^24p^4$	35 Br $4s^24p^5$	36 Kr $4s^24p^6$
37 Rb $5s^1$	38 Sr $5s^2$	39 Y $4d^15s^2$	40 Zr $4d^25s^2$	41 Nb $4d^45s^1$	42 Mo $4d^55s^1$	43 Tc $4d^55s^2$	44 Ru $4d^75s^1$	45 Rh $4d^85s^1$	46 Pd $4d^{10}$	47 Ag $4d^{10}5s^1$	48 Cd $4d^{10}5s^2$	49 In $5s^25p^1$	50 Sn $5s^25p^2$	51 Sb $5s^25p^3$	52 Te $5s^25p^4$	53 I $5s^25p^5$	54 Xe $5s^25p^6$
55 Cs $6s^1$	56 Ba $6s^2$	*	72 Hf $5d^26s^2$	73 Ta $5d^36s^2$	74 W $5d^46s^2$	75 Re $5d^56s^2$	76 Os $5d^66s^2$	77 Ir $5d^76s^2$	78 Pt $5d^96s^1$	79 Au $5d^{10}6s^1$	80 Hg $5d^{10}6s^2$	81 Tl $6s^26p^1$	82 Pb $6s^26p^2$	83 Bi $6s^26p^3$	84 Po $6s^26p^4$	85 At $6s^26p^5$	86 Rn $6s^26p^6$
87 Fr $7s^1$	88 Ra $7s^2$	**	104 Rf $6d^27s^2$	105 Db $6d^37s^2$	106 Sg $6d^47s^2$	107 Bh $6d^57s^2$	108 Hs $6d^67s^2$	109 Mt $[6d^77s^2]$	110 Ds $[6d^87s^2]$	111 Rg $[6d^97s^2]$	112 Cn $[6d^{10}7s^2]$	113 Nh $7s^27p^1$	114 Fl $[7s^27p^2]$	115 Mc $[7s^27p^3]$	116 Lv $[7s^27p^4]$	117 Ts $[7s^27p^5]$	118 Og $[7s^27p^6]$

	57 La $5d^16s^2$	58 Ce $4f^15d^16s^2$	59 Pr $4f^36s^2$	60 Nd $4f^46s^2$	61 Pm $4f^56s^2$	62 Sm $4f^66s^2$	63 Eu $4f^76s^2$	64 Gd $4f^75d^16s^2$	65 Tb $4f^96s^2$	66 Dy $4f^{10}6s^2$	67 Ho $4f^{11}6s^2$	68 Er $4f^{12}6s^2$	69 Tm $4f^{13}6s^2$	70 Yb $4f^{14}6s^2$	71 Lu $5d^16s^2$
*															
**	89 Ac $6d^17s^2$	90 Th $6d^27s^2$	91 Pa $5f^26d^17s^2$	92 U $5f^36d^17s^2$	93 Np $5f^46d^17s^2$	94 Pu $5f^67s^2$	95 Am $5f^77s^2$	96 Cm $5f^76d^17s^2$	97 Bk $5f^97s^2$	98 Cf $5f^{10}7s^2$	99 Es $5f^{11}7s^2$	100 Fm $5f^{12}7s^2$	101 Md $5f^{13}7s^2$	102 No $5f^{14}7s^2$	103 Lr $7s^27p^1$

Figure 7.11 Periodic table showing the configuration of the valence electrons for each element. (Note 1: Core electrons are not shown. Note 2: Some elements vary slightly from the expected configuration due to subtle quantum mechanical effects. Configurations in brackets are theoretical.)

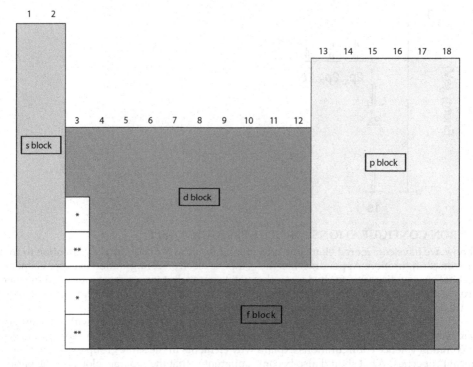

Figure 7.12 The different blocks of the periodic table, based on the subshell in which the highest energy valence electrons are found.

The number of groups (columns) in each block is related to the number of electrons that can fit into each subshell. The s block contains two groups because the s subshell can hold two electrons. The p block consists of six groups because the p subshell can hold six electrons. The d block is ten groups across because the d subshell can hold ten electrons. And the f block contains 14 groups because the f subshell can hold 14 electrons.

PROBLEM 7.10
How would the periodic table change if there were three possible spin quantum number values for electrons ($-\frac{1}{2}$, 0, and $+\frac{1}{2}$) instead of just two ($-\frac{1}{2}$ and $+\frac{1}{2}$)?

ELECTRONIC STABILITY

Given the shell structure for electrons, there are certain magic numbers of electrons that correspond to stable electron configurations. Those numbers are 2, 10, 18, 36, 54, 86, and 118. These numbers correspond to the elements He, Ne, Ar, Kr, Xe, Rn, and Og. Looking at Figure 7.4, we can see that for two electrons ($1s^2$), this completes shell 1. Neon has ten electrons, which completes the second shell ($1s^2 2s^2 2p^6$). In the case of argon, with 18 electrons, the 3s and 3p subshells are filled, but not the 3d subshell. Argon is, however, electronically stable. If we look at Xe, Rn, and Og, we see a similar phenomenon: they are all chemically stable but do not have completely filled shells. Electronic stability, then, comes from filling the highest energy s and p subshells. Since it takes eight electrons (two for s and six for p) to fill the valence s and p subshells, this is often referred to as the octet rule.[11] It is important to be careful, however, as d-block and f-block elements can have eight or more valence electrons but are not stable. The eight electrons must occupy the s and p subshells.

Only the noble gases (group 18) have stable electron configurations. Therefore, all other elements have unstable electron configurations, and they will gain or lose electrons to try to achieve stability. In terms of losing or gaining electrons, atoms will do whichever is easier (so a main-group atom with three valence electrons will lose three electrons rather than try to gain five). If we look at the elements in group 1 (Figure 7.11) we can see that they will tend to lose their one valence electron (becoming preferentially 1+ ions) to achieve a noble gas configuration. For example, caesium atoms have [Xe]$6s^1$ configurations and they will tend to lose the one valence electron to

1	2	3	4	5	6	7	8	9	10	11	12	13	14	15	16	17	18
H 1+/1−																	He 0
Li 1+	Be 2+											B 3+	C variable	N 3−	O 2−	F 1−	Ne 0
Na 1+	Mg 2+											Al 3+	Si variable	P 3−	S 2−	Cl 1−	Ar 0
K 1+	Ca 2+	Sc 3+	Ti variable	V variable	Cr variable	Mn variable	Fe variable	Co variable	Ni variable	Cu variable	Zn 2+	Ga 3+	Ge variable	As variable	Se 2−	Br 1−	Kr 0
Rb 1+	Sr 2+	Y 3+	Zr 4+	Nb variable	Mo variable	Tc variable	Ru variable	Rh variable	Pd variable	Ag 1+	Cd 2+	In variable	Sn variable	Sb variable	Te variable	I 1−	Xe 0
Cs 1+	Ba 2+	*	Hf 4+	Ta variable	W variable	Re variable	Os variable	Ir variable	Pt variable	Au variable	Hg variable	Tl variable	Pb variable	Bi variable	Po variable	At variable	Rn 0
Fr 1+	Ra 2+	**	Rf unknown	Db unknown	Sg unknown	Bh unknown	Hs unknown	Mt unknown	Ds unknown	Rg unknown	Cn unknown	Nh unknown	Fl unknown	Mc unknown	Lv unknown	Ts unknown	Og unknown

*	La 3+	Ce 3+	Pr 3+	Nd 3+	Pm 3+	Sm 3+	Eu 3+	Gd 3+	Tb 3+	Dy 3+	Ho 3+	Er 3+	Tm 3+	Yb 3+	Lu 3+
**	Ac 3+	Th variable	Pa variable	U variable	Np variable	Pu variable	Am 3+	Cm 3+	Bk 3+	Cf 3+	Es 3+	Fm 3+	Md 3+	No variable	Lr 3+

Figure 7.13 Periodic table showing the common charge each element tends to take on in ionic compounds (if known). 0 means that monatomic ions of this element do not naturally form.

become Cs^+ and have a configuration of [Xe]. If there are multiple valence electrons, the highest energy valence electrons (those farthest to the right in standard notation or closest to the top in an orbital energy diagram) are lost first. In contrast, the halogens (elements in group 17) all have one electron fewer than a noble gas, and so they will tend to gain one electron to achieve a noble gas configuration. For example, fluorine has the configuration $[He]2s^22p^5$, and when it gains an electron (to become F^-) the configuration of the fluoride anion is $[He]2s^22p^6$, which is the same as that for neon. With this information in hand, then, we can see how the common charges (Figure 7.13) that each element adopts when it ionizes are what they are.

PROBLEM 7.11
Using Figure 7.11 and Figure 7.13, answer the following questions.

a.

  i)   Why do all group-3 elements commonly form 3+ ions?

  ii) In Chapter 3 we discussed that the lanthanoids caused a lot of frustration for Mendeleev. Why would he be led to think that they all belonged to group 3?

b. Transition metals can form several different ionic charges, but a 2+ ion is a common charge for most transition metals. Why would 2+ ions commonly form for transition metals?

PROBLEM 7.12
Write the electron configuration for each of the following ions.

a. Fe^{3+}

b. P^{3-}

c. Sn^{2+}

d. I^-

e. Sc^{3+}

PROBLEM 7.13

For each of the following configurations: (i) identify if it is a ground-state, excited-state, or impossible electron configuration and explain your answer; and (ii) if it is a ground-state configuration, identify whether it is a stable ground state or unstable ground state and explain.

a. $[Ne]3s^23p^1$

b. $[Ar]3d^64s^2$

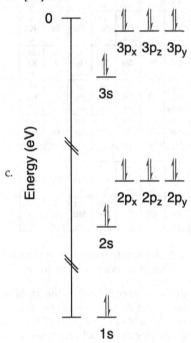

c.

d. $[Xe]4f^{14}5d^{10}6s^26p^5$

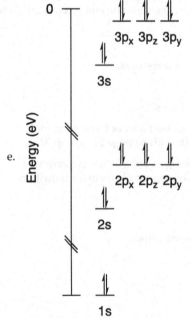

e.

f. $[Kr]4d^85s^1$

g. $1s^22s^22p^63s^23p^63d^94s^24p^6$

NOTES

1. This is a publication of de Broglie thesis in full: de Broglie, L. Recherches Sur la Théorie des Quanta. *Ann. Phys.* (Paris), **1925**, *10* (3), 22–128. DOI: 10.1051/anphys/192510030022.

2. (i) Heisenberg, W. Über den anschaulichen Inhalt der quantentheoretischen Kinematik und Mechanik. *Z. Phys.*, **1927**, *43* (3-4), 172–198. DOI: 10.1007/BF01397280.

   (ii) Kennard, E.H. Zur Quantenmechanik einfacher Bewegungstypen. *Z. Phys.*, **1927**, *44* (4–5), 326–352. DOI: 10.1007/BF01391200.

3. (i)   Heisenberg, W. Über quantentheoretische Umdeutung kinematischer und mechanischer Beziehungen. *Z. Phys.*, **1925**, *33* (1), 879–893. DOI: 10.1007/BF01328377.

   (ii) Born, M.; Jordan, P. Zur Quantenmechanik. *Z. Phys.*, **1925**, *34* (1), 858–888. DOI: 10.1007/BF01328531.

   (iii) Born, M.; Heisenberg, W.; Jordan, P. Zur Quantenmechanik. II. *Z. Phys.*, **1926**, *35* (8-9), 557–615. DOI: 10.1007/BF01379806.

4. Schrödinger, E. An Undulatory Theory of the Mechanics of Atoms and Molecules. *Phys. Rev.*, **1926**, *28* (6), 1049–1070. DOI: 10.1103/PhysRev.28.1049.

5. Dirac, P.A.M. On the Theory of Quantum Mechanics. *Proc. R. Soc. Lond. A*, **1926**, *112* (762), 661–677. DOI: 10.1098/rspa.1926.0133.

6. Bohr introduced n in his model of the atom (see Chapter 6). Arthur Sommerfeld refined Bohr's model and introduced the quantum numbers k (now ℓ) and m (now m_ℓ) to try to refine Bohr's model (Sommerfeld, A. *Atombau und Spektrallinien*. Braunschweig: Friedr. Vieweg & Sohn, 1921.). The Bohr–Sommerfeld model still suffered from unresolved contradictions and paradoxes, but the quantum number convention has continued and been given a more rigorous, truly quantum mechanical foundation.

7. Bohr, N. Der Bau der Atome und die physikalischen und chemischen Eigenschaften der Elemente. *Z. Physik*, **1922**, *9* (1), 1-67. DOI: 10.1007/BF01326955.

8. (i)   Pauli, W. Über die Gesetzmässigkeiten des anomalen Zeemaneffektes. *Z. Phys.*, **1923**, *16*, 155–164. DOI: 10.1007/BF01327386.

   (ii) Pauli, W. Über den Einfluss der Geschwindigkeitsabhängigkeit der Elektronenmasse auf den Zeemaneffekt. *Z Phys.*, **1925**, *31*, 373–385. DOI: 10.1007/BF02980592.

   (iii) Pauli, W. Über den Zusammenhang des Abschlusses der Elektronengruppen im Atom mit der Komplexstruktur der Spektren. *Z. Phys.*, **1925**, *31*, 765–783. DOI: 10.1007/BF02980631.

9. Hund, F. Zur Deutung verwickelter Spektren, I. *Z. Phys.*, **1925**, *33*, 345–371. DOI: 10.1007/BF01328319.

10. All reported configurations conform with those reported by NIST Atomic Spectra Database: Kramida, A.; Ralchenko, Y.; Reader, J.; NIST ASD Team. *NIST Atomic Spectra Database* (ver. 5.10), [Online]. Available: https://physics.nist.gov/asd. [2023, April 28]. National Institute of Standards and Technology, Gaithersburg, MD. DOI: https://doi.org/10.18434/T4W30F.

11. Langmuir, I. The Arrangement of Electrons in Atoms and Molecules. *J. Am. Chem. Soc.*, **1919**, *41* (6), 868–934. DOI: 10.1021/ja02227a002.

8 Periodic Trends

The arrangement of electrons around a nucleus and the quantum mechanical rules that govern electron configuration were investigated in Chapter 7. In this chapter, we will turn our attention to some of the properties of atoms – atomic radius, ionization energy, electron affinity, and electronegativity – and how those properties show repeating (periodic) patterns. These periodic trends have their origin in the electron configuration of each element and are manifestations of Coulomb's law.

COULOMB'S LAW

Charged particles experience electrostatic (coulombic) attraction or repulsion according to Coulomb's law (Equation 8.1).[1] The force is proportional to the product of the charges (Q_1 and Q_2) and inversely proportional to the interparticle distance squared (r^2). If the product of the charges is negative, then the force is attractive; if the product of the charges is positive, then the force is repulsive.

$$F_{electrostatic} = k_e \frac{Q_1 Q_2}{r^2} \tag{8.1}$$

$F_{electrostatic}$ is the electrostatic force (N).

k_e is the Coulomb constant ($8.988 \times 10^9 \frac{Nm^2}{C^2}$).
Q_1 is the charge of ion one (C).
Q_2 is the charge of ion two (C).
r is the interparticle distance (m).

Because of their equal and opposite charge, an electron (-1.602×10^{-19} C) and a proton (1.602×10^{-19} C) are attracted to each other electrostatically. As more protons are added to the nucleus and more electrons are added to the atom, we can apply a modified version of Coulomb's law to model the attraction of a valence electron to the nucleus (Equation 8.2):

$$F_{electrostatic} = k_e \frac{-e^2 Z_{eff}}{r^2} \tag{8.2}$$

$F_{electrostatic}$ is the electrostatic force (N).

k_e is the Coulomb constant ($8.988 \times 10^9 \frac{Nm^2}{C^2}$)
e is the elementary charge of an electron and a proton (1.602×10^{-19} C).
Z_{eff} is the effective charge of the nucleus (unitless).
r is the distance between the nucleus and electron (m).

The modification of the numerator accounts for two factors. First, the attraction between a valence electron and the nucleus is always attractive, and so the sign has been fixed as negative. Second, the charge of the nucleus is the product of the elementary charge (e) and the atomic number (Z). The full eZ magnitude of the nucleus is only felt by the 1s electrons. Those 1s electrons, and each successive shell of core electrons, shield the valence electrons from the nucleus. This means that the valence electrons feel the attraction of the effective nuclear charge (Z_{eff}), where the effective nuclear charge is the full nuclear charge less a shielding constant (S):

$$Z_{eff} = Z - S \tag{8.3}$$

Z_{eff} is the effective nuclear charge.
Z is the nuclear charge.
S is the shielding constant.

The degree to which core electrons shield the nucleus from a valence electron can be approximated following a set of rules,[2] calculated through computational chemistry,[3] or measured through X-ray spectroscopy. Figure 8.1 shows the effective nuclear charge found from X-ray spectroscopy.[4]

Aside from the effective nuclear charge (Z_{eff}), the other important term of Equation 8.2 to discuss is the distance between a valence electron and the nucleus (r). The distance between an electron and the nucleus is dependent upon the shell number, and the shell radius, the electron occupies. The shell radius can be calculated using the electron localization function (ELF).[5] Table 8.1 shows the radius in picometers (1 pm is 1×10^{-12} m) for shells 1–3 of an iron atom.[6]

DOI: 10.1201/9781003479338-8

1	2	3	4	5	6	7	8	9	10	11	12	13	14	15	16	17	18
1 **H** 1.00																	2 **He**
3 **Li** 1.26	4 **Be** 1.66											5 **B** 1.56	6 **C** 1.82	7 **N** 2.07	8 **O** 2.00	9 **F** 2.26	10 **Ne**
11 **Na** 1.84	12 **Mg** 2.25											13 **Al** 1.99	14 **Si** 2.32	15 **P** 2.63	16 **S** 2.62	17 **Cl** 2.93	18 **Ar**
19 **K** 2.26	20 **Ca** 2.68	21 **Sc** 2.77	22 **Ti** 2.83	23 **V** 2.82	24 **Cr** 2.82	25 **Mn** 2.96	26 **Fe** 3.04	27 **Co** 3.04	28 **Ni** 3.00	29 **Cu** 3.01	30 **Zn** 3.32	31 **Ga** 2.66	32 **Ge** 3.05	33 **As** 3.40	34 **Se** 3.39	35 **Br** 3.73	36 **Kr**
37 **Rb** 2.77	38 **Sr** 3.23	39 **Y** 3.45	40 **Zr** 2.55	41 **Nb** 3.56	42 **Mo** 3.61	43 **Tc** 3.66	44 **Ru** 3.68	45 **Rh** 3.70	46 **Pd** 3.13	47 **Ag** 3.73	48 **Cd** 4.06	49 **In** 3.26	50 **Sn** 3.67	51 **Sb** 3.98	52 **Te** 4.07	53 **I** 4.38	54 **Xe**
55 **Cs** 3.21	56 **Ba** 3.71	*	72 **Hf** 4.46	73 **Ta** 4.57	74 **W** 4.60	75 **Re** 4.57	76 **Os** 4.74	77 **Ir** 4.91	78 **Pt** 4.88	79 **Au** 4.94	80 **Hg** 5.25	81 **Tl** 4.02	82 **Pb** 4.43	83 **Bi** 4.39	84 **Po** 4.72	85 **At** 4.96	86 **Rn**
87 **Fr**	88 **Ra**	**	104 **Rf**	105 **Db**	106 **Sg**	107 **Bh**	108 **Hs**	109 **Mt**	110 **Ds**	111 **Rg**	112 **Cn**	113 **Nh**	114 **Fl**	115 **Mc**	116 **Lv**	117 **Ts**	118 **Og**

*	57 **La** 3.84	58 **Ce** 3.87	59 **Pr** 3.79	60 **Nd** 3.81	61 **Pm** 3.83	62 **Sm** 3.86	63 **Eu** 3.88	64 **Gd** 4.04	65 **Tb** 3.93	66 **Dy** 3.96	67 **Ho** 3.99	68 **Er** 4.02	69 **Tm** 4.04	70 **Yb** 4.46	71 **Lu**
**	89 **Ac**	90 **Th** 4.65	91 **Pa** 4.65	92 **U** 4.65	93 **Np**	94 **Pu**	95 **Am**	96 **Cm**	97 **Bk**	98 **Cf**	99 **Es**	100 **Fm**	101 **Md**	102 **No**	103 **Lr**

Figure 8.1 Effective nuclear charge (Z_{eff}) of each element from X-ray spectroscopic data.

Table 8.1 Shell Number and Shell Radius (pm) for an Iron Atom Calculated Using ELF

Shell Number	Shell Radius in an Iron Atom (pm)
1	5.00
2	22.23
3	106.36

PROBLEM 8.1

What will happen to the attractive force ($F_{electrostatic}$) between the nucleus and a valence electron as Z_{eff} increases?

PROBLEM 8.2

What will happen to the force ($F_{electrostatic}$) between the nucleus and a valence electron as r increases?

PROBLEM 8.3

Considering lithium to fluorine, what is changing to cause Z_{eff} to increase?

PROBLEM 8.4

Considering hydrogen to caesium (Figure 8.1), the change in Z_{eff} is relatively small. Why does Z_{eff} for lithium (1.26) show only a 26% increase compared to hydrogen, while Z shows a 200% increase?

PROBLEM 8.5

Given the relationship between shell number and radius (Table 8.1), how does the attractive force ($F_{electrostatic}$) between the nucleus and a valence electron in shell 3 compare to the same nucleus and an electron in shell 2?

PROBLEM 8.6

Given the relationship between shell number and radius (Table 8.1), what would you predict for the trend in atomic size for an atom in period 1 compared to an atom in period 2 compared to an atom in period 3? Explain.

ATOMIC RADIUS (r)

If we consider an atom to be a picometer-scale sphere, atoms will have both a measurable radius and volume. This concept was first developed from studying X-ray diffraction data of crystals. Bragg proposed that the crystals consisted of hard-sphere atoms that were closely packed, and so the internuclear distance in crystals would be the sum of atomic radii.[7] This idea was revived and expanded by Slater,[8] who developed the idea of a covalent atomic radius (r_{cov}) that represents the atomic size when bonded to another atom.[9,10] Another atomic radius value is the van der Waals radius (r_{vdw}). The van der Waals radius represents the closest another nonbonded atom can approach and is the balance point between intermolecular attraction (Chapter 12) and electron–electron repulsion.[11] Figure 8.2 shows the van der Waals radii, quantum mechanically calculated radii, and covalent radii of atoms 1–92.

As seen in Figure 8.2, the van der Waals radii are significantly larger than the covalent radii of the same atom. This is because bonded atoms require being close together to allow for orbital overlap and mixing (Chapter 9), whereas the van der Waals radii reflect the point at which two nonbonded atoms will start to repel due to Pauli-exclusion-based electron–electron repulsion. It should be noted that there is not a direct relationship between a particular shell number, or shell radius, and the covalent or the van der Waals radii; however, the empirical van der Waals radii do correspond to quantum mechanically calculated radii based on electron density parameters.[12] The quantum mechanical results are strongly correlated with an atom's valence shell radius.

Considering Figure 8.2, a regular (periodic) pattern can be seen, particularly with the covalent radii. Figure 8.3 reorganizes the elements to show them by group and period. With this reformatting of data, it is easier to see the increase in atomic size, within a group, as one moves from period

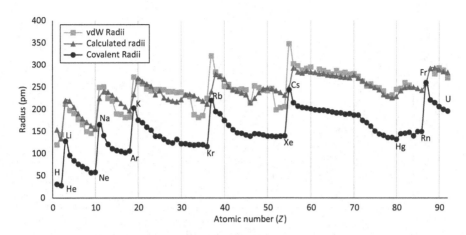

Figure 8.2 Van der Waals (vdW), calculated, and covalent atomic radii of elements 1–92.

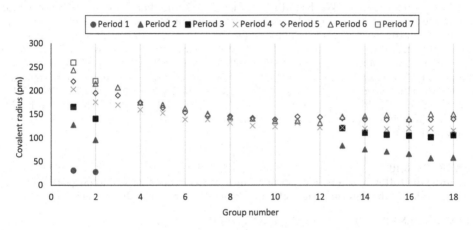

Figure 8.3 Covalent radii by group number of elements 1–88, omitting the f-block elements.

1 to period 7. The general pattern of atomic radius diminution moving from group 1 to group 18, within a period, is evident in both Figure 8.2 and Figure 8.3.

PROBLEM 8.7
Consider the elements in group 1, Figure 8.3. Using the electron configuration of each element, provide an explanation for the trend $r_{cov}(H) < r_{cov}(Li) < r_{cov}(Na) < r_{cov}(K) < r_{cov}(Rb) < r_{cov}(Cs) < r_{cov}(Fr)$.

PROBLEM 8.8
Consider the elements in period 4, Figure 8.3. Using the electron configuration of each element and Coulomb's law, provide an explanation for the general decrease in covalent radius from potassium to krypton.

So far, only the radii of neutral atoms have been considered. As seen in Figure 8.4, however, the radius of an atom changes significantly upon the loss of electrons to form a cation. The decrease in radius is due to changes in the electron configuration and the change of shielding that comes with the loss of valence electrons. There is very little change upon gaining electrons to form an anion. Consider Figure 8.4 to answer the problems that follow.

PROBLEM 8.9
Provide an explanation for why cations are smaller than their corresponding neutral atoms. Think about how Z_{eff} and r might change in going from the neutral atom to the ion.

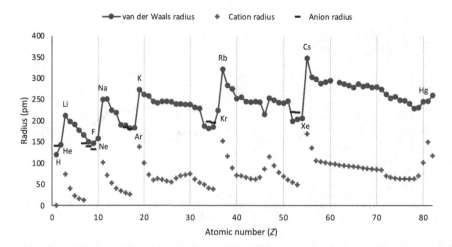

Figure 8.4 Atomic (van der Waals) radii of elements 1–82, and their cationic and anionic radii. All values are for a coordination number of 6, for transition metals the radii are for low-spin states and are either for the most common oxidation number or for the oxidation state of the element as it appears in a naturally occurring oxide. (Data from (i) Shannon, R.D.; Prewitt, C.T. Effective Ionic Radii in Oxides and Fluorides. *Acta Cryst.*, **1969**, B25, 925–946. DOI: 10.1107/S0567740869003220. (ii) Shannon, R.D. Revised Effective Ionic Radii and Systemic Studies of Interatomic Distances in Halides and Chalcogenides. *Acta Cryst.*, **1976**, A32, 751–767. DOI: 10.1107/S0567739476001551. (iii) Marcus, Y. Ionic Radii in Aqueous Solutions. *Chem. Rev.*, **1988**, 88 (8), 1475–1498. DOI: 10.1021/cr00090a003. (iv) Libowitz, G.G. *The Solid-State Chemistry of Binary Metal Hydrides.* W.A. Benjamin Inc.: New York, 1965.)

PROBLEM 8.10
Arrange the following ions in order of atomic radius (from smallest to largest): Ca^{2+}, Cl^-, S^{2-}, K^+. What is the determining factor (Z_{eff} or shell number) in this order?

IONIZATION ENERGY (E_i) AND ELECTRON AFFINITY (E_{ea})

Figure 8.4 shows the change in radius when an atom becomes an ion. Aside from changing the radius, there is also an energy change associated with losing or gaining an electron. To remove an electron from an atom, i.e., ionize an atom, energy must be added. Ionization energy (E_i) is the term for the energy needed to overcome the electrostatic attraction ($F_{electrostatic}$) and remove one electron from an atom to form a cation (Equation 8.4). For an atom with multiple electrons, successive ionization energy values can be measured where the first ionization energy is the energy it takes to form a cation from a neutral atom, the second ionization energy is the energy it takes to form a dictation from a cation, and so forth.

$$X + E_i \rightarrow X^+ + 1\ e^- \tag{8.4}$$

When an electron is added to an atom, energy is released, which would be reported as a negative value. Electron affinity (E_{ea}) is the term for the energy released when an electron is added to an atom to form an anion:

$$X + 1\ e^- \rightarrow X^- + E_{ea} \tag{8.5}$$

To measure electron affinity, it is easier to measure the amount of energy necessary to remove an electron from an anion (Equation 8.6). Therefore, the values for electron affinity are reported as positive values. While the values are positive, a better sense of electron affinity comes from thinking of it in the context of Equation 8.5, where electron affinity measures how much more stable the atom becomes when an electron is added.

$$X^- + E_{ea} \rightarrow X + 1\ e^- \tag{8.6}$$

The first ionization energy and electron affinity values for elements 1–92 are shown in Figure 8.5.[13] A larger ionization energy means that it is harder to remove an electron from the atom. A larger electron affinity means that the atom is more stable when an electron is added. An atom

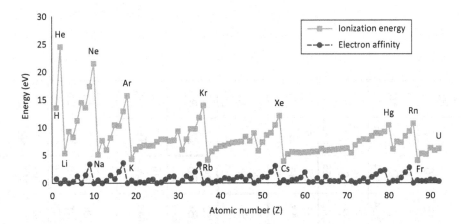

Figure 8.5 First ionization energy (E_i) and electron affinity (E_{ea}) for elements 1–92.

with an electron affinity of zero means that the anion was unstable and so the energy required to remove the electron could not be measured.

PROBLEM 8.11
Notice that E_i(H) < E_i(He). Provide an explanation for this.

PROBLEM 8.12
Provide an explanation for this trend in ionization: E_i(H) > E_i(Li) > E_i(Na) > E_i(K) > E_i(Rb) > E_i(Cs).

PROBLEM 8.13
Moving from lithium to neon, we cover the entire second period of elements. Can you make a generalization for the overall trend for E_i within a period?

PROBLEM 8.14
Does the generalized trend you proposed in Problem 8.13 hold up for the other periods in Figure 8.5?

PROBLEM 8.15
If we consider E_{ea}, is there really a trend (that is, could you draw a linear line) for elements 3–10 (period 2)?

PROBLEM 8.16
We can, however, analyze E_{ea} values for different groups.

a. The elements of which group have the largest electron affinity values? Why are their electron affinity values so high?

b. Why do group 1 and group 11 elements have an appreciable electron affinity?

c. Why do groups 2, 12, and 18 have an electron affinity of zero?

ELECTRONEGATIVITY (χ)

Ionization energy and electron affinity are useful concepts when discussing atoms in a vacuum, however, chemists tend to focus much of their time on atoms within molecules. In this context, electronegativity (χ) is a more useful concept because it is the power of an atom to attract electrons to itself.[14,15] Linus Pauling first proposed the concept of electronegativity, and the scale first proposed by Pauling remains one of the most widely used (Figure 8.6).[16]

1	2	3	4	5	6	7	8	9	10	11	12	13	14	15	16	17	18
1 H 2.20																	2 He
3 Li 0.98	4 Be 1.57											5 B 2.04	6 C 2.55	7 N 3.04	8 O 3.44	9 F 3.98	10 Ne
11 Na 0.93	12 Mg 1.31											13 Al 1.61	14 Si 1.90	15 P 2.19	16 S 2.58	17 Cl 3.16	18 Ar
19 K 0.82	20 Ca 1.00	21 Sc 1.36	22 Ti 1.54	23 V 1.63	24 Cr 1.66	25 Mn 1.55	26 Fe 1.83	27 Co 1.88	28 Ni 1.91	29 Cu 1.90	30 Zn 1.65	31 Ga 1.81	32 Ge 2.01	33 As 2.18	34 Se 2.55	35 Br 2.96	36 Kr 3.0
37 Rb 0.82	38 Sr 0.95	39 Y 1.22	40 Zr 1.33	41 Nb 1.6	42 Mo 2.16	43 Tc 1.9	44 Ru 2.2	45 Rh 2.28	46 Pd 2.20	47 Ag 1.93	48 Cd 1.69	49 In 1.78	50 Sn 1.96	51 Sb 2.05	52 Te 2.30	53 I 2.66	54 Xe 2.6
55 Cs 0.79	56 Ba 0.89	*	72 Hf 1.3	73 Ta 1.5	74 W 2.36	75 Re 1.9	76 Os 2.0	77 Ir 2.20	78 Pt 2.28	79 Au 2.54	80 Hg 2.00	81 Tl 2.04	82 Pb 2.33	83 Bi 2.02	84 Po 2.0	85 At 2.2	86 Rn
87 Fr 0.7	88 Ra 0.9	**	104 Rf	105 Db	106 Sg	107 Bh	108 Hs	109 Mt	110 Ds	111 Rg	112 Cn	113 Nh	114 Fl	115 Mc	116 Lv	117 Ts	118 Og

*	57 La 1.1	58 Ce 1.12	59 Pr 1.13	60 Nd 1.14	61 Pm 1.2	62 Sm 1.17	63 Eu 1.1	64 Gd 1.20	65 Tb 1.2	66 Dy 1.22	67 Ho 1.23	68 Er 1.24	69 Tm 1.25	70 Yb 1.1	71 Lu 1.27
**	89 Ac 1.1	90 Th 1.0	91 Pa 1.5	92 U 1.38	93 Np 1.36	94 Pu 1.28	95 Am 1.30	96 Cm	97 Bk	98 Cf	99 Es	100 Fm	101 Md	102 No	103 Lr

Figure 8.6 Pauling electronegativity (χ_P) values.

Pauling electronegativity (χ_P) values are based on the strength of a bond (in electronvolt) between two atoms (the dissociation energy, E_d; Chapter 15), generically AB, and the homodiatomic bond strengths AA and BB (Equation 8.6). Note that the Pauling electronegativity value is unitless and so the value inside the radical in Equation 8.6 is multiplied by the $eV^{-1/2}$ to cancel units. To develop this scale, hydrogen was arbitrarily set to 2.20 and values greater than hydrogen were chosen for elements that have a greater tendency to pull electrons toward themselves than hydrogen.

$$|\chi_A - \chi_B| = \sqrt{E_d(AB) - \frac{E_d(AA) + E_d(BB)}{2}}\left(eV^{-1/2}\right) \tag{8.6}$$

χ_A and χ_B are the electronegativity values for atom A and atom B, respectively.
E_d is the dissociation energy for the bond between the listed atoms in parentheses.

While Pauling's electronegativity scale is the most widely used, and the scale that will be employed throughout the rest of this book, it is not immediately clear from the formal definition how electronegativity relates to any periodic trend. It is therefore worth considering other electronegativity scales. The Allred–Rochow electronegativity (χ_{AR}) scale defines electronegativity as the charge experienced by an electron on the surface of an atom (Equation 8.7), which is the modified form of Coulomb's law (Equation 8.2) stripped of the Coulomb constant and the elementary charge terms:[17]

$$\chi_{AR} = \frac{Z_{eff}}{r^2} \tag{8.7}$$

Z_{eff} is the effective nuclear charge.
r is the covalent radius of the atom (pm).

Given the familiarity and widespread use of Pauling's scale, it is common to linearly transform other electronegativity scales so that they are comparable to Pauling values (Equation 8.8). The Allred–Rochow electronegativity values are transformed using Equation 8.8. Allred–Rochow values calculated from Z_{eff} (Figure 8.1) and r_{cov} (Figure 8.2) are shown in Figure 8.7.

$$\chi_{AR} = 5963\frac{Z_{eff}}{r^2} + 0.7212 \tag{8.8}$$

The Mulliken electronegativity (χ_M) scale defines the electronegativity average of ionization energy and electron affinity:[18]

$$\chi_M = \frac{E_i + E_{ea}}{2} \tag{8.9}$$

The Mulliken electronegativity values are also known as absolute electronegativity (or electroaffinity) values and are often reported with the unit of electronvolt. These electronvolt values are transformed using Equation 8.10 and are shown in Figure 8.8:

$$\chi_M = 0.374\left(\frac{E_i + E_{ea}}{2}\right) + 0.744 \tag{8.10}$$

The Allen electronegativity (χ_M) scale, also called configuration energy, defines electronegativity as the average energy (kJ/mol) of a valence electron in a free atom:[19]

$$\chi_A = \frac{n_s\varepsilon_s + n_p\varepsilon_p}{n_s + n_p} \tag{8.11}$$

ε_s is the valence electron energy of s-orbital electrons.
ε_p is the valence electron energy of p-orbital electrons.
n_s is the number of s-orbital electrons.
n_p is the number of p-orbital electrons.

These average, one-electron energy values are multiplied by 1.75×10^{-3} to transform the energy values (kJ/mol) into values that are comparable to the Pauling scale. The Allen electronegativity values are shown in Figure 8.9.

This book, and most chemists, only employ the Pauling scale (χ_P). However, it is worth noting the commonality of the results and trends (Figure 8.10) of these different scales. This tells us two things: First, electronegativity can be derived from multiple different sets of data. Second,

1	2	3	4	5	6	7	8	9	10	11	12	13	14	15	16	17	18
1 H																	2 He
3 Li 1.18	4 Be 1.80											5 B 2.04	6 C 2.60	7 N 3.17	8 O 3.46	9 F 3.87	10 Ne
11 Na 1.12	12 Mg 1.40											13 Al 1.53	14 Si 1.84	15 P 2.09	16 S 2.14	17 Cl 2.40	18 Ar
19 K 1.05	20 Ca 1.24	21 Sc 1.29	22 Ti 1.38	23 V 1.44	24 Cr 1.59	25 Mn 1.63	26 Fe 1.76	27 Co 1.86	28 Ni 1.88	29 Cu 1.75	30 Zn 2.05	31 Ga 1.79	32 Ge 1.98	33 As 2.15	34 Se 2.12	35 Br 2.27	36 Kr
37 Rb 1.06	38 Sr 1.23	39 Y 1.29	40 Zr 1.22	41 Nb 1.51	42 Mo 1.63	43 Tc 1.73	44 Ru 1.75	45 Rh 1.82	46 Pd 1.69	47 Ag 1.78	48 Cd 1.89	49 In 1.69	50 Sn 1.85	51 Sb 1.95	52 Te 2.00	53 I 2.07	54 Xe
55 Cs 1.04	56 Ba 1.20	*	72 Hf 1.59	73 Ta 1.66	74 W 1.77	75 Re 1.92	76 Os 2.08	77 Ir 2.19	78 Pt 2.29	79 Au 2.31	80 Hg 2.52	81 Tl 1.86	82 Pb 1.96	83 Bi 1.92	84 Po 2.16	85 At 2.04	86 Rn
87 Fr	88 Ra	**	104 Rf	105 Db	106 Sg	107 Bh	108 Hs	109 Mt	110 Ds	111 Rg	112 Cn	113 Nh	114 Fl	115 Mc	116 Lv	117 Ts	118 Og

*	57 La 1.26	58 Ce 1.28	59 Pr 1.27	60 Nd 1.28	61 Pm 1.30	62 Sm 1.31	63 Eu 1.31	64 Gd 1.35	65 Tb 1.34	66 Dy 1.36	67 Ho 1.37	68 Er 1.39	69 Tm 1.39	70 Yb 1.48	71 Lu
**	89 Ac	90 Th	91 Pa	92 U	93 Np	94 Pu	95 Am	96 Cm	97 Bk	98 Cf	99 Es	100 Fm	101 Md	102 No	103 Lr

Figure 8.7 Allred–Rochow electronegativity (χ_{AR}) values transformed into values comparable to χ_P values.

1	2	3	4	5	6	7	8	9	10	11	12	13	14	15	16	17	18
1 **H** 2.85																	2 **He** 4.77
3 **Li** 1.29	4 **Be** 1.91											5 **B** 1.77	6 **C** 2.51	7 **N** 2.89	8 **O** 2.99	9 **F** 4.06	10 **Ne** 4.20
11 **Na** 1.23	12 **Mg** 1.60											13 **Al** 1.37	14 **Si** 1.95	15 **P** 2.27	16 **S** 2.50	17 **Cl** 3.27	18 **Ar** 3.11
19 **K** 1.08	20 **Ca** 1.32	21 **Sc** 1.43	22 **Ti** 1.46	23 **V** 1.53	24 **Cr** 1.56	25 **Mn** 1.56	26 **Fe** 1.68	27 **Co** 1.77	28 **Ni** 1.82	29 **Cu** 1.85	30 **Zn** 1.93	31 **Ga** 1.37	32 **Ge** 1.88	33 **As** 2.20	34 **Se** 2.15	35 **Br** 3.01	36 **Kr** 2.79
37 **Rb** 1.04	38 **Sr** 1.24	39 **Y** 1.39	40 **Zr** 1.49	41 **Nb** 1.61	42 **Mo** 1.64	43 **Tc** 1.63	44 **Ru** 1.74	45 **Rh** 1.78	46 **Pd** 1.83	47 **Ag** 1.83	48 **Cd** 1.85	49 **In** 1.33	50 **Sn** 1.75	51 **Sb** 1.98	52 **Te** 2.22	53 **I** 2.70	54 **Xe** 2.44
55 **Cs** 0.99	56 **Ba** 1.17	*	72 **Hf** 1.45	73 **Ta** 1.64	74 **W** 1.79	75 **Re** 1.66	76 **Os** 1.95	77 **Ir** 2.14	78 **Pt** 2.24	79 **Au** 2.33	80 **Hg** 2.12	81 **Tl** 1.38	82 **Pb** 1.62	83 **Bi** 1.71	84 **Po** 2.10	85 **At** 2.44	86 **Rn** 2.18
87 **Fr** 1.02	88 **Ra** 1.22	**	104 **Rf**	105 **Db**	106 **Sg**	107 **Bh**	108 **Hs**	109 **Mt**	110 **Ds**	111 **Rg**	112 **Cn**	113 **Nh**	114 **Fl**	115 **Mc**	116 **Lv**	117 **Ts**	118 **Og**

*	57 **La** 1.32	58 **Ce** 1.33	59 **Pr** 1.37	60 **Nd** 1.56	61 **Pm** 1.24	62 **Sm** 1.26	63 **Eu** 1.39	64 **Gd** 1.35	65 **Tb** 1.48	66 **Dy** 1.35	67 **Ho** 1.36	68 **Er** 1.37	69 **Tm** 1.52	70 **Yb** 1.34	71 **Lu** 1.25
**	89 **Ac** 1.20	90 **Th** 1.46	91 **Pa** 1.37	92 **U** 1.39	93 **Np**	94 **Pu**	95 **Am**	96 **Cm**	97 **Bk**	98 **Cf**	99 **Es**	100 **Fm**	101 **Md**	102 **No**	103 **Lr**

Figure 8.8 Mulliken electronegativity (χ_M) values transformed into values comparable to χ_P values.

1	2	3	4	5	6	7	8	9	10	11	12	13	14	15	16	17	18
1 H 2.30																	2 He 4.16
3 Li 0.91	4 Be 1.58											5 B 2.05	6 C 2.54	7 N 3.07	8 O 3.61	9 F 4.19	10 Ne 4.79
11 Na 0.87	12 Mg 1.29											13 Al 1.61	14 Si 1.92	15 P 2.25	16 S 2.59	17 Cl 2.87	18 Ar 3.24
19 K 0.73	20 Ca 1.03	21 Sc 1.19	22 Ti 1.38	23 V 1.53	24 Cr 1.65	25 Mn 1.75	26 Fe 1.80	27 Co 1.84	28 Ni 1.88	29 Cu 1.85	30 Zn 1.59	31 Ga 1.76	32 Ge 1.99	33 As 2.21	34 Se 2.42	35 Br 2.69	36 Kr 2.97
37 Rb 0.71	38 Sr 0.96	39 Y 1.12	40 Zr 1.32	41 Nb 1.41	42 Mo 1.47	43 Tc 1.51	44 Ru 1.54	45 Rh 1.56	46 Pd 1.58	47 Ag 1.87	48 Cd 1.52	49 In 1.67	50 Sn 1.82	51 Sb 1.98	52 Te 2.16	53 I 2.36	54 Xe 2.58
55 Cs 0.66	56 Ba 0.88	*	72 Hf 1.16	73 Ta 1.34	74 W 1.47	75 Re 1.60	76 Os 1.65	77 Ir 1.68	78 Pt 1.72	79 Au 1.92	80 Hg 1.76	81 Tl	82 Pb	83 Bi	84 Po	85 At	86 Rn
87 Fr	88 Ra	**	104 Rf	105 Db	106 Sg	107 Bh	108 Hs	109 Mt	110 Ds	111 Rg	112 Cn	113 Nh	114 Fl	115 Mc	116 Lv	117 Ts	118 Og

*	57 La	58 Ce	59 Pr	60 Nd	61 Pm	62 Sm	63 Eu	64 Gd	65 Tb	66 Dy	67 Ho	68 Er	69 Tm	70 Yb	71 Lu 1.09
**	89 Ac	90 Th	91 Pa	92 U	93 Np	94 Pu	95 Am	96 Cm	97 Bk	98 Cf	99 Es	100 Fm	101 Md	102 No	103 Lr

Figure 8.9 Allen electronegativity (χ_A) values transformed into values comparable to χ_P values.

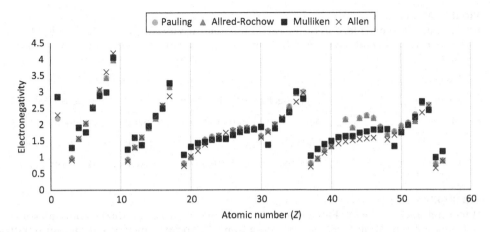

Figure 8.10 Comparison of Pauling, Allred–Rochow, Mulliken, and Allen electronegativity values for elements 1–56. Note that all values for Allred–Rochow, Mulliken, and Allen electronegativity scales have been transformed into values comparable to χ_P values.

electronegativity is widely applicable to a breadth of atomic and molecular phenomena, as shall be seen throughout future chapters. For this reason, electronegativity has been proposed as the third dimension (after periods and groups) of the periodic table.[20]

PROBLEM 8.17
Consider the electrostatic potential maps of lithium hydride, hydrogen, and hydrogen fluoride.

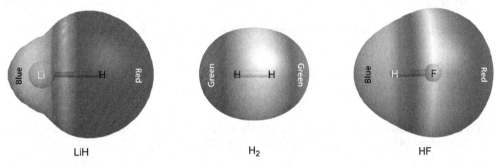

Red indicates areas of high electron density, blue indicates areas of low electron density, and green indicates a middling amount of electron density.

Using the electronegativity values in Figure 8.6 and your knowledge of what electronegativity means, can you explain the differences in color for H in each of the above compounds?

PROBLEM 8.18
Look at the electronegativity values in Figure 8.6.

a. LiF and RbCl are example ionic compounds. What is the difference in electronegativity ($\Delta\chi_P$) for Li and F? For Rb and Cl?

b. Now consider CH_4 and N_2O_4, example molecular compounds. What is the difference in electronegativity ($\Delta\chi_P$) for C and H? For N and O?

c. Can you make any generalization about the magnitude of the electronegativity differences ($\Delta\chi_P$) for the ionic compounds versus the molecular compounds?

PROBLEM 8.19
For each pair of elements, determine their electronegativity difference ($\Delta\chi_P$) and indicate which atom should have high electron density and which should have low electron density.

a. C and H $\Delta\chi_P$_____ high-electron density atom____ low-electron density atom____

b. Fe and H $\Delta\chi_P$_____ high-electron density atom____ low-electron density atom____

c. Br and Al $\Delta\chi_P$_____ high-electron density atom____ low-electron density atom____

d. C and F $\Delta\chi_P$_____ high-electron density atom____ low-electron density atom____

e. Fr and F $\Delta\chi_P$_____ high-electron density atom____ low-electron density atom____

PROBLEM 8.20
When sodium fluoride (NaF) forms from sodium and fluorine, the final compound consists of a sodium cation (Na^+) and fluoride anion (F^-). Sodium fluoride will not form with a sodide anion (Na^-) and fluorine cation (F^+).

a. Provide an explanation why sodium is always the cation and fluorine is always the anion in sodium fluoride based on the electronegativity values (χ) of sodium and fluorine.

b. A more rigorous consideration looks at the energetics of the processes involved. Using Figure 8.4, provide an explanation based on energy why sodium will always form a cation and fluorine will always form an anion when sodium and fluorine react together to make sodium fluoride.

NOTES

1. Coulomb, C.A. Premier mémoire sur l'éctricité et le magnétisme. *Histoire de l'Academie Royale des Sc.*, **1785**, 569–577.

2. Slater, J.C. Atomic Shielding Constants. *Phys. Rev.*, **1930**, 36 (1), 57–64. DOI: 10.1103/physrev.36.57.

3. Koseki, S.; Fedorov, D.G.; Schmidt, M.W.; Gordon, M.S. Spin–Orbit Splittings in the Third-Row Transition Elements: Comparison of Effective Nuclear Charge and Full Breit–Pauli Calculations. *J. Phys. Chem. A.*, **2001**, 105 (35), 8262–8268. DOI: 10.1021/jp011677r.

4. Husain, M.; Batra, A. Electronegativity Scale from X-Ray Photoelectron Spectroscopic Data. *Polyhedron*, **1989**, 8 (9), 1233–1237. DOI: 10.1016/S0277-5387(00)81146-8.

5. Becke, A.D.; Edgecombe, K.E. A Simple Measure of Electron Localization in Atomic and Molecular Systems. *J. Chem. Phys.*, **1990**, 92 (9), 5397–5403. DOI: 10.1063/1.458517.

6. The ELF results for iron were for iron in its $^5D_4$ ground state: Kahout, M.; Savin, A. Atomic Shell Structure and Electron Numbers. *Int. J. Quantum Chem.*, **1996**, 60 (4), 875–882. DOI: 10.1002/(SICI)1097-461X(1996)60:4<875::AID-QUA10>3.0.CO;2-4.

7. Bragg, W.L. XVIII. The Arrangement of Atoms in Crystals. *Lond. Edinb. Dublin Philos. Mag. J. Sci.*, **1920**, 40 (236), 169–189. DOI: 10.1080/14786440808636111.

8. Slater, J.C. Atomic Radii in Crystals. *J. Chem. Phys.*, **1964**, 41 (10), 3199–3204. DOI: 10.1063/1.1725697.

9. Cordero, B.; Gómez, V.; Platero-Prats, A.E.; Revés, M; Echeverría, J.; Cremades, E.; Barragán, F.; Alvarez, S. Covalent Radii Revisited. *Dalton Trans.*, **2008**, 21, 2832–2838. DOI: 10.1039/B801115J.

10. Bańkowski, Z. Covalent Radii of Noble Gas Atoms and Position of Hydrogen in the Periodic System of Elements. *Nature*, **1966**, 209, 71–72. DOI: 10.1038/209071a0.

11. (i) Bondi, A. Van der Waals Volumes and Radii. *J. Phys. Chem.*, **1964**, *68* (3), 441–451. DOI: 10.1021/j100785a001.

(ii) Alvarez, S. A Cartography of the Van der Waals Territories. *Dalton Trans.*, **2013**, *42*, 8617–8636. DOI: 10.1039/c3dt50599e.

12. Rahm, M.; Hoffmann, R.; Ashcroft, N.W. Atomic and Ionic Radii of Elements 1–96. *Chem. Eur. J.*, **2016**, *22* (41), 14625–14632. DOI: 10.1002/chem.201602949.

13. Sansonetti, J.E.; Martin, W.C.; Young, S.L. *Handbook of Basic Atomic Spectroscopic Data*. NIST Standard Reference Database 108. DOI: 10.18434/T4FW23.

14. Pauling, L. The Nature of the Chemical Bond. IV. The Energy of Single Bonds and the Relative Electronegativity of Atoms. *J. Am. Chem. Soc.*, **1932**, *54* (9), 3570-3582. DOI: 10.1021/ja01348a011.

15. "Electronegativity." IUPAC. *Compendium of Chemical Terminology*, 2nd ed. Compiled by A.D. McNaught and A. Wilkinson. Blackwell Scientific Publications, Oxford, 1997. Online version (2019–) created by S.J. Chalk. DOI: 10.1351/goldbook.E01990.

16. (i) The data most widely cited today are the revised Pauling values given by Allred: Allred, A.L. Electronegativity Values from Thermochemical Data. *J. Inorg. Nucl.*, **1961**, *17* (3–4), 215–221. DOI: 10.1016/0022-1902(61)80142-5.

(ii) Tellurium value from: Huggins, M.L. Bond Energies and Polarities. *J. Am. Chem. Soc.*, **1953**, *75* (17), 4123–4126. DOI: 10.1021/ja01113a001.

(iii) Noble gas values from: Allen, L.C.; Huheey, J.E. The Definition of Electronegativity and the Chemistry of the Noble Gases. *J. Inorg. Nucl. Chem.*, **1980**, *42* (10), 1523–1524. DOI: 10.1016/0022-1902(80)80132-1.

(iv) All other values with only one decimal place are from: Gordy, W.; Thomas, W.J.O. Electronegativity of the Elements. *J. Chem. Phys.*, **1956**, *24* (2), 439–444. DOI: 10.1063/1.1742493.

17. Allred, A.L.; Rochow, E.G. A Scale of Electronegativity Based on Electrostatic Force. *J. Inorg. Nucl.*, **1958**, *5* (4), 264–268. DOI: 10.1016/0022-1902(58)80003-2. This system was first proposed by Gordy: Gordy, W. A New Method of Determining Electronegativity from Other Atomic Properties. *Phys. Rev.*, **1946**, *69* (11–12), 604–607. DOI: 10.1103/PhysRev.69.604.

18. Note that Mulliken used "particular 'valence states'" when calculating E_i and E_{ea}. Here the E_i and E_{ea} from Figure 8.4 are used: Mulliken, R.S. A New Electroaffinity Scale; Together with Data on Valence States and on Valence Ionization Potentials and Electron Affinities. *J. Chem. Phys.*, **1934**, *2* (11), 782–793. DOI: 10.1063/1.1749394.

19. Allen, L.C. Electronegativity Is the Average One-Electron Energy of the Valence-Shell Electrons in Ground-State Free Atoms. *J. Am. Chem. Soc.*, **1989**, *111* (25), 9003–9014. DOI: 10.1021/ja00207a003.

20. Sun, X. Redefinition of Electronegativity as the Average Valence Electron Energy: The Third Dimension of the Periodic Table. *Chem. Educator*, **2000**, *5* (2), 54–57. DOI: 10.1007/s00897990363a.

9 Bonding

In this chapter, we will move our attention from the scale of a singular atom to atoms that are connected – bonded – together. This will expand our focus from the picometer (1×10^{-12} m) scale to the nanometer (1×10^{-9} m) scale.

The principles, theories, and rules discussed in Chapter 7 for the electron configuration of an atom give us a string of alphanumeric characters that represent how the electrons are distributed in an atom. For example, the electron configuration of carbon is $1s^2 2s^2 2p^2$. This can be abbreviated using the noble gas notation to highlight only the valence electrons of carbon: $[He]2s^2 2p^2$. Each of these configurations, while rigorously sound in terms of theory and experiment, lacks a certain fluidity and facility in its presentation. It is for this reason that Lewis dot symbols, developed by G.N. Lewis in 1916,[1] are still so widely employed by chemists. Consider the Lewis dot symbol for carbon (Figure 9.1).

The fact that carbon has four valence electrons is more readily apparent to both a practicing chemists and a learner with this symbology. If we consider a periodic table of Lewis dot symbols (Figure 9.2), the patterns and group trends in electron configuration become much more apparent.

Looking at the Lewis dot symbols (Figure 9.2), recall that atoms are at their most stable (lowest energy) when they have filled valence shells. Main group elements, aside from hydrogen and helium, typically follow the octet rule,[2] which states that a stable atom has eight valence electrons. Specifically, those eight valence electrons correspond to filled s and p subshells. For hydrogen and helium, there is no 1p subshell and so they only need a duet of electrons to fill the 1s subshell and to be stable. Transition metals require 12 electrons to achieve stability, which corresponds to filling the valence s and d subshells.[3] Looking at the main group elements in Figure 9.2, we can see that only group 18 elements have an octet of electrons, which is why the noble gas elements naturally exist as monatomic gases. For all other elements, they can only achieve stability through forming di- or polyatomic molecules or through forming compounds.

Atoms in molecules and compounds are held together by bonds. Bonds can be thought of as the energy, or force, holding atoms together. More specifically, when ions and atoms are brought together, forming ionic bonds or covalent bonds, respectively, they release energy (Figure 9.3). To break ions and atoms in a compound apart, energy (dissociation energy, E_d) must be added to break the bond. This chapter will consider the nature of both ionic and covalent bonding.

IONIC BONDS

An ionic bond refers to the electrostatic attraction experienced between a positively charged ion (cation) and a negatively charged ion (anion). Ions, and hence ionic bonds, are formed in cases where the interacting atoms have a large electronegativity difference. The electronegativity difference is so large that when two atoms interact the electrons are not shared but transferred from one to the other. The atom that loses electrons, called a cation, empties its valence shell leaving a stable, noble-gas core. The atom that gains electrons fills its valence shell to achieve a noble-gas configuration and an octet.[4] This exchange of electrons to form cations and anions is a redox reaction (Chapter 19). The electrostatic attraction between a cation and an anion is called an ionic bond,[5] which is described mathematically by Coulomb's law (Chapter 8).[6]

Although we will discuss ionic bonds and covalent bonds separately in this chapter, they exist as part of a continuum of bonding. Very few bonds are entirely ionic and only the bonds between atoms of the same element are entirely covalent. The amount of ionic character in a bond can be estimated using the Pauling electronegativity (χ_P) values of the atoms involved:[7]

$$f_i(AB) = 1 - e^{\left(-\frac{(\chi_A - \chi_B)^2}{4}\right)} \tag{9.1}$$

$f_i(AB)$ is the fraction ionic character of the bond between atom A and atom B.
χ_A is the Pauling electronegativity value of atom A.
χ_B is the Pauling electronegativity value of atom B.

Figure 9.1   The Lewis dot symbol for carbon showing the valence configuration (left) and for a hybridized (Chapter 10) carbon atom (right) that highlights the tendency of carbon to make four bonds.

DOI: 10.1201/9781003479338-9

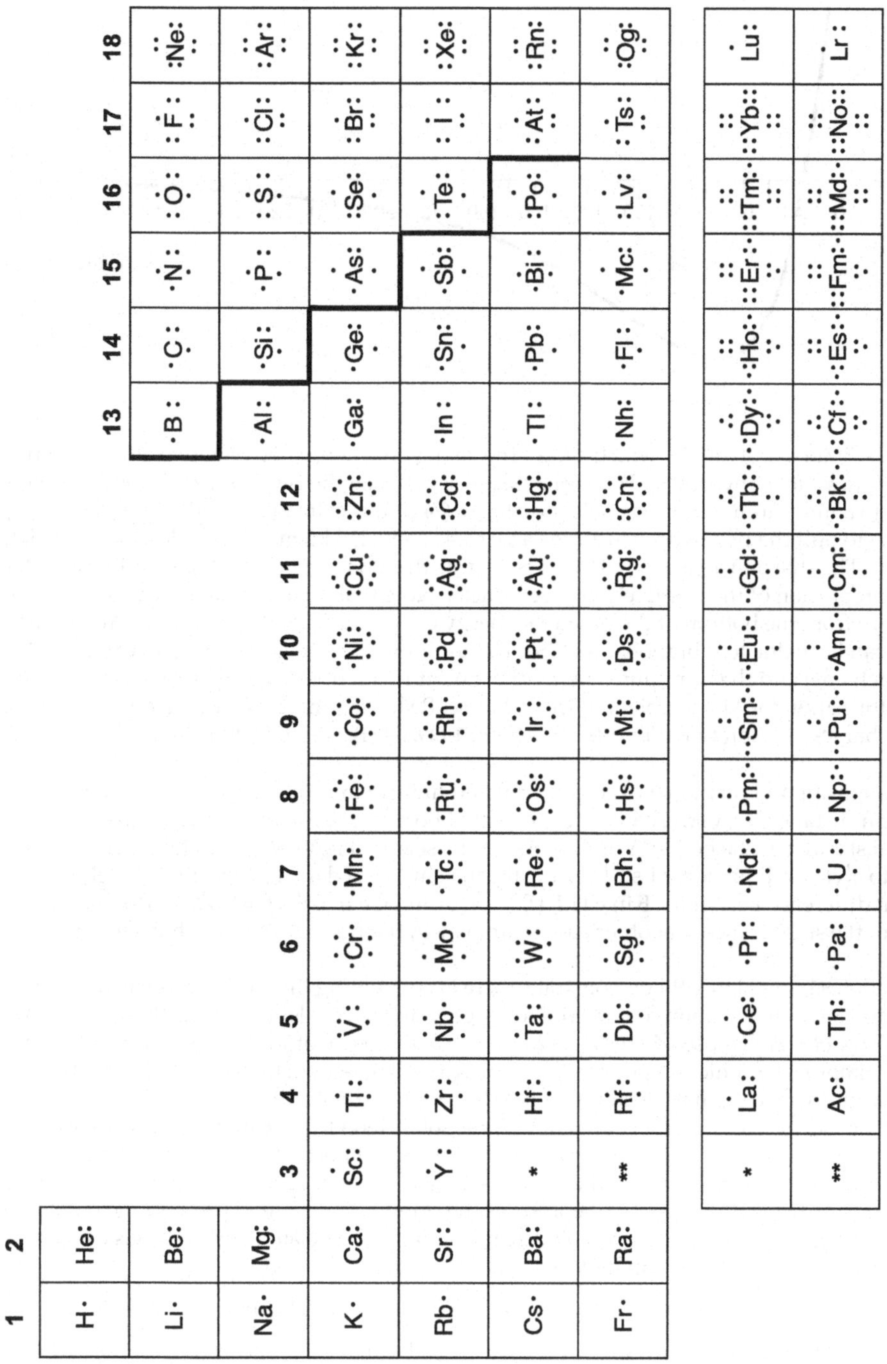

Figure 9.2 Lewis dot symbols for the elements.

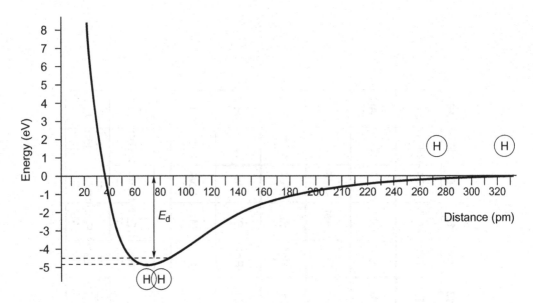

Figure 9.3 Morse potential – which shows the change in energy between two atoms as a function of distance – for dihydrogen. When the distance between the two nuclei is 74 pm, the energy decreases to a minimum. The internuclear distance (74 pm) is called the bond length; the energy it would take to pull the two hydrogen atoms apart – 4.52 eV (436 kJ/mol) – is called the dissociation energy (E_d). The sharp increase in potential energy as the atoms get closer together is due to the coulombic repulsion of the positively charged nuclei. Note that the dissociation energy (E_d) does not measure from the bottom of the potential energy curve but from the minimum energy that the molecule can have due to vibrational motion. The difference between the electronic potential minimum (black line) and the minimum energy the molecule can have is called the zero-point energy (ZPE). (Information for Morse potential from: Morse, P.M. Diatomic Molecules According to the Wave Mechanics. II. Vibrational Levels. *Phys. Rev.*, *1929*, *34* (1), 57–64. DOI: 10.1103/PhysRev.34.57.)

The relationship in Equation 9.1 is theoretical and highly simplified as it does not consider the internuclear distance between A and B, the extent of bonding between the two atoms, or what other atom(s) may be attached to A or to B. As can be seen in Figure 9.3, despite not taking these factors into account, Equation 9.1 still closely fits the empirical data. It is worth highlighting the significant difference between CF in CH_3F ($f_i = 0.253$) and CF in CF_4 ($f_i = 0.457$). These specific data underscore the significance that other atoms can play in the fraction of ionic character between two atoms.

As Equation 9.1 and Figure 9.4 demonstrate, there is no dividing line between covalent bonds and ionic bonds. However, it is common practice to roughly divide covalent bonds as those where $\Delta\chi_P < 1.7$ and ionic bonds as those where $\Delta\chi_P > 1.7$, where a difference of 1.7 corresponds to 50% ionic character.[8] Empirically, ionic compounds are compounds that tend to form crystal lattices (Figure 9.5). Crystal lattices are regular repeating unit cells of ions and not a discrete set of bonded atoms – a hallmark of molecular or covalent compounds. Ionic solids tend to have high melting points (hundreds to thousands of degrees Celsius) and tend to be solids at normal temperature and pressure.

The electrostatic attraction among the ions of a crystal lattice is reflected in the lattice enthalpy ($\Delta_{lattice}H°$). The lattice enthalpy, which represents the stabilization gained through the bonding of oppositely charged ions, can be represented as the enthalpy associated with gaseous cations and anions coming together to form an ionic solid:

$$Na^+(g) + Cl^-(g) \rightarrow NaCl(s) \quad \Delta_{lattice}H° = -788 \text{ kJ/mol}$$

To determine the lattice enthalpy, we can use the Born–Fajans–Haber cycle (Figure 9.6),[9] which considers the total energy change associated with going from the elements that make up the ionic compound, here Na(s) and $Cl_2(g)$, and breaking down the overall process into individual, quantifiable steps.

The magnitude of the lattice enthalpy is affected by the bond strength (E_d), the ionization energy (E_i), and the electron affinity (E_{ea}) of the elements involved (Equation 9.2).

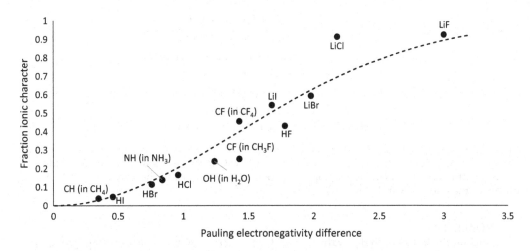

Figure 9.4 Plot of fraction ionic character (f_i) against the Pauling electronegativity difference ($\Delta\chi_p$). The dashed line plots Equation 9.1, and individual data come from experimental and computational sources. (Data from (i) Dailey, B.P.; Townes, C.H. The Ionic Character of Diatomic Molecules. *J. Chem. Phys.*, **1955**, *23* (1), 118–123. DOI: 10.1063/1.1740508. (ii) Pilcher, G.; Skinner, H.A. Valence-States of Boron, Carbon, Nitrogen and Oxygen. *J. Inorg. Nucl. Chem.*, **1962**, *24* (8), 937–952. DOI: 10.1016/0022-1902(62)80211-5.)

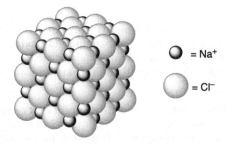

Figure 9.5 An example crystal lattice of sodium chloride. (Based on Figure 8 from: Barlow, W. XXVI. Geometrische Untersuchung über eine mechanische Ursache der Homogenität der Structur und der Symmetrie; mit besonderer Anwendung auf Krystallisation und chemische Verbindung. *Z. Kristallogr. Cryst. Mater.*, **1898**, *29* (1–6), 433–588. DOI: 10.1524/zkri.1898.29.1.433.)

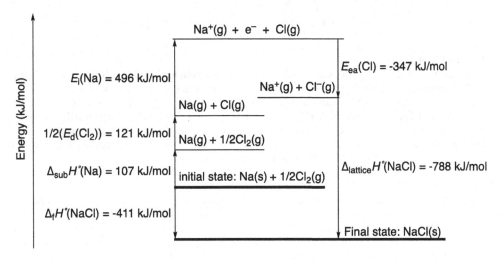

Figure 9.6 Example Born–Fajans–Haber cycle diagram for sodium chloride.

$$\Delta_{\text{lattice}}H° = \Delta_{\text{sub}}H° + E_d + E_i + E_{ea} - \Delta_f H°$$ (9.2)

$\Delta_{\text{lattice}}H°$ is the lattice enthalpy (kJ/mol).

$\Delta_{\text{sub}}H°$ is the enthalpy of sublimation (kJ/mol).

E_d is the dissociation energy (kJ/mol).

E_i is the ionization energy (kJ/mol).

E_{ea} is the electron affinity (kJ/mol).

$\Delta_f H°$ is the enthalpy of formation (kJ/mol).

While not a part of the lattice enthalpy calculation when using the Born–Fajans–Haber cycle, the lattice enthalpy is also a manifestation of Coulomb's law (Chapter 8). As such, we can consider trends in lattice energy based on the magnitude of ion charges (Q) and the size of ions, which influences the internuclear distance (r).

PROBLEM 9.1

Consider the lattice enthalpy values of MgO ($\Delta_{\text{lattice}}H° = -3929$ kJ/mol) and MgF_2 ($\Delta_{\text{lattice}}H° = -2922$ kJ/mol).

a. Provide an explanation of why magnesium oxide has a larger lattice enthalpy value than magnesium fluoride.

b. The boiling point of MgO is 3600 °C and the boiling point of MgF_2 is 2260 °C. Provide an explanation for the difference in boiling points.

PROBLEM 9.2

F, Cl, Br, and I are called halogens, which means "salt maker" in Greek. These are so named because halogens typically react to make ionic compounds (salts). Salts that contain Cl^-, Br^-, or I^- are very soluble in water, while salts that contain F^- are less soluble in water. In the context of lattice energy, explain why salts with fluoride are typically less soluble in water.

PROBLEM 9.3

Ionic liquids are ionic compounds whose melting point is at or below room temperature. Below is an example compound that is an ionic liquid (its melting point is –71 °C). Propose an explanation of why this ionic compound, 1-butyl-3-methylimidazolium tetrafluoridoborate, has such a low melting point.

1-butyl-3-methylimdazolium tetrafluoridoborate

PROBLEM 9.4

You find two bottles (bottle A and bottle B) labeled as chromium chloride. Remembering IUPAC nomenclature and knowing chromium's common charges, you know that chromium chloride could mean chromium(II) chloride ($CrCl_2$) or chromium(III) chloride ($CrCl_3$). You determine the melting point of the chemical in each bottle. Bottle A's compound melts at 1152 °C and bottle B's compound melts at 824 °C. Using this information and your knowledge of ionic compounds, identify which compound ($CrCl_2$ or $CrCl_3$) is in each bottle and briefly explain your reasoning.

COVALENT BONDS

A covalent[10] bond refers to a region of relatively high electron density between nuclei that arises at least partly from the sharing of electrons and gives rise to an attractive force and a characteristic internuclear distance (bond length).[11] Covalent compounds, and hence covalent bonds, are formed in cases where the interacting atoms have no electronegativity difference or only a moderate one ($\Delta \chi_P < 1.7$), and the electrons are shared between the atoms rather than being transferred. In cases of unequal sharing of electrons ($0.4 < \Delta \chi_P < 1.7$), the covalent bond – shared electrons – is strengthened by electrostatic attraction, partial ionic bonding. Bonds that involve a mixture of covalent and ionic bonding are polar covalent bonds.

Unlike ionic compounds, covalent compounds consist of atoms with a defined structure. These well-defined structures, molecules, are represented with Lewis structures and the bonding described by quantum mechanical theory. The following sections will first describe how to draw Lewis structures and then how to understand the bonding of atoms through quantum mechanical theory.

Lewis Structures

The most common model for depicting molecules was developed by G.N. Lewis in 1916. Lewis's model for showing covalent bonding in molecules predates the development of modern valence bond theory, but the results from valence bond theory and the results from properly drawn Lewis structures will often match quite closely. For this reason, Lewis structures are the medium chemists use most often to think about, to communicate, and to represent the structure of molecules.[12] Table 9.1 shows three examples of Lewis structures.

The assumptions with Lewis structures are the same as those with valence bond theory: a pair of electrons can be localized on a single atom (one center), which is called a lone pair and is drawn as two dots next to the atomic symbol (Table 9.1); a pair of electrons can be shared between two atoms (two centers), which is called a two-center–two-electron (2c-2e) bond and is drawn as a line connecting the two atomic symbols (Table 9.1). There can be up to six pairs of electrons, three lines, connecting two main group atoms (dinitrogen in Table 9.1), and four,[13] five,[14] and six bonds[15] between two atoms are possible in transition metal bonding.

The following paragraphs will focus on the development of a regularized system for drawing Lewis structures from a chemical formula. It should be noted that while the system presented below will give results that match closely with those that come from natural bond orbital analysis, a full computational analysis is still necessary to appreciate the subtleties of bonding in most structures. To start, let's consider the Lewis structure of water (Table 9.1). To draw any Lewis structure, the first step is determining the total number of valence electrons in the system. In this work, Figure 9.2 will be useful.

Each hydrogen atom has one valence electron:	$2H(1 e^-) = 2 e^-$
The oxygen atom has six valence electrons:	$+1O(6 e-) = 6 e-$
The total number of valence electrons is:	$8 e^-$ total

The second step is to determine the number of electrons needed to satisfy the valence shell of each atom. Each hydrogen atom needs 2 electrons total, each main group atom needs 8 electrons total, and each transition metal needs 12 electrons.

Each hydrogen atom needs two electrons:	$2H(2 e^-) = 4 e^-$
The oxygen atom needs eight electrons:	$+1O(8 e-) = 8 e-$
The total number of electrons required is:	$12 e^-$ needed

There are not enough valence electrons for each atom to have a full valence shell alone, and so the atoms must share electrons to complete their valence shells. The next step is to find the difference between the number of electrons needed and the total number of electrons, which is the number of electrons shared.

Table 9.1 Name, Chemical Formula, and Lewis Structure of Three Example Molecules

Name	Water	Ethene	Dinitrogen
Chemical formula	H_2O	C_2H_4	N_2
Lewis structure	H-Ö-H	H H | | H-C=C-H	:N≡N:

$$12 \text{ e}^- \text{ needed}$$
$$\underline{- 8 \text{ e}^= \text{ total}}$$
$$4 \text{ e}^- \text{ shared}$$

A bond, which is drawn as a line in Lewis structures, represents a shared pair of electrons. And so, the four shared electrons in water represent two bonds.

$$4\text{e}^- \left(\frac{1 \text{ bond}}{2\text{e}^-} \right) = 2 \text{ bonds}$$

The final mathematical step is to determine the number of unshared electrons. By subtracting the number of shared electrons from the total number of valence electrons, we can find the number unshared. In the case of water, this gives us four unshared electrons, which will make two lone pairs.

$$8 \text{ e}^- \text{ total}$$
$$\underline{- 4 \text{ e– shared}}$$
$$4 \text{ e}^- \text{ unshared}$$

Now, the results from above are translated into a structure. First, the least electronegative, non-hydrogen atom is placed in the middle and the other atoms are arranged around that central atom. Here the only non-hydrogen atom is oxygen, which goes in the middle, and the hydrogen atoms are arranged around the central atom (Figure 9.7).

Lines, or bonds, are then drawn to connect the peripheral atoms to the central atom (Figure 9.8). Then we take stock of how many electrons are around each atom.

With just the lines drawn, each hydrogen atom is connected to one bond. Each bond represents two electrons, which means that the valence shell of each hydrogen atom is full. For the oxygen atom, it is connected to two lines, which means it only has four electrons around it. To satisfy the valence shells of any atoms left unfilled, we add unshared electrons as unshared (lone) pair, and we start at the most electronegative atoms first. Here, the two lone pair are drawn on the central oxygen atom, which will satisfy the oxygen atom's valence shell. That is, two lines (four electrons) and two lone pairs (four electrons) give a full octet. This completes the Lewis structure of water (Figure 9.9).

For polyatomic ions, the same approach holds, but the charge must be considered when calculating the valence electron total. For each negative charge, one additional electron is added to the total number of valence electrons. For example, hydroxide (HO^-) has eight valence electrons: one electron from the hydrogen atom, six electrons from the oxygen atom, and one electron represented by the negative charge. For each positive charge, one electron is subtracted from the total number of valence electrons. For example, oxidanium (H_3O^+) has eight valence electrons: one from each hydrogen atom (three total), six from the oxygen atom, and one electron is subtracted because of the 1+ charge. Otherwise, mathematical steps proceed as above, which gives the structures of hydroxide and oxidanium seen in Figure 9.10. Note the bracket with the charge to the right. This is a way of denoting that the entire structure has a charge. We will see a way of formally designating where this charge is below.

A final type of molecule that will be considered here is the less common, odd-electron species, which is referred to as a radical. Notice all the above examples have even numbers of electrons. If we consider NO (an important signaling molecule in biology), calculation of the number of bonds and unshared electrons finds that NO should have five shared electrons, two and a half bonds,

H O H

Figure 9.7   Arrangement of atoms in water.

H−O−H

Figure 9.8   Partial Lewis structure of water with bonds drawn between connected atoms.

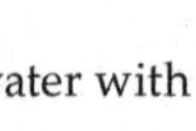

H−Ö−H

Figure 9.9   Complete Lewis structure of water with bonds and lone pair.

$$H-\overset{\displaystyle ..}{\underset{\displaystyle ..}{O}}: \Big]^{-} \qquad \overset{\displaystyle ..}{\underset{\displaystyle |}{\underset{H}{H-O-H}}} \Big]^{+}$$

Figure 9.10 The Lewis structures of hydroxide (left) and oxidanium (right).

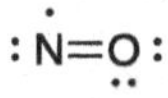

$$: \overset{\displaystyle .}{N} = \overset{\displaystyle ..}{O} :$$

Figure 9.11 Lewis structures of nitric oxide.

and six unshared electrons. What this means is that there are two bonds and the half bond is a single, unshared electron (also called a radical electron). A Lewis structure for nitric oxide is shown in Figure 9.11.

Looking at Figure 9.11, notice that we have placed the unshared pairs around oxygen (the more electronegative atom) to satisfy its octet. The single, unpaired electron is placed on the nitrogen, which is left with less than a full octet. Odd-electron (radical) species like NO are typically highly reactive and do not persist for very long.

PROBLEM 9.5
Draw Lewis structures for each of the following chemical formulae: CH_4, NH_3, HF, NH_2^-, NH_4^+, HS^-, BH_4^-, CH_3.

Bond Length

Although Lewis structures are incredibly useful in showing information about bonding, it is worth highlighting information that is not included. In a Lewis structure, all atoms and bonds are drawn the same size and length. As shown in Table 9.2, the length of the H–X bond varies substantially from HF to HI due to the increase in atomic radius of the halogen atoms from fluorine to iodine. The length of a bond is correlated with several properties, one of which is dissociation energy (E_d). Dissociation energy is the amount of energy it takes to pull apart two atoms that are bonded together (Figure 9.3). As can be seen in Table 9.2, shorter bonds tend to be stronger bonds and longer bonds tend to be weaker bonds. There is no standardized system for showing differences in bond length in Lewis structures and the viewer is expected to know trends in atomic size.

Formal Charge

In Figure 9.10, hydroxide (HO^-) is drawn using a bracket notation to show that there was an overall negative charge. Chemists find it useful to formally identify specific atoms as carrying the charge(s). There are several systems to bookkeep electrons in molecules. In Lewis structures, formal charge is most used. Formal charge is a bookkeeping tool that chemists use to gauge how many electrons are

Table 9.2 H–X Lewis Structures, Optimized Structures, Bond Lengths, and Dissociation Energies

H–X Lewis Structure	Computed Structure	H–X Bond Length (pm)	E_d (kJ/mol)
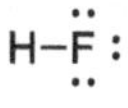 $H-\overset{..}{\underset{..}{F}}:$		92	569
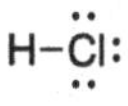 $H-\overset{..}{\underset{..}{Cl}}:$		127	431
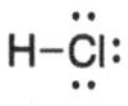 $H-\overset{..}{\underset{..}{Cl}}:$		141	364
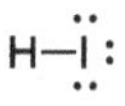 $H-\overset{..}{\underset{..}{I}}:$		161	297

around a given atom compared to the neutral atom by itself.[16] The word "formal" in formal charge is short for formalism, that is, an agreed-upon method for determining a bookkeeping charge value. Formal charge should not be confused with actual charge (see below). The formula for formal charge (FC) is shown in Equation 9.3. It is important to note that the formal charge must be calculated for each atom in a molecule and that all nonzero formal charges must be included next to the atom. The sum of all individual formal charges must equal the molecular charge.

$$FC = e_v - e_{lp} - b \tag{9.3}$$

e_v is the number of valence electrons.
e_{lp} is the number of unshared electrons.
b is the number of bonds.

With this information in hand, the formal charge for the hydrogen atom in hydroxide would be 0 (1 – 0 – 1), and the oxygen atom would be 1– (6 – 6 – 1). Typically, when the formal charge is 0 then the formal charge is omitted from the structure. For a formal charge of 1+ this is denoted by a plus sign inscribed in a circle and a charge of 1– is denoted by a negative sign inscribed in a circle. A complete Lewis structure always includes all nonzero formal charges, and the complete Lewis structures for hydroxide and oxidanium are shown in Figure 9.12.

PROBLEM 9.6
For each of the following molecules, calculate the formal charge on each atom and indicate any formal charges by appropriately drawing in the formal charge(s).

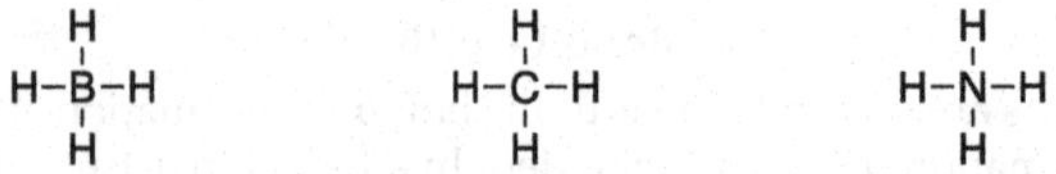

Formal charges are useful for several reasons (they significantly help in predicting the reactivity of a molecule), and formal charges are useful for trying to decide among possible Lewis structures for a given molecular formula. Consider the formula $AlCl_4^-$. One could potentially draw the Lewis structures in Figure 9.13.

By using formal charges, the best Lewis structure can be chosen (Figure 9.14).

Looking at the Lewis structures with formal charges (Figure 9.14), the first Lewis structure is the correct and best structure for aluminium tetrachloride. There are two important takeaways from this. First, chemical structures tend to be more stable when atoms are arranged radially from a central atom rather than linearly. Second, the Lewis structure with fewer formal charges tends to be the better structure.

Figure 9.12 The Lewis structures of hydroxide (left) and oxidanium (right) showing formal charges.

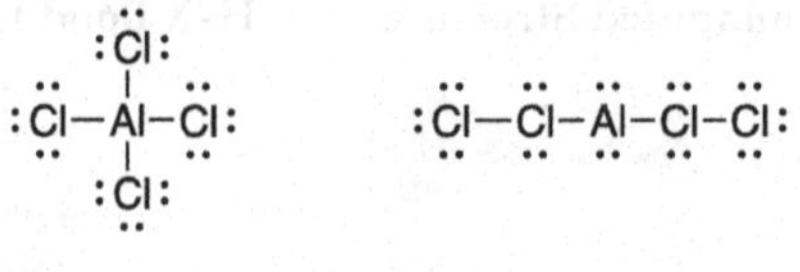

Figure 9.13 Possible Lewis structures of aluminium tetrachloride.

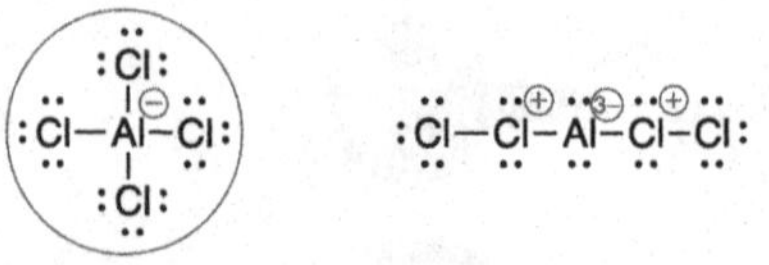

Figure 9.14 Possible Lewis structures of aluminium tetrachloride with formal charges and best structure circled.

PROBLEM 9.7

For the following series of Lewis structures drawn for the formula CH_4O, correctly assign all formal charges to each atom and then choose which Lewis structure is the best depiction of the molecule.

H–C̈–O–H   H–C̈–Ö–H   H–C–O–H

PROBLEM 9.8

Draw a Lewis structure for each of the following. Include all nonzero formal charges: CO, HCN, CH_2O, $SOCl_2$.

Actual Charge

In calculating formal charge, the key assumption is that all bonds are perfectly covalent. We will see in Chapter 19 another system for bookkeeping electrons is oxidation numbers, which assumes that all bonds are purely ionic. As seen in Figure 9.4, most bonds fall in between those two extremes, but we use formal charge and oxidation numbers because they allow us to calculate a "charge" on each atom. However, the actual charge distribution within a molecule is heavily influenced by the electronegativity differences between atoms, which is not accounted for by formal charge, and the nature of the covalent bonding, which is not accounted for by oxidation numbers. Ideally, we would want to know the actual, empirical charges on each atom. There is no universal way to measure the charge on every atom in a molecule. We can, however, approximate the empirical charge with computational analysis. One method (the details are beyond the scope of this document) is natural population analysis (NPA),[17] which is a result of a natural bond orbital (NBO) calculation.[18] In Figure 9.15, the formal charges (0 for hydrogen and 1+ for nitrogen), oxidation numbers (+1 for hydrogen and –3 for nitrogen), and NPA results (+0.495 for hydrogen and –0.979 for nitrogen) are shown for the ammonium cation. What is most apparent is that formal charge has no relationship to the actual distribution of electron density. Nitrogen is more electronegative than hydrogen and so it should be (and is) more electron-rich. This is apparent in the oxidation numbers, charges, and electrostatic potential map (red is electron-rich and blue is electron-poor). We will deal with oxidation numbers in more detail in Chapter 19, but it is worth noting that they provide a better sense of distribution but overestimate the magnitude of charges in a molecule.

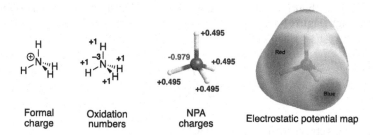

Formal charge   Oxidation numbers   NPA charges   Electrostatic potential map

Figure 9.15 Formal charge, oxidation numbers, NPA charges, and electrostatic potential for ammonium.

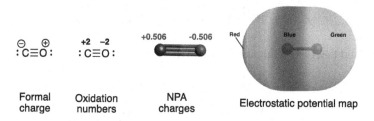

Formal charge   Oxidation numbers   NPA charges   Electrostatic potential map

Figure 9.16 Formal charge, oxidation numbers, NPA charges, and electrostatic potential for carbon monoxide.

In Figure 9.16, the formal charges (1– for carbon and 1+ for oxygen), oxidation numbers (+2 for carbon and –2 for oxygen), and NPA results (+0.506 for carbon and –0.506 for oxygen) for carbon monoxide are shown. Again, formal charge does not give a good sense of the actual distribution of charge, but oxidation numbers do while overestimating the magnitude of the charges.

Although formal charge does not indicate actual charge, many chemists, especially organic chemists, use it extensively because of the ease with which one can calculate the charge, and it can be used to predict reactivity patterns.

Contributing Structures

We will now turn our attention to some more nuanced Lewis structure examples. Let's start by considering formate (HCO_2^-). Calculation of the number of bonds and unshared electrons shows that formate should have four bonds and ten unshared electrons. Carbon is the least electronegative element and so it will go in the center with oxygen and hydrogen atoms surrounding it. Drawing in the bonds, lone-pair electrons, and formal charges, we could draw formate as shown in Figure 9.17.

We have a model – Lewis structures are model structures of the real molecular species – in hand for formate. The model indicates that the two oxygen atoms are nonequivalent. One oxygen is negatively charged, the other is neutral. One oxygen is making a double bond to carbon, and one is making a single bond, which we would expect to show up as differences in bond lengths since double bonds are shorter than single bonds.

As with any good scientific model, it must be tested against empirical results. X-ray crystallographic analysis of formate shows that the two oxygen atoms are equivalent: the C-O bond lengths are both 127 pm.[19] What does this say about our Lewis structure model of formate? These results tell us that the original Lewis structure model has limitations. Specifically, the assumption is that all electrons are localized, that is, they are either on a single atom as a lone pair or they are between two atoms as a bond. If we go back and consider the Lewis structure that was drawn above, it was arbitrarily decided, by the author, that the double bond would be drawn between the top oxygen atom and the carbon; however, it would be equally acceptable to draw either of the Lewis structures shown in Figure 9.18.

Both Lewis structures in Figure 9.18 are equally correct to draw (in that they all follow the rules of drawing good Lewis structures). Each Lewis structure differs from the other only in terms of where we drew the lone-pair electrons and the double bond. However, each of these Lewis structures is – by itself – a bad depiction of formate. In these situations, where multiple Lewis structures are acceptable to draw for a molecule, to get an accurate depiction of the molecule we must consider all of them. We call all the valid Lewis structures for a single molecule contributing structures.[20,21,22] To show their relationship as contributing structures, we draw a double-headed arrow in between each (Figure 9.19).[23]

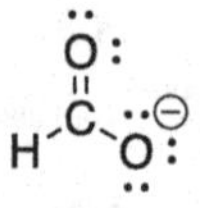

Figure 9.17 A Lewis structure for formate (HCO_2^-).

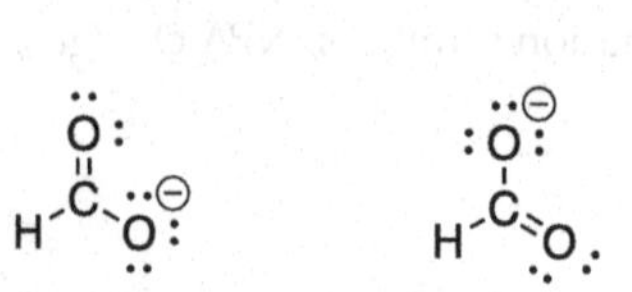

Figure 9.18 Valid Lewis structures of formate.

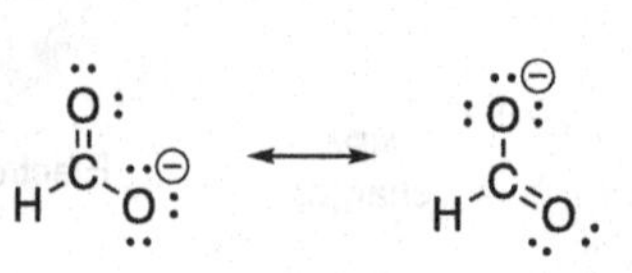

Figure 9.19 Two contributing structures of formate.

What contributing structures indicate is that the electrons in a molecule are spread among multiple atoms, delocalized. It should be stressed that formate, and any molecule for which there are contributing structures, is not oscillating back and forth between the different forms, each Lewis structure is itself a poor approximation by chemists to describe a more complex reality. Contributing structures must be included whenever an arbitrary choice is made in a Lewis structure drawing, here the choice of whether to draw the double bond between the top oxygen or the right oxygen. If you make an arbitrary choice, then the other option must be drawn and considered.

PROBLEM 9.9
For each of the following chemical formulae, draw all contributing structures.

Trioxygen (O_3)

Boron trifluoride (BF_3)

Dinitrogen monoxide (N_2O)

Chlorofluorocarbonyl (COFCl)

Hypercoordinate Molecules

When in the center of a molecule, the relatively small atoms of the second period (B, C, N, O, and F) can, at most, coordinate with four areas of electron density (bonded atoms or lone pair), e.g., BH_4^-, $CHCl_3$, and NF_4^+. As atoms increase in size in periods three and beyond, the central atom can be coordinated to more than four different areas of electron density (bonded atoms or lone pair), e.g., PF_5, SF_6, and IF_7. We will term these compounds hypercoordinate:[24] they are coordinated to more than four areas of electron density. Let us start by drawing a Lewis structure for krypton difluoride (KrF_2). Calculation of the number of bonds and unshared electrons finds that krypton difluoride should have one bond and 20 unshared electrons. Krypton is the least electronegative atom and so it will go in the center with the fluorine atoms surrounding it. We then draw in the bond to the central krypton, complete the remaining octets by adding in lone-pair electrons, and add in any nonzero formal charges (Figure 9.20).

In Figure 9.20, the fluorine on the left side was arbitrarily chosen to not have a line drawn. To fully account for the structure of KrF_2, all contributing structures (Figure 9.21) need to be drawn.

What we can gather from the contributing structures of KrF_2 is that the bonds between krypton and fluorine are both covalent and ionic. In addition, krypton has three lone pairs and two attached atoms, for a total of five areas of electron density, more than four areas of electron density around itself (something that is not possible for the smaller atoms of period two). In compounds like krypton difluoride, we see an example of a three-center–four-electron (3c-4e) bond,[25] which will be considered further below. The nature of the bonding in these compounds has been debated since the beginning of modern chemistry. Langmuir and others have favored the view that the octet rule generally holds true for most main group elements and that the bonding is best represented as shown in Figure 9.21.[26] Lewis, however, proposed that these hypercoordinate compounds are also hypervalent (Figure 9.22) and that these compounds involve the central atom having an expanded octet.[27] This debate has continued for nearly a century, but both

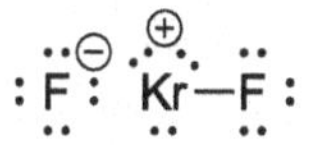

Figure 9.20 Krypton difluoride Lewis structure.

Figure 9.21 Krypton difluoride contributing structures.

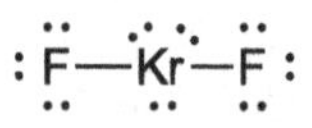

Figure 9.22 Krypton difluoride hypervalent structure.

computational and theoretical evidence support the octet-centric view proposed by Langmuir (Figure 9.21) over expanded octet structures.[28] While not supported by theory or experiment, hypervalent structures are commonly shown – because they avoid formal charges[29] – in books and online.

PROBLEM 9.10

Draw the Lewis structures for each of the following chemical formulae: SO_3, PF_5, SF_6, PO_4^{3-}, ClO_2^-, and I_3^-. Check the Lewis structures you have drawn in this problem against the results found online, where hypervalent structures are almost exclusively presented.

QUANTUM MECHANICAL UNDERSTANDING OF BONDING

Lewis structures provide a way to visualize the bonding in covalent compounds. As we have seen, however, electrons in atoms are best described through a quantum mechanical approach (Chapter 7). Lewis structures do not convey how the configuration of electrons changes as atoms combine into compounds nor do they explain the change in energy (Figure 9.3) that is observed when atoms form bonds. With the Lewis structure model as our guide, we will see in the following sections how quantum mechanics strengthens and challenges our understanding of bonds and bonding. In turn, we will consider two complementary quantum approaches: valence bond (VB) theory and molecular orbital (MO) theory.

Valence Bond Theory

In valence bond theory the key assumption, as with Lewis structures, is that electron pairs are localized on a single atom (one center) or between two atoms (two centers). Electron pairs localized to a single center are called nonbonding electrons, or lone pair in the language of Lewis structures. Electron pairs localized between two centers are called a shared pair or a bond pair. Recall that when a bond forms, the formation of a bond is the result of stabilization (lowering the energy) of the electrons involved (Figure 9.3). For our first example of VB theory, let us consider the simplest molecule dihydrogen (H_2). Here (Figure 9.23) we have two hydrogen atoms; each hydrogen has

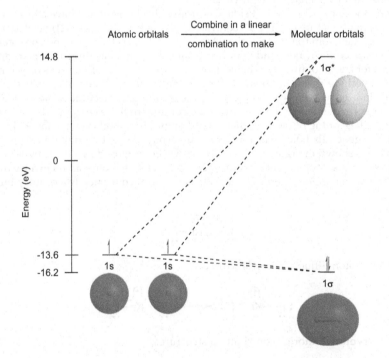

Figure 9.23 σ-orbital mixing in dihydrogen (H_2). Orbital images and energy values calculated at the CCSD(T)/cc-pVTZ level of theory and basis set.

one electron located in the 1s orbital. Bonding is the result of overlap between two singly occupied atomic orbitals to form two new molecular orbitals.[30] The lower energy orbital is a bonding (stabilizing) orbital with orbital density between the two atoms. The higher energy orbital is an antibonding (destabilizing) orbital with a node between the two atoms. Since the electron density is along the internuclear axis, we call these orbitals σ orbitals. The bonding orbital is termed σ and the antibonding orbital is labeled as σ*.

Now, with our molecular orbital diagram, we add electrons following the rules we saw for orbital filling: the Aufbau principle (electrons are filled from the lowest energy orbital to the highest), Hund's rule (equal energy orbitals are filled singly before being doubly filled), and the Pauli exclusion principle (at most there can be two, spin-paired electrons per orbital). If we put the two hydrogen-atom electrons into the diagram in Figure 9.23, then we can see that the electrons are now lower in energy combined in dihydrogen than they were on two separate hydrogen atoms. This lowering of electron energy when shared by the two hydrogen atoms in dihydrogen is what we call a bond.

At this point, we can introduce a formal definition for bond order, that is, the number of bonds between two atoms. The bond order (p_{rs}) between two atoms (r and s) is half the difference between the number of electrons in bonding orbitals and the number of electrons in antibonding (*) orbitals. For dihydrogen, $p_{HH} = \frac{1}{2}(2 - 0) = 1.0$, that is, a single bond exists between the two atoms of dihydrogen. In contrast, dihelium would have the same molecular orbital energy diagram but there are now four electrons that fill the σ and σ* orbitals. For dihelium $p_{HeHe} = \frac{1}{2}(2 - 2) = 0$, no bond exists between the two atoms of dihelium, and no stable molecule forms when two helium atoms combine.

PROBLEM 9.11
Which has a larger ionization energy (E_i): H or H_2? Explain your answer.

PROBLEM 9.12
The dihelium cation (He_2^+) is a stable polyatomic ion with a dissociation energy (E_d) of 2.47 eV. Using Figure 9.23, explain why He_2^+ is stable in contrast to unstable He_2.

Single, σ bonds are generally the lowest energy bond that can form between two atoms. Also, the σ orbital density lies along the internuclear axis and is spherically symmetric, which means that σ bonds can freely rotate. Before considering other types of bonding, we will first look at σ bonding in a slightly more complicated example: hydrogen fluoride (HF).

The bonding in HF comes from the overlap of the hydrogen 1s orbital with a valence fluorine orbital. The valence orbitals of fluorine are 2s and $2p_x$, $2p_y$, and $2p_z$. When a main group element with p orbitals bonds, the valence s orbital hybridizes (mixes) to some extent with the p orbitals. The nature of the hybridization depends on the molecular shape (Chapter 10) and the energy of the electrons involved in the two atoms that bond. For HF, the 2s orbital mixes with the $2p_z$ orbital to make two sp hybrid orbitals: one that is 78.5% s and 21.5% p (s^4p) and the other that is 21.5% s and 78.5% p (sp^4) (Figure 9.24). The higher energy sp^4 orbital will mix with the hydrogen 1s orbital to produce the σ bond. The remaining orbitals on fluorine – the s^4p hybrid orbital and the unhybridized $2p_x$ and $2p_y$ orbitals – will contain the fluorine lone pair. Since the lone pair are in two different types of orbitals, the three lone pair are non-equivalent energetically.

When looking at a Lewis structure, any single bond between two atoms is a σ bond. If two atoms are connected by a double (or triple) bond, then there exists both a σ bond and a π bond (or π bonds) between the two atoms. Let's consider the bonding in ethene (Figure 9.25). Each carbon is connected to each hydrogen atom by a σ bond. The two carbon atoms are connected by a double bond, which means there is one σ bond and one π bond.

Because there is already a σ bond between the carbon atoms, and the σ bond orbital lies along the internuclear axis, the π bond must form through some other type of orbital overlap. Specifically, π bonds form through the facial overlap of parallel p orbitals. Figure 9.26 shows the p-orbital mixing that results in the creation of a π bond. Note that a singular p orbital has orbital density above and below a nodal plane at the atom. This means that a singular π bond, which is formed from p orbitals, has orbital density above and below a nodal plane that exists along the internuclear axis.

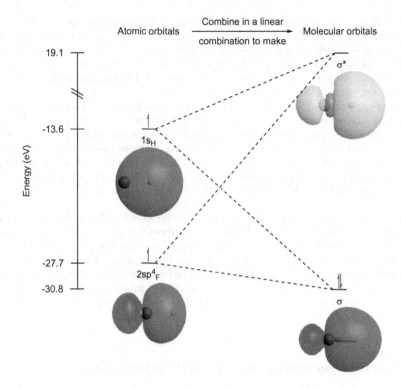

Figure 9.24 σ-orbital mixing in hydrogen fluoride (HF). Orbital images and energy values calculated at the CCSD(T)/cc-pVTZ level of theory and basis set.

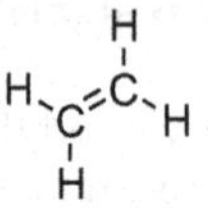

Figure 9.25 Lewis structure of ethylene.

PROBLEM 9.13
Unlike σ bonds, π bonds cannot rotate. Compare Figure 9.26 with Figure 9.24. Provide an explanation of why π bonds cannot rotate.

PROBLEM 9.14
Answer the following questions about dichlorine (Cl_2).

a. Draw a Lewis structure for dichlorine and identify the type of bond that exists between the chlorine atoms.

b. Dichlorine (Cl_2) has a dissociation energy (E_d) of 2.52 eV, but when one electron is added (Cl_2^-) the dissociation energy decreases to 1.26 eV.

   i. Where (into which molecular orbital) does this added electron go?

   ii. Why does this decrease the dissociation energy?

c. The bond lengths for these two are 201 pm and 264 pm. Which bond length goes with which molecular entity? Explain.

d. What would adding one more electron do to the bond energy and the bond length? Explain.

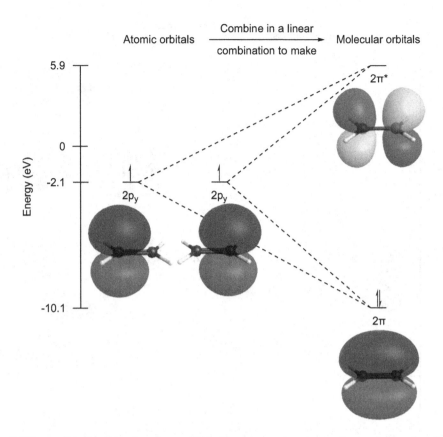

Atomic orbitals $\xrightarrow[\text{combination to make}]{\text{Combine in a linear}}$ Molecular orbitals

Figure 9.26 π-orbital mixing in ethene. Orbital images and energy values calculated at the CCSD(T)/cc-pVTZ level of theory and basis set.

Valence bond theory compliments our chemical sense that we can glean from Lewis structures while also grounding, validating, and deepening our understanding of bonding from a true quantum mechanical perspective. We will now consider molecular orbital theory, an alternative – but not contradictory – approach that provides us with a new and different way to consider bonding.

MOLECULAR ORBITAL THEORY

Molecular orbital theory is a quantum mechanical theory that does not approach chemical structure with Lewis structures in mind. Instead, molecular orbital theory was based on developing a quantum framework for molecular bonding based on spectroscopy – the interaction of light and molecules.[31] Given this different starting point, MO theory does not assume that electrons are localized, either to one atom or between two atoms, nor does it assume that only valence electrons are relevant to bonding. In fact, molecular orbital theory considers the interaction of all orbitals in bonding, and electrons can be delocalized over the entire molecule.

Figure 9.27 shows the molecular orbital diagram for dioxygen (O_2). This diagram is more complicated than the orbital diagrams we saw for VB theory because we are considering the entire set of orbitals and their combinations. There are some observations worth noting. First, as we saw in valence bond theory, bonding orbitals (those not denoted by asterisk) are lower in energy than the atomic orbitals that combine to produce them. For example, the $2p_z$ orbitals (–16.5 eV) combine to produce the σ_{2pz} bonding orbital (–17.9 eV). Second, σ orbitals form from s orbitals and from directly overlapping p orbitals, while π bonds form from perpendicular, facially overlapping p orbitals. Also, as we saw above, σ bonds (–17.9 eV) are lower in energy than π bonds (–17.4 eV). Finally, the bond order for dioxygen is 2.0 ($p_{OO} = \frac{1}{2}(10 - 6) = 2.0$).

Since molecular orbital theory looks at all the orbital interactions for a system, rather than the overlap of individual overlapping orbitals, one result is more immediately clear from the MO

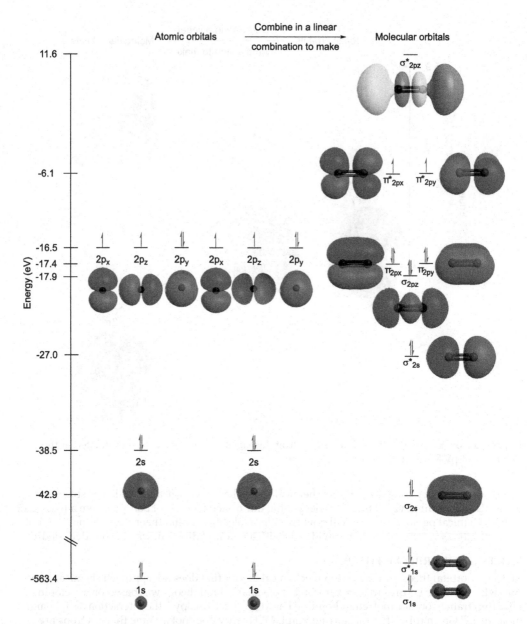

Figure 9.27 Full molecular orbital treatment of the bonding in dioxygen showing the combination of atomic orbitals (left) to produce molecular orbitals (right). Orbital images and energy values calculated at the CCSD(T)/cc-pVTZ level of theory and basis set. Note that the interaction of the electron spin and the orbital angular momentum leads to slightly (<1 eV) different energy values for filled or partially filled orbitals that are degenerate when unfilled: p, d, and f. The importance and effects of spin–orbit coupling are beyond the scope of this chapter and the values in the diagrams shown here do not include the effects of spin–orbit coupling.

diagram (Figure 9.27): the ground state for dioxygen has two unpaired electrons. This means that dioxygen is paramagnetic, i.e., it will align with a magnetic field, in contrast to dihydrogen (Figure 9.23) where the electrons are paired and the molecule is diamagnetic, i.e., it will align against a magnetic field.[32] With deeper analysis, valence bond theory can also explain the paramagnetism of dioxygen, but MO theory can do so more readily.

PROBLEM 9.15
Use MO theory to explain the following bond length data (in picometers [pm]) and dissociation energy (E_d) data (in eV). Can you provide a trend for how bond order relates to bond length and bond dissociation energy?

O_2^+ bond length = 112 pm; E_d = 6.48 eV

O_2 bond length = 121 pm; E_d = 5.16 eV

O_2^- bond length = 128 pm; E_d = 4.55 eV

O_2^{2-} bond length = 149 pm; E_d = 2.21 eV

ELECTRON DELOCALIZATION (MO THEORY) AND CONJUGATION (VB THEORY)

So far, we have only considered small molecules with two atoms, which obscures how electrons can be delocalized across the entire system. We will reconsider the delocalized π bonding in formate (HCO_2^-) (Figure 9.17). A π bond exists between the central carbon atom and the right oxygen atom (Figure 9.28), which is made up of the overlap of a carbon p orbital and an oxygen p orbital.

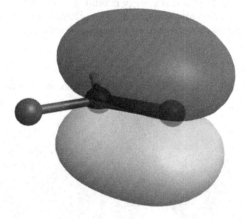

Figure 9.28 π bonding orbital between carbon and the right oxygen atom.

Figure 9.29 A $n_O \rightarrow \pi^*_{CO}$ donor–acceptor interaction between carbon and the left oxygen atom.

In formate, the left oxygen atom donates a p-orbital lone pair that is accepted by the π^*_{CO} orbital (Figure 9.29). This donor–acceptor interaction creates a partial π bond between the left oxygen atom and the carbon atom. And because electron density is added to the π^*_{CO} orbital, the amount of π bonding between the carbon atom and the right oxygen atom is reduced. This donor–acceptor interaction is called conjugation. We show this conjugation with contributing structures (Figure 9.19).

When we reconsider the three-center–four-electron bonding in krypton difluoride (KrF_2) (Figure 9.20) we see a similar pattern. A σ bond exists between the central krypton atom and the right fluorine atom (Figure 9.30), which is made up of the overlap of a krypton p orbital and a fluorine p orbital.

In krypton difluoride, the left fluorine atom donates a p-orbital lone pair that is accepted by the σ^*_{KrF} orbital (Figure 9.31). This donor–acceptor interaction creates a partial covalent bond between the left fluorine atom and the krypton atom. And because electron density is added to the σ^*_{KrF} orbital, the amount of covalent bonding between the krypton atom and the right fluorine atom is reduced. We show this conjugation with contributing structures (Figure 9.21).

When considered through the lens of molecular orbital theory, the bonding in formate is the result of three p_z orbitals overlapping (one each from carbon and the two oxygen atoms) to produce three molecular orbitals (ψ_1–ψ_3) (Figure 9.32). The lowest energy orbital (ψ_1) is a bonding orbital that shows a two-electron bond extending across all three atoms. The distribution of electrons across more than two atoms is called delocalization. The other occupied orbital (ψ_2) is nonbonding (nb), which shows electron density on the two end oxygen atoms and no bonding between any of the three atoms.

Figure 9.33 shows an analogous set of molecular orbitals for krypton difluoride, which shows the delocalization of σ bonding. The lowest energy orbital (ψ_1) is a bonding orbital that shows a two-electron bond extending across all three atoms. The other occupied orbital (ψ_2) is nonbonding (nb), which shows electron density on the two end fluorine atoms and no bonding between any of the three atoms.

BONDING SUMMARY

Only noble gases have stable electron configurations. All other atoms must combine to produce

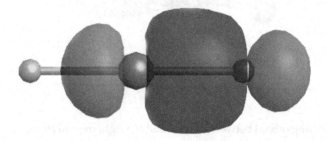

Figure 9.30 A σ bonding orbital between krypton and the right fluorine atom.

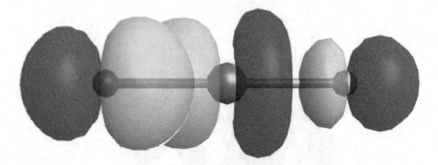

Figure 9.31 A $n_F \rightarrow \sigma^*_{KrF}$ donor–acceptor interaction between krypton and the left fluorine atom.

electronically stable species. For atoms with large electronegativity differences, electrons are

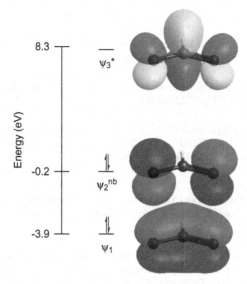

Figure 9.32 Three molecular orbitals of formate that form from overlapping p_z orbitals from the carbon atom and both oxygen atoms. Orbital images and energy values calculated at the CCSD(T)/cc-pVTZ level of theory and basis set.

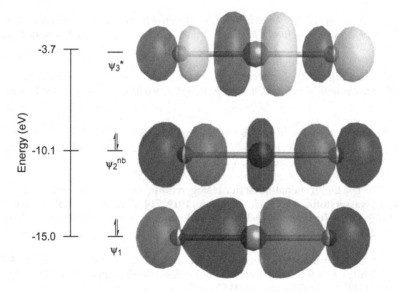

Figure 9.33 Three molecular orbitals of krypton difluoride that form from overlapping p_z orbitals from the krypton atom and both fluorine atoms. Orbital images and energy values calculated at the CCSD(T)/cc-pVTZ level of theory and basis set.

transferred from the metal (or metalloid) to the nonmetal atom. The resulting ions are attracted electrostatically to produce crystal lattices. For atoms with small electronegativity differences, electrons are shared between the atoms. This spreads the electrons out across two (localized) or more (delocalized) atoms, which decreases their energy and stabilizes the ensemble. The distribution of electrons impacts how molecules interact and will be considered in detail in Chapter 12.

NOTES

1. Lewis, G.N. The Atom and the Molecule. *J. Am. Chem. Soc.*, **1916**, *38* (4), 762-785. DOI: 10.1021/ja02261a002.

2. Langmuir, I. The Arrangement of Electrons in Atoms and Molecules. *J. Am. Chem. Soc.*, **1919**, *41* (6), 868–934. DOI: 10.1021/ja02227a002.

3. This corresponds to filling the s and d subshells. The p subshell is not substantially involved in transition-metal bonding: Landis, C.R.; Weinhold, F. Valence and Extra-Valence Orbitals in Mian Group and Transition Metal Bonding. *J. Comput. Chem.*, **2007**, *28* (1), 198–203. DOI: 10.1002/jcc.20492.

4. This model of ionic bonding, electrons transfer to form octet-abiding ions, was first proposed by Kossel: Kossel, W. Über Molekülbildung als Frage des Atombaus. *Ann. Phys.*, **1916**, *354* (3), 229–362. DOI: 10.1002/andp.19163540302.

5. "Ionic bond." IUPAC. *Compendium of Chemical Terminology*, 2nd ed. Compiled by A.D. McNaught and A. Wilkinson. Blackwell Scientific Publications, Oxford, 1997. Online version (2019–) created by S.J. Chalk. DOI: 10.1351/goldbook.IT07058.

6. Coulomb, C.A. Premier mémoire sur l'éctricité et le magnétisme. *Histoire de l'Academie Royale des Sc.*, **1785**, 569–577.

7. Pauling, L. *The Nature of the Chemical Bond and the Structure of Molecules and Crystals. An Introduction to Modern Structural Chemistry*, 3rd ed. Cornell University Press: Ithaca, NY, 1960. p. 98.

8. Pauling, L. *The Nature of the Chemical Bond and the Structure of Molecules and Crystals. An Introduction to Modern Structural Chemistry*, 3rd ed. Cornell University Press: Ithaca, NY, 1960. p. 100.

9. (i) Born, M. Die Elektronenaffinität der Halogenatome. *Ber. Dtsch. Phys. Ges.*, **1919**, 679–685. https://hdl.handle.net/2027/uc1.c058625598?urlappend=%3Bseq=3%3Bownerid=13510798903545560-7 (Accessed 03/26/23).

   (i) Fajans, K. Die Elektronenaffinität der Halogenatome und die Ionisierungsarbeit der Halogenwasserstoffe. *Ber. Dtsch. Phys. Ges.*, **1919**, 714–722. https://hdl.handle.net/2027/uc1.c058625598?urlappend=%3Bseq=3%3Bownerid=13510798903545560-7 (Accessed 03/26/23).

   (ii) Haber, F. Betrachtungen zur Theorie der Wärmetönung. *Ber. Dtsch. Phys. Ges.*, **1919**, 750–768. https://hdl.handle.net/2027/uc1.c058625598?urlappend=%3Bseq=3%3Bownerid=13510798903545560-7 (Accessed 03/26/23).

10. Langmuir introduced the term "covalence," which is commonly used instead of the more transparent term "electron-pair bond": Langmuir, I. The Arrangement of Electrons in Atoms and Molecules. *J. Am. Chem. Soc.*, **1919**, *41* (6), 868–934. DOI: 10.1021/ja02227a002.

11. "Covalent bond." IUPAC. *Compendium of Chemical Terminology*, 2nd ed. Compiled by A.D. McNaught and A. Wilkinson. Blackwell Scientific Publications, Oxford, 1997. Online version (2019–) created by S.J. Chalk. DOI: 10.1351/goldbook.C01384.

12. (i) Lewis, G.N. The Atom and the Molecule.*J. Am. Chem. Soc.*, **1916**, *38* (4), 762-785. DOI: 10.1021/ja02261a002.

   (ii) Lewis, G.N. The Magnetochemical Theory. *Chem. Rev.*, **1924**, *1* (2), 231-248. DOI: 10.1021/cr60002a003.

(iii) Lewis, G.N. The Chemical Bond. *J. Chem. Phys.*, **1933**, *1* (1), 17-28. DOI: 10.1063/1.1749214. Cotton, F.A. Metal-Metal Bonding in $[Re_2X_8]^{2-}$ Ions and Other Metal Atom Clusters. *Inorg. Chem.*, **1965**, *4* (3), 334–336. DOI: 10.1021/ic500025a016.

13. Cotton, F.A. Metal-Metal Bonding in $[Re_2X_8]^{2-}$ Ions and Other Metal Atom Clusters. *Inorg. Chem.*, **1965**, *4* (3), 334–336. DOI: 10.1021/ic500025a016.

14. Nguyen, T.; Sutton, A.D.; Brynda, M.; Fettinger, J.C.; Long, G.J.; Power, P.P. Synthesis of a Stable Compound with Fivefold Bonding Between Two Chromium(I) Centers. *Science*, **2005**, *310* (5749), 844–847. DOI: 10.1126/science.1116789.

15. Borin, A.C.; Gobbo, J.P.; Roos, B.O. A Theoretical Study of the Binding and Electronic Spectrum of the Mo_2 Molecule. *Chem. Phys.*, **2008**, *343* (2–3), 210–216. DOI: 10.1016/j.chemphys.2007.05.028.

16. Formal charge comes from Langmuir's concept of residual charge: Langmuir, I. Types of Valence. *Science*. **1921**, *54*, 59–67. DOI: 10.1126/science.54.1386.59.

17. Reed, A. E.; Weinstock, R. B.; Weinhold, F. Natural Population Analysis. *J. Chem. Phys.* **1985**, *83* (2), 735–746. DOI: 10.1063/1.449486.

18. Glendening, E. D.; Badenhoop, J. K.; Reed, A. E.; Carpenter, J. E.; Bohmann, J. A.; Morales, C. M.; Landis, C. R.; Weinhold, F. A. *NBO 6.0*, Theoretical Chemistry Institute: University of Wisconsin, Madison, 2013.

19. Zachariasen, W.H. The Crystal Structure of Sodium Formate, $NaHCO_2$. *J. Am. Chem. Soc.*, **1940**, *62* (5), 1011–1013. DOI: 10.1021/ja01862a007.

20. "Contributing structure." IUPAC. *Compendium of Chemical Terminology*, 2nd ed. Compiled by A.D. McNaught and A. Wilkinson. Blackwell Scientific Publications, Oxford, 1997. Online version (2019–) created by S.J. Chalk. DOI: 10.1351/goldbook.C01309.

21. This concept is most often termed "resonance" following the name given by Pauling:

    (i) Pauling, L. The Nature of the Chemical Bond. III. The Transition from One Extreme Bond Type to Another. *J. Am. Chem. Soc.*, **1932**, *54* (3), 988–1003. DOI: 10.1021/ja01342a022.

    (ii) Pauling, L.; Sherman, J. The Nature of the Chemical Bond. VI. The Calculation from Thermochemical Data of the Energy of Resonance of Molecules Among Several Electronic Structures. *J. Chem. Phys.*, **1933**, *1* (8), 606–617. DOI: 10.1063/1.1749335.

22. Ingold's development of similar ideas, along with his development of electron-pushing arrows, led him to term this phenomenon "mesomerism": Ingold, C.K. Principles of an Electronic Theory of Organic Reactions. *Chem. Rev.*, **1934**, *15* (2), 225–274. DOI: 10.1021/cr60051a003.

23. The double-headed arrow notation was introduced in Germany: Eistert, B. Zur Schreibweise chemischer Formeln und Reaktionsabläufe. *Ber. Dtsch. Chem. Ges.*, **1938**, *71* (2), 237–240. DOI: 10.1002/cber.19380710206.

24. Hypercoordinate simply means that there are more than four atoms attached to the central atom. Following the recommendation of Paul von Rague Schleyer (Letters: Hypervalent compounds. *Chem. Eng. New.*, **1984**, *62* (22), 4. Hypercoordinate is the term that shall be used in this text, but the terms hypervalent and expanded octet are other terms that are also used for these molecules.

25. (i) Rundle, R.E. Electron Deficient Compounds. *J. Am. Chem. Soc.*, **1947**, *69* (6), 1327–1331. DOI: 10.1021/ja01198a028.

(ii) Rundle, R.E. Electron Deficient Compounds. II. Relative Energies of "Half-Bonds." *J. Chem. Phys.*, **1949**, *17*, 671–675. DOI: 10.1063/1.1747367.

(iii) Pimentel, G.C. The Bonding of Trihalide and Bifluoride Ions by the Molecular Orbital Methods. *J. Chem. Phys.*, **1951**, *19*, 446–448. DOI: 10.1063/1.1748245.

26. Langmuir, I. Types of Valence. *Science*. **1921**, *54*, 59-67. DOI: 10.1126/science.54.1386.59.

27. Lewis, G.N. Valence and the Structure of Atoms and Molecules. The New York Catalog Co.: New York, 1923.

28. (i) Suidan, L.; Badenhoop, J.K.; Glendening, E.D.; Weinhold, F. Common Textbook and Teaching Misrepresentations of Lewis Structures. *J. Chem. Educ.*, **1995**, *72* (7), 583–586. DOI: 10.1021/ed072p583.

    (ii) Collins, G.A.D.; Cruickshank, D.W.J.; Breeze, A. Bonding in Krypton Difluoride. *J. Chem. Soc., Faraday Trans. 2*, **1974**, *70*, 393–397. DOI: 10.1039/F29747000393.

    (iii) Kutzelnigg, W. Chemical Bonding in Higher Main Group Elements. *Angew. Chem. Int. Ed.*, **1984**, *23* (4), 272–295. DOI: 10.1002/anie.198402721.

    (iv) Reed, A.E.; Weinhold, F. On the Role of d Orbitals in SF_6. *J. Am. Chem. Soc.*, **1986**, *108* (13), 3586–3593. DOI: 10.1021/ja00273a006.

29. While IUPAC recommendations and conventions are closely followed in this book, the recommended depiction of hypervalent structures (Brecher, J. Graphical Representation Standards for Chemical Structure Diagrams. *Pure Appl. Chem.*, **2008**, *80* (2), 277–410. DOI: 10.1351/pac200880020277) will not be observed because hypervalent structures obfuscate key details of bonding.

30. (i) The original development of VB theory: Heitler, W.; London, F. Wechselwirkung neutraler Atome und homöopolare Bindung nach der Quantenmechanik. *Z. Phys.*, **1927**, *44* (6-7), 455–472. DOI: 10.1007/BF01397394.

    (ii) A modern VB system for quantum mechanical calculations: Natural Bond Orbital – NBO 7. E. D. Glendening, J, K. Badenhoop, A. E. Reed, J. E. Carpenter, J. A. Bohmann, C. M. Morales, P. Karafiloglou, C. R. Landis, and F. Weinhold, Theoretical Chemistry Institute, University of Wisconsin, Madison (2018).

31. (i) Hund, F. Zur Deutung einiger Erscheinungen in den Molekelspektren, *Z. Phys.*, **1926**, *36* (9–10), 657–674. DOI: 10.1007/BF01400155.

    (ii) Mulliken, R.S. The Assignment of Quantum Numbers for Electrons in Molecules. I. *Phys. Rev.*, **1928**, *32* (2), 186–222. DOI: 10.1103/physrev.32.186.

    (iii) Lennard-Jones, J.E. The Electronic Structure of Some Diatomic Molecules, *Trans. Faraday Soc.*, **1929**, *25*, 668–686. DOI: 10.1039/TF9292500668.

32. (i) "Paramagnetic." IUPAC. *Compendium of Chemical Terminology*, 2nd ed. Compiled by A.D. McNaught and A. Wilkinson. Blackwell Scientific Publications, Oxford, 1997. Online version (2019–) created by S.J. Chalk. DOI: 10.1351/goldbook.P04404.

    (ii) "Diamagnetic." *IUPAC Compendium of Chemical Terminology*, 2nd ed. Compiled by A.D. McNaught and A. Wilkinson. Blackwell Scientific Publications, Oxford, 1997. Online version (2019–) created by S.J. Chalk. DOI: 10.1351/goldbook.d01668.

10 Molecular Shapes

Lewis structures were not developed to depict correct molecular shapes. It is, thus, completely acceptable when drawing a Lewis structure to depict a carbon atom with four bonds connected to it forming a square planar shape as shown for methane (Figure 10.1).

From experimentation, however, it is known that methane is not flat, as the Lewis structure in Figure 10.1 would suggest. Methane has a distinctly three-dimensional shape (Figure 10.2).

To properly account for this three-dimensional shape, we need to employ two new ways of drawing bonds. Looking at the methane molecular model (Figure 10.2), we can see that two hydrogen atoms lie in the plane of the page. By convention, if something is in the plane of the page, we use a regular line (Figure 10.3).

The other two hydrogen atoms in the methane molecular model (Figure 10.2) are not in the plane of the page. One hydrogen atom is coming out of the plane at the viewer. To show this, we employ a wedge bond (⬤), which starts at the central atom and broadens and boldens as it comes toward the atom that is meant to be above the plane of the page (Figure 10.4).

The last hydrogen atom in the methane molecular model (Figure 10.2) is going back, into the plane of the page away from the viewer. To show this, a dashed bond (⋯⋯) is used, which starts at the central atom and broadens but remains dashed as it goes toward the atom that is meant to be behind the plane of the page. In this way, we can provide a Lewis structure that fully accounts for the three-dimensional shape of the actual molecule. Compare, now, the model with the wedge-dash Lewis structure (Figure 10.5).

For methane, we had a model to look at, but how would we predict the shape of a molecule without having a model already at hand? One of the most used systems for predicting molecular shape is valence shell electron pair repulsion (VSEPR),[1] where the shape is predicted based on counting the number of areas of electron density (attached atoms or lone pairs) around a central atom. Table 10.1 shows examples of counting the areas of electron density around an atom. This system works well for main group compounds, but it should be noted that for transition metals, the number of unshared electrons around the metal atom does not directly affect the molecular shape.

$$
\begin{array}{c}
\text{H} \\
| \\
\text{H—C—H} \\
| \\
\text{H}
\end{array}
$$

Figure 10.1 Lewis structure of methane.

Figure 10.2 Three-dimensional model of methane. The black sphere represents the carbon atom, and the white spheres represent hydrogen atoms.

$$
\begin{array}{c}
\text{H} \\
| \\
\text{H}^{\diagup}\text{C}
\end{array}
$$

Figure 10.3 The carbon atom and two hydrogen atoms of methane that all lie in the same plane. These are drawn with regular straight, lines showing the bonds are flat in the plane of the page.

$$
\begin{array}{c}
\text{H} \\
| \\
\text{H}^{\diagup}\text{C}_{\blacktriangleright} \\
\text{H}
\end{array}
$$

Figure 10.4 The wedge bond employed in the drawing of methane to represent the hydrogen atom that comes up, out of the plane of the page.

DOI: 10.1201/9781003479338-10

Figure 10.5 The molecular model (left) and three-dimensional Lewis structure (right) of methane.

VSEPR theory then predicts a shape based on the total number of areas of electron density around the central atom (Table 10.2).

VSEPR is based upon the assumption that only the repulsion of electrons around an atom is responsible for the molecular geometry of that atom. It ignores the impact of important factors such as bond polarization, delocalization, orbital mixing, H-bonding, and ring strain. It is worth noting that VSEPR predictions for oxygen and other atoms with multiple lone pairs strongly diverge from those results from quantum mechanics. As such, the VSEPR predictions should be treated as predictions and not as fact.

As the shape of a molecule is considered, review the atomic orbitals of carbon (Chapter 7, Figure 7.8). If we consider the valence orbitals (those involved in bonding), the s orbital is spherically symmetric, and the p orbitals point along the x, y, and z axes. How then can we understand the linear, trigonal planar, and tetrahedral shapes carbon adopts (Table 10.2) given these orbitals? In molecular orbital theory, the three-dimensional shape arises from the overlap of s and p orbitals to produce delocalized molecular orbitals of the lowest energy; however, to use this approach

Table 10.1 Counting Areas of Electron Density

Lewis Structure	Number of Connected Atoms	Number of Lone Pairs	Total Number of Areas of Electron Density
:O=C=O:	:O=C=O: = 2	:O=C=O: = 0	2
⊖: C≡O :⊕	⊖: C≡O :⊕ = 1	⊖: C≡O :⊕ = 1	2
⊖:O–C–O:⊖ (O above)	⊖:O–C–O:⊖ = 3	⊖:O–C–O:⊖ = 0	3
⊖:O–S=O: (⊕)	⊖:O–S=O: = 2	⊖:O–S=O: = 1	3
H–C–H (H above, H below)	H–C–H = 4	H–C–H = 0	4
⊖:O–S–O:⊖ (O top, ⊕ bottom)	⊖:O–S–O:⊖ = 3	⊖:O–S–O:⊖ = 1	4
F–S–F (F top ⊖, F bottom, ⊕)	F–S–F = 4	F–S–F = 1	5
:F–Xe::F: (⊕)	:F–Xe::F: = 2	:F–Xe::F: = 3	5
F–W–F (F surrounding)	F–W–F = 6	F–W–F = 0	6

Table 10.2   VSEPR Reference for Two to Six Areas of Electron Density

Areas of Electron Density around Central Atom	Electron Geometry	Generic Model[a]	Example
2	Linear	180° A–X–A	:C≡O: ; H–C≡N: ; :O=C=O:
3	Trigonal planar	A–X–A, A 120°	F=B(–F)(–F) (borate); O=C(–O⁻)(–O⁻) (carbonate); O=S(–O)(–O) (sulfite)
4	Tetrahedral	109.5° A–X(–A)(–A)–A	CH₄ (H–C with four H); NH₃ (H–N(–H)–H); H₂O (H–O(–H))
5	Trigonal bipyramidal	90° A–X(–A)(–A)–A, 120°	SF₄ type (S with four F)
6	Octahedral	90° A–X(–A)(–A)(–A)–A	WF₆ type (W with six F)

[a] Ideal angles are shown. Lone pairs cause a compression of those idealized angles.

would require a thorough grounding in molecular symmetry and group theory, which is beyond the scope of this chapter. Working from a valence bond theory, chemists have developed the concept of hybridized orbitals, which involves the mixing of two different types of atomic orbitals (often s and p) to produce a new set of one-electron, atomic hybrid orbitals (Figure 10.6).[2] These orbital hybridization designations correspond to the type and number of orbitals mixed: one s orbital and three p orbitals make four sp^3 orbitals. The exact type of hybridization that occurs depends on the geometry of the compound. A simplified approach to hybridization assumes that a tetrahedral atom is sp^3 hybridized, a trigonal planar atom is sp^2 hybridized, and a linear atom is sp hybridized. Figures 10.6, 10.7, and 10.8 show the three common electronic geometries of carbon atoms, their hybridization designation, and the types of orbitals present in each case.

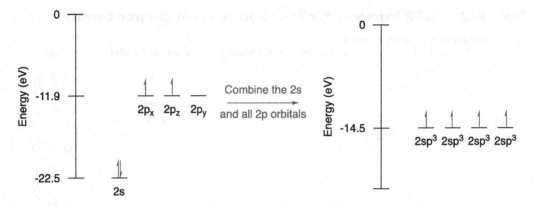

Figure 10.6 Diagram of sp³ hybridization in tetrahedral carbon atoms.

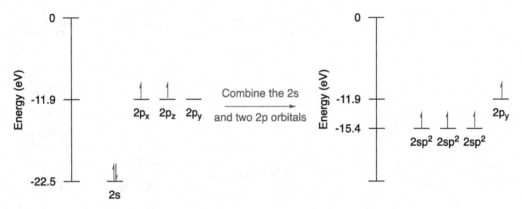

Figure 10.7 Diagram of sp² hybridization in trigonal planar carbon atoms.

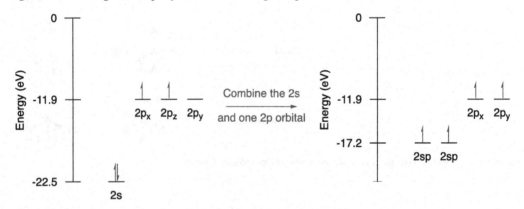

Figure 10.8 Diagram of sp hybridization in line carbon atoms.

PROBLEM 10.1
Identify the electron geometry and hybridization of each carbon atom.

a.
$$\text{H–N–C–N–H}$$
with O: above C, H below each N

b. H–N=C=O:

c.
$$\text{H–C–C–H}$$
with H H above and H H below

Let us consider the hybridization of carbon atoms a bit further. Table 10.3 shows three hydrocarbons that contain sp³-, sp²-, and sp-hybridized carbon atoms. There are two ways to remember the hybrid orbitals associated with each type of carbon. One can strictly memorize that tetrahedral

Table 10.3   The Electron Geometry Carbon Can Adopt in Ethane, Ethylene, and Acetylene and the Corresponding Orbitals of One Carbon Atom in Each Overlaid on the Corresponding Molecular Models

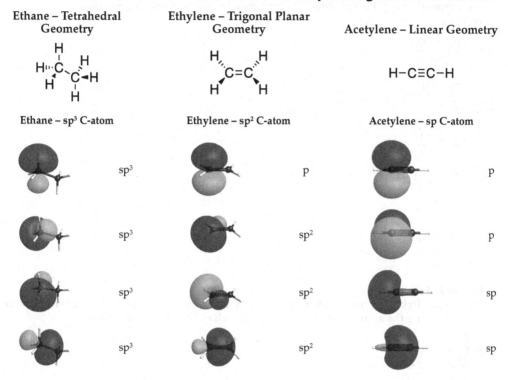

is sp^3, trigonal planar is sp^2, and linear is sp; however, it is more efficient to have a tool one can draw upon. The best way to determine the hybridization of a carbon atom is to use the sp^{SN-1} mnemonic, where SN is the steric number or the number of areas of electron density around an atom.

An sp^3-hybridized carbon atom can make four single bonds (with the orbitals lying along the bonding axis) and will adopt a tetrahedral geometry: the four C-atom hybrid orbitals in ethane are all sp^3 orbitals creating four tetrahedrally oriented single bonds.

An sp^2-hybridized carbon atom can make two single bonds and one double bond. The single bonds use the hybridized orbitals, while the double bond uses both the hybridized orbital and the unhybridized p orbital. The molecule will adopt a trigonal planar geometry with the second bond of the double bond existing above and below the molecular plane.

An sp-hybridized carbon atom can make two double bonds (or one single and one triple bond) and will adopt a linear geometry with the bonds using the p orbitals perpendicular to each other and the molecular axis. The two hybrid orbitals in acetylene are arranged linearly and 180° apart.

In the discussion of hybridization above, we focused on carbon for three reasons. First, carbon is the atom for which the simplified approach to hybridization works well, which is why hybridization is widely used in organic chemistry. Second, a simplified prediction of the hybridization for atoms with one or more lone pairs can vary substantially from the computational results. This text will only ask you to predict hybridization for carbon atoms, which will be a routine part of future organic chemistry study.

PROBLEM 10.2

For each of the following Lewis structures, count the number of areas of electron density and redraw the Lewis structure (using Table 10.2) to appropriately show the correct three-dimensional shape. Provide the electron geometry shape name and identify the bond angle(s).

a. $:\ddot{C}l-\ddot{N}-\ddot{C}l:$
   $\quad\quad :\ddot{C}l:$

b. $:N{\equiv}\overset{\oplus}{N}-\overset{\ominus}{\ddot{N}}:$

c. $:\ddot{F}-\ddot{O}-\ddot{F}:$

d. $:\ddot{I}-\ddot{I}::\ddot{I}:^{\ominus}$

e. $:\ddot{F}-\overset{:\ddot{F}::\ddot{F}:}{\underset{:\ddot{F}:}{\overset{\diagdown}{P}_{\oplus}}}:\ddot{F}:^{\ominus}$

f. $\overset{:\overset{\ominus}{\ddot{O}}:}{\underset{:\underset{\ominus}{\ddot{O}}:}{\overset{\ominus}{\ddot{O}}-\overset{2\oplus}{S}-\overset{\ominus}{\ddot{O}}:}}$

PROBLEM 10.3

What is the hybridization of each carbon atom in the following molecule? Redraw the molecule showing the appropriate bond angles and shape of each carbon atom.

$$\begin{array}{c} H \\ | \\ H-C-H \quad :\ddot{O}\ H \\ | \quad\quad\quad || \ | \\ :N-C{\equiv}C-C-C-H \\ | \quad\quad\quad | \\ H-C-H \quad\quad H \\ | \\ H \end{array}$$

PROBLEM 10.4

For each of the following chemical formulae, draw a Lewis structure that shows the correct three-dimensional shape of the molecule.

a. SF_6

b. COS

c. BF_3

d. ClF_3

e. PO_4^{3-}

f. SO_2Cl_2

NOTES

1. Gillespie, R.J.; Nyholm, R.S. Inorganic Stereochemistry. *Q. Rev. Chem. Soc.*, **1957**, *11*, 339–380. DOI: 10.1039/QR9571100339.

2. Pauling, L. The Nature of the Chemical Bond. Application of Results Obtained from the Quantum Mechanics and from a Theory of Paramagnetic Susceptibility to the Structure of Molecules. *J. Am. Chem. Soc.*, **1931**, *53* (4), 1367–1400. DOI: 10.1021/ja01355a027.

11 Gases

Our study of chemistry began with the nucleus (Chapter 4) and then slowly expanded outward to include electrons (Chapters 6 and 7) and then bonding between atoms (Chapter 9). In this chapter and in Chapters 12 and 13, our focus will expand to consider not just a single molecular entity but an ensemble. Specifically, this chapter will focus on gases, that is, an ensemble of molecular entities (atoms or molecules) without definite shape or volume and where each molecular entity is widely separated from the others (Figure 11.1). Because the particles are widely separated (gases are termed dispersed phases), we can simplify our analysis by assuming that there are no particle–particle interactions (Chapter 12). When studying liquids and solids (Chapter 13), the particles are substantially closer together; liquids and solids are termed condensed phases (Figure 11.1), and hence particle–particle interactions cannot be assumed not to exist. In fact, as we shall see, particle–particle interactions are the reason why, for example, water is a liquid and glucose is a solid at normal temperature and pressure.

GAS PARTICLES AND ENERGY

We will begin our study of gases by considering the physical behavior of the molecular entities that make up a gas. We will make several inferences about the behavior of gases that collectively are described as the kinetic theory of gases.[1] First, based on the density of a gas (argon has a density of 1.661 g/L at normal temperature and pressure) and the van der Waals radius of an argon atom (0.183 nm, Chapter 8) we can infer that the particles that make up a gas are much smaller than the distances between the particles (Figure 11.1). While it would be more realistic to account for the size of a gas particle, we can simplify our analysis by assuming that a generalized, hypothetical gas – we will use the term ideal gas – is composed of zero-dimensional point masses. We can see the general validity of this assumption if we consider the mean free path,[2] that is, the average distance a gas particle can move before it collides with another particle. For air at normal temperature (25.0 °C, 298 K) and pressure (1.00 atm, 101325 Pa)[3], the mean free path (66.35 nm)[4] is 181 times greater than the diameter (0.366 nm) of an argon atom.

PROBLEM 11.1
We can approximate the change in mean free path (l) in air using:

$$l = 2.22 \times 10^{-10} \frac{\text{m}}{\text{K}} T$$

l is the mean free path (m).
T is the temperature (K).

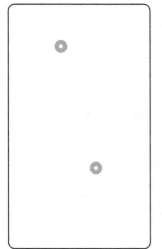
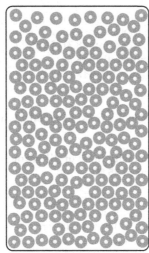

Figure 11.1 Density of argon atoms in a sample of gaseous argon (left), liquid argon (middle), and solid argon (right). The atoms occupy 0.064%, 57%, and 64% of the space, respectively. (Data from van Witzenburg, W. and Stryland, J.C. Density Measurements of Compressed Solid and Liquid Argon. *Can. J. Phys.*, **1968**, *46* (7), 811–816. DOI: 10.1139/p68-102.)

DOI: 10.1201/9781003479338-11

Argon gas condenses into liquid argon at −186 °C. What is the mean free path (l) at −185 °C? How does this compare to the van der Waals diameter (0.366 nm) of an argon atom?

The second inference made about gases is that the particles of a gas move in every possible direction with a distribution of velocity (v) magnitudes, and their average kinetic energy (E_k) is proportional to the absolute temperature (T). Initially, this inference was conjecture, but additional support came with the development of our understanding of heat (the transfer of thermal energy) and temperature (the measure of an object's thermal energy).[5,6,7] What is most important to know is that temperature provides a macroscopic measurement of particles' mean kinetic energy. If we assume that particles of gas are zero dimensional, this means that the only motion possible is translation through space. The kinetic energy of the gas particles can be calculated with Equation 11.1:

$$\overline{E_k} = \frac{1}{2} n M \overline{v}^2 \tag{11.1}$$

$\overline{E_k}$ is the mean kinetic energy of the gas particles (J).
n is the amount of substance (mol).
M is the molar mass (kg/mol).
$\overline{v}$ is the mean velocity (m/s).

Because the mean kinetic energy of a gas depends only on temperature,[8] we can calculate this average internal energy of a gas using Equation 11.2. Note that this expression incorporates the gas constant R, which is the macroscopic form of a fundamental constant relating temperature and energy.

$$\overline{E_k} = \frac{3}{2} n R T \tag{11.2}$$

$\overline{E_k}$ is the mean kinetic energy of the gas particles (J).
n is the amount of substance (mol).
R is the gas constant (8.314 J/(mol K)).
T is the temperature (K).

Now with Equation 11.1 and Equation 11.2, we can find the root-mean-square velocity (v_{rms}), the quadratic mean, for the gas particles in a sample:

$$\sqrt{\overline{v}^2} = v_{rms} = \sqrt{\frac{3RT}{M}} \tag{11.3}$$

$\overline{v}$ is the mean velocity (m/s).
v_{rms} is the root-mean-square velocity (m/s).
R is the gas constant (8.314 J/(mol K)).
T is the temperature (K).
M is the molar mass (kg/mol).

PROBLEM 11.2
What is the root-mean-square velocity (m/s) of argon gas at −185 °C? At 25 °C? At 323 °C?

PROBLEM 11.3
What is the root-mean-square velocity (m/s) of dihydrogen gas at 25 °C? Of dioxygen gas at 25 °C? Of sulfur hexafluoride gas at 25 °C?

While the root-mean-square velocity is a useful property of a gas, it is only the quadratic mean velocity of a particular gas at a given temperature. An actual sample of gas will have a distribution of velocity magnitudes (Figure 11.2).[9] The calculation of this distribution is beyond the scope of this book, but it is worth noting that at low temperatures the root-mean-square velocity is lower and the distribution, or spread, of velocity magnitudes is narrower. As temperature increases, the root-mean-square velocity increases and the spread of velocity magnitudes widens.

Because gas particles are in constant random motion, we can conclude that gas particles collide frequently with one another and with the walls of the container. Given the mean free path in air and the v_{rms} of argon at 298 K (421 m/s), this corresponds to roughly 1×10^{10} collisions per

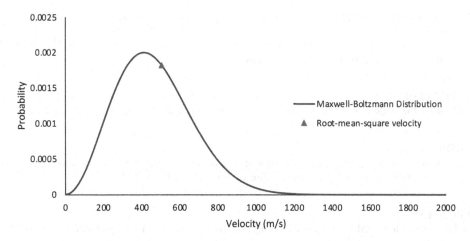

Figure 11.2 Maxwell–Boltzmann distribution of gas particles in air (0.028 97 kg/mol) at normal temperature (25.0 °C, 298 K) and pressure (1.00 atm, 101 325 Pa) with a root-mean-square velocity of 506.6 m/s.

second.[10] The third inference about ideal gases is that these collisions are perfectly elastic and follow Newton's laws. Pressure, then, is the macroscopic result of the frequent random collisions of the gas particles with the container. Pressure is, by definition, the amount of force per area (N/m²), which is the definition of the unit pascal (1 Pa = 1 N/m²). We can also think of pressure as the amount of energy per volume (J/m³), and the internal energy, mean kinetic energy, of a gas can be calculated as the product of pressure and volume:[11]

$$\overline{E_k} = \frac{3}{2}pV \tag{11.4}$$

$\overline{E_k}$ is the mean kinetic energy of the gas particles (J).
p is the pressure (Pa).
V is the volume (m³).

PROBLEM 11.4
Two rigid containers of gas are 1.00 L in size, which will have a greater amount of kinetic energy: the container with a pressure of 101 kPa or the container with a pressure of 202 kPa (note that 1000 L = 1 m³)?

IDEAL GAS LAW

The macroscopic variables that define a gas, as seen above, are pressure (p), volume (V), amount of substance (n), and temperature (T). In combination, the amount of substance and temperature (Equation 11.2) along with pressure and volume (Equation 11.4) relate to the mean kinetic energy of a gas. As such, we can relate these macroscopic variables to each other and arrive at a general equation describing a gas:

$$pV = nRT \tag{11.5}$$

p is pressure (Pa or atm).
V is volume (m³ or L).
n is the amount of gas (mol).
R is the gas constant (8.314 $\frac{Pa\,m^3}{mol\,K}$ or 0.082 06 $\frac{L\,atm}{mol\,K}$).
T is the temperature (K).

Equation 11.5 is called the ideal gas law, which allows for the determination of any one variable of a gas, given the other three variables. There are important limitations on the ideal gas law. This law works only for gases that are at low to moderate pressure and moderate to high temperature.

At high pressure the assumption that gas particles are zero-dimensional point masses diverges from the observed behavior, and at low temperatures the assumption that collisions are perfectly elastic also no longer holds. The importance of these discrepancies between an ideal gas and the behavior of a real gas will be investigated in Chapter 12.

PROBLEM 11.5
Calculate the pressure (atm) exerted by 1.55 g of Xe gas at 25 °C in a 560 mL container.

PROBLEM 11.6
Once filled with 4.4 g of CO_2 gas, a flask of unknown volume is found to have a pressure of 0.961 atm. If the flask is 27 °C, what is the volume of the flask (L)?

PROBLEM 11.7
A 4.23 g sample of unknown gas exerts a pressure of 0.965 atm in a 1.00 L container at 445.7 K.

a. Calculate the amount of gas (mol) present in the container.

b. Calculate the molar mass (kg/mol) given the mass (4.23 g) and your answer to part a.

PROBLEM 11.8
Two gas containers at 22.5 °C are connected by a valve. Container A is 9.2 L and the pressure inside is 1.75 atm. Container B is 5.4 L and the pressure inside is 0.82 atm.

a. Calculate the amount of gas (mol) present in each container.

b. When the valve is opened, what is the total volume of the container AB? What is the total amount of gas inside the container AB?

c. Using your answers in part b, what will be the final pressure (atm) after the valve is opened?

SPECIFIC RELATIONSHIPS: The Gas Laws

While the ideal gas law relates all macroscopic variables that define the state of a gas (p, V, n, and T), our focus is often on how changing one variable affects another variable. Historically, these individual gas laws developed independently of the kinetic theory of gases presented above, but the ideal gas law was first derived from the compilation of these individual laws. Each law is named for the scientist who is credited with its discovery, but knowledge of these names should not be the focus of your attention throughout this section.

Our first gas law is Boyle's law,[12] which considers the relationship between pressure and volume at a constant amount of gas (n) and a constant temperature (T). With a constant amount of gas and a constant temperature, the product of pressure and volume for a gas is a constant value. As such, the product of pressure and volume under two different conditions will be equal (Equation 11.6). What this means is that as volume increases the pressure decreases, and as volume decreases the pressure increases. We can see this graphically in Figure 11.3.

$$p_1 V_1 = p_2 V_2 \qquad (11.6)$$

p_1 is the pressure (Pa or atm) for condition 1.
V_1 is the volume (m³ or L) for condition 1.
p_2 is the pressure (Pa or atm) for condition 2.
V_2 is the volume (m³ or L) for condition 2.

PROBLEM 11.9
When inhaling, a person's chest cavity increases in volume, and when exhaling their chest cavity decreases in volume.

a. Provide an explanation for how increasing our chest cavity volume helps us inspire, that is, get air into our lungs.

b. Provide an explanation as to why decreasing our chest cavity volume pushes air out of our lungs.

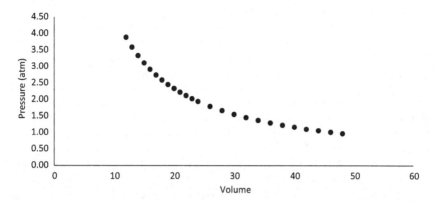

Figure 11.3 Plot of gas pressure (atm) versus volume for air at normal temperature. Note that these data come from Boyle's original paper and the author could not determine the units for volume used by Boyle.

The next gas law is Charles's law, which was published by Gay-Lussac, but Gay-Lussac credited unpublished work by Charles.[13] Charles and Gay-Lussac studied the relationship between volume and temperature at a constant pressure (*p*) and a constant amount of gas (*n*). With a constant pressure and a constant amount of gas, the quotient of volume and temperature for a gas is a constant value. As such the quotient of volume and temperature under two different conditions will be equal (Equation 11.7). What this means is that as temperature increases, the volume increases, and as temperature decreases, the volume decreases. We can see this graphically in Figure 11.4.

$$\frac{V_1}{T_1} = \frac{V_2}{T_2}$$
(11.7)

V_1 is the volume (m^3 or L) for condition 1.
T_1 is the temperature (K) for condition 1.
V_2 is the volume (m^3 or L) for condition 2.
T_2 is the temperature (K) for condition 2.

PROBLEM 11.10
Consider the data in Figure 11.4 for 1.0 mol helium gas at 1.0 atm showing the relationship between volume (*V*) and temperature (*T*). The trendline for these data and the linear fit equation are shown. Determine the *x* intercept of these data. What is the significance of this temperature?

There is a similar relationship between pressure and temperature as there is between volume and temperature. This gas law is often Gay-Lussac's law, but it is more properly attributed to

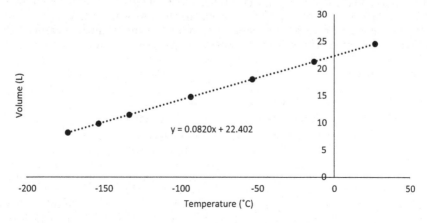

Figure 11.4 Plot of gas volume (L) versus temperature (°C) for 1.0 mol helium gas at 1.0 atm.

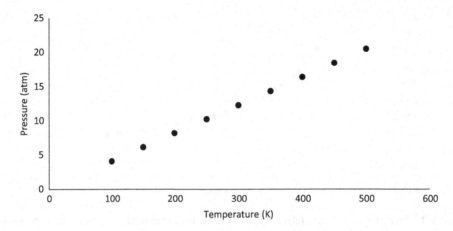

Figure 11.5 Plot of gas pressure (atm) versus temperature (K) for 5.0 mol dinitrogen gas in a 10.0 L fire extinguisher.

Amonton.[14] With a constant volume and a constant amount of gas, the quotient of pressure and temperature for a gas is a constant value. As such, the quotient of pressure and temperature under two different conditions will be equal (Equation 11.8). What this means is that as temperature increases, the pressure increases, and as temperature decreases, the pressure decreases. We can see this graphically in Figure 11.5.

$$\frac{p_1}{T_1} = \frac{p_2}{T_2}$$ (11.8)

p_1 is the pressure (Pa or atm) for condition 1.
T_1 is the temperature (K) for condition 1.
p_2 is the pressure (Pa or atm) for condition 2.
T_2 is the temperature (K) for condition 2.

PROBLEM 11.11
Consider the data in Figure 11.5 for 5.0 mol dinitrogen gas in a 10.0 L fire extinguisher showing the relationship between pressure (p) and temperature (T). Fire extinguishers are great tools for helping to escape from fires, or to put out a small fire, but fire extinguishers put into a fire can be quite dangerous. Consider Figure 11.5 and provide an explanation as to why a fire extinguisher dropped into a fire is dangerous.

Avogadro's law considers the amount of gas (n) and the volume of gas (V),[15] and it is perhaps one of the most important laws in terms of the effect that it had not only on our understanding of gases but also on linking the atomic world to the macroscopic. Avogadro's law states that equal volumes of gases, at equal pressure and temperature, have the same number of particles. Or, in other words, the quotient of volume (V) to the amount of gas (n) is constant:

$$\frac{V_1}{n_1} = \frac{V_2}{n_2}$$ (11.9)

V_1 is the volume (L) for condition 1.
n_1 is the amount of gas (mol) for condition 1.
V_2 is the volume (L) for condition 2.
n_2 is the amount of gas (mol) for condition 2.

Avogadro's law provided a scientific basis for determining the number of particles in a mole, now called the Avogadro constant (N_A) in his honor. While the value was first precisely determined after the Millikan oil drop experiment (Chapter 3), Loschmidt determined the size of air molecules,[16] which allowed a determination of the number of particles in a sample of gas, a forerunner of the Avogadro constant. Avogadro's law also provided the basis for rationally determining relative atomic mass (A_r) values in 1858.[17]

PROBLEM 11.12
In a 1.0 L container is 1.31 g of dioxygen gas and in a separate 1.0 L container is 0.0826 g of dihydrogen gas. Both containers are at 1.0 atm and 25 °C.

a. What can we say about the number of particles in the two containers?

b. What is the relative mass of dihydrogen gas to dioxygen gas (m_{H2}/m_{O2})?

c. Now, if we were to set that atomic mass of oxygen as 16.000, what is the relative atomic mass of hydrogen? How does this compare with the value in Appendix 2?

GAS MIXTURES

This chapter has only investigated pure gases, but gases are often found in mixtures. A great example is the air around you, which is a mixture of nitrogen, oxygen, argon, water, and carbon dioxide. To conclude this chapter, we will consider two important aspects of gas mixtures: pressure and effusion.

For mixtures of gases, the ideal gas law still holds true. An important ramification of this is Dalton's law of partial pressures: the total pressure of a gas mixture is the sum of the partial pressures of each gas (Equation 11.10).[18] Because gas particles experience perfectly elastic collisions, when behaving ideally, only the amount of gas particles, not their identity, affects the pressure.

$$p_{Total} = p_A + p_B \qquad (11.10)$$

p_{Total} is the total pressure (Pa or atm).
p_A is the partial pressure of gas A (Pa or atm).
p_B is the partial pressure of gas B (Pa or atm).

PROBLEM 11.13
A sample of air in a 10.0 L container has the following partial pressures of each gas: p_{O2} = 20.9 kPa, p_{N2} = 78.1 kPa, p_{Ar} = 0.97 kPa, p_{H2O} = 1.28 kPa, p_{CO2} = 0.05 kPa. What is the total pressure of the sample?

The partial pressure of a gas can be determined from the concentration of that gas:

$$p_A = p_{Total}\, y_A \qquad (11.11)$$

p_A is the partial pressure of gas A (Pa or atm).
p_{Total} is the total pressure (Pa or atm).
y_A is the mole fraction of gas A.

PROBLEM 11.14
In the lungs, the normal mole fraction of CO_2 (y_{CO2}) is 0.046. What is the partial pressure (kPa) of CO_2 in your lungs (total pressure is 101.3 kPa)?

While the identity of a gas does not affect how it contributes to the total pressure of a system, gas identity can and does affect the root-mean-square velocity of a gas, which depends on the molar mass (Equation 11.3). Lighter gases are faster moving, and heavier gases are slower moving. When a container has a small hole, the gases will escape at a different rate depending on their root-mean-square velocity (Equation 11.12). This equation is called Graham's law of effusion.[19]

$$\frac{v_{rms}(\text{lighter gas})}{v_{rms}(\text{heavier gas})} = \sqrt{\frac{M_{\text{heavier gas}}}{M_{\text{lighter gas}}}} \qquad (11.12)$$

v_{rms}(lighter gas) is the root-mean-square velocity (m/s) of the lighter gas.
v_{rms}(heavier gas) is the root-mean-square velocity (m/s) of the heavier gas.
$M_{\text{heavier gas}}$ is the molar mass (kg/mol) of the heavier gas.
$M_{\text{lighter gas}}$ is the molar mass (kg/mol) of the lighter gas.

Let's look at an example of a container with dihydrogen gas (M = 0.002 0160 kg/mol) and dioxygen gas (M = 0.031 998 kg/mol). This will give a ratio of root-mean-square velocity magnitudes of 3.9840, which means that dihydrogen molecules will effuse about four times faster than dioxygen molecules. This difference in effusion will lead to enrichment of the lighter gas, in this example the dihydrogen, outside the container and enrichment of the heavier gas, here dioxygen, inside the container.

PROBLEM 11.15
Natural uranium is a mixture of fissionable uranium-235 (0.72% abundance, m_{U-235} = 235.04 u) and non-fissionable uranium-238 (99.3% abundance, m_{U-238} = 238.05 u). One method of creating enriched uranium (increasing the uranium-235 content) is to take advantage of the differential rate of effusion of uranium hexafluoride (UF_6) gas.

a. What is the ratio of root-mean-square velocity magnitudes for $^{235}UF_6$ and $^{238}UF_6$?

b. Given the ratio calculated in part a, how effective do you think effusion is as a method of enrichment?

NOTES

1. This analysis comes from the Kinetic Theory of Gases first proposed by Bernoulli and later expanded during the 19th century: Bernoulli, D. *Hydrodynamica, sive De Viribus et Motibus Fluidorum Commentarii*. Joh. Henr. Decker für Johannes Reinhold Dulsecker: Straßburg, 1738.

2. Clausius, R. Ueber die Art der Bewegung, welche wir Wärme nennen. *Ann. Phys.*, **1857**, *176* (3), 353–380. DOI: 10.1002/andp.18571760302.

3. Note that pascal (Pa) is the SI unit of pressure, where 1 Pa is 1 N/m^2. Non-SI units that are commonly used are atmosphere (1.0 atm is 101325 Pa) and bar (1 bar is 100 000 Pa). The standard state pressure ($p°$) since 1982 has been a pressure of 100 000 Pa (1 bar or 0.987 atm). Normal state pressure is defined by the National Institute of Standards and Technology (NIST) as 1 atm (101 325 Pa or 1.01 bar).

4. Jennings, S.G. The Mean Free Path in Air. *J. Aerosol Sci.*, **1988**, *19* (2), 159–166. DOI: 10.1016/0021-8502(88)90219-4.

5. (i) Carnot, S. Réflexions sur la Puissance Motrice du Feu et sur les Machines Propres a Développer cette Puissance. Bachelier: Paris, 1824.

    (ii) Clapeyron, E. Mémoire sur la Puissance Motrice de la Chaleur. *Journal de l'École Royale Polytechnique*, **1834**, *14*, 153–190.

6. Joule, J.P. III. On the Mechanical Equivalent of Heat. *Phil. Trans. R. Soc.*, **1850**, *140*, 61–82. DOI: 10.1098/rstl.1850.0004.

7. Clausius, R. *The Mechanical Theory of Heat* (Trans. W.R. Browne), MacMillan and Co.: London, 1879.

8. Joule, J.P. LIV. On the Changes of Temperature Produced by the Rarefaction and Condensation of Air. *Lond. Edinb. Dublin Philos. Mag. J. Sci. Series 3*, **1845**, *26* (174), 369–383. DOI: 10.1080/14786444508645.

9. (i) Maxwell, J.C. V. Illustrations of the Dynamical Theory of Gases.–Part I. On the Motions and Collisions of Perfectly Elastic Spheres. *Lond. Edinb. Dublin Philos. Mag. J. Sci. Series 4*, **1860**, *19* (124), 19–32. DOI: 10.1080/14786446008642818.

    (ii) Maxwell, J.C. II. Illustrations of the Dynamical Theory of Gases.–Part II. On the Process of Diffusion of Two or More Kinds of Moving Particles Among One Another. *Lond. Edinb. Dublin Philos. Mag. J. Sci. Series 4*, **1860**, *20* (130), 21–37. DOI: 10.1080/14786446008642902.

    (iii) Boltzmann, L. Weitere Studien über das Wärmegleichgewicht unter Gasmolekülen. *Sitz.-Ber. Akad. Wiss. Wien (II)*, **1872**, *66*, 275–370. DOI: 10.1007/978-3-322-84986-1_3.

(iv) Boltzmann, L. Über die Beziehung zwischen dem zweiten Hauptsatz der mechanischen Wärmetheorie und der Warscheinlichkeitsrechnung respective den Sätzen über das Wärmegleichgewicht. *Sitz.-Ber. Akad. Wiss. Wien (II)*, **1877**, *76*, 373–435. DOI: 10.1017/CBO9781139381437.011.

10. Maxwell, J.C. V. Illustrations of the Dynamical Theory of Gases.–Part I. On the Motions and Collisions of Perfectly Elastic Spheres. *Lond. Edinb. Dublin Philos. Mag. J. Sci. Series 4*, **1860**, *19* (124), 19–32. DOI: 10.1080/14786446008642818.

11. The reason for the 3/2 term comes from the full physics derivation of this expression, which can be found in: *The Feynman Lectures on Physics*, Volume 1. Chapter 39. The Kinetic Theory of Gases. Gottlieb, M. and Pfeiffer, R., California Institute of Technology. https://www.feynmanlectures.caltech.edu/I_39.html (Accessed 07/03/2023).

12. Boyle, R. A *Defence of the Doctrine Touching the Spring and Weight of the Air*. Thomas Robinson: London, 1662.

13. Gay-Lussac, J.L. Recherches sur la dilatation des gaz et des vapeurs. *Ann. Chim. Phys.*, **1802**, *43*, 137–75.

14. Amontons, G. Moyen de substituer commodement l'action du feu, a la force des hommes et des cheveaux pour mouvoir les machines. *Mem. Acad. R. Sci. Paris*. **1699**, *3*, 112–126.

15. Avogadro, A. Essai d'une manière de determiner les masses relatives des molecules elementaries des corps, et les proportions selon lesquelles ells entrent dans ces cominaisons. *J. Phys. Chim. Hist. Nat. Arts.*, **1811**, *73*, 58–76.

16. Loschmidt was the first to estimate the size of air molecules, which allows for an estimation of the number of particles in a mole. For this work, the Avogadro constant is occasionally symbolized as L rather than N_A: Loschmidt, J. Zur Grösse der Luftmoleküle. *Sitz.-Ber. K. Akad. Wiss. Wien*, **1865**, *52* (2), 395–413.

17. Cannizzaro, S. *Sketch of a Course of Chemical Philosophy*. Chicago: The University of Chicago Press, 1911. Translated from: *Il Nuovo Cimento*, **1858**, *7*, 321–366.

18. Dalton, J. Essay IV. On the Expansion of Elastic Fluids by Heat. *Mem. Proc. Manch. Lit. Philos. Soc.*, **1802**, *5* (2), 595–602.

19. Graham, T. XXVII. On the Law of the Diffusion of Gases. *Lond. Edinb. Dublin Philos. Mag. J. Sci. Series 3*, **1833**, *2* (9), 175–190. DOI: 10.1080/14786443308648004.

12 The van der Waals Equation and Intermolecular Forces

The physical properties of a gas are related through the ideal gas law:

$$pV = nRT \tag{12.1}$$

p is pressure (Pa or atm).
V is volume (m^3 or L).
n is the amount of gas (mol).
R is the gas constant (8.314 $\frac{\text{Pa m}^3}{\text{mol K}}$ or 0.082 06 $\frac{\text{L atm}}{\text{mol K}}$).
T is the temperature (K).

The ideal gas law holds for most gases at low to moderate pressures and moderate to high temperatures. Specifically, the ideal gas law works for gases that adhere to the assumptions of the kinetic molecular theory.[1] These assumptions include:

1) A gas is composed of particles whose size is much smaller than the distances between them (so gas particles can be treated as zero-dimensional point masses).

2) Gas particles move randomly at various speeds in every possible direction.

3) Gas particles collide frequently with one another and with the walls of the container. These collisions are perfectly elastic.

4) The average kinetic energy of gas particles is proportional to the absolute temperature.

While these assumptions are usually true for most conditions, there are two assumptions that ignore the physical reality of gas particles: (1) gas particles can be treated as zero-dimensional point masses and (2) collisions between the particles and the container are perfectly elastic. Since gases are matter, we know they must have both mass and volume, which means that the first assumption will break down (especially when gas particles are forced close together at high pressures). And particles always interact when they are in close contact, which means that at high pressures and low temperatures (when particles are closer together and moving more slowly), the interaction between particles will become more significant. A handful of approaches have been developed to address the nonzero particle size and particle interactions of gases. We will focus on the van der Waals equation:[2]

$$\left(p + \frac{n^2a}{V^2}\right)(V - nb) = nRT \tag{12.2}$$

p is pressure (atm).
V is volume (L).
n is the amount of gas (mol).
a is the pressure correction term ($\frac{\text{atm L}^2}{\text{mol}^2}$).

b is the volume correction term ($\frac{\text{L}}{\text{mol}}$).

R is the gas constant (0.082 06 $\frac{\text{L atm}}{\text{mol K}}$).

T is the temperature (K).

The volume correction term b accounts for the fact that gas particles have nonzero particle size. The units provide a measure of what volume (L) a mole (mol) of gas particles occupies. As such, a bigger chemical species would have a larger b value (L/mol). The b value also provides a means of determining which chemical species is larger because the larger the b value, the larger the chemical species. For example, let's consider the b values for the halogens (Table 12.1). We can see that the b values increase as we move down the group 17 elements from fluorine (smallest, top of group) to iodine (largest, bottom of group). This quantitative measure follows the expected size progression from the periodic trends (recall that as you go down a group, atoms increase in size due to the larger shell number).

DOI: 10.1201/9781003479338-12

Table 12.1 *b* Values of Halogens

Halogen	F_2	Cl_2	Br_2	I_2
b value (L/mol)	0.0290	0.0542	0.0591	0.0731

PROBLEM 12.1
What makes a gas ideal? What makes a gas real?

PROBLEM 12.2
Consider the following gases: Ar, SF_6, H_2, CO_2, and CH_4. In terms of their structure (not looking up the values), rank them in order of smallest to biggest *b* value. Explain your answer.

The pressure correction term *a* accounts for the fact that particle–particle collisions are not perfectly elastic. The units of the *a* value are $\dfrac{atm\,L^2}{mol^2}$, which is a complex set of units, but it is not, at first, clear what they mean. Let's unpack and rearrange these units a bit. First, the powers are expanded to show the individual terms:

$$\frac{(L\,atm)L}{mol\,mol} \tag{12.3}$$

Then the numerator and denominator are both multiplied by 1/L, and we arrive at a more tangible set of units:

$$\frac{L\,atm}{mol\left(\dfrac{mol}{L}\right)} \tag{12.4}$$

The numerator (L atm) is a unit of energy (1 L atm = 101.325 J), and the denominator has the unit of moles, which is the amount of substance, and mol/L, which is the amount of concentration:

$$\frac{energy}{(amount)(amount\ concentration)} \tag{12.5}$$

Thus, the *a* value is a measure of energy (normalized per particle amount and amount concentration) necessary to change the distance between particles. This attractive energy is the energy that holds particles together via intermolecular forces (IMFs), which consist of van der Waals interactions and hydrogen bonds. Hydrogen bonds and van der Waals interactions are stabilizing attraction between molecular entities. Specifically, this means that molecular entities are in a lower energy state when they are in close proximity and energy must be added to disperse condensed phases and molecular aggregates.[3]

NONCOVALENT (VAN DER WAALS) INTERACTIONS

We will utilize these *a* and *b* values to provide insights into the size of gas particles and their non-covalent interactions. Consider Figure 12.1, which shows the correlation between *b* and *a* values for a sampling of chemical species. The specific chemicals and their *b* and *a* values are provided in Table 12.2.

Looking at Figure 12.1, the *b* and *a* values show significant correlation ($R^2 = 0.9705$): the larger a molecular entity is, the larger the strength of its intermolecular forces.[4] This general trend is the most universal of intermolecular forces: dispersion interactions (London interactions).[5] Note that dispersion interactions are often also called van der Waals interactions, but dispersion is only one type of van der Waals interactions along with dipole–dipole interactions (Keesom interactions), which are considered below, and dipole–dispersion interactions (Debye force), which are not considered here.

(London) Dispersion Interactions

Dispersion interactions are attractive interactions between molecules that arise because the dispersion of electrons on the surface leads to polarization as two molecular entities approach one another. That means that some areas are electron-rich (indicated with ∂^-, meaning partial negative charge) and some areas are electron-poor (indicated with ∂^+, meaning partial positive

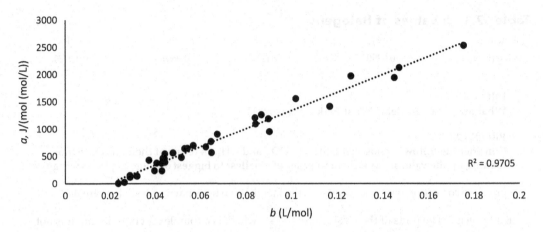

Figure 12.1 Correlation of b and a values for an average sample of chemical species.

Table 12.2 Chemical Species, with Their b and a Values (Represented in Figure 12.1)

Chemical	$b\left(\dfrac{L}{mol}\right)$	$a\left(\dfrac{J}{mol\left(\dfrac{mol}{L}\right)}\right)$	Chemical	$b\left(\dfrac{L}{mol}\right)$	$a\left(\dfrac{J}{mol\left(\dfrac{mol}{L}\right)}\right)$
He	0.024	3.5	H_2Se	0.048	559.6
H_2	0.027	24.9	PH_3	0.052	475.8
NO	0.029	147.9	HI	0.053	639.2
F_2	0.029	118.7	Cl_2	0.054	642.7
O_2	0.032	140.0	SO_2	0.057	695.6
Ar	0.032	137.3	Rn	0.062	668.8
Kr	0.040	235.5	C_2H_6	0.065	564.4
HCl	0.041	374.9	C_3H_8	0.090	950.9
CO_2	0.043	370.6	$CHCl_3$	0.102	1554.3
CH_4	0.043	233.0	C_4H_{10}	0.117	1411.5
H_2S	0.043	460.4	CCl_4	0.126	1968.9
HBr	0.044	456.0	C_5H_{12}	0.146	1938.3
NO_2	0.044	543.1	$SiCl_4$	0.147	2123.8
N_2O	0.044	390.3	C_6H_{14}	0.175	2530.1

Figure 12.2 Model of dispersion interaction in chlorine showing the formation of an induced dipole in neighboring chlorine molecules and the resulting attraction (hash marks).

charge).[6] These partially negative and partially positive areas then attract each other as shown in Figure 12.2.

The attraction is the result of an electrostatic force that is described by the London equation (Equation 12.6). Note that since the interaction energy is inversely proportional to distance (r), the attraction is most intense as particles approach during a collision.

$$E_{\text{dispersion}} = \frac{-3\alpha_1\alpha_2}{2r^6} \frac{E_i(1)E_i(2)}{E_i(1)+E_i(2)} \tag{12.6}$$

$E_{\text{dispersion}}$ is the stabilizing energy of the dispersion interaction (kJ/mol).
α_1 is the polarizability volume of molecular entity one (m^3).
α_2 is the polarizability volume of molecular entity two (m^3).
r is the distance between molecular entities (m).
$E_i(1)$ is the ionization energy of molecular entity one (kJ/mol).
$E_i(2)$ is the ionization energy of molecular entity two (kJ/mol).

Considering Equation 12.6, it is worth considering the polarizability volume (α) further, as this is the core of the dispersion interaction. Consider methane (CH_4), which has a van der Waals volume of 2.58×10^{-29} m^3. The polarizability volume (α) can be thought of as the size of the deformation that the electron cloud can undergo. For methane, the polarizability volume (α) is 2.6×10^{-30} m^3, in other words, the polarizability volume is 10% of the total molecular volume. This 10% value is a typical ratio of polarizability volume to molecular volume. In methane, this leads to an attractive – stabilizing as shown by the negative value – dispersion energy of –9.8 kJ/mol.[7] As molecules increase in size, their molecular volumes increase, which means that there is a greater polarizability volume or deformation ability of the electron cloud. This leads to stronger stabilizing interactions. For example, tetrachloromethane (CCl_4) has a van der Waals volume of 8.54×10^{-29} m^3 and a polarizability volume of 1.051×10^{-29} m^3 (12% of the total volume). The greater polarizability volume, which is a function of greater size, corresponds to a greater dispersion interaction energy of –32.6 kJ/mol.[4] Note that for isomers, molecules with the same formula but different structures, linear isomers have greater polarizability volumes (and hence greater dispersion interaction energy) than more spherical isomers.[8] We can see this in the comparison of C_4H_{10} isomers butane, which is linear and has a dispersion interaction energy –11.7 kJ/mol, and 2-methylpropane, which is more spherical and has a dispersion interaction energy –7.5 kJ/mol.[9] It is important to keep in mind that dispersion interactions are, with limited exceptions that will be discussed below, the most common form of intermolecular attraction interaction. All other intermolecular forces that will be considered are add-ons that some molecules have and others do not.

If we look at the binary compounds of group 14, elements with hydrogen (CH_4, SiH_4, GeH_4, and PbH_4) have an exceptionally good correlation between size (b) and strength of their intermolecular forces (a) (Figure 12.3). This can be attributed to the increasing size of the gas molecule and its ability to engage in stabilizing dispersion interactions.

The strength of the intermolecular forces between particles is directly related to the boiling point of a chemical species; thus, it is not surprising that the boiling of these compounds is also well correlated to the b value (Figure 12.4). The reason for this relationship is that the greater the strength of the intermolecular forces between molecules in the liquid phase, the more energy will be required to separate them into gas particles. In the case of boiling point, this means greater thermal energy. Liquid particles are in close contact allowing for strong electrostatic forces between particles, whereas gas particles are separated by great distances and the electrostatic forces between noncolliding particles are negligible.

While dispersion interactions are universal, there are some molecules that have a weaker dispersion interaction than their size (b values) would suggest. Consider the chemical species highlighted in Figure 12.5 with X's and listed in Table 12.3. All these chemical species fall below the general trend for other compounds.

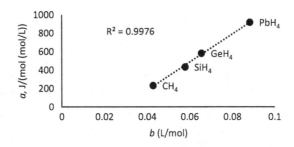

Figure 12.3 Correlation of b and a values for binary compounds of group-14 elements with hydrogen.

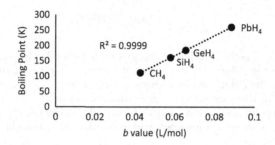

Figure 12.4 Correlation of b values and boiling point for binary compounds of group-14 elements with hydrogen.

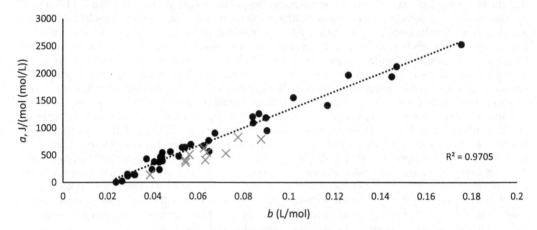

Figure 12.5 Correlation of b and a values for an average sample of molecules (black circles) and with some low-dispersion chemicals (X).

Table 12.3   Low-Dispersion Chemical Species' b and a Values

Chemical	$b \left(\dfrac{L}{mol} \right)$	$a \left(\dfrac{J}{mol\left(\dfrac{mol}{L}\right)} \right)$
BF_3	0.054	403.3
F_3N	0.055	362.7
CH_3F	0.056	507.5
CH_2F_2	0.063	626.6
CF_4	0.063	409.4
CHF_3	0.064	544.9
SiF_4	0.072	532.9
C_2H_5F	0.078	827.8
SF_6	0.088	796.1

The chemicals highlighted in Figure 12.5 and listed in Table 12.3 all have one thing in common: fluorine in their chemical formula. The highly electronegative fluorine atoms are responsible for the reduction in attractive dispersion interactions for these compounds. Fluorine is so highly electronegative (4.0 on the Pauling scale) that it holds onto its electrons more tightly than any other element that forms compounds, which reduces the polarizability of these molecules. For example, tetrafluoromethane (CF_4) has a van der Waals volume of 5.01×10^{-29} m^3 and a polarizability volume of 2.9×10^{-30} m^3 (6% of the total volume).[10] The reduced polarizability volume, which is a function of fluorine's electronegativity, corresponds to a reduced dispersion interaction energy of –14.7 kJ/mol.[11] In general, the presence of fluorine atoms in a chemical structure reduces the ability of electrons to disperse and form ∂^- and ∂^+ regions (see Figure 12.2), which reduces the stabilizing attraction between molecules. An everyday example of utilizing low-surface energy materials like this poly(tetrafluoroethylene), which is applied to make nonstick cooking materials (under the brand name Teflon), and in stain- and water-repellent clothing (an example brand is Gore-Tex).

PROBLEM 12.3
Considering the intermolecular forces involved, can you provide an explanation for the difference in the boiling point between these two compounds?

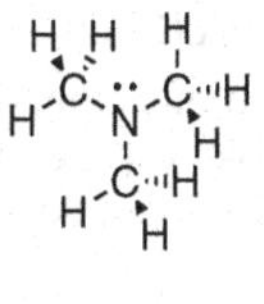

C_3H_9N
$b = 0.1$ L/mol
boiling point = 7 °C

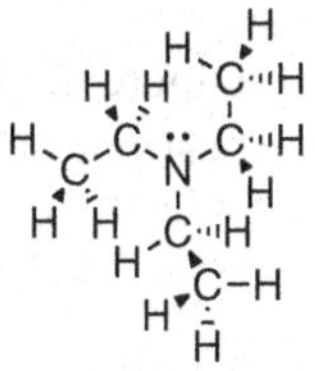

C_6H_7N
$b = 0.2$ L/mol
boiling point = 89 °C

PROBLEM 12.4
Sulfur hexafluoride (SF_6) has a b value roughly equivalent to that of propane (C_3H_8). Sulfur hexafluoride's boiling point is 10 K lower than that of propane. Provide an explanation for this effect.

PROBLEM 12.5
Consider the following set of data looking at fluorinated ethane compounds. Provide an explanation for the trend in boiling points.

Name	Fluoroethane	1,1,1-Trifluoroethane	1,1,1,2,2,2-Hexafluoroethane
Structure	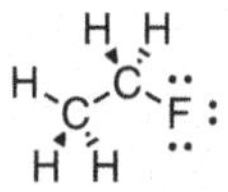	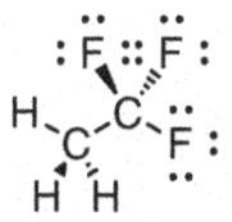	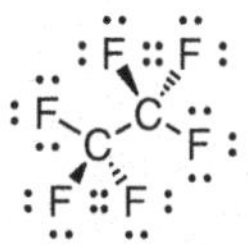
b (L/mol)	0.076 61	0.094 17	0.098 92
Boiling point (K)	236	226	195

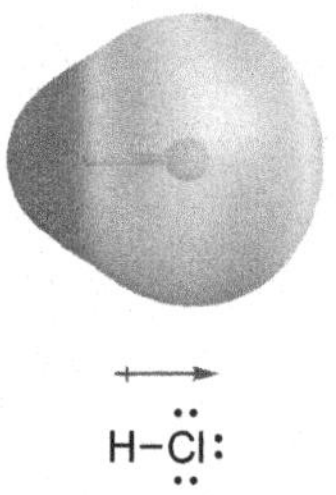

$H-\overset{..}{\underset{..}{Cl}}:$

Figure 12.6 Electrostatic potential map (top) and Lewis structure with dipole moment arrow (bottom) of HCl. In an electrostatic potential map, areas of high electron density are shown in red or with the symbol ∂^-, and areas of low electron density are shown in blue or with the symbol ∂^+, and a middling electron density is shown in green and no symbol is shown.

Table 12.4 Trend in the Dipole Moment (*p*), and the Dipole Length (*l*$_p$) for Hydrogen Halides

Hydrogen Halide	Pauling Electronegativity of Halogen Atom	Dipole Moment (C m)	Dipole Moment (D)	Dipole Length (pm)
HF	3.98	6.07×10^{-30}	1.82	37.9
HCl	3.16	3.60×10^{-30}	1.08	22.5
HBr	2.96	2.73×10^{-30}	0.82	17.1
HI	2.66	1.47×10^{-30}	0.44	9.2

Dipole–Dipole (Keesom) Interactions

Dipole–dipole interactions are a stronger,[12] more fixed version of dispersion interactions. Whereas dispersion relies on the random fluctuation of electrons, dipole–dipole interactions occur between molecules that have a fixed, unequal distribution of charge (polar molecules). Consider, for example, HCl in Figure 12.6.

There is a high electronegativity difference ($\Delta\chi_P$ = 0.96) between chlorine (χ_P = 3.16) and hydrogen (χ_P = 2.20). For covalent molecules containing bonded atoms with electronegativity differences ($\Delta\chi_P$) of 0.4 to 1.7, there will be a polar bond between the atoms; greater than 1.7 will lead to electron transfer and the formation of an ionic bond. We can see the result of this polar bond in HCl in Figure 12.6, where the H is left electron-poor (blue), and the Cl is rendered electron-rich (red). In other words, the HCl molecule has an electric dipole moment (*p*). The value of the dipole moment in HCl is 3.60×10^{-30} Cm or 1.08 D, where D is the non-SI unit debye, which is more convenient for reporting dipole moments. The dipole moment can also conveniently be represented in SI units as a dipole length (*l*$_p$). The dipole length of HCl is 22.5 pm. The dipole moment and dipole length for the hydrogen halides are shown in Table 12.4.

The HCl molecules will then interact through attraction of the opposite ends of the dipole (electron-poor to the electron-rich) in what is called a dipole–dipole interaction (Figure 12.7).

Most molecules with polar bonds will have a fixed, unequal distribution of charge (an electric dipole) and will engage in dipole–dipole interactions. There are a small number of molecules that have polar bonds but are not polar. These molecules are highly symmetric and thus there is no electron-poor end and no electron-rich end, so they cannot engage in dipole–dipole interactions.

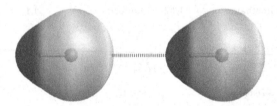

Figure 12.7 Dipole–dipole interaction (hash marks) between HCl molecules.

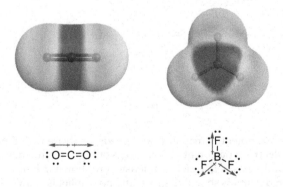

Figure 12.8 Electrostatic potential maps (top) and Lewis structures with dipole moment arrows (bottom) with polar bonds, but which are nonpolar due to their high symmetry.

These molecules can still engage in dispersion interactions. A couple of examples are shown in Figure 12.8.

PROBLEM 12.6
Methane (CH_4) and carbon dioxide (CO_2) are similar in terms of b values. Methane boils at 111.6 K, while carbon dioxide sublimates at 194.7 K. Draw the structures, consider the distribution of charge, and propose an explanation for this difference in behavior. Note: This involves a type of intermolecular force that was not explicitly covered but an extension of the noncovalent interactions covered.

PROBLEM 12.7
Considering the intermolecular forces involved, can you provide an explanation for the difference in the boiling point for the two compounds below?

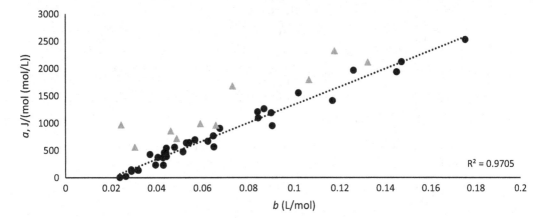

C_3H_9N
$b = 0.1$ L/mol
boiling point = 7 °C

C_3H_7NO
$b = 0.1$ L/mol
boiling point = 153 °C

HYDROGEN BONDS (H-BONDS)

While dispersion interactions are nearly universal, they are not the only important intermolecular interaction. We will now turn our attention to an intermolecular interaction that is much more exclusive: hydrogen bonds (H-bonds).[13] First, let's see the significance of this interaction on the graph of b and a values, in Figure 12.9, where molecules with hydrogen-bonding ability are shown as triangles and appear significantly above the trend for typical chemical species (circles).

Looking at Figure 12.9, we can see that all the molecules that can hydrogen bond have significantly higher a values (significantly stronger intermolecular forces) than we would expect from their size alone (b values). This substantial increase in a values for these molecules is due to their ability to hydrogen bond. Let's look at the chemical formulae of these molecules in Table 12.5.

Hydrogen bonding is a concept that has attracted significant debate since it was first proposed. This is in part because a hydrogen bond is more complicated than a purely electrostatic, noncovalent interaction and involves a degree of covalency (bonding). We will explore hydrogen bonding using the definition proposed by Weinhold et al.,[14] where a hydrogen bond is defined as "a fractional chemical bond due to partial intermolecular A–H⋯:B ↔ A:⋯H–B⁺ delocalization (partial 3-center/4-electron proton-sharing between Lewis bases), arising most commonly from quantum mechanical $n_B \rightarrow \sigma^*_{AH}$ donor-acceptor interaction." Let's unpack this terminology a bit in terms of

Figure 12.9 Correlation of b and a values for an average sample of molecules (circles) and with some molecules that engage in hydrogen bonding (triangles).

Table 12.5 Hydrogen-Bonding Chemical Species' b and a Values

Chemical	$b \left(\dfrac{L}{mol} \right)$	$a \left(\dfrac{J}{mol \left(\dfrac{mol}{L} \right)} \right)$
HF	0.024	969.2
H_2O	0.030	561.0
N_2H_4	0.046	857.2
CH_3NH_2	0.049	720.0
CH_3OH	0.066	959.8
CH_3CO_2H	0.106	1794.5
C_6H_5OH	0.118	2323.4
C_4H_9OH	0.132	2117.7

$$:\!F\!-\!H\cdots:\!\overset{\ominus}{F}\!: \quad\longleftrightarrow\quad \overset{\ominus}{:}\!F:\cdots H\!-\!F\!:$$

Figure 12.10 Contributing structures for the hydrogen bonding between hydrogen fluoride and fluoride.

the strongest hydrogen bond, i.e., −188 kJ/mol of stabilization energy, between hydrogen fluoride and fluoride (Figure 12.10).

If we consider the contributing structures (Figure 12.10), we can see that the hydrogen bond in the hydrogen fluoride–fluoride pair (F–H···F⁻) consists of a partial bond from each fluorine to the central hydrogen atom. The bond order between each fluoride and the central hydron (H⁺) is 0.5. This interaction consists of three atomic nuclei (the three centers are F, H, and F); and four electrons, two from the HF bond and two from the fluorine lone pair (Chapter 9). The fluoride is donating a pair of electrons and is defined as a Lewis base (this will be considered further in Chapter 20). In terms of orbital interactions, this is formally defined as the donation of electrons from the fluoride nonbonding lone pair (n_F) into the HF antibonding orbital (ω^*_{FH}).

Hydrogen bonds are possible for any molecular entity that contains an A–H bond, but the most significant hydrogen bonding – energy values of 10–20 kJ/mol – occurs when hydrogen atoms are attached to N, O, or F interacting with an N, O, or F lone pair. These are the types of moieties that

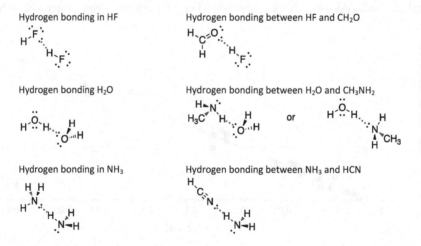

Figure 12.11 Examples of hydrogen bonds between two of the same molecular entities (left) and two different molecular entities (right).

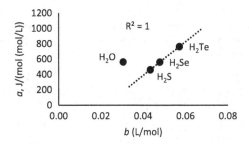

Figure 12.12 Correlation of b and a values for binary compounds of group-16 elements and hydrogen.

are most often considered in hydrogen bonding. An example of the types of intermolecular hydrogen bonds involving N, O, and F is shown in Figure 12.11.

The effect of hydrogen bonding in a gas can be dramatic by creating much larger a values than one would expect based on size (b value). Figure 12.12 shows the a and b values of the group-16 elements with hydrogen (H_2O, H_2S, H_2Se, and H_2Te). Water (H_2O) is the only member of this series that forms hydrogen bonds of any significant strength, and as a result, water's a value is much larger than it would be without the hydrogen bond. The impact of hydrogen bonding on a values or molecular properties is largest for small molecular entities. This is because the hydrogen bond is created by a portion of the structure rather than the whole molecule like dispersion forces. If the molecular entity is larger, then the impact of a single hydrogen bond from one small region of the molecule will not be as impactful.

The increase in interaction strength, due to the presence of a hydrogen bond in addition to dispersion, results in significantly higher boiling points. While hydrogen bonds are important in the gas phase, they are even more important in condensed phases such as solids or liquids where the molecular entities are in proximity. The result is a dramatic increase in the boiling point for water relative to other group-16 elements with hydrogen (H_2S, H_2Se, and H_2Te) (Figure 12.13).

PROBLEM 12.8
Considering the intermolecular forces involved, can you provide an explanation for the difference in the boiling point for the two isomers below?

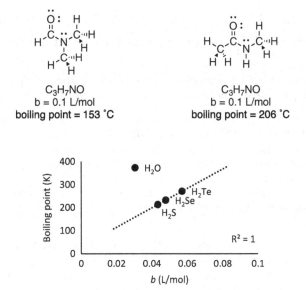

Figure 12.13 Correlation of b values and boiling points for binary compounds of group-16 elements and hydrogen.

PROBLEM 12.9

Consider the following alcohols, their structures, and their miscibility in water (miscible means that they will always form a homogeneous mixture, regardless of amounts). Provide an explanation for the trend you see.

miscible   miscible   miscible

73 g/L   22 g/L

PROBLEM 12.10

Surfactants are large molecules that have pieces with dissimilar intermolecular forces. Most commonly, surfactants have a hydrophilic (water loving) and hydrophobic (water fearing) component. Consider tetraethylene glycol monodecyl ether.

Tetraethylene glycol monodecyl ether

This is made from tetraethylene glycol and decane.

tetraethylene glycol   decane

Both tetraethylene glycol and decane have very similar b values (0.3 L/mol). The two a values are 51.6 $\frac{atm\,L^2}{mol^2}$ and 56.9 $\frac{atm\,L^2}{mol^2}$. Assign the a value to the correct structure and explain your answer.

Reconsider the structure of tetraethylene glycol monodecyl ether. Using your answer in part a, identify which part of the molecule is hydrophobic and which part is hydrophilic. Explain your answer.

THE IMPORTANCE OF INTERMOLECULAR FORCES

Dispersion, dipole–dipole interactions, and hydrogen bonds are the three most common types of intermolecular forces. Intermolecular forces are central to both chemistry and biology. Because they govern the way molecules interact, intermolecular forces dictate the phase behavior of chemicals and their miscibility and solubility. Intermolecular forces also drive, among other things, the assembly of cell membranes, the secondary and tertiary structure of proteins, and the binding of drugs to protein targets.

NUCLEAR, ELECTROSTATIC, BONDING, AND INTERMOLECULAR INTERACTIONS

Intermolecular forces are the last type of stabilizing interaction between particles that will be investigated in this book. We have discussed nuclear binding energy, bonding, hydrogen bonding, and van der Waals interactions. While these different interactions differ in their binding energy, how much energy it takes to pull two interacting particles apart from one another, they are all stabilizing, attractive interactions between particles (Figure 12.14). That is, the particles are all lower in energy when engaged in these interactions, and it requires added energy to disrupt these forces.

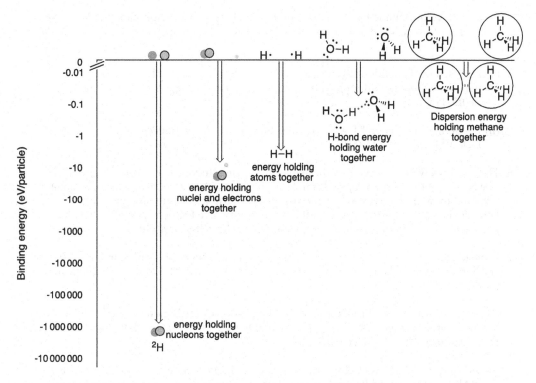

Figure 12.14 Binding energy comparison among nuclear binding, proton–electron binding, covalent bonding, hydrogen bonding, and dispersion interactions (as an example van der Waals interaction).

When nuclear binding energy is changed, we call that a nuclear reaction (Chapter 5). When bonds are changed, we call that a chemical reaction (Chapter 19 and Chapter 20). And when intermolecular forces are changed, we call that a physical change (Chapter 13).

ACKNOWLEDGMENTS

Dr. Brian Esselman (University of Wisconsin–Madison) collaborated on this chapter providing immeasurable help editing and running all computational chemistry calculations.

NOTES

1. Bernoulli, D. *Hydrodynamica, sive De Viribus et Motibus Fluidorum Commentarii.* Joh. Henr. Decker für Johannes Reinhold Dulsecker: Straßburg, **1873**.

2. van der Waals, J.D. Thesis: Over de Continuïteit van den Gas- en Vloeistoftoestand. University of Leiden, **1873**.

3. (i) Jones, J.E. On the Determination of Molecular Fields.–I. From the Variation of the Viscosity of a Gas with Temperature. *Proc. R. Soc. Lond. A*, **1924**, *106* (738), 441–462. DOI: 10.1098/rspa.1924.0081.

   (ii) Jones, J.E. On the Determination of Molecular Fields.–II. From the Equation of State of a Gas. *Proc. R. Soc. Lond. A*, **1924**, *106* (738), 463–477. DOI: 10.1098/rspa.1924.0082.

   (iii) Lennard-Jones, J.E. Cohesion. *Proc. Phys. Soc.*, **1931**, *43* (5), 461–482. DOI: 10.1088/0959-5309/43/5/301.

4. Note that b values (L/mol) are being used as an analog for surface area, the parameter upon which polarizability volume more directly correlate. For example, butane (a cylindrical molecular) and 2-methylpropane (a spherical molecule) are isomers of C_4H_{10} and both have

the same *b* value. Butane has a boiling point of 272 K and 2-methylpropane has a boiling point of 261 K, which is due to the smaller surface area of a sphere compared to a cylinder when both have the same volume.

5. (i)   Eisenschitz, R.; London F. Über das Verhältnis der van der Waalschen Kräfte zu den homöopolaren Bindungskräften. *Z. Physik*, **1930**, *60* (7-8), 491–527. DOI: 10.1007/BF01341258.

    (ii) London, F. The General Theory of Molecular Forces. *Trans. Faraday Soc.*, **1937**, *33*, 8b–26. DOI: 10.1039/TF937330008B.

6. The ∂ symbol was first used to indicate partial charges: Ingold, C.K.; Ingold, E.H. CLXIX.–The Nature of the Alternating Effect in Carbon Chains. Part V. A Discussion of Aromatic Substitution with Special Reference to the Respective Roles of Polar and non-Polar Dissociation; and a Further Study of the Relative Directive Efficiencies of Oxygen and Nitrogen. *J. Chem. Soc.*, **1926**, *129*, 1310–1328. DOI: 10.1039/JR9262901310.

7. Israelachvili, J.N. 5 – Interactions Involving the Polarization of Molecules. In *Intermolecular and Surface Forces*, 3rd ed. Academic Press: Burlington, MA, **2011**, 91–106. DOI: 10.1016/B978-0-12-375182-9.10005-3.

8. Gussoni, M.; Rui, M.; Zerbi, G. Electron and Relaxation Contribution to Linear Molecular Polarizability. An Analysis of the Experimental Values. *J. Mol. Struct.*, **1998**, *447* (3), 163–215. DOI: 10.1016/S0022-2860(97)00292-5.

9. (i)   Tsuzuki, S.; Honda, K.; Uchimaru, T.; Mikami, M. Magnitude of Interaction Between *n*-Alkane Chains and Its Anisotropy: High-Level ab Initio Calculations of *n*-Butane, *n*-Pentane, and *n*-Hexane Dimers. *J. Phys. Chem. A*, **2004**, *108* (46), 10311–10316. DOI: 10.1021/jp048403z.

    (ii) Jalkanen, J.-P.; Pakkanen, T.A.; Yang, Y.; Rowley, R.L. Interaction Energy Surfaces of Small Hydrocarbon Molecules. *J. Chem. Phys.*, **2003**, *118* (12), 5474–5483. DOI: 10.1063/1.1540106.

10. Gussoni, M.; Rui, M.; Zerbi, G. Electron and Relaxation Contribution to Linear Molecular Polarizability. An Analysis of the Experimental Values. *J. Mol. Struct.*, **1998**, *447* (3), 163–215. DOI: 10.1016/S0022-2860(97)00292-5.

11. William E. Acree, Jr., James S. Chickos, Phase Transition Enthalpy Measurements of Organic and Organometallic Compounds. In *NIST Chemistry WebBook, NIST Standard Reference Database Number 69*, P.J. Linstrom and W.G. Mallard, Eds., National Institute of Standards and Technology, Gaithersburg MD. DOI: 10.18434/T4D303.

12. Keesom, W.M. The Second Virial Coefficient for Rigid Spherical Molecules Whose Mutual Attraction Is Equivalent to That of a Quadruplet Placed at Its Center. *Proc. Sect. Sci. K. Ned. Akad. Wet. Amst.*, **1915**, *18*, 636–646.

13. (i)   Lewis, G.N. *Valence and the Structure of Atoms and Molecules.* Chemical Catalog Co.: New York, 1923.

    (ii) Pauling, L. The Shared-Electron Chemical Bond. *Proc. Natl. Acad. Sci. USA.*, **1928**, *14* (4), 359–362. DOI: 10.1073/pnas.14.4.359.

14. Weinhold, F.; Klein, R.A. What Is a Hydrogen Bond? Resonance Covalency in the Supramolecular Domain. *Chem Educ. Res. Pract.* **2014**, *51* (3), 276–285. DOI: 10.1039/C4RP00030G.

13 States, Phases, and Physical Changes

This book has, so far, investigated the structure of atoms and molecules. This chapter represents a bridge as we will transition from covering the states of molecular entities (solids, liquids, and gases) to changes in state (physical changes). We covered gases in detail (Chapter 11), and now solids and liquids will receive greater attention.

STATE VERSUS PHASE

Before diving into this topic, we first need to establish a distinction in two common terms: state and phase. The three common states of matter (or states of aggregation) are solids, liquids, and gases. The states of matter differ depending on the density of particle packing, the arrangement and degree of order, and the type(s) of particle motion.[1] In contrast to a state, a phase is a system that is uniform in physical state and in chemical composition.[2] Let us consider vinaigrette as an example (Figure 13.1). A mixture of oil and water, like a vinaigrette salad dressing, is in the liquid state. Within the liquid vinaigrette, there are two phases, which differ in chemical composition: an oil phase (olive oil) and an aqueous phase (water and acetic acid).

> ### PROBLEM 13.1
> For each of the following, identify the state(s) of matter (solid, liquid, or gas) and whether the material is a singular phase or more than one phase.
>
> a. A pencil
>
> b. Partially melted gallium metal

SOLID, LIQUID, AND GAS PROPERTIES

Now that we have firmly established our terminology, let's reconsider solids, liquids, and gases in terms of the density of particle packing, the arrangement and degree of order, and the type(s) of particle motion. Both solids and liquids are condensed phases of matter, which means they cannot be easily compressed and there is very little space between particles (Figure 13.2). In contrast, gases are dispersed phases with significant distance between the particles (Figure 13.2), which makes gases compressible. Unlike liquids or gases, solids are distinguished by having long-range order in the arrangement of particles. The persistence of long-range order is because solid particles vibrate (move around a fixed point) but do not easily rotate (spin around their center of mass) nor do they translate (move through three-dimensional space). In contrast, liquids and gases are fluid phases that take on the shapes of their containers as the particles not only vibrate but also rotate and translate. The greater extent of motion for liquids and gases is observable in terms of the lack of long-range order (lack of regular, repeating arrangement) among the particles (Figure 13.2). The phase that a given chemical has at normal temperature and pressure (25.0 °C, 101 kPa) is dictated by the strength of the intermolecular forces present in the compound.

> ### PROBLEM 13.2
> Consider Figure 13.2. Rank the mean kinetic energy $(\overline{E_k})$ of a solid (ice) versus a liquid (water) versus a gas (steam) from highest to lowest. Explain your answer.

Solids, liquids, and gases are the most common phases that one encounters. There are several other phases that matter can exist in, including Bose–Einstein condensate, superfluid, supercritical fluid, and plasma. We will not consider all these phases here, but it is worth highlighting two from this list. The first is plasma, which is a gas that is so high energy that electrons are no longer bound to their nuclei. Plasma is dispersed phases with free ions and free electrons. The other phase to note, which we will consider further below, is a supercritical fluid. A supercritical fluid has the density of a fluid but the energy and cavity-filling properties of a gas. Supercritical fluids have gained increasing attention for their applications as solvents in chemical reactions, decaffeinating coffee, and for dry cleaning.

PHASE CHANGES

The state of matter for a substance can be changed (Figure 13.3) by altering the pressure and/or by altering the temperature. The amount of thermal energy, or the change in pressure, needed to affect the phase of a substance depends on the strength of the interactions (bonds, Chapter 9, or

DOI: 10.1201/9781003479338-14

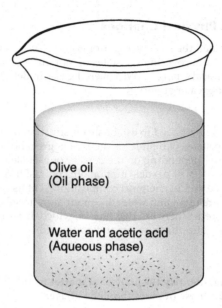

Figure 13.1   Diagram of an olive oil vinaigrette with olive oil (top) and water and acetic acid (bottom). A vinaigrette is in the liquid state and contains an oil phase (top) and an aqueous phase (bottom).

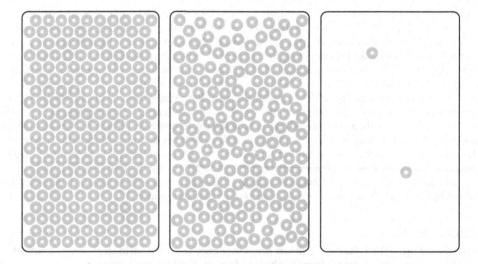

Figure 13.2   Density of argon atoms in a sample of solid argon (left), liquid argon (middle), and gaseous argon (right). The atoms occupy 64%, 57%, and 0.064% of the space, respectively. (Data from van Witzenburg, W. and Stryland, J.C. Density Measurements of Compressed Solid and Liquid Argon. Can. J. Phys., **1968**, 46 (7), 811–816. DOI: 10.1139/p68-102.)

intermolecular forces, Chapter 12) between the particles. The pressure–temperature impact on the phase behavior of a substance is represented by a pressure–temperature phase diagram (Figure 13.3).[3] Moving from solid to liquid to gas, at constant pressure, requires increasing the mean kinetic energy of the particles by increasing the temperature. We are considering first-order phase transitions,[4] that is, a phase transition where a plot of heat, volume, or internal energy versus temperature shows a clear discontinuity (Figure 13.4). This discontinuity corresponds to the phase transition. For a plot of heat energy versus temperature, the magnitude of the discontinuity is proportional to the strength of the intermolecular forces between the particles. For the example of carbon dioxide, the jump in heat corresponds to the energy required to disrupt the van der Waals interactions between particles.

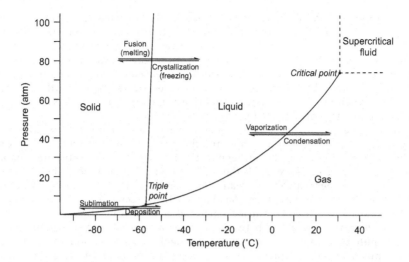

Figure 13.3 Pressure–temperature diagram for carbon dioxide showing the phase change terminology. The triple point is at –56.57 °C and 5.11 atm. The critical point is at 30.98 °C and 72.79 atm.

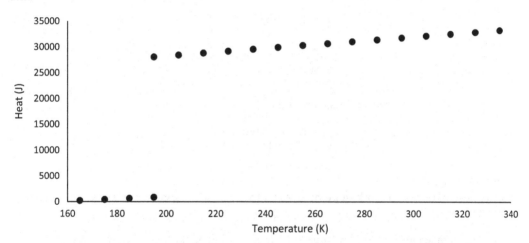

Figure 13.4 Thermal energy transfer (heat) absorbed by one mole of carbon dioxide at 1 atm as it is warmed from 165 K to 335 K. The normal (1 atm) sublimation temperature is 195 K and the enthalpy of sublimation ($\Delta_{sub}H°(CO_2, 1 \text{ atm})$) is 27.2 kJ/mol. (Data from Bryson III, C.E.; Cazcarra, V; Levenson, L.L. Sublimation Rates and Vapor Pressures of Water, Carbon Dioxide, Nitrous Oxide, and Xenon. J. Chem. Eng. Data, **1974**, 19 (2), 107–110. DOI: 10.1021/je60061a021.)

PROBLEM 13.3
If we compare carbon dioxide (sublimates at –78 °C at 1 atm) and carbon disulfide (boils at 46 °C at 1 atm), we can see that carbon disulfide must be given more thermal energy to vaporize.

a. What can we say about the strength of the intermolecular forces between carbon dioxide molecules versus the strength of the intermolecular forces between carbon disulfide?

b. Can you provide an explanation for the differences in intermolecular forces?

PROBLEM 13.4
The enthalpy of fusion ($\Delta_{fus}H°$) for water is 6.01 kJ/mol and the enthalpy of vaporization ($\Delta_{vap}H°$) for water is 40.65 kJ/mol. Provide an explanation of why more heat is required to vaporize water than is required to melt water. Consider the particles (Figure 13.2) and how they change upon vaporization versus melting.

Aside from changing temperature, altering the amount of thermal energy, the state of a substance can also be changed by changing the pressure. Changing temperature is something that most people have lived experience with, but most people typically do not have lived experience with altering the pressure of a substance. Let's consider what's at play in changing pressure. To reduce the pressure, a vacuum is used. Vacuums mechanically reduce the amount of gas present in a container, which will in turn reduce the pressure (force per area) on a sample. A sufficient decrease in pressure can cause a substance to melt and to vaporize as the molecular entities are no longer held together by the force of gas particles hitting the surface.

While decreasing the pressure can only be done with a vacuum, there are two different ways to increase pressure, depending on the sample and the pressures desired. One way to increase pressure is by adding an inert gas, like nitrogen or argon, to a container. As the amount of gas in the container increases, the pressure inside the container will also increase. It should be noted that pressurizing a container with gas can lead to explosive results, if not careful. The other way to increase pressure is through the application of a force using some sort of mechanical press. For example, a diamond anvil cell can pressurize a material up to 7.7×10^6 atm.

In terms of the intermolecular forces between particles, a material with strong intermolecular forces will require very little added pressure to condense from a gas, and it will, conversely, require the attainment of very low pressures to vaporize. A substance with weak intermolecular forces will require high pressures to condense and only a small reduction in pressure to vaporize. The relationship between pressure and the boiling point – the temperature at which a liquid will convert to gas – is expressed by the Clausius–Clapeyron equation:[5]

$$\ln\left(\frac{p_2}{p_1}\right) = \frac{-\Delta_{vap}H^{\circ}}{R}\left(\frac{1}{T_2} - \frac{1}{T_1}\right) \tag{13.1}$$

p_2 is the pressure for condition two (Pa or atm).
p_1 is the pressure for condition one (Pa or atm).
$\Delta_{vap}H^{\circ}$ is the enthalpy of vaporization (J/mol).
R is the gas constant ($\frac{8.314\,J}{mol\,K}$).

T_2 is the temperature for condition two (K).
T_1 is the temperature for condition one (K).

PROBLEM 13.5
The normal (1.00 atm) boiling point of water ($\Delta_{vap}H^{\circ} = 43.9$ kJ/mol) is 100.0 °C.

a. What is the boiling point (°C) of water on the top of Mount Everest (0.31 atm)?

b. Will it take more time or less time to cook something on the top of Mount Everest? Explain.

SUPERCRITICAL FLUIDS AND THE VAN DER WAALS EQUATION

For chemicals in a liquid state, there are always two phases present. Most of the molecular entities are in the liquid phase and a minority of the particles are also in the gas (vapor) phase. The amount of vapor above a liquid is quantified as the vapor pressure. Altogether, then, a liquid is really a two-phase, fluid system (liquid and vapor). The vapor phase will be predominant at lower pressures, or higher temperatures, and the liquid phase will be predominant at higher pressures, or lower temperatures (Figure 13.3). Once a substance exceeds its critical point, the critical temperature (T_c) and the critical pressure (p_c), it becomes a supercritical fluid. A supercritical fluid is a single phase that cannot be vaporized with higher temperatures nor liquefied with higher pressures. A supercritical fluid (SCF) has a density close to that of a liquid, but it fills a container like a gas. While a supercritical fluid has interesting applications, as noted above, the critical point temperature and pressure also allow for the determination of a (Equation 13.2) and b values (Equation 13.3) of the van der Waals equation.[6] Recall from Chapter 12 that the a value is a measure of the strength of a particle's intermolecular forces (a greater a value corresponds to stronger intermolecular forces); the b value is a measure of a particle's size (a greater b value corresponds to a larger particle).

$$a = \frac{27}{64}\frac{\left(RT_c\right)^2}{p_c} \tag{13.2}$$

a is the van der Waals pressure correction term $(\dfrac{\text{atm}\,L^2}{\text{mol}^2})$.

R is the gas constant $(0.082\,06\ \dfrac{L\,\text{atm}}{\text{mol}\,K})$.

T_c is the critical temperature (K).

p_c is the critical pressure (atm).

$$b = \frac{RT_c}{8p_c} \tag{13.3}$$

b is the volume correction term $(\dfrac{L}{\text{mol}})$.

R is the gas constant $(0.082\,06\ \dfrac{L\,\text{atm}}{\text{mol}\,K})$.

T_c is the critical temperature (K).

p_c is the critical pressure (atm).

PROBLEM 13.6

The critical point of methane is −82.3 °C and 45.79 atm. The critical point of ammonia is 132.4 °C and 111.3 atm.

a. Calculate the a and b values for methane and for ammonia.

b. Which substance (methane or ammonia) has stronger intermolecular forces between particles? Based on their structures, provide an explanation for the difference.

PHASE AND PHYSICAL CHANGE

The phase of a substance is the result of the balance of intermolecular forces, temperature, and pressure. A physical change is the change in phase of a substance, not the chemical formula, by altering its temperature and/or pressure. Physical changes can be understood through quantifying the macroscopic variables of temperature and pressure. In the following chapters, we will move on to studying chemical reactions, which can involve a change in temperature and/or pressure but must involve a change in the chemical formula of the substance.

NOTES

1. Barker, J.A.; Henderson, D. What Is "Liquid"? Understanding the States of Matter. *Rev. Mod. Phys.*, **1976**, *48* (4), 587–671. DOI: 10.1103/RevModPhys.48.587.

2. "Phase." IUPAC. *Compendium of Chemical Terminology*, 2nd ed. Compiled by A.D. McNaught and A. Wilkinson. Blackwell Scientific Publications, Oxford, 1997. Online version (2019–) created by S.J. Chalk. DOI: 10.1351/goldbook.P04528.

3. Pressure–temperature diagrams were first introduced by Gibbs: Gibbs, J.W. A Method of Geometrical Representation of the Thermodynamic Properties of Substances by Means of Surfaces. *Trans. Conn. Acad. Arts Sci.* **1873**, 2, 382–404.

4. Sauer, T. A Look Back at the Ehrenfest Classification. *Eur. Phys. J. Special Topics*, **2017**, *226*, 539–549. DOI: 10.1140/epjst/e2016-60344-y.

5. (i) Clapeyron, E. Mémoire sur la Puissance Motrice de la Chaleur. *Journal de l'École Royale Polytechnique*, **1834**, *14*, 153–190.

    (ii) Clausius, R. Ueber die bewegende Kraft der Wärme und die Gesetze, welches ich daraus für die Wärmelehre selbst ableiten lassen. *Ann. Phys.*, **1850**, *155* (4), 10.1002/andp.1850.1550403.

6. Berberan-Santos, M.N.; Bodunov, E.N.; Pogliani, L. The Van der Waals Equation: Analytical and Approximate Solutions. *J. Math. Chem.*, **2008**, *43* (4), 1437–1457. DOI: 10.1007/s10910-007-9272-4.

14 Chemical Kinetics

Chemical kinetics is the study of the speed of chemical reactions. The term "kinetic" may be familiar from physics, where it means motion, and the speed of an object in motion is reported in m/s. In a chemical reaction, chemists are still interested in the speed at which one or more chemicals react to produce a new chemical, which is reported as mol/(L s) (or M/s). Chemical kinetics can be divided into two aspects. The focus of this chapter is macroscopic kinetics,[1] which considers how the speed of a chemical reaction can depend on changes in the amount concentration of reactant(s) and/or product(s), changes in surface area, and temperature. While this macroscopic understanding can give a chemist knowledge that allows them to control the reaction's progress, the power of kinetics comes from how this macroscopic understanding feeds into the other aspect of kinetics: microscopic kinetics.[2] Microscopic kinetics considers how the nanoscale behavior and elementary steps of atoms and molecules produce the observable macroscopic phenomena that can be measured in a lab. The knowledge chemists have gained from microscopic kinetics is the basis for a significant amount of modern organic chemistry, inorganic chemistry, and biochemistry because chemists have learned how and why bonds are broken and made, and how both enzymes and chemists can manipulate these processes.

MACROSCOPIC KINETICS: Rate of Reaction

Macroscopic kinetics focuses on describing, analyzing, quantifying, and defining how reactions progress through time. To get a sense of how we describe and quantify reactions, first consider a hypothetical reaction where a reactant (R) is becoming a product (P) with a 1:2 stoichiometric ratio:

$$R \rightarrow 2\,P$$

The first step in a macroscopic kinetics analysis is to quantify the reaction progress by determining the amount concentration (and change in amount concentration) of each chemical over time (Figure 14.1). Note that the end state of the reaction, where the ratio of the amount concentration of products to the amount concentration of reactants is constant, is called equilibrium. Equilibrium is something that will be investigated further in future chapters. For now, it is useful to know that at equilibrium the reaction has achieved a dynamic state where the relative amount concentration values do not change, but the reactant is still becoming a product and the product is becoming a reactant.

Although a full analysis of the entire reaction time course will provide a richer understanding of the reaction, it is common to study only the initial changes in amount concentration. During the initial phase of the reaction, the changes are both at their fastest and can be approximated as linear (Figure 14.2).

The linear trend lines in Figure 14.2 have slopes with values of mol/(L s). That is to say, the top trend line would represent the rate of disappearance (negative value) of reactant R and the bottom trend line would represent the rate of appearance (positive value) of product P.

PROBLEM 14.1
Determine the initial rates (mol/(L s)) of change in amount concentration for each of the following datasets.

a. Reaction: $Br_2(g) + CH_4(g) \xrightarrow{light} HBr(g) + CH_3Br(g)$. See also Table 14.1.

b. Reaction: $2\,NaN_3(s) \xrightarrow{330°C} 2\,Na(s) + 3\,N_2(g)$. See also Table 14.2.

c. Reaction: $(CH_3)_3CBr(sln) + NaOH(sln) \xrightarrow{water,ethanol} (CH_3)_3COH(sln) + NaBr(sln)$.[3] See also Figure 14.3.

The trend lines shown in Figure 14.2 represent the rate of change for individual components of the reaction, but not necessarily the rate of reaction itself. The rate of reaction is how fast (mol/(L s)) the overall reaction happens, which is always a positive quantity. The relationship between the rate of reaction and the individual component rates for this reaction is shown in Equation 14.1:

DOI: 10.1201/9781003479338-13

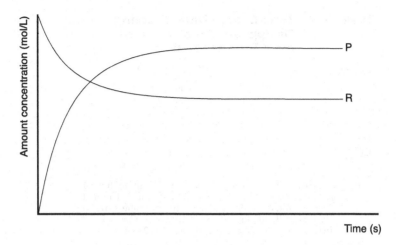

Figure 14.1 Reaction time course plot for the reaction of R → 2 P.

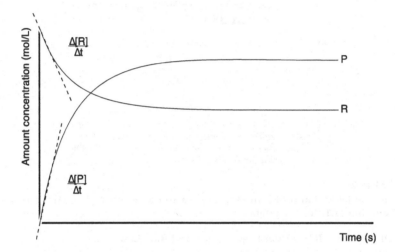

Figure 14.2 Initial rate analysis (linear fits shown as dashed lines) looking at the change in reactant amount concentration ($\Delta[R]$) over time and the change in product amount concentration ($\Delta[P]$) over the change in time.

$$\text{rate of reaction} = -\frac{1}{1}\left(\frac{\Delta[R]}{\Delta t}\right) = +\frac{1}{2}\left(\frac{\Delta[P]}{\Delta t}\right) \tag{14.1}$$

The rate of reaction is the negative of the rate of disappearance of the reactant R, and the rate of reaction is one half the rate of appearance of the product. The negative sign for the disappearance of reactant R corrects for the fact that the rate of reaction is always positive (and the rate of disappearance of reactants is always negative). The ratios $\frac{1}{1}$ and $\frac{1}{2}$ consider the stoichiometric relationship between the reaction (the numerator is always one because there is only one reaction) and the individual components: there is one R for every one reaction, hence the $\frac{1}{1}$ in front of the rate of disappearance of R; there are two P for every one reaction, hence the $\frac{1}{2}$ in front of the rate of appearance of P.

Table 14.1 Time Course Data Showing Disappearance of Bromine over Time at 25 °C

Time (s)	$[Br_2(g)]$ (mol/L)
0	1.49×10^{-3}
60	1.44×10^{-3}
180	1.35×10^{-3}
300	1.26×10^{-3}
420	1.17×10^{-3}
600	1.04×10^{-3}

Source: Kistiakowsky, G.B.; Van Artsdalen, E.R. Bromination of Hydrocarbons. I. Photochemical and Thermal Bromination of Methane and Methyl Bromine. Carbon Hydrogen Bond Strength in Methane. *J. Chem. Phys.* **1944,** 12 (12), 469–478. DOI: 10.1063/1.1723896.

Table 14.2 Time Course Data Showing Appearance of Nitrogen over Time at 330 °C

Time (s)	$[N_2(g)]$ (mol/L)
0	0
1	6.60×10^{-4}
2	1.131×10^{-3}
3	1.467×10^{-3}

Source: Walker, R.F. Thermal Decomposition of Sodium Azide: Crystal Size Effects, Topochemistry and Gas Analyses. *J. Phys. Chem. Solids.* **1968,** 29 (6), 985-1000. DOI: 10.1016/0022-3697(68)90235-7.

PROBLEM 14.2
Reconsider Tables 14.1 and 14.2 and Figure 14.3 and determine the initial rate of reaction (using Equation 14.1) for each dataset.

MACROSCOPIC KINETICS: Order Dependence and Rate Law

Knowledge of the initial rate of reaction is an important first piece in a kinetics analysis of a reaction. The next step is to determine how the rate of reaction is impacted by changing the amount concentration (or for heterogeneous reactions the surface area) of each reactant that is present.[4] Note that the amount concentration of product can also influence the reaction and in practice it is often similarly analyzed, but such reactions are beyond the scope of this chapter.

Consider the reaction of sodium iodide with butyl chloride, C_4H_9Cl, to produce butyl iodide, C_4H_9I, and sodium chloride (the phase notation (sln) is to indicate chemicals dissolved in solution; here acetone is the solvent).

$$NaI(sln) + C_4H_9Cl(sln) \xrightarrow[\text{acetone } 40° C]{} C_4H_9I(sln) + NaCl(s)$$

The initial rate of reaction is determined as before, but now there are several experiments where the initial amount concentration of each reactant is varied. The results of the analysis of the reaction of sodium iodide with butyl chloride are summarized in Table 14.3.[5]

Reviewing the data in Table 14.3, the initial amount concentration of sodium iodide and butyl chloride both influence the initial rate of reaction. The relationship between the initial amount concentration and initial rate of reaction – this is called order dependence – is quantified by relating the factor change in amount concentration and the factor change in rate. For example, consider experiments 1 and 2 in Table 14.3. The amount concentration of sodium iodide is constant (when considering order dependence, only one variable at a time can be considered) and the amount concentration of butyl chloride doubles ($\frac{0.166\,\text{mol/L}}{0.083\,\text{mol/L}} = 2$) and the rate of reaction also doubles

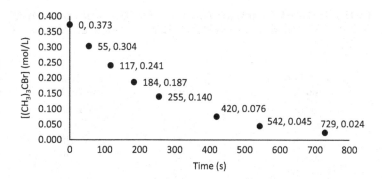

Figure 14.3 Time course data showing disappearance of $(CH_3)_3CBr$ over time. (Data from Bateman, L.C.; Cooper, K.A.; Hughes, E.D.; Ingold, C.K. 178. Mechanism of Substitution at a Saturated Carbon Atom. Part XIII. Mechanisms Operative in the Hydrolysis of Methyl, Ethyl, Isopropyl, and tert.-Butyl Bromides in Aqueous Solutions. J. Chem. Soc. (Resumed). **1940**, 925–935. DOI: 10.1039/JR9400000925.)

Table 14.3 Initial Rate of Reaction Analysis for the Reaction of Sodium Iodide with Butyl Chloride at 25 °C

Experiment	[NaI] (mol/L)	[C₄H₉Cl] (mol/L)	Rate of Reaction (mol/(L s))
1	0.083	0.083	1.1×10^{-7}
2	0.083	0.166	2.1×10^{-7}
3	0.083	0.332	4.3×10^{-7}
4	0.166	0.166	4.2×10^{-7}

$(\dfrac{2.1\times10^{-7}\,\text{mol}/(\text{L}\,\text{s})}{1.1\times10^{-7}\,\text{mol}/(\text{L}\,\text{s})} = 2)$. What this means is there is a first-order dependence of the rate of reaction

on the amount concentration of butyl chloride. In more mathematical terms, we would say there is a linear relationship between the rate and the amount concentration of butyl chloride. This is confirmed when considering experiments 2 and 3, where the amount concentration of butyl chloride doubles and again the rate doubles. Together, then, we can say that the (reaction) rate is directly proportional to the amount concentration of butyl chloride:

$$\text{rate} \propto \left[C_4H_9Cl\right]$$

Now if the relationship between the rate and the amount concentration of sodium iodide (Table 14.3) is considered, the amount concentration of sodium iodide doubles between experiments 2 and 4 $(\dfrac{0.166\,\text{mol}/\text{L}}{0.083\,\text{mol}/\text{L}} = 2)$ and the rate of reaction also doubles $(\dfrac{4.2\times10^{-7}\,\text{mol}/(\text{L}\,\text{s})}{2.1\times10^{-7}\,\text{mol}/(\text{L}\,\text{s})} = 2)$, which

means that the rate also has a first-order dependence (is directly proportional) to the amount concentration of sodium iodide.

$$\text{rate} \propto \left[C_4H_9Cl\right]\left[NaI\right]$$

Last, the goal is to find the rate law. The rate law is an equation relating the rate of reaction to the amount concentration of reactant, with the appropriate exponent. To go from a proportional relationship to an equation, a constant is added, the rate constant, k.

$$\text{rate} = k[C_4H_9Cl][NaI]$$

To find the value of the rate constant, the values from a particular experiment are added in for the rate and amount concentration values. If the values from experiment 1 (Table 14.3) are used, we can then solve for k to find that k has a value of 1.6×10^{-5} L/(mol s):

Table 14.4 **Initial Rate of Reaction Analysis for the Reaction of Quinolone ($C_{11}H_{13}NO_2$) and Maleimide ($C_4H_3NO_2$) with an Iridium Catalyst at 25 °C**

Experiment	[$C_{11}H_{13}NO_2$(sln)] (mol/L)	[$C_4H_3NO_2$(sln)] (mol/L)	[Iridium Catalyst] (mol/L)	Rate of Reaction (mol/L)
1	0.0396	0.103	7.00×10^{-4}	1.70×10^{-6}
2	0.0496	0.103	7.00×10^{-4}	2.03×10^{-6}
3	2.00×10^{-5}	0.052	2.80×10^{-4}	1.17×10^{-6}
4	2.00×10^{-5}	0.075	2.80×10^{-4}	1.17×10^{-6}
5	2.00×10^{-5}	0.103	4.60×10^{-4}	1.17×10^{-6}
6	2.00×10^{-5}	0.103	6.20×10^{-4}	1.23×10^{-6}

$$1.1 \times 10^{-7} \frac{mol}{L\,s} = k[0.083\ mol\,/\,L][0.083\ mol\,/\,L]$$

$$k = 1.6 \times 10^{-5} \frac{L}{mol\,s}$$

Altogether, the full rate law for the reaction of sodium iodide with butyl chloride is:

$$rate = 1.6 \times 10^{-5} \frac{L}{mol\,s}[C_4H_9Cl][NaI]$$

PROBLEM 14.3
For the following dataset (Table 14.4), determine the rate law for the reaction.

$$\text{Reaction: } C_{11}H_{13}NO_2(sln) + C_4H_3NO_2(sln) \xrightarrow{\text{Iridium catalyst}} C_{15}H_{16}N_2O_4(sln).[6]$$

MACROSCOPIC KINETICS: Reaction Order
Once the rate law for a reaction is known, we can determine the overall reaction order. Considering the rate law above, the reaction is first order in C_4H_9Cl, first order in NaI, and second-order (or bimolecular) overall. The overall order of a reaction is the sum of the exponents for each component of the reaction. We will consider zeroth-order, first-order, and second-order reactions in more detail.

Consider a hypothetical reaction of A converting into B:

$$A \rightarrow B$$

This reaction was investigated in a macroscopic kinetics analysis to interrogate the dependence of the rate of reaction on the amount concentration of A. The rate versus amount concentration data are summarized in Figure 14.4.

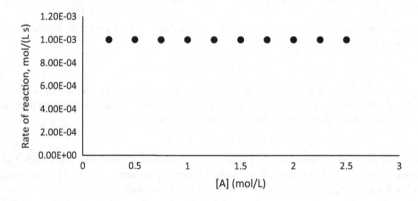

Figure 14.4 Rate of reaction compared to the amount concentration of A (mol/L).

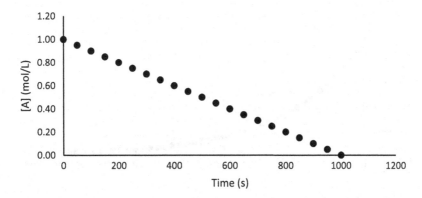

Figure 14.5 Zeroth-order reaction time course.

The rate of reaction is independent of the amount concentration of A and so the rate law for the reaction is zeroth order, where the rate is only dependent upon the rate constant:

$$\text{rate} = k[A]^0$$

$$\text{rate} = k$$

A zeroth-order reaction is one where the system will not proceed any faster or slower by changing the amount of reactant(s) present. This situation is not uncommon for heterogeneous reactions (where different phases are present) and the phase boundary can become saturated. Since the rate is independent of the amount concentration of A, the time course for a zeroth-order reaction is a linear line (Figure 14.5).

Now, consider a hypothetical reaction of C converting into D:

$$C \rightarrow D$$

This reaction was investigated in a macroscopic kinetics analysis to interrogate the dependence of the rate of reaction on the amount concentration of C. The rate versus amount concentration data are summarized in Figure 14.6.

The rate of reaction shows a linear dependence of the amount concentration of C and so the rate law for the reaction is first order (unimolecular), where the rate is linearly dependent upon the amount concentration of C:

$$\text{rate} = k[C]$$

A first-order reaction will increase or decrease in rate in direct response to the amount of C that is present. Since the rate is dependent upon the amount concentration of C, the time course for a first-order reaction shows an exponential decrease as the amount concentration of C decreases over time and slows the rate of reaction (Figure 14.7).

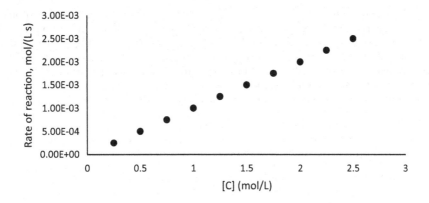

Figure 14.6 Rate of reaction compared to the amount concentration of C (mol/L).

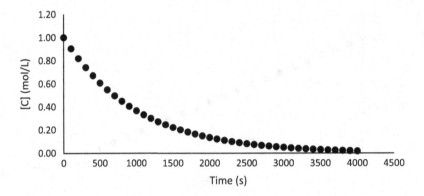

Figure 14.7 First-order reaction time course.

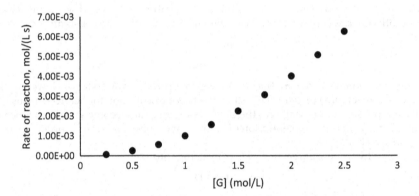

Figure 14.8 Rate of reaction compared to the amount concentration of G (mol/L).

Finally, consider a hypothetical reaction of G converting into H:

$$G \rightarrow H$$

This reaction was investigated in a macroscopic kinetics analysis to interrogate the dependence of the rate of reaction on the amount concentration of G. The rate versus amount concentration data are summarized in Figure 14.8.

The rate of reaction shows a square or quadratic dependence of the amount concentration of G and so the rate law for the reaction is second order (bimolecular), where the rate is dependent upon the amount concentration of G squared:

$$\text{rate} = k[G]^2$$

A second-order reaction will significantly increase or decrease in rate in a squared response to the amount of G that is present. Since the rate is dependent upon the square of the amount concentration of G, the time course for a second-order reaction shows an inverse square decrease as the amount concentration of G decreases over time and slows the rate of reaction (Figure 14.9).

More common are reactions that are first-order overall (unimolecular), second-order overall (bimolecular), or third-order overall (termolecular). Reactions can be more complicated – high order or even non-integer orders – but that falls outside the scope of this chapter.

MACROSCOPIC KINETICS: The Integrated Rate Law

Consider the solvolysis (breaking of a molecule with a solvent) of $(CH_3)_3CBr$, *tert*-butyl bromide, in aqueous solution to produce $(CH_3)_3COH$, *tert*-butyl alcohol:

$$(CH_3)_3\,CBr(sln) + NaOH(sln) \xrightarrow[\text{water, ethanol}]{} (CH_3)_3\,COH(sln) + NaBr(sln)$$

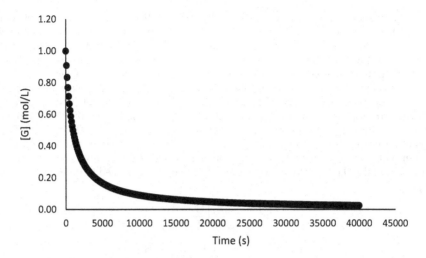

Figure 14.9 Second-order reaction time course.

After a kinetics analysis of this reaction, the rate law is found to be first-order overall and only depends on the amount concentration of *tert*-butyl bromide, $(CH_3)_3CBr$:

$$\text{rate} = k[(CH_3)_3CBr]$$

While this form of the rate law is useful for knowing how fast the reaction will occur, a chemist may want to know what the amount concentration of *tert*-butyl bromide, $(CH_3)_3CBr$, will be at a particular point. For this, the integrated rate law is used:

$$[(CH_3)_3CBr]_t = [(CH_3)_3CBr]_0 e^{-kt}$$
$$k \text{ is rate constant (1/s).}$$
$$t \text{ is time (s).}$$

There are integrated rate laws that can be found for other reaction orders, but the first-order integrated rate law is the most used type of integrated rate law.

PROBLEM 14.4
Radioactive decay is a first-order reaction. Fluorine-18 is a radioisotope used in positron-emission tomography (PET) scans. If the rate constant for ^{18}F to decay is 1.05×10^{-4} 1/s, how long will it take for a dose of fluorine-18 to decay by half?

$$[^{18}F]_t = [^{18}F]_0 e^{-kt}$$

MACROSCOPIC KINETICS: Rate Constant and Temperature Dependence
If a glow stick is cooled and then put in the freezer, it will dim. Conversely, when a glow stick is placed in a bath of hot water, it glows brighter. In this example, the speed of a reaction is being modulated by temperature. Affecting a reaction with temperature is intuitive because people have more day-to-day experience with the phenomenon. Food is stored in the refrigerator or freezer to keep it longer (slowing the rate of decay and decomposition), and food is heated up to speed up the denaturation and condensation reactions that people find tasty.

   While affecting temperature and rate is intuitive, there is no obvious place for temperature in the rate laws above. The rate constant, however, is related to temperature according to the Arrhenius equation:[7]

$$k = Ae^{-E_a/RT} \tag{14.2}$$

k is the rate constant (units depend on reaction order).
A is the pre-exponential factor (units match k).
E_a is the activation energy (kJ/mol).
R is the gas constant (0.008 314 kJ/(mol K)).
T is temperature (K).

With this, it can be seen how the rate of reaction – specifically the rate constant – varies with temperature. One of the key points to note is the activation energy (E_a) term. Activation energy is the minimum amount of energy necessary to start a reaction. Think of the room that you are in as you read this. If it were to be filled with a significant amount of natural gas, aside from the difficulty this might cause to you and your ability to breathe, you might also worry about the natural gas catching fire or blowing up. Specifically, you would worry about anything that could cause or create a spark. That spark would provide the initial activation energy to start the conflagration. Without that spark, without the activation energy, the reaction would never happen. The activation energy can vary substantially from reactions that can happen with the touch of a feather (the explosive decomposition of nitrogen triiodide) to those like a flame that require hundreds of kilojoules per mole to begin. To find the activation energy, the rate constant is measured at several temperatures. Equation 14.3 (a ratio of Arrhenius equations) can be used to determine the activation energy:

$$\ln\left(\frac{k_2}{k_1}\right) = \frac{E_a}{R}\left(\frac{1}{T_1} - \frac{1}{T_2}\right)$$

(14.3)

k_1 is the rate constant at temperature one.
k_2 is the rate constant at temperature two.
E_a is the activation energy (kJ/mol).
R is the gas constant (0.008 314 kJ/(mol K)).
T_1 is temperature one (K).
T_2 is temperature two (K).

PROBLEM 14.5
The hydrolysis of *tert*-butyl bromide, $(CH_3)_3CBr$, is studied in 90% aqueous acetone.[8] At 25.0 °C, the rate constant is found to be 1.30×10^{-5} 1/s, and at 50.0 °C the rate constant is found to be 1.93×10^{-4} 1/s. What is the activation energy (kJ/mol) for this reaction?

⤳ MACROSCOPIC KINETICS: Catalysts
As we have seen, the rate of reaction can be increased by increasing the concentration, for those species that show up in the rate law, and by increasing the temperature (Equation 14.3). While both of these approaches work, both have their limitations. Increasing the concentration to increase the rate of reaction has an inherent limit: the solubility limit of the chemical in the solution. In addition, using a higher concentration of reactant could be problematic in terms of cost if the reactant is expensive and in terms of toxicity or danger if the chemical is hazardous. Finally, increasing the temperature to increase the rate of reaction has a limit: the boiling point of the solvent. And using a higher temperature increases the energetic cost, reducing the energy efficiency of the reaction.

A catalyst is any substance that increases the rate of a reaction without modifying the overall change in reaction energy (Chapter 17).[9] Catalysts can be classified as homogeneous, in which only one phase is involved, and heterogeneous, in which the reaction occurs at or near an interface between phases. Catalysts are important because they increase the rate of reaction without changing the temperature of the reaction. This means that using a catalyst can make a process more energy efficient. In addition, catalysts are typically effective in substoichiometric amounts. This means that catalysts do not substantially add to the material cost of a reaction, and if the catalyst can be recovered, which is often true of heterogeneous catalysts, then using catalysts also does not impact the atom economy of a reaction.

Catalysts increase the rate of reaction by decreasing the activation energy of the reaction. For example, the isomerization (Figure 14.10) of *cis*-but-2-ene to *trans*-but-2-ene is exergonic ($\Delta U < 0$). The activation energy for this rotation is calculated (at B3LYP/6-311+G(2d,p) level of theory and basis set) as 264 kJ/mol. This barrier is large enough that at room temperature this isomerization does not occur. With the addition of a diiodine catalyst, the activation energy decreases to 55 kJ/mol,[10] and the reaction happens rapidly at room temperature.

Two important points should be made about catalysts. First, catalysts do not show up in a balanced equation. If a catalyst is consumed as a reactant during one step, then it must be produced as a product in a later step. For this reason, catalysts are written above the reaction arrow to show their presence in a reaction (Figure 14.11). Second, although a catalyst does not show up in the overall reaction stoichiometry, it does appear in the rate law, and partial reaction orders are

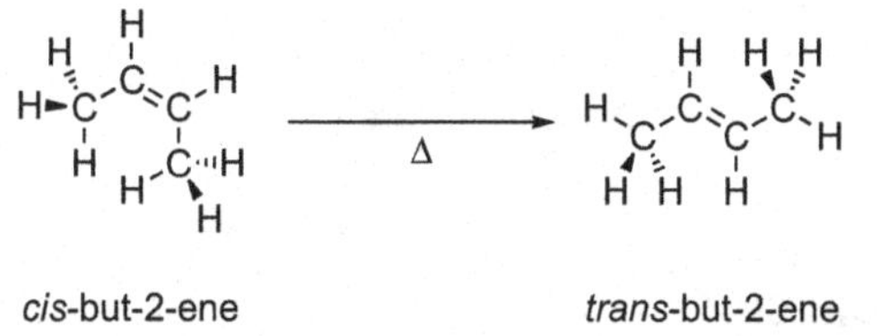

Figure 14.10 The isomerization of *cis*-but-2-ene (left) to *trans*-but-2-ene (right).

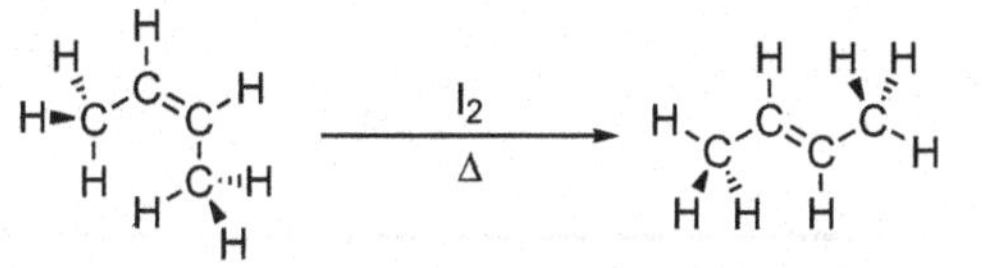

Figure 14.11 The isomerization of *cis*-but-2-ene to *trans*-but-2-ene with a diiodine catalyst.

possible for catalysts. For example, the uncatalyzed rate law for the isomerization reaction is rate = $k_{uncat}[cis$-but-2-ene$]$, and the catalyzed rate law is rate = $k_{cat}[cis$-but-2-ene$][I_2]^{1/2}$.[11]

Given the significant impact that catalysts can have on the rate of reaction by decreasing the activation energy, it is perhaps unsurprising that catalysts are ubiquitous. In cells, catalysts are called enzymes. In internal combustion engine cars, catalysts in the catalytic converters help minimize smog-producing chemicals. And finally, in industry, catalysts are widely used to help make processes more profitable by reducing process time, minimizing energy costs, and reducing material expenditure.

MICROSCOPIC KINETICS

Most of this chapter is dedicated to macroscopic kinetics, the description of and quantification of the rate of reaction, and the factors that affect the rate of reaction. In this last section, microscopic kinetics will be briefly considered. Microscopic kinetics takes the results of macroscopic kinetics and tries to develop an understanding of how the reaction is happening at the atomic/molecular scale.

To begin, let's reconsider the reaction of sodium iodide and butyl chloride, C_4H_9Cl:

$$NaI(sln) + C_4H_9Cl(sln) \xrightarrow[\text{acetone } 40°C]{} C_4H_9I(sln) + NaCl(s)$$

This reaction involves two chemicals (NaI and C_4H_9Cl) reacting together. The macroscopic kinetics analysis from above provided us with a sense of how the rate of reaction is affected by these two chemicals:

$$rate = \left(1.6 \times 10^{-5} \, L/(mol \, s)\right)[C_4H_9Cl][NaI]$$

If the overall reaction stoichiometry is compared with the rate law, the stoichiometric coefficients of NaI and C_4H_9Cl (1 and 1) match the exponents for those terms in the rate law (1 and 1). A reasonable inference could be then that these two chemicals are reacting in a single, bimolecular step. That is, it could be inferred that the reaction has a single step where both chemicals come together. This is shown in a reaction coordinate diagram (Figure 14.12).

In Figure 14.12, the activation energy (E_a) can be seen as the energy it takes for the reactants to get over the energetic barrier on the way to the products. The highest energy point is called the transition state (where bonds are being broken and being made). There is only a single step (only a single energetic peak) between the reactants and products (the relative energy of the reactants and products will be investigated in future chapters). To understand how the atoms and molecules arrange themselves and how bonds are made and broken over the course of this step would require further study.

Now let's contrast the example above with the reaction of *tert*-butyl bromide undergoing solvolysis in the presence of sodium hydroxide:

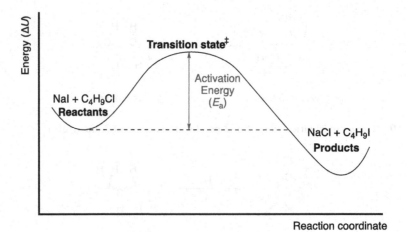

Figure 14.12 Reaction coordinate diagram showing the reaction of NaI and C_4H_9Cl.

$$(CH_3)_3 CBr(sln) + NaOH(sln) \xrightarrow[\text{water, ethanol}]{} (CH_3)_3 COH(sln) + NaBr(sln)$$

The macroscopic kinetics analysis from above provided us with a sense of how the rate of reaction is affected by these two chemicals:

$$\text{rate} = k[(CH_3)_3CBr]$$

If the rate law is compared with the overall reaction, one of the reactants, NaOH, does not show up in the rate law. What this indicates is that this reaction cannot happen in a single, elementary step, but it must be made up of multiple steps. A possible inference, then, is that the rate law corresponds to a single, unimolecular step, in a multistep reaction, and that it corresponds to the slowest, rate-determining step in that multistep reaction. Figure 14.13 shows a plausible reaction coordinate diagram for this multistep reaction.

In Figure 14.13, there are two new pieces compared to Figure 14.12. The first is there is a second energetic maximum (a second transition state), which implies two steps to the overall reaction. There is also a new, high-energy, local minimum, which is referred to as an intermediate. An intermediate is a chemical species that is formed over the course of the reaction but reacts further and is not present at the end of the reaction. In this proposed reaction coordinate diagram, the first step is the rate-determining step (the absolute highest energy peak). This hypothetical reaction

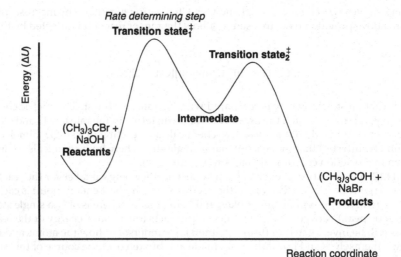

Figure 14.13 Proposed reaction coordinate diagram showing the reaction of $(CH_3)_3CBr$ and NaOH in aqueous acetone.

156

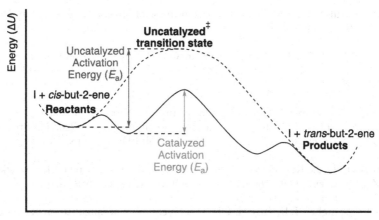

Figure 14.14 Reaction coordinate diagram showing the reaction of uncatalyzed isomerization of *cis*-but-2-ene to *trans*-but-2-ene (dashed curve) and the catalyzed isomerization (solid curve). For this diagram, the active catalyst, an iodine atom, is shown rather than the diiodine precatalyst.

coordinate diagram assumes that the *tert*-butyl bromide first reacts by itself (high energy) to form an intermediate before that intermediate more rapidly goes on to form the product. This proposed diagram is based on several inferences made by comparing the rate law to the overall reaction stoichiometry. To support or refute this proposed reaction coordinate diagram (reaction mechanism), further research would be necessary. But in these two small vignettes, however, hopefully it is clear how the macroscopic kinetics analyses can begin to provide a microscopic kinetics' understanding of reactions.

⤳ MICROSCOPIC KINETICS: Catalysts

To conclude our discussion of kinetics, we will consider the effect of a catalyst in the context of a reaction coordinate diagram (Figure 14.14). There are several items that are important to note. The first is the activation energy is substantially lower than the uncatalyzed reaction. Second, the reaction mechanism is often substantially different with a catalyst. In Figure 14.14, for example, we can see that the uncatalyzed reaction is a single step, while the catalyzed reaction is a three-step reaction.[12] Finally, the catalyst and substrate[13] typically react to form a catalyst–substrate complex, which is lower in energy than the initial free catalyst and free substrate.

ACKNOWLEDGMENTS

A special thank you to Dr. Anna Dunn (https://orcid.org/0000-0003-2852-7755) for her thoughtful reviews and criticism of this chapter.

NOTES

1. "Macroscopic kinetics." IUPAC. *Compendium of Chemical Terminology*, 2nd ed. Compiled by A.D. McNaught and A. Wilkinson. Blackwell Scientific Publications, Oxford, 1997. Online version (2019–) created by S.J. Chalk. DOI: 10.1351/goldbook.M03677.

2. "Microscopic kinetics." IUPAC. *Compendium of Chemical Terminology*, 2nd ed. Compiled by A.D. McNaught and A. Wilkinson. Blackwell Scientific Publications, Oxford, 1997. Online version (2019–) created by S.J. Chalk. DOI: 10.1351/goldbook.M03915.

3. The phase notation (sln) is to indicate chemicals dissolved in solution, where here acetone is the solvent.

4. (i) Waage, P.; Guldberg, C.M. Studier over Affiniteten. *Forh. Vidensk.-Selsk. Kristiania*. **1864**, 35–41. (Translation by H.I. Abrash: *J. Chem. Educ.*, **1986**, *63* (12), 1044–1047. DOI: 10.1021/ed063p1044.).

(ii) Waage, P.; Guldberg, C.M. Forsøg til Bestemmelse af Lovene for Affiniteten. *Forh. Vidensk.-Selsk. Kristiania.* **1864**, 92–94.

5. Pace, R.D.; Regmi, Y. The Finkelstein Reaction: Quantitative Reaction Kinetics of an S_N2 Reaction Using Nonaqueous Conductivity. *J. Chem. Educ.* **2006**, *83* (9), 1344–1348. DOI: 10.1021/ed083p1344.

6. Swords, W.B.; Chapman, S.J.; Hofstetter, H.; Dunn, A.L.; Yoon, T.P. Variable Temperature LED–NMR: Rapid Insights into a Photocatalytic Mechanism from Reaction Progress Kinetic Analysis. *J. Org. Chem.* **2022**, *87* (17), 11776–11782. DOI: 10.1021/acs.joc.2c01479.

7. Arrhenius, S. Über die Reaktionsgeschwindigkeit bei der Inversion von Rohrzucker durch Säuren. *Z. Phys. Chem.* **1889**, *4U* (1), 226–248. DOI: 10.1515/zpch-1889-0416.

8. Bateman, L.C.; Hughes, E.D.; Ingold, C.K. 184. Mechanism of Substitution at a Saturated Carbon Atom. Part XIX. A Kinetic Demonstration of Unimolecular Solvolysis of Alkyl Halides. (Section A) Kinetics of, and Salt Effects in, the Hydrolysis of *tert.*-Butyl Bromide in Aqueous Acetone. *J. Chem. Soc. (Resumed).* **1940**, 960–966. DOI: 10.1039/JR9400000960.

9. "Catalyst." IUPAC. *Compendium of Chemical Terminology*, 2nd ed. Compiled by A.D. McNaught and A. Wilkinson. Blackwell Scientific Publications, Oxford, 1997. Online version (2019–) created by S.J. Chalk. DOI: 10.1351/goldbook.C00876.

10. Hepperle, S.; Li, Q.; East, A.L.L. Mechanism of Cis/Trans Equilibration of Alkenes via Iodine Catalysis. *J. Phys. Chem. A.*, **2005**, *109* (48), 10975–10981. DOI: 10.1021/jp053727o.

11. Dickinson, R.G.; Lotzkar, H. The Kinetics of the Thermal Isomerization of Cinnamic Acid Catalyzed by Iodine. *J. Am. Chem. Soc.*, **1937**, *59* (3), 472–475. DOI: 10.1021/ja01282a013.

12. Benson, S.W.; Egger, K.W.; Golden, D.M. Iodine-Catalyzed Isomerization of Olefins. III. Kinetics of the Geometrical Isomerization of Butene-2 and the Rate of Rotation About a Single Bond. *J. Am. Chem. Soc.*, **1965**, *87* (3), 468–476. DOI: 10.1021/ja01081a013.

13. A substrate is any chemical that the catalyst acts on.

15 Enthalpy

Chemists are very interested in knowing the thermal energy flow associated with a reaction (the heat, q) of a chemical or physical process. While heat can be calculated from calorimetry, typically chemists are most interested in knowing the heat per mole at constant pressure of a reaction, enthalpy $(\Delta_r H°)$,[1] for several reasons. First, enthalpy relates to the change in energy for the system $(\Delta U = n\Delta H - P\Delta V$; see Chapter 2) and since the magnitude of heat in a reaction is often significantly greater than any accompanying work, then the change in energy (ΔU) can be approximated as essentially the same as enthalpy $(\Delta U \approx n\Delta H)$. Second, knowing the change in enthalpy, which is normalized per mole of reaction, for many reactions allows chemists to start developing an understanding of what, on the nanoscale, is driving the macroscopic thermal energy changes that are being observed.

EXPERIMENTAL ENTHALPY: Calorimetry

To experimentally find the enthalpy of a reaction or a process, a chemist must conduct a calorimetry (literally "heat measure") experiment. In a calorimetry experiment, the reaction or process takes place within a closed, insulated system: the calorimeter. It is often the case that water, whose temperature is being monitored, surrounds the reaction (or the reaction is taking place in water). The water is referred to as a bath.

Consider a common coffee-cup calorimeter experiment, where an aqueous reaction is happening inside an insulated coffee cup. In this example, the neutralization reaction of aqueous hydrogen chloride reacting with sodium hydroxide is considered:[2]

$$HCl(aq) + NaOH(aq) \rightarrow NaCl(aq) + H_2O(l)$$

For this experiment, a Styrofoam coffee cup contains aqueous hydrogen chloride (61.0 g of H_2O and 1.49 g of HCl) at 16.8 °C. To that is added a sodium hydroxide solution (62.0 g of H_2O and 1.63 g of NaOH) also at 16.8 °C. The final temperature of the mixture is 21.3 °C. To find the enthalpy of this reaction, we first need to calculate the amount of heat of the water bath (q_{bath}) using Equation 15.1. Here the heat capacity (c_p) of the solutions is assumed to be the same as the heat capacity of pure water $(4.184 \frac{J}{g\ °C})$.

$$q_{bath} = mc_p\Delta T \tag{15.1}$$

$$q_{bath} = (126.2\ g)\left(4.184\frac{J}{g\ °C}\right)(21.3\ °C - 16.8\ °C)$$

$$q_{bath} = 2400\ J$$

It is important to note that this corresponds to the heat of the water bath (not the reaction). The heat of the reaction is the objective. The heat of the bath (q_{bath}) is equal and opposite the heat of reaction (q_{rxn}):

$$q_{bath} = -q_{rxn} \tag{15.2}$$

$$q_{rxn} = -q_{bath} = -2400\ J$$

Next, we need to find the amount (mol) of one of the reactants. The 1.49 g of HCl (0.03646 kg/mol) would correspond to 0.0409 mol HCl. Now we take the ratio of heat (q) to moles (n):

$$\frac{q_{rxn}}{n} = \frac{-2400\ J}{0.0409\ mol\ HCl} = -58\ 000\frac{J}{mol\ HCl}$$

Finally, to find the enthalpy of reaction $(\Delta_r H°)$, we need to convert this from $\frac{J}{mol\ HCl}$ to $\frac{J}{mol\ rxn}$

(usually reported as just J/mol, the rxn is omitted). The balanced reaction is used to relate the amount of HCl (1 mol HCl) to the amount of reaction (1 mol rxn). Overall, this gives us an enthalpy of reaction $(\Delta_r H°)$ for this neutralization of –58 000 J/mol:

$$\Delta_r H° = -58\ 000\frac{J}{mol\ HCl}\left(\frac{1\ mol\ HCl}{1\ mol\ rxn}\right) = 58\ 000\frac{J}{mol\ rxn}$$

DOI: 10.1201/9781003479338-15

In the above example, the step of going from mole of HCl to mole of reaction seems superfluous. Let us consider this a bit further in a reaction with more complicated stoichiometry:

$$2 \, Al(s) + 3 \, Cl_2(g) \rightarrow 2 \, AlCl_3(s)$$

In a calorimetry experiment, the following heat per mole values are found: $-352.1 \dfrac{kJ}{mol \, Al}$, $-352.1 \dfrac{kJ}{mol \, AlCl_3}$, and $-234.7 \dfrac{kJ}{mol \, Cl_2}$. As can be seen, depending on which chemical we choose, the heat per mole can vary. To avoid choosing a particular chemical in each reaction, the value is normalized to the moles of reaction by using the stoichiometric coefficients:

$$-352.1 \dfrac{kJ}{mol \, Al} \left(\dfrac{2 \, mol \, Al}{1 \, mol \, rxn} \right) = -704.2 \, kJ / mol$$

$$-352.1 \dfrac{kJ}{mol \, AlCl_3} \left(\dfrac{2 \, mol \, AlCl_3}{1 \, mol \, rxn} \right) = -704.2 \, kJ / mol$$

$$-234.7 \dfrac{kJ}{mol \, Cl_2} \left(\dfrac{3 \, mol \, Cl_2}{1 \, mol \, rxn} \right) = -704.1 \, kJ / mol$$

Now, normalized to moles (the "of reaction" is implicit), the stoichiometry of the reaction does not matter in the reporting of enthalpy values.

Although coffee-cup calorimetry can be a useful technique for quickly, if crudely, determining the enthalpy of reaction, high-precision calorimetry requires more significant setup. While there are a variety of calorimetry setups, the most important consideration is that we must consider the temperature change of the calorimeter (and the heat associated with that). To measure the heat absorbed by the calorimeter, we use heat capacity (C) of the calorimeter, which relates the amount of heat energy to the temperature increase (J/°C). This is different than a *specific* heat capacity, which is for the energy per degree *per gram* (J/(g °C)). This adds one additional step to the calculation of q_{bath} (C is always provided).

$$q_{bath} = q_{water} + q_{calorimeter} \tag{15.3}$$

$$q_{bath} = mc_p(H_2O(l))\Delta T + C\Delta T$$

PROBLEM 15.1
Find the enthalpy of reaction ($\Delta_r H°$) in each of the following calorimetry experiments.

a. In a calorimeter ($C = 1686.2$ J/°C) that contains 2257.0 g of water ($c_p(H_2O(l)) = 4.184 \dfrac{J}{g \, °C}$), 0.8106 g of benzene ($C_6H_6$) is reacted with oxygen, which causes the temperature to rise by 3.054 °C.[3] What is the enthalpy of reaction ($\Delta_r H°$)? The balanced equation is:

$$2 \, C_6H_6(l) + 15 \, O_2(g) \rightarrow 12 \, CO_2(g) + 6 \, H_2O(g)$$

b. In a coffee-cup calorimetry experiment, conducted by the author, 2.1981 g of ammonium chloride (NH_4Cl, 0.053 49 kg/mol) is added to 86.6883 g water ($c_p(H_2O(l)) = 4.184 \dfrac{J}{g \, °C}$). The temperature goes from 16.8 °C to 15.0 °C. What is the enthalpy of reaction ($\Delta_r H°$)? The balanced equation is:

$$NH_4Cl(s) \rightarrow NH_4Cl(aq)$$

FINDING ENTHALPY WITHOUT DIRECT EXPERIMENTATION
Calorimetry is how chemists calculate the enthalpy of reaction, which given the work and equipment required, would limit the utility of enthalpy. Given the importance of enthalpy to

understanding the energy change of a system, chemists have developed three different ways of calculating the enthalpy of a reaction without doing a calorimetry experiment every time: dissociation energy (E_d),[4] standard formation enthalpy ($\Delta_f H°$), and Hess's law.[5] Each of these methods will be considered in turn.

Dissociation Energy (E_d)

The thermal energy change of a process comes from the breaking (endothermic) and making (exothermic) of bonds and intermolecular forces. The overall reaction enthalpy, then, is the sum of endothermic (breaking) steps and exothermic (making) steps. Therefore, if we know the strength of a bond, measured as how much energy it takes to break (dissociate) the bond, we can determine the energy associated with a chemical change where that bond is made or broken. An example of bond dissociation energy values is shown in Table 15.1, and a more complete list is included in Appendix 6.

To use dissociation energy values to calculate an enthalpy of reaction, we need to know what bonds are being broken and what bonds are being made. The enthalpy of reaction then is the sum of bonds broken (endothermic) minus the sum of bonds made (exothermic). Since it is the bonds of reactants that are broken and the bonds of products that are formed, this can be formulated as Equation 15.4:

$$\Delta_r H° = \sum n E_d(\text{reactants}) - \sum n E_d(\text{products}) \tag{15.4}$$

Here an example of calculating reaction enthalpy using bond dissociation energy values will be considered.

Example 15.1 Finding $\Delta_r H°$ Using E_d Values

Objective: Find the enthalpy of reaction for the reaction of hydrogen and fluorine to make hydrogen fluoride.

$$H_2(g) + F_2(g) \rightarrow 2\ HF(g)$$

To solve: Using Equation 15.4, the known information is substituted into the equation.

$$\Delta_r H° = \sum n E_d(\text{reactants}) - \sum n E_d(\text{products})$$

$$\Delta_r H° = 1E_d(H–H) + 1E_d(F–F) - 2E_d(H–F)$$

First, there are the two reactant dissociation energy values.
→ $1E_d(H–H)$ is 1 (the stoichiometry coefficient) multiplied by the E_d of the H–H bond in H_2.
→ $1E_d(F–F)$ is 1 (the stoichiometry coefficient) multiplied by the E_d of the F–F bond in F_2.

Then the product dissociation energy value is subtracted.
→ $2E_d(H–F)$ is 2 (the stoichiometry coefficient) multiplied by the E_d of the H–F bond in HF.

Finally, values are substituted into the equation (Table 15.1).
$\Delta_r H° = 1(436\ \text{kJ/mol}) + 1(159\ \text{kJ/mol}) - 2(570.\ \text{kJ/mol})$
$\Delta_r H° = -545\ \text{kJ/mol}$

Table 15.1 Selected Dissociation Energy (E_d) Values

Bond	E_d (kJ/mol)	Bond	E_d (kJ/mol)
H–H	436	I–I	152
O=O	498	H–F	570.
F–F	159	H–Cl	431
C–Cl	350.	H–Br	366
Cl–Cl	243	H–I	298
H–O	497	H–S	381
H–C	439	H–N	450
Al–Cl	502	Cu–Cl	378

PROBLEM 15.2
For each reaction below, calculate $\Delta_r H°$ using E_d values.

a. $2 H_2(g) + O_2(g) \rightarrow 2 H_2O(l)$

b.
$$\begin{array}{cccc}
\text{H} & & & \text{H} \\
| & & & | \\
\text{H–C–H} & \text{:Cl–Cl:} & \longrightarrow & \text{H–C–Cl:} \quad \text{H–Cl:} \\
| & & & | \\
\text{H} & & & \text{H}
\end{array}$$

c. $2 Al(s) + 3 CuCl_2(aq) \rightarrow 2 AlCl_3(aq) + 3 Cu(s)$

Standard Formation Enthalpy ($\Delta_f H°$); Also Colloquially Called Heat of Formation

An ideal system for calculating enthalpy would be based on quantifying the absolute amount of thermal energy in a chemical. Then we could sum all the thermal energy values of reactants and products and find the difference. A core problem is that zero thermal energy would mean absolutely zero atomic and molecular motion. Although atoms stop translating (moving through Cartesian space) at absolute zero, they never stop vibrating (oscillating around a fixed point). This means that there is no definable zero thermal energy point. The scientific community has defined an arbitrary zero point: the enthalpy of formation ($\Delta_f H°$) for an element in its most stable state is zero.

$$\Delta_f H°(C(s, graphite)) = 0 \text{ kJ/mol}$$

$$\Delta_f H°(N_2(g)) = 0 \text{ kJ/mol}$$

$$\Delta_f H°(O_2(g)) = 0 \text{ kJ/mol}$$

When a reaction is carried out, using this as our standard, we can then determine the enthalpy of formation ($\Delta_f H°$) for any chemical and chemists create tables (like Table 15.2) of these enthalpy of formation values ($\Delta_f H°$). A longer list of enthalpy of formation values can be found in Appendix 7.

Then Equation 15.5 is used to find the enthalpy of a reaction ($\Delta_r H°$):

$$\Delta_r H° = \sum n\Delta_f H°(products) - \sum n\Delta_f H°(reactants) \qquad (15.5)$$

Here an example of calculating reaction enthalpy using standard enthalpy of formation values will be considered.

Example 15.2 Finding $\Delta_r H°$ Using $\Delta_f H°$ Values

Consider the chemical reaction for the combustion of methane:

$$CH_4(g) + 2 O_2(g) \rightarrow CO_2(g) + 2 H_2O(g)$$

To solve: Using Equation 15.5, start by plugging in known information.

$$\Delta_r H° = \sum n\Delta_f H°(products) - \sum n\Delta_f H°(reactants)$$

$$\Delta_r H° = [2\Delta_f H°(H_2O(g)) + 1\Delta_f H°(CO_2(g))] - [1\Delta_f H°(CH_4(g)) + 1\Delta_f H°(O_2(g))]$$

Table 15.2   Selected Enthalpy of Formation ($\Delta_f H°$) Values

Substance	$\Delta_f H°$ (kJ/mol)	Substance	$\Delta_f H°$ (kJ/mol)
$CO_2(g)$	−393.5	$CH_4(g)$	−74.6
$UO_2(s)$	−1085.0	$NH_4^+(aq)$	−132.5
$NO(g)$	91.3	$NO_3^-(aq)$	−207.4
$H_2O(l)$	−285.8	$UF_4(s)$	−1914.2
$H_2O(g)$	−241.8	$HF(aq)$	−321.1
$HF(l)$	−299.0	$HF(g)$	−273.3
$NH_4NO_4(s)$	−365.2	$O_2(g)$	0
$Cu(s)$	0	$Al(s)$	0
$CuCl_2(aq)$	−269.6	$AlCl_3(aq)$	−1032.6

First, there are the $\Delta_f H°$ values of the two products:

→ The $2\Delta_f H°(H_2O(g))$ is 2 (the stoichiometry coefficient) multiplied by the $\Delta_f H°$ value of water gas.
→ The $1\Delta_f H°(CO_2(g))$ is 1 (the stoichiometry coefficient) multiplied by the $\Delta_f H°$ value of carbon dioxide gas.

Then $\Delta_f H°$ values of the two reactants are subtracted:
→ The $1\Delta_f H°(CH_4(g))$ is 1 (the stoichiometry coefficient) multiplied by the $\Delta_f H°$ value of methane gas.
→ The $1\Delta_f H°(O_2(g))$ is 1 (the stoichiometry coefficient) multiplied by the $\Delta_f H°$ value of oxygen gas.

Now the values from Table 15.2 are plugged in and numerically solved for.

$$\Delta_r H° = [2(-241.8 \text{ kJ/mol}) + 1(-393.5 \text{ kJ/mol})] - [1(-74.6 \text{ kJ/mol}) + 1(0 \text{ kJ/mol})]$$
$$\Delta_r H° = -802.5 \text{ kJ/mol}$$

PROBLEM 15.3

For each equation, calculate $\Delta_r H°$ using $\Delta_f H°$ values. Notice the standard enthalpy of formation can be used not just for chemical reactions but also for phase changes and for the dissociation of ionic compounds in water.

a. $UO_2(s) + 4 HF(aq) \rightarrow UF_4(s) + 2 H_2O(l)$

b. $HF(g) \rightarrow HF(l)$

c. $NH_4NO_3(s) \rightarrow NH_4^+(aq) + NO_3^-(aq)$

d. $2 Al(s) + 3 CuCl_2(aq) \rightarrow 2 AlCl_3(aq) + 3 Cu(s)$

Hess's Law

Hess's law is the last way that we can find an unknown reaction enthalpy without doing work in the laboratory. Hess's law is both very useful and the least straightforward of the three because it requires a bit of puzzle solving. To illustrate Hess's law, an example will be considered at length.

Example 15.3 Finding $\Delta_r H°$ Using Hess's Law

To illustrate, let's consider the following reaction:

$$C_2H_4(g) + 6 F_2(g) \rightarrow 2 CF_4(g) + 4 HF(g)$$

The reaction enthalpy for that reaction is the objective, but the information known is the reaction enthalpy values for several reactions:

$$H_2(g) + F_2(g) \rightarrow 2 HF(g) \; \Delta_r H° = -537 \text{ kJ/mol}$$
$$C(s) + 2 F_2(g) \rightarrow CF_4(g) \; \Delta_r H° = -680. \text{ kJ/mol}$$
$$2 C(s) + 2 H_2(g) \rightarrow C_2H_4(g) \; \Delta_r H° = 52.3 \text{ kJ/mol}$$

Our challenge, then, is to come up with a way that these reactions can be added together to get the overall reaction desired. To do that, we need to look at each reaction in turn.

Let's consider the first reaction. We need to ask: Which chemical (or chemicals) show up in the overall reaction? Does that chemical show up on the correct side (reactant or product) in the individual reaction as it does in the overall reaction? And what is the coefficient of that reactant or product compared to the overall reaction above?

$$H_2(g) + F_2(g) \rightarrow 2 HF(g) \; \Delta_r H° = -537 \text{ kJ/mol}$$

Analysis: HF (a product in the overall reaction) shows up in the product side of this reaction just as it does in the overall reaction. The only thing that is amiss is that this has a stoichiometric coefficient of 2, whereas in the overall it has a coefficient of 4. To remedy this, we need to multiply the equation through by 2 (which also means multiplying the enthalpy value by 2).

Modified reaction and enthalpy:

$$2\ H_2(g) + 2\ F_2(g) \rightarrow 4\ HF(g)\ \Delta_rH° = -1074\ kJ/mol$$

For the second reaction provided:

$$C(s) + 2\ F_2(g) \rightarrow CF_4(g)\ \Delta_rH° = -680\ kJ/mol$$

Analysis: CF_4 (a product in the overall reaction) shows up in the product side of this reaction just as it does in the overall reaction. The only thing that is amiss is that this has a stoichiometric coefficient of 1, whereas in the overall it has a coefficient of 2. To remedy this, we need to multiply the equation through by 2 (which also means multiplying the enthalpy value by 2).

Modified reaction and enthalpy:

$$2\ C(s) + 4\ F_2(g) \rightarrow 2\ CF_4(g)\ \Delta_rH° = -1360\ kJ/mol$$

For the last reaction provided:

$$2\ C(s) + 2\ H_2(g) \rightarrow C_2H_4(g)\ \Delta_rH° = 52.3\ kJ/mol$$

Analysis: C_2H_4 (a reactant in the overall reaction) shows up in the product side of this reaction; however, the coefficient (1) is what it needed to be. So, the reaction is reversed, which means changing the sign of the enthalpy value.

Modified reaction and enthalpy:

$$C_2H_4(g) \rightarrow 2\ C(s) + 2\ H_2(g)\ \Delta_rH° = -52.3\ kJ/mol$$

Now we need to consider all the modified reactions and investigate whether they can add to the overall reaction:

$$2\ H_2(g) + 2\ F_2(g) \rightarrow 4\ HF(g)\ \Delta_rH° = -1074\ kJ/mol$$

$$2\ C(s) + 4\ F_2(g) \rightarrow 2\ CF_4(g)\ \Delta_rH° = -1360.\ kJ/mol$$

$$C_2H_4(g) \rightarrow 2\ C(s) + 2\ H_2(g)\ \Delta_rH° = -52.3\ kJ/mol$$

Now if these are added up (and add up the enthalpy values too):

$$2\ H_2(g) + 2\ F_2(g) + 2\ C(s) + 4\ F_2(g) + C_2H_4(g) \rightarrow 2\ C(s) + 2\ H_2(g) + 4\ HF(g) +$$
$$2\ CF_4(g)\ \Delta_rH° = -2486.3\ kJ/mol$$

If done correctly, then there should be terms on the product and reactant side, which can cancel and consolidate to simplify. Here we can (1) consolidate $2\ F_2(g) + 4\ F_2(g)$ to $6\ F_2(g)$, (2) cancel the $2\ H_2(g)$ that is on the left and the right sides, and (3) cancel the 2 C(s) that is on the left and right sides:

$$6\ F_2(g) + C_2H_4(g) \rightarrow 2\ CF_4(g) + 4\ HF(g)\ \Delta_rH° = -2486\ kJ/mol$$

Now the reaction matches the overall reaction. This means that the $\Delta_rH°$ value is the $\Delta_rH°$ for the overall reaction. If we did not arrive at the overall reaction, then we would need to reevaluate the changes made to the individual reactions before adding.

PROBLEM 15.4
For each equation, calculate $\Delta_rH°$ using the individual reaction enthalpy values provided.

a. Overall reaction: $Ca^{2+}(aq) + 2\ OH^-(aq) + CO_2(g) \rightarrow CaCO_3(s) + H_2O(l)$

Individual reactions:

$CaCO_3(s) \rightarrow CaO(s) + CO_2(g)\ \Delta_rH° = 178.3\ kJ/mol$

$CaO(s) + H_2O(l) \rightarrow Ca(OH)_2(s)\ \Delta_rH° = -65.2\ kJ/mol$

$Ca(OH)_2(s) \rightarrow Ca^{2+}(aq) + 2\ OH^-(aq)\ \Delta_rH° = -16.7\ kJ/mol$

b. Overall reaction: $P_4(s) + 6\ Cl_2(g) \rightarrow 4\ PCl_3(l)$

Individual reactions:

$P_4(s) + 10\ Cl_2(g) \rightarrow 4\ PCl_5(s)\ \Delta_rH° = -1774.0\ kJ/mol$

$PCl_3(l) + Cl_2(g) \rightarrow PCl_5(s)\ \Delta_rH° = -123.8\ kJ/mol$

c. Overall reaction: $C_2H_4(g) + H_2O(l) \rightarrow C_2H_5OH(l)$

Individual reactions:

$C_2H_4(g) + 3\ O_2(g) \rightarrow 2\ CO_2(g) + 2\ H_2O(l)\ \Delta_rH° = -1411.1$ kJ/mol

$C_2H_5OH(l) + 3\ O_2(g) \rightarrow 2\ CO_2(g) + 3\ H_2O(l)\ \Delta_rH° = -1367.5$ kJ/mol

MEANING OF ENTHALPY

Throughout this chapter, we have discussed the importance of enthalpy as it relates to internal energy (ΔU) and how to calculate enthalpy experimentally (calorimetry) and theoretically (dissociation energy, standard enthalpy of formation, and Hess's law). It is important to briefly discuss enthalpy on its own terms.

First, enthalpy indicates whether a reaction is energetically more stable before a reaction ($\Delta_rH° > 0$) or after a reaction ($\Delta_rH° < 0$). As such, an endothermic reaction ($\Delta_rH° > 0$) is referred to as a reactant-favored process because the system energy is lower before the reaction than it is after, and the reaction will tend to not move forward – toward products – unless acted upon external factors. In contrast, an exothermic reaction ($\Delta_rH° < 0$) is referred to as a product-favored process because the system energy is lower after the reaction than it is before, and the reaction will tend to move forward – toward products – so long as the initial activation energy is provided.

Second, enthalpy provides a means of comparing different reactions as a means of trying to gain insight into what is occurring at the atomic scale. Consider the phase change of lithium halide salts dissolving and dissociating in water:

$$LiF(s) \rightarrow Li^+(aq) + F^-(aq) \qquad \Delta_rH° = 4.9 \text{ kJ/mol}$$

$$LiCl(s) \rightarrow Li^+(aq) + Cl^-(aq) \qquad \Delta_rH° = -37.1 \text{ kJ/mol}$$

$$LiBr(s) \rightarrow Li^+(aq) + Br^-(aq) \qquad \Delta_rH° = -48.9 \text{ kJ/mol}$$

$$LiI(s) \rightarrow Li^+(aq) + I^-(aq) \qquad \Delta_rH° = -63.3 \text{ kJ/mol}$$

In changing from fluoride to iodide, there is a shift from a weakly endothermic dissociation (LiF) to a moderately exothermic dissociation (LiI). From these sets of data, the following questions could be asked: Is this trend a manifestation of changing ionic radii (Chapter 8)? Does this change in reaction enthalpy stem from differences in lattice energy (Chapter 9)? Why does the fluoride ion, which would produce the strongest hydrogen bond (Chapter 12), dissolving in water produce an endothermic reaction? These and other questions would require further experimentation to definitively answer, but enthalpy provides quantitative data that can be used to help provide an answer.

PROBLEM 15.5
Calculate the enthalpy of reaction for each of the following using Appendix 7. What trend(s) do you notice? What topics do the trend(s) suggest should be investigated further to try to understand the data?

a. $2\ Al(s) + 3\ Cu(NO_3)_2(aq) \rightarrow 2\ Al(NO_3)_3(aq) + 3\ Cu(s)$

b. $Al(s) + 3\ AgNO_3(aq) \rightarrow Al(NO_3)_3(aq) + 3\ Ag(s)$

c. $Al(s) + Au(NO_3)_3(aq) \rightarrow Al(NO_3)_3(aq) + Au(s)$

⤳ ENTHALPY AND ENERGY

Making and breaking bonds or intermolecular forces is a key source of and sink for energy. The energy obtained from the combustion of hydrocarbons comes from breaking C–C and C–H bonds and making C=O and O–H bonds. This energy can be used to drive the pistons of an internal combustion engine and to boil water by disrupting the intermolecular forces between water molecules, which drives steam turbines and makes electricity. Hydrocarbons have substantial amounts of chemical potential energy, but that comes at the cost of greenhouse gas emissions. The transition to a sustainable energy economy requires that all chemical potential energy used has its ultimate origins in renewable and net-zero-carbon resources, e.g., solar, wind, water, geothermal, biomass, and (potentially) fusion.

NOTES

1. Enthalpy was a name suggested by K. Onnes for Gibb's "heat function" concept: Dalton, J.P. Researches on the Joule-Kelvin-Effect, Especially at Low Temperatures. I. Calculations for Hydrogen. *Proceedings of the Section of Sciences* (Koninklijke Akademie van Vetenschappen), **1909**, *11* (part 2), 863–873.

2. Adapted from: Richards, T.W.; Rowe, A.W. The Heats of Neutralization of Potassium, Sodium, and Lithium Hydroxides with Hydrochloric, Hydrobromic, Hydriodic and Nitric Acids, at Various Dilutions. *J. Am. Chem. Soc.*, **1922**, *44* (4), 684–707. DOI: 10.1021/ja01425a004.

3. Adapted from: Richards, T.W. Davis, H.S. The Heats of Combustion of Benzene, Toluene, Aliphatic Alcohols, Cyclohexanol, and Other Carbon Compounds. *J. Am. Chem. Soc.*, **1920**, *42* (8), 1599–1617. DOI: 10.1021/ja01453a011.

4. Szwarc, M. The Determination of Bond Dissociation Energies by Pyrolytic Methods. *Chem. Rev.*, **1950**, *47* (1), 75–173. DOI: 10.1021/cr60146a002.

5. Hess, G.H. *Bulletin scientifique, Académie imperiale des sciences (St. Petersburg)*, **1840**, *8*, 257–272.

16 Entropy

The first law of thermodynamics provides an understanding of how a system's internal energy can change, but it does not give a sense of direction to that change. That is, the first law does not provide an understanding of whether endothermic or exothermic processes are favored. It does not explain whether one should expect to find that exoworkic changes are favored and endoworkic changes are disfavored or vice versa.

THE SECOND LAW OF THERMODYNAMICS

Although the first law does not provide us with a sense of direction or favorability of a process, our lived experience does. If you make a cup of coffee (or tea or hot cocoa) and let it sit out for long enough, the once-hot beverage will become room temperature. Similarly, if you get a cup of ice water and let it sit out long enough, the water will eventually become uniformly room temperature. A cup of room-temperature water will never, on its own, become hot water (nor will it become cold water). The sense of how heat will tend to flow is one formulation of the second law of thermodynamics:[1] no process is possible where the only result is heat is transferred from a colder reservoir to a hotter one.[2] That is a slightly more archaic way to say it is impossible for a room-temperature object, on its own, to become hotter (or to become colder). The natural direction of a process, happening without added heat or work, is for energy to disperse or spread out. Matter also tends to disperse or spread out, which is a manifestation of the tendency for energy to spread out. Processes that concentrate energy, and by extension matter, will not happen on their own and require heat or work to occur.

REACTION NOTATION: Bidirectional Arrows

Before considering some reactions and processes, this is a useful moment to update the notation that is commonly used for reactions. It is most common to represent a chemical reaction and use a unidirectional arrow, $\rightarrow$, to show the transformation of reactants (the left side) to products (the right side). As we move forward in considering the energy of a reaction (and subsequently its equilibrium), we will start to use bidirectional arrows, $\rightleftarrows$,[3] to show the transformation of reactants (the left side) to products (the right side). What this notation helps to highlight is that a reaction can proceed from left to right (turning reactants into products) as we have considered to date, and a reaction could also proceed from right to left (turning products into reactants). Since both directions of change are possible, then, we will need to consider which side is favored. If energetically, in terms of the change in internal energy (ΔU) or in terms of the dispersal of energy, the right side of the reaction is favored, we say that it is a product-favored process. If the left side of the reaction is favored, we would say that it is a reactant-favored process. Consider the example of nitrogen triiodide decomposition:

$$2\,NI_3(s) \rightleftarrows N_2(g) + 3\,I_2(g) \qquad \Delta_r H° = \text{-290 kJ/mol} \qquad (16.1)$$

The decomposition reaction is highly exothermic, meaning that the internal energy of the system is lower after the reaction than before the reaction. The lowering of the internal energy makes the system more stable and so the right side is favored enthalpically. The spread of energy (and by extension, matter) also favors the right side as the concentrated solid is now highly dispersed as a gas. Altogether, then, we can say that this is a product-favored reaction.

PROBLEM 16.1

For each of the following, identify only whether energy and/or matter is more dispersed on the left side/before the right side/after (do not consider enthalpy). Then identify and explain whether you think this is a product-favored or reactant-favored process.

a. $NaCl(s) \rightleftarrows NaCl(aq)$

b. $H_2O(g) \rightleftarrows H_2O(l)$

c. $2\,C_6H_6(l) + 15\,O_2(g) \rightleftarrows 12\,CO_2(g) + 6\,H_2O(g)$

d. $Pb(NO_3)_2(aq) + Na_2SO_4(aq) \rightleftarrows PbSO_4(s) + 2NaNO_3(aq)$

e. $2\,NaCl(s) \rightleftarrows 2\,Na(s) + Cl_2(g)$

f. A container of 1.0 mol neon gas goes from 5.0 L in volume to 10.0 L in volume.

g. A 50 mM KBr solution is diluted to 25 mM by the addition of more water.

DOI: 10.1201/9781003479338-16

h. A refrigerator gets colder on the inside than it is on the outside.

i. A transport protein maintains more H^+(aq) in the intermembrane space of a mitochondrion than there is in the matrix.

ENTROPY

The spread or dispersal of energy is a useful framework for thinking about the second law of thermodynamics; however, it does not provide a quantifiable system for comparing processes and their (dis)favorability. To quantify the spread of energy (and matter) we use entropy.[4] Entropy (S) is a measure of the dispersal of energy and has units of J/(mol K). Note that units are energy per mole kelvin, which considers that the spread of energy is related to and affected by temperature. If the change in entropy ($\Delta_r S°$) is positive ($\Delta_r S° > 0$), then the spread of energy is increasing (we say "entropy increases"), which favors the products (the right side of an equation or the end stage of a process). If the change in entropy ($\Delta_r S°$) is negative ($\Delta_r S° < 0$), then the spread of energy is decreasing (we say "entropy decreases"), which favors the reactants (the left side of an equation or the initial stage of a process).

THE THIRD LAW OF THERMODYNAMICS

In quantifying and measuring entropy, we have a significant advantage over enthalpy (where scientists have had to define an arbitrary zero point, $\Delta_f H°$ of an element in its most stable state). As temperature decreases, the entropy of a thermodynamic system approaches a constant – that is independent of all other thermodynamic state variables – as its absolute temperature approaches zero. This is the third law of thermodynamics.[5] For a pure element in a perfectly crystalline state at absolute zero (0 K), the entropy would be 0 J/(mol K). With this information in hand, the standard molar entropy ($S°$) has been measured for many substances (example values can be seen in Appendix 7). And to calculate entropy we use Equation 16.2, which should look very familiar for how we calculate enthalpy ($\Delta_r H°$) using standard enthalpy of formation values ($\Delta_f H°$).

$$\Delta_r S° = \sum n S°(\text{products}) - \sum n S°(\text{reactants}) \qquad (16.2)$$

PROBLEM 16.2
For each of the following, calculate $\Delta_r S°$ using Appendix 7. How do your results compare with your predictions?

a. $NaCl(s) \rightleftarrows NaCl(aq)$

b. $H_2O(g) \rightleftarrows H_2O(l)$

c. $2 C_6H_6(l) + 15 O_2(g) \rightleftarrows 12 CO_2(g) + 6 H_2O(g)$

d. $Pb(NO_3)_2(aq) + Na_2SO_4(aq) \rightleftarrows PbSO_4(s) + 2NaNO_3(aq)$

e. $2 NaCl(s) \rightleftarrows 2 Na(s) + Cl_2(g)$

PROBLEM 16.3
When we calculate $\Delta_r S°$, we arrive at the net or total entropy change for a process. This can, however, obscure factors at the atomic level that are contributing to an increase in entropy and those that are contributing to a decrease in entropy. The model below is an idealized atom-scale view of what happens as NaCl dissolves in water.

a. Referring to the model, provide an example of how entropy is increasing in the system.

b. Referring to the model, provide an example of how entropy is decreasing in the system.

c. When some ionic compounds dissolve, the value for $\Delta_r S° > 0$. What can you say about the two effects discussed in parts a and b?

d. When some ionic compounds dissolve, the value for $\Delta_r S° < 0$. What can you say about the two effects discussed in parts a and b?

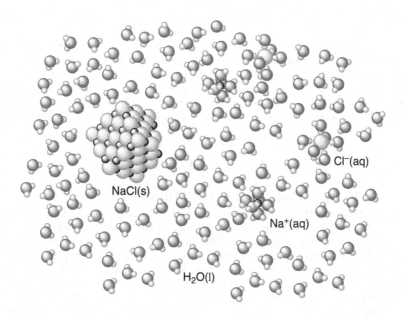

NaCl(s)

Cl⁻(aq)

Na⁺(aq)

H₂O(l)

NOTES

1. The idea that later became the second law was first expressed by S. Carnot: Carnot, S. Réflexions sur la Puissance Motrice du Feu et sur les Machines Propres a Développer cette Puissance. Bachelier: Paris, 1824.

2. (i) Clausius, R. Ueber die bewegende Kraft der Wärme und die Gesetze, welche sich daraus für die Wärmelehre selbst arbeiten lassen. *Ann. Phys.* (Leipzig), **1850**, *155* (4), 500–524. DOI: 10.1002/andp.18501550306.

   (ii) Clausius, R. Ueber eine veränderte Form des zweiten Hauptsatzes der mechanischen Wärmetheorie. *Ann. Phys.* (Leipzig), **1854**, *169* (12), 481–506. DOI: 10.1002/andp.18541691202.

3. The use of bidirectional arrows is first attributed to van't Hoff, J.H. Études de Dynamique Chimique. Frederik Muller & Co.: Amsterdam, 1884. Bidirectional arrows with full arrow heads (⇄) are for reactions not at equilibrium (but proceeding in both directions), while bidirectional arrows with half arrow heads (⇌) are for reactions at equilibrium (Chapter 18), per IUPAC convention: *Quantities, Units, and Symbols in Physical Chemistry*, 3rd ed., prepared for publication by E.R. Cohen, T. Cvitas, J.G. Frey, B. Holmstrom, K. Kuchitsu, R. Marquardt, I. Mills, F. Pavese, M. Quack, J. Stohner, H. Strauss, M. Takami, and A.J. Thor, RSC Publishing, 2007.

4. Clausius, R. Ueber verschiedene für die Anwendung bequeme Formen der Hauptgleichungen der mechanischen Wärmetheorie. *Ann. Phys.* (Leipzig), **1865**, *201* (7), 353–400. DOI: 10.1002/andp.18652010702.

5. Nernst, W. *The New Heat Theorem: Its Foundations in Theory and Experiment.* Methuen & Co. Ltd.: London, 1926.

17 Gibbs Energy

So far in discussing the energetics of reactions, we have considered the flow of thermal energy (enthalpy, $\Delta_r H°$) and the dispersal or spread of energy (entropy, $\Delta_r S°$), and we have considered what makes a reaction reactant favored ($\Delta_r H° > 0$ or $\Delta_r S° < 0$) or product favored ($\Delta_r H° < 0$ or $\Delta_r S° > 0$). While we have considered these individually, the flow of thermal energy and the dispersal of energy are not happening in isolation. Rather both processes happen at the same time, and both have an impact on the outcome of a reaction. Consider the hydration of ethylene with water:

$$C_2H_4(g) + H_2O(l) \rightleftharpoons C_2H_5OH(aq) \qquad \Delta_r H° = -54.9 \text{ kJ/mol}$$

$$\Delta_r S° = -140.8 \text{ J/(mol K)} \tag{17.1}$$

DETERMINING THE FAVORABILITY OF A REACTION

The hydration of ethylene with water is product-favored enthalpically ($\Delta_r H° < 0$) and reactant-favored entropically ($\Delta_r S° < 0$). In terms of this reaction, the question then becomes: Is this a product-favored or a reactant-favored reaction? We need means of identifying whether a reaction is likely to proceed on its own, is product favored, or if it requires energy to proceed, is reactant favored. To do that, we use the composite quantity Gibbs energy ($\Delta_r G°$) shown in Equation 17.2.[1,2] $\Delta_r G°$ provides a way of determining whether the reaction is product favored and will spontaneously proceed, once the activation energy has been provided, to make products ($\Delta_r G° < 0$) or if the reaction is reactant favored and needs energy continuously supplied to make products ($\Delta_r G° > 0$).

$$\Delta_r G° = \Delta_r H° - T\Delta_r S° \tag{17.2}$$

$\Delta_r G°$ is Gibbs energy (in kJ/mol).
$\Delta_r H°$ is the reaction enthalpy (heat flow in kJ/mol).
$\Delta_r S°$ is the reaction entropy (spread of energy and matter in kJ/(mol K); this must be converted from the J/(mol K) it is normally reported in).
T is the temperature (K).

Using Equation 17.2 for the hydration of ethylene, we find that at room temperature (298.15 K), the reaction has a Gibbs energy of −12.9 kJ/mol. Note that the entropy value must be converted to kJ/(mol K) or the enthalpy must be converted to J/mol.

$$\Delta_r G° = -54.9 \text{ kJ} / \text{mol} - \left(298.15 \text{ K}\right)\left(-0.1408 \text{ kJ} / \left(\text{mol K}\right)\right)$$

$$= -12.9 \text{ kJ} / \text{mol}$$

The negative value (an exergonic reaction) means that this reaction is product favored. It should be noted that this reaction proceeds *very* slowly. The fact that Gibbs energy tells us a reaction is favored does not tell us anything about the kinetics – specifically the activation energy – of a reaction. In addition, since the equation to calculate Gibbs energy from enthalpy and entropy includes temperature (Equation 17.2), the temperature at which a reaction takes place plays a significant role in whether it is reactant favored or product favored.

PROBLEM 17.1
For each reaction, calculate $\Delta_r G°$ at 77.0 K (the temperature of liquid nitrogen), 298.15 K (room temperature), and 737 K (the surface temperature of Venus).

a. $CH_3OH(g) + H_2(g) \rightleftharpoons CH_4(g) + H_2O(g)$

$\Delta_r H° = -116.0$ kJ/mol, $\Delta_r S° = 4.6$ J/(mol K)

b. $Mg(OH)_2(s) \rightleftharpoons MgO(s) + H_2O(g)$

$\Delta_r H° = 81.0$ kJ/mol, $\Delta_r S° = 152.6$ J/(mol K)

c. $2 CH_4(g) \rightleftharpoons C_2H_6(g) + H_2(g)$

$\Delta_r H° = 64.9$ kJ/mol, $\Delta_r S° = -12.9$ J/(mol K)

d. $CO(g) + 2 H_2(g) \rightleftharpoons CH_3OH(g)$

$\Delta_r H° = -90.1$ kJ/mol, $\Delta_r S° = -219.2$ J/(mol K)

DOI: 10.1201/9781003479338-17

e. $H_2(g) + Br_2(g) \rightleftharpoons 2\ HBr(g)$

   $\Delta_r H° = -103.7\ kJ/mol,\ \Delta_r S° = 21.2\ J/(mol\ K)$

EFFECT OF TEMPERATURE ON THE FAVORABILITY OF A REACTION

Considering the answers to the calculations above, you should be able to see the relationships shown in Table 17.1. For an exothermic reaction where entropy increases ($\Delta_r H° < 0$ or $\Delta_r S° > 0$), the reaction is product favored at all temperatures, and for an endothermic reaction where entropy decreases ($\Delta_r H° > 0$ or $\Delta_r S° < 0$), the reaction is reactant favored at all temperatures. For reactions where enthalpy and entropy favor different sides of the reaction, the temperature comes into play. For an exothermic reaction where entropy decreases ($\Delta_r H° < 0$ or $\Delta_r S° < 0$), the reaction is product favored at low temperatures and reactant favored at high temperatures. In an endothermic reaction where entropy increases ($\Delta_r H° > 0$ or $\Delta_r S° > 0$), the reaction is reactant favored at low temperatures and product favored at high temperatures.

The terms "high temperature" and "low temperature" are relative because they will vary for each reaction. We can determine the temperature (in kelvins) at which a reaction will invert from product favored to reactant favored (or from reactant favored to product favored, depending on the specific reaction) by using Equation 17.3:

$$T = \frac{\Delta_r H°}{\Delta_r S°} \tag{17.3}$$

In using Equation 17.3, the first step is to convert $\Delta_r H°$ to J/mol, or $\Delta_r S°$ to kJ/(mol K), so that the units appropriately cancel. If we consider the reaction from Equation 17.1, we found that the reaction was product favored at room temperature. Based on the sign of its enthalpy (−54.9 kJ/mol) and entropy (−140.8 J/(mol K)), this reaction should be product favored at low temperatures and reactant favored at high temperatures. The inversion temperature, using Equation 17.3 and converting entropy to −0.1408 kJ/(mol K), shows that the inversion happens at 390 K (117 °C):

$$T = \frac{-54.9\ kJ\,/\,mol}{-0.1408\ kJ\,/\,(mol\ K)} = 390.\ K\left(or\ 117°C\right)$$

It should be noted that one could try to find the inversion temperature of an exothermic reaction where entropy increases ($\Delta_r H° < 0$ or $\Delta_r S° > 0$) or for an endothermic reaction where entropy decreases ($\Delta_r H° > 0$ or $\Delta_r S° < 0$), but the result will be a negative temperature in kelvins. What this tells us is that there is no possible temperature at which these reactions will change from product favored to reactant favored or vice versa.

PROBLEM 17.2
For each reaction, determine the inversion temperature in Celsius (°C).

a. $CH_3OH(g) + H_2(g) \rightleftharpoons CH_4(g) + H_2O(g)$

   $\Delta_r H° = -116.0\ kJ/mol,\ \Delta_r S° = 4.6\ J/(mol\ K)$

b. $Mg(OH)_2(s) \rightleftharpoons MgO(s) + H_2O(g)$

   $\Delta_r H° = 81.0\ kJ/mol,\ \Delta_r S° = 152.6\ J/(mol\ K)$

c. $2\ CH_4(g) \rightleftharpoons C_2H_6(g) + H_2(g)$

   $\Delta_r H° = 64.9\ kJ/mol,\ \Delta_r S° = -12.9\ J/(mol\ K)$

d. $CO(g) + 2\ H_2(g) \rightleftharpoons CH_3OH(g)$

   $\Delta_r H° = -90.1\ kJ/mol,\ \Delta_r S° = -219.2\ J/(mol\ K)$

Table 17.1 Relationship between the Sign of Entropy and the Temperature at Which a Reaction Is Reactant or Product Favored

	$\Delta_r H° < 0$	$\Delta_r H° > 0$
$\Delta_r S° < 0$	Reactant favored at high temperatures Product favored at low temperatures	Reactant favored at all temperatures Product favored at no temperature
$\Delta_r S° > 0$	Reactant favored at no temperature Product favored at all temperatures	Reactant favored at low temperatures Product favored at high temperatures

e. $H_2(g) + Br_2(g) \rightleftarrows 2 HBr(g)$

$\Delta_r H° = -103.7$ kJ/mol, $\Delta_r S° = 21.2$ J/(mol K)

STANDARD GIBBS ENERGY OF FORMATION ($\Delta_f G°$)

Although Gibbs energy is a composite of enthalpy and entropy – and calculating Gibbs energy from enthalpy and entropy gives us the most amount of information about the reaction and the effect of temperature – measuring or calculating the enthalpy and entropy of reaction and then finding Gibbs energy can add a significant amount of time to finding the answer to whether a reaction will happen at room temperature. Another more immediate way to calculate Gibbs energy (only at 298 K) is to use the standard Gibbs energy of formation ($\Delta_f G°$). These values are found with the same assumption as the standard enthalpy of formation ($\Delta_f H°$), which is that standard Gibbs energy of formation of any pure element in its most stable state is 0. The standard Gibbs energy of formation values can then be collated (see Appendix 7), and the Gibbs energy of reaction ($\Delta_r G°$) can be calculated using Equation 17.4, which should look familiar in form to how we have calculated other thermodynamic quantities.

$$\Delta_r G° = \sum n \Delta_f G°(\text{products}) - \sum n \Delta_f G°(\text{reactants}) \tag{17.4}$$

Reconsidering the hydration of ethylene (Equation 17.1) and using Appendix 7, we can recalculate the Gibbs energy of reaction using $\Delta_f G°$ values:

$$\Delta_r G° = \sum n \Delta_f G°(\text{products}) - \sum n \Delta_f G°(\text{reactants})$$

$$= 1\Delta_f G°(C_2H_5OH(l)) - [1\Delta_f G°(C_2H_4(g)) + 1\Delta_f G°(H_2O(l))]$$

$$= -174.8 \text{ kJ/mol} - [68.4 \text{ kJ/mol} + -237.1 \text{ kJ/mol}]$$

$$= -6.1 \text{ kJ/mol}$$

The value does, and usually will, differ somewhat from the calculation of Gibbs energy using enthalpy and entropy. The values should be comparable, but they will differ because of the difference in reference points for standard molar entropy ($S°$) and enthalpy of formation ($\Delta_f H°$) versus the reference point of standard Gibbs energy of formation ($\Delta_f G°$).

PROBLEM 17.3

Using the standard Gibbs energy of formation values ($\Delta_f G°$) in Appendix 7, calculate the Gibbs energy of reaction ($\Delta_r G°$) for each of the following.

a. $3 C_2H_2(g) \rightleftarrows C_6H_6(l)$

b. $2 H_2O_2(l) \rightleftarrows 2 H_2O(l) + O_2(g)$

c. $Pb(s) + PbO_2(s) + 2 H_2SO_4(aq) \rightleftarrows 2 PbSO_4(s) + 2 H_2O(l)$

d. $2 KMnO_4(aq) + 6 HCl(aq) + 5 CH_2O(g) \rightleftarrows 5 HCO_2H(aq) + 2 MnCl_2(s) + 2 KCl(aq) + 3 H_2O(l)$

e. $B_2O_3(s) + 3 Mg(s) \rightleftarrows 2 B(s) + 3 MgO(s)$

GIBBS ENERGY AND REACTION ANALYSIS

While Gibbs energy values ($\Delta_r G°$) are useful in determining whether a reaction is reactant favored or product favored, Gibbs energy is also unto itself a useful quantity. For reactions where the Gibbs energy of reaction is positive (endergonic reactions, $\Delta_r G° > 0$), the energy of the products is higher than the energy of the reactants (Figure 17.1), and the value of Gibbs energy represents the minimum amount of energy necessary to make the reaction happen. The value $\Delta_r G°$ only provides the relative difference in energy between the reactants and products and does not account for activation energy (E_a), so the true energy necessary to make an endergonic reaction happen is the value Gibbs energy plus the activation energy.

For reactions where the Gibbs energy of reaction is negative (exergonic reactions, $\Delta_r G° < 0$), the energy of the products is lower than the energy of the reactants (Figure 17.2), and the value of Gibbs energy represents the maximum amount of energy available (or free) to be used to do work.

A useful analogy to think about for endergonic and exergonic reactions is rolling up or rolling down a hill. For an endergonic reaction, this would be analogous to rolling up a hill. As you lay at the bottom of the hill you must input energy to flip yourself over (the activation energy, E_a) and you must input energy for every amount of progress you make up the hill. In contrast, in an exergonic reaction, it is analogous to rolling down a hill. As you lay at the top of the hill, you must

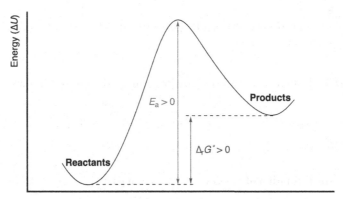

Figure 17.1 Endergonic ($\Delta_r G° > 0$) reaction showing the relative energy of products and reactants, the magnitude of the Gibbs energy of reaction ($\Delta_r G°$), and the magnitude of the activation energy (E_a).

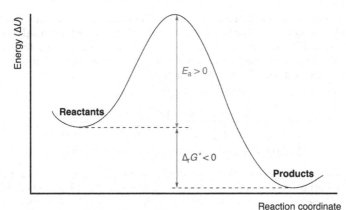

Figure 17.2 Exergonic ($\Delta_r G° < 0$) reaction showing the relative energy of products and reactants, the magnitude of the Gibbs energy of reaction ($\Delta_r G°$), and the magnitude of the activation energy (E_a).

input energy to flip yourself over (the activation energy, E_a), but once you start rolling the energy released by rolling down the hill keeps you going without you having to do anything.

⤳ GIBBS ENERGY AND COUPLED REACTIONS

Chemists often consider the ability to use available, exergonic ($\Delta_r G° < 0$) Gibbs energy to provide energy for endergonic ($\Delta_r G° > 0$) reactions. When an exergonic reaction is used to drive an endergonic reaction, it is said that the reactions are "coupled"; this can help to make reactions more energy efficient. The trade-off is a potential increased cost in mass as more chemical matter is used and the atom economy can decrease. Consider the following set of questions that investigate coupled reactions.

PROBLEM 17.4

Metals are often dug out of the ground as ores, compounds with oxygen, sulfur, and other elements. To isolate the pure metal, many metallurgical reactions and processes involve transformations that will produce a metal oxide. To isolate the metal, the metal oxide could be heated until it decomposes. Consider an example of iron(III) oxide – the ore is known as hematite – decomposing:

$$2\ Fe_2O_3(s) \rightleftarrows 4\ Fe(s) + 3\ O_2(g)$$

Using the values in Appendix 7, calculate $\Delta_r G°$ for the decomposition of iron(III) oxide to pure iron. Is this a product-favored or reactant-favored process? Can you offer an explanation as to why the reaction needs to be heated?

PROBLEM 17.5
Calculate the temperature iron(III) oxide must be heated to (in °C) to spontaneously decompose into iron and oxygen.

PROBLEM 17.6
Now consider the formation of carbon dioxide from solid, graphite carbon (coke):

$$C(s, graphite) + O_2(g) \rightleftharpoons CO_2(g)$$

What is $\Delta_r G°$ for the oxidation of coke to carbon dioxide? Is this a product-favored or reactant-favored process?

PROBLEM 17.7
Now consider the formation of water from hydrogen gas:

$$2 H_2(g) + O_2(g) \rightleftharpoons 2 H_2O(l)$$

What is $\Delta_r G°$ for the oxidation of hydrogen to water? Is this a product-favored or reactant-favored process?
What are the advantages of oxidizing hydrogen over oxidizing coke?

PROBLEM 17.8
In industrial metallurgy, it is often the case that the reduction of a metal oxide to pure metal is accomplished by heating that metal oxide with coke. Increasingly, companies are looking to switch to heating the metal oxide with hydrogen. If you take the chemical equations for the decomposition of iron(III) oxide and the formation of water and add them, you get a new combined (coupled) reaction:

$$2 Fe_2O_3(s) \rightleftharpoons 4 Fe(s) + 3 O_2(g)$$
$$6 H_2(g) + 3 O_2(g) \rightleftharpoons 6 H_2O(l)$$

$$6 H_2(g) + 2 Fe_2O_3(s) \rightleftharpoons 4 Fe(s) + 6 H_2O(l), overall$$
$$3 H_2(g) + Fe_2O_3(s) \rightleftharpoons 2 Fe(s) + 3 H_2O(l), simplified$$

In coupling a reaction there is a different overall reaction, but one that still produces pure iron, the desired product. The new $\Delta_r G°$ for this process can be calculated as you did above by using the $\Delta_f G°$ values, but since we are adding the two reactions together, we can find $\Delta_r G°$ by adding the $0.5\Delta_r G°$ from Problem 17.4 and $1.5\Delta_r G°$ from Problem 17.7. What is $\Delta_r G°$ for this coupled reaction?

PROBLEM 17.9
Considering your answer to Problem 17.8, can you explain why metallurgy couples the oxidation of hydrogen with the reduction of a metal oxide (like iron(III) oxide)?

PROBLEM 17.10
Problem 17.8 states that the metal oxide is heated together with hydrogen. Given your answer to Problem 17.8 and the value of $\Delta_r G°$, can you provide an explanation as to why metal oxide and hydrogen must be heated? (There are two reasons for this.)

PROBLEM 17.11
Consider the hydrolysis of ATP in water:

$$ATP(aq) + H_2O(l) \rightleftharpoons ADP(aq) + HPO_4{}^{2-}(aq) \quad \Delta_r G° = -30.5 \text{ kJ/mol}$$

Looking at the $\Delta_r G°$ value, can you provide an explanation for how the body uses ATP (in a general sense)? That is, from an energetic standpoint, what is ATP doing? Consider the metallurgy example and the role of dihydrogen.

GIBBS AND REACTION CONTROL

Coupling reactions are a useful means of effecting a desired chemical transformation, which means that coupling reactions provide us (along with temperature) a means of controlling chemical reactions. Up until this point, one could easily be forgiven for thinking that chemistry is nothing more than understanding the how and why of atoms, molecules, and change. That is a substantial portion of what is taught in chemistry. But now with $\Delta_r G°$, we can predict and

Figure 17.3 The structure and standard Gibbs energy of formation values of methanol and iodomethane.

understand the choice that goes into a chemist's thinking when they decide to make a new chemical. Let's look at the following two chemicals: methanol and iodomethane (Figure 17.3).

Methanol is a common chemical that can be made industrially from carbon monoxide and hydrogen or biosynthetically from methane. Iodomethane is a much more useful reactant in chemical synthesis and is typically produced from methanol. We will consider how we can effect this chemical transformation below. Consider the questions on the following pages that investigate the transformation of methanol to iodomethane. First, consider the reaction of methanol becoming iodomethane (Figure 17.4).

You can see that we have an unbalanced reaction, which we need to first balance (Figure 17.5).

Note that the ions are indicated as being aqueous because most reactions need some sort of solvent to allow the chemical to mix and water is the solvent with the most readily available data.

PROBLEM 17.12
What is $\Delta_r G°$ for the reaction to make iodomethane from methanol and iodide?

PROBLEM 17.13
Are we going to succeed in making iodomethane this way at 25 °C?

PROBLEM 17.14
Given the $\Delta_f H°$ and $S°$ values for each of the following structures, what are $\Delta_r H°$ and $\Delta_r S°$? Is this a reaction that could be made to happen by changing the temperature to be higher or lower than 25 °C?

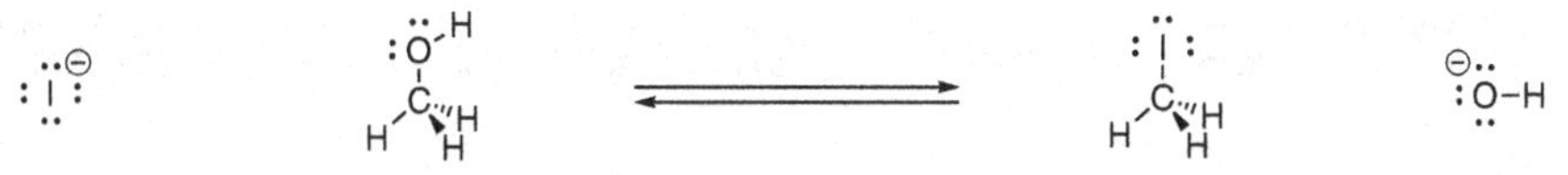

iodide(aq)
$\Delta_f G° = -51.6$ kJ/mol
$\Delta_f H° = -55.2$ kJ/mol
$S° = 111.3$ J/(mol K)

methanol(aq)
$\Delta_f G° = -175.3$ kJ/mol
$\Delta_f H° = -245.9$ kJ/mol
$S° = 133.1$ J/(mol K)

iodomethane(l)
$\Delta_f G° = 13.4$ kJ/mol
$\Delta_f H° = -15.5$ kJ/mol
$S° = 163.2$ J/(mol K)

hydroxide(aq)
$\Delta_f G° = -157.2$
$\Delta_f H° = -230.0$ kJ/mol
$S° = -10.8$ J/(mol K)

Even if we can change the temperature and affect the reaction outcome, chemists typically like to keep reactions as close to room temperature as possible, as heating or cooling a reaction can

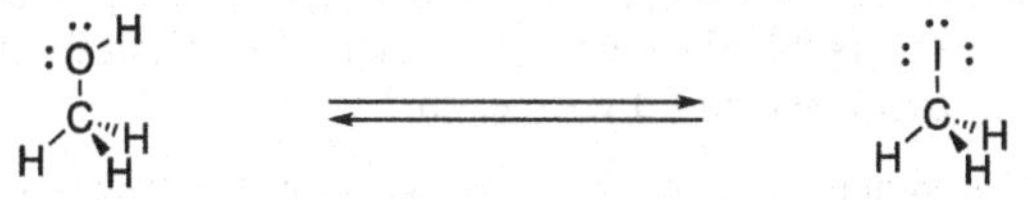

methanol(aq)
$\Delta_f G° = -175.3$ kJ/mol

iodomethane(l)
$\Delta_f G° = 13.4$ kJ/mol

Figure 17.4 Unbalanced reaction of methanol becoming iodomethane.

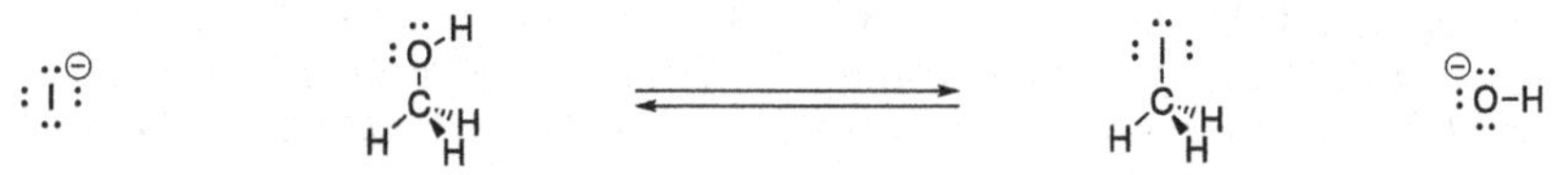

iodide(aq)
$\Delta_f G° = -51.6$ kJ/mol

methanol(aq)
$\Delta_f G° = -175.3$ kJ/mol

iodomethane(l)
$\Delta_f G° = 13.4$ kJ/mol

hydroxide(aq)
$\Delta_f G° = -157.2$ kJ/mol

Figure 17.5 Balanced reaction of methanol becoming iodomethane.

present its own challenges. Given all this information, then, the only option is to look at a different reactant other than iodide, though realistically, iodide does not exist all by itself, as a bottle of negative charge would explode, and thus some other byproduct other than just hydroxide. Looking in the stockroom, a chemist finds that they have NaI(aq), HI(aq), P(s, red), and I_2(s).

PROBLEM 17.15
Calculate $\Delta_r G°$ for each reaction. Which reaction will work to make iodomethane at room temperature and which will not?

sodium iodide(aq)
$\Delta_f G° = -313.5$ kJ/mol

methanol(aq)
$\Delta_f G° = -175.3$ kJ/mol

iodomethane(l)
$\Delta_f G° = 13.4$ kJ/mol

sodium hydroxide(aq)
$\Delta_f G° = -419.1$ kJ/mol

a. $\Delta_r G° =$

hydrogen iodide(aq)
$\Delta_f G° = -51.6$ kJ/mol

methanol(aq)
$\Delta_f G° = -175.3$ kJ/mol

iodomethane(l)
$\Delta_f G° = 13.4$ kJ/mol

water(l)
$\Delta_f G° = -237.1$ kJ/mol

b. $\Delta_r G° =$

iodine(s)
$\Delta_f G° = 0$ kJ/mol

phosphorus(s, red)
$\Delta_f G° = -12.2$ kJ/mol

methanol(aq)
$\Delta_f G° = -175.3$ kJ/mol

iodomethane(l)
$\Delta_f G° = 13.4$ kJ/mol

phosphoric acid(l)
$\Delta_f G° = -1123.6$ kJ/mol

water(l)
$\Delta_f G° = -237.1$ kJ/mol

c. $\Delta_r G° =$

⇝ PROBLEM 17.16
Consider the reactions in Problem 17.15.

a.   Which reaction is (or reactions are) best in terms of Gibbs energy?

b.   Which reaction is (or reactions are) best in terms of atom economy?

c.   Which reaction is (or reactions are) better in terms of chemical hazards (use the internet to find the safety data sheets [SDS] for each chemical other than methanol and iodomethane and identify the hazards associated with each)?

d.   Is there a single best reaction, from a green chemistry perspective, to produce iodomethane?

NOTES

1. Gibbs energy is also commonly referred to as Gibbs free energy though this terminology has been deprecated. "Gibbs energy (function) *G*." IUPAC. *Compendium of Chemical Terminology*, 2nd ed. Compiled by A.D. McNaught and A. Wilkinson. Blackwell Scientific Publications, Oxford, 1997. Online version (2019–) created by S.J. Chalk. DOI: 10.1351/goldbook.G02629.

2. Gibbs's original presentation was more closely in line with the first law of thermodynamics: $\Delta_r G° = \Delta_r U° + P\Delta V - T\Delta_r S°$. In: Gibbs, J.W. A Method of Geometrical Representation of the Thermodynamic Properties of Substances by Means of Surfaces. *Trans. Conn. Acad. Arts Sci.* **1873**, 2, 382–404.

18 Equilibrium

As we have seen, $\Delta_r G°$ provides a measure for whether a reaction is product-favored ($\Delta_r G° < 0$) energetically or whether a reaction is reactant-favored ($\Delta_r G° > 0$) energetically. Although $\Delta_r G°$ provides a useful understanding of the energy changes associated with a reaction, as a field chemistry is frequently more interested in the amounts of those chemicals.

EQUILIBRIUM AND GIBBS ENERGY

Let us start the discussion of equilibrium by reconsidering the hydration of ethylene with water:

$$C_2H_4(g) + H_2O(l) \rightleftharpoons C_2H_5OH(aq) \qquad \Delta_r G°(298.15 \text{ K}) = -12.9 \text{ kJ/mol} \qquad (18.1)$$

Equilibrium is a dynamic state where the overall amount concentration of chemicals is not changing, but there is an ongoing interconversion of reactants to products and products to reactants.[1] To measure the amount of each chemical, we utilize the idea of a unitless equilibrium constant (K). The equilibrium constant tells us whether there is a higher amount concentration of products ($K > 1$, a product-favored reaction) or a higher amount concentration of reactants ($K < 1$, a reactant-favored reaction). While we can measure the actual amount concentrations (or actual pressure) of each chemical (see below), we can also relate K to $\Delta_r G°$ using Equation 18.2 and Equation 18.3:[2]

$$K = e^{-\left(\frac{\Delta_r G°}{RT}\right)} \qquad (18.2)$$

K is the ratio of product amount concentration to reactant amount concentration (unitless).
$\Delta_r G°$ is the Gibbs energy of reaction (kJ/mol).
R is the gas constant (0.008 314 kJ/(mol K)).
T is the temperature (K).

$$\Delta_r G° = -RT \ln K \qquad (18.3)$$

K is the ratio of product amount concentration to reactant amount concentration (unitless)
$\Delta_r G°$ is the Gibbs energy of reaction (kJ/mol).
R is the gas constant (0.008 314 kJ/(mol K)).
T is the temperature (K).

Using Equation 18.2 for the hydration of ethylene we find that at room temperature (298.15 K), the reaction has a thermodynamic standard temperature (298.15 K) equilibrium constant (K) of 200. Since the equilibrium constant is dependent upon temperature, the temperature always needs to be clearly specified for a given value.

$$K(298.15 \text{ K}) = e^{-\left(\frac{-12.9 \text{ kJ/mol}}{(0.008314 \text{ J/(mol K)})(298.15 \text{ K})}\right)} = 200$$

The value (200) is much greater than 1, which indicates that the amount of product ($C_2H_5OH(aq)$) is 200 times greater than the amount of reactant ($C_2H_4(g)$).[3] In general, for $K \gg 1$, the amount of products at equilibrium is greater than the amount of reactants, while for $K \ll 1$ the amount of reactants at equilibrium is greater than the amount of products. For K near 1, especially for complex reactions, the exact amount of products and reactants will need to be calculated, which will be explored further below.

PROBLEM 18.1
Using Appendix 7, calculate $\Delta_r G°$ for this reaction (25 °C) and then convert that value to the equilibrium constant (K).

$$H_2O(l) \rightleftharpoons H^+(aq) + OH^-(aq)$$

PROBLEM 18.2
Convert the $\Delta_r G°$ values to equilibrium constants (K) at 298 K.

a. $ATP(aq) + H_2O(l) \rightleftharpoons ADP(aq) + HPO_4^{2-}(aq)$ $\Delta_r G° = -30.5 \text{ kJ/mol}$

b. $H_2(g) + CO_2(g) \rightleftharpoons H_2O(g) + CO(g)$ $\Delta_r G° = 28.62 \text{ kJ/mol}$

DOI: 10.1201/9781003479338-18

PROBLEM 18.3
Convert each equilibrium constant (*K*) to $\Delta_r G°$ at 298 K.

a. $N_2O_4(g) \rightleftharpoons 2\,NO_2(g)$ $K = 0.0063$

b. $CH_3CO_2H(l) + C_2H_5OH(aq) \rightleftharpoons CH_3CO_2C_2H_5(l) + H_2O(l)$ $K = 0.4$

c. $CuCl_2(aq) + Fe(s) \rightleftharpoons Cu(s) + FeCl_2(aq)$ $K = 2.4 \times 10^{26}$

IMPORTANCE OF EQUILIBRIUM AND GIBBS ENERGY

"Chemistry," declared Roger Kornberg in an interview, "is the queen of all sciences. Our best hope of applying physical principles to the world around us is at the level of chemistry. In fact, if there is one subject which an educated person should know in the world it is chemistry." Kornberg won the 2006 Nobel Prize in chemistry for his work on transcription, which involved unraveling the more than a dozen complicated proteins involved in the copying of DNA into RNA. He would know how important chemistry is in uncovering the details of a ubiquitous life process.[4]

I must therefore inevitably take my cue from Kornberg and ask the following question: *What equation would you regard as the most important one in science?* For most people the answer to this question would be easy: Einstein's famous mass-energy formula, $E = mc^2$. Some people may cite Newton's inverse square law of gravitation. And yet it should be noted that both equations are virtually irrelevant for most practicing physicists, chemists, and biologists. They are familiar to the public mainly because they have been widely publicized and are associated with two very famous scientists. There is no doubt that both Einstein and Newton are supremely important for understanding the universe, but they both suffer from the limitations of reductionist science that preclude the direct application of the principles of physics to the everyday workings of life and matter.

Take Einstein's formula for instance. About the only importance it has for most physical scientists is the fact that it is responsible for the nuclear processes that have forged the elements in stars and supernovae. Chemists deal with reactions that involve not nuclear processes but the redistribution of electrons. Except in certain cases, Einstein therefore does not figure in chemical or biological processes. Newton's gravitational formula is equally distant. Chemical reactions involve the attraction and repulsion of charges, which are processes governed by the electromagnetic force. This force is stronger than the gravitational force by a factor of 10^{36}, an unimaginable number. Thus, gravity is too weak for chemists and biologists to bother with it in their work. The same goes for many physicists who deal with atomic and molecular interactions.

Instead, here are two equations that have a far greater and more direct relevance to the work done by most physical and biological scientists. The equations lie at the boundary of physics and chemistry, and both are derived from a science whose basic truths are so permanently carved in stone that Einstein thought they would never, ever need to be modified. That science is thermodynamics, and the equations we are talking about involve the most basic variables in thermodynamics. They apply without exception to every important physical and chemical process you can think of, from the capture of solar energy by plants and solar cells to the combustion of fuel inside trucks and human bodies to the union of sperm and egg.

Two thermodynamic quantities govern molecular behavior, and indeed the behavior of all matter in the universe. One is *enthalpy*, usually denoted by the symbol *H*, and roughly representing the quantity of energy and the strength of interactions and bonds between different atoms and molecules. The other is *entropy*, usually denoted by the symbol *S*, and roughly representing the quality of energy and the disorder in any system. Together the enthalpy and entropy make up the *Gibbs energy*, *G*, which roughly denotes the amount of useful work that can be extracted from any living or non-living system. In practical calculations, what we are concerned with are changes in these quantities rather than their absolute values, so each one of them is prefaced by the symbol Δ indicating change. The celebrated second law of thermodynamics states that the entropy of a spontaneous process always increases, and it is indeed one of the universal facts of life, but that is not what we are concerned with here.

Think about what happens when two molecules – of any kind – interact with each other. It need not even be an actual reaction; it can simply be the binding of two molecules to one another by strong (hydrogen bond) or weak (van der Waals interactions) forces. The interaction is quantified by an *equilibrium constant*, *K*, which is simply the ratio of the concentrations of the products of the reaction to the starting material (reactants). The bigger the equilibrium constant, the more the amount of the products. *K* thus tells us how much of a reaction has been completed, how much reactant has been converted to product. Our first great equation relates this equilibrium constant to the Gibbs energy of the interaction through the following formula:

$$\Delta_r G^\circ = -RT \ln K$$

or, in other words

$$K = e^{-\left(\frac{\Delta_r G^\circ}{RT}\right)}$$

Here ln is the natural logarithm base e, R is a fundamental constant called the gas constant, T is the ambient temperature, and $\Delta_r G^\circ$ is the energy change. This equation tells us two major things and one minor thing. The minor thing is that reactions can be driven toward reactants or toward products by temperature increases, and exponentially so (but that is not the same as speeding them up; this goal is the domain of *kinetics*, not thermodynamics). But the major things are what is critical here. First, the equation says that the Gibbs energy in a product-favored reaction with a favorable (greater than 1) equilibrium constant is always going to be negative; the more negative it is, the better. And that is what you find. The Gibbs energy change for many of biology's existential reactions, like the coupling of biological molecules with ATP (the "energy currency" of the cell), the process of electron transfer mediated by chlorophyll, and the oxidation of glucose to provide energy, is indeed negative. Life has also worked out clever little tricks to couple reactions with positive (unfavorable) $\Delta_r G^\circ$ changes to those with negative $\Delta_r G^\circ$ values to give an overall favorable energy profile.

The second feature of the equation is a testament to the wonder that is life, and it never ceases to amaze me. It attests to what scientists and philosophers have called "fine-tuning" the fact that evolution has somehow succeeded in minimizing the error inherent in life's processes, in carefully reining in the operations of life within a narrow window. Look again at that expression. It says that $\Delta_r G^\circ$ is related to K not linearly but exponentially. That is a dangerous proposition because it means that even a tiny change in $\Delta_r G^\circ$ will correspond to a large change in K. How tiny? No bigger than 12.6 kJ/mol.

A brief digression to appreciate how small this value is. Energies in chemistry are usually expressed as kilojoules per mole. A bond between two carbon atoms is about 335 kJ/mol. A bond between two nitrogen atoms is 946 kJ/mol, indicating why nitrogen can be converted to ammonia by breaking this bond only at very high temperatures and pressures and in the presence of a catalyst. A hydrogen bond – the noncovalent interaction that holds together biological molecules like DNA and proteins – is anywhere between 8.4 and 41.8 kJ/mol.

Thus, 12.6 kJ/mol is a fraction of the typical energy of a bond. It takes just a little jiggling around to overcome this energy barrier; if you ask a chemist to predict or optimize a reaction within this range, they will be extremely uncomfortable. The exponential, highly sensitive dependence of K on $\Delta_r G^\circ$ means that changing $\Delta_r G^\circ$ from close to zero to 12.6 kJ/mol will translate to changing K from 1:99.98 in favor of products to 99.98:1 in favor of reactants (remember that K is a ratio). It is not even chemistry; it is a simple mathematical truth. Thus, a tiny change in $\Delta_r G^\circ$ can all but completely shift a chemical reaction from favoring products to favoring reactants. Naturally, this will be very bad if the goal of a reaction is to create products that are funneled into the next chemical reaction. Little changes in the Gibbs energy can therefore radically alter the flux of matter and energy in life's workings. But this does not happen. Evolution has fine-tuned life so well that it has remained a game played within a 12.6 kJ/mol energy window for more than 2.5 billion years. It is so easy for this game to quickly spiral out of hand, but it does not. It does not happen for the trillions of chemical transactions that trillions of cells execute every day in every single organism on this planet.

And it does not happen for a reason; cells would have a very hard time modulating their key chemical reactions if the Gibbs energies involved in those reactions had been too large. Life would be quickly put into a death trap if every time it had to react, fight, move, or procreate it had to suddenly change Gibbs energies for each of its processes by hundreds of kilojoules per mole. There are lots of bonds broken and formed in biochemical events, of course, and as we saw before, these *bond* energies can easily amount to hundreds of kJ/mol. But the tendency of the reactants or products containing those bonds to accumulate is governed by these tiny changes in Gibbs energy which nudge a reaction one way or another. In one sense then, life is optimizing small changes (in Gibbs energy of reactions) between two large numbers (bond energies). This is always a balancing act on the edge of a cliff, and life has managed to be successful in it for billions of years.

Thus, we all hum along smoothly, beneficiaries of a 12.6 kJ/mol energy window, going about our lives even as we are held hostage to the quirks of thermodynamic optimization, walking along an exponential energy precipice. And all because

$$K = e^{-\left(\frac{\Delta_r G^\circ}{RT}\right)}$$

REACTION QUOTIENT (Q)

So far, the equilibrium constant (K) has been considered as an outgrowth of Gibbs energy for a reaction. Now, we are going to consider reactions, the amounts of chemicals involved, and their equilibria without reference to $\Delta_r G^\circ$. For a generic reaction b B + d D $\rightleftharpoons$ e E + f F, we can define a reaction quotient (Q):[5]

$$Q = \frac{a_E^e a_F^f}{a_B^b a_D^d} \qquad (18.4)$$

There is a new piece of notation here. Let us dissect one of the terms in Equation 18.4: a_B^b. The term a_B formally means the *activity* of chemical B.[6] You should think of this as the "actual amount" of chemical B; also note that a is unitless, so Q is also unitless. The superscript b means that we are raising this term to the "b" power, where that represents the stoichiometric coefficient of that chemical in the balanced equation. When we talk about actual amounts or activities, what this looks at is the idea that a measured concentration, in terms of amount concentration or pressure, can be less than we might expect, which can be due to intermolecular interactions or aggregation.

In practice, we tend to treat activity as though it were just the amount concentration, which it is for dilute solutions. If we are looking at the amount of each chemical in terms of the amount concentration (mol/L), then Q is called Q_c (the subscript c means amount concentration). We can, then, also calculate Q_c directly by measuring the amount concentrations of each chemical:

$$Q_c = \frac{[E]^e [F]^f}{[B]^b [D]^d} \qquad (18.5)$$

If it is a gas-phase reaction and we measure the amounts of the chemicals in terms of pressures, then Q is called Q_p (the subscript p means pressure). We can, then, also calculate Q_p directly by measuring the partial pressure of each gas (p_i) at equilibrium:

$$Q_p = \frac{p_E^e p_F^f}{p_B^b p_D^d} \qquad (18.6)$$

At its heart, we need to remember that the basic premise of a reaction quotient (Q) is that it is the ratio of the amount of products (usually measured in amount concentration) over the amount of reactants (also usually measured in amount concentration). It is worth noting that by itself, Q does not tell us anything. Q is useful in that it tells us where a reaction is relative to its equilibrium constant (K), which is dictated by the energy of the reaction. Let us investigate this idea of the reaction quotient by looking at the interconversion of nitrogen dioxide and dinitrogen tetroxide:

$$2\ NO_2(g) \rightleftharpoons N_2O_4(g) \qquad\qquad K_c = 170 \text{ at } 25\ ^\circ C$$

Now consider a situation where we have 1.0×10^{-2} mol/L N_2O_4 and 5.0×10^{-3} mol/L NO_2 at 25 °C. Is this reaction at equilibrium? If not, which way will the reaction go (to make more products or to make more reactants)? To solve this problem, we need to calculate the value of Q_c.

$$Q = \frac{[N_2O_4]}{[NO_2]^2} = \frac{[1.0 \times 10^{-2}]}{[5.0 \times 10^{-3}]^2} = 4.0 \times 10^2$$

What does this indicate? This tells us that the ratio of products to reactants with these concentrations is 400. If the equilibrium ratio is 170, that means that there are more products (and fewer reactants) under these initial conditions than there will be once the reaction reaches equilibrium. So, the reaction will convert product (N_2O_4) to reactant (NO_2) until the equilibrium ratio of 170 is reached.

If we start with all products and no reactant ($Q = \infty$), then the reaction will convert product to reactant until the equilibrium ratio is reached. If we start with all reactants and no product ($Q = 0$), then the reaction will convert reactant to product until the equilibrium ratio is reached. However, what if we have a random mix of product and reactant? How do we know which way the reaction is going to proceed? Or is the reaction already at equilibrium? The answer to these questions is the reason that Q is most useful. If a reaction is at equilibrium then the reaction quotient is the equilibrium constant ($Q = K$). If a reaction is not at equilibrium ($Q \neq K$), then the reaction will proceed to produce more reactant from product, when $Q > K$, or to produce more product from reactant, when $Q < K$. A useful mnemonic is to write K first and then Q with the correct inequality symbol. Using this approach when $K < Q$, the reaction will proceed toward the left (think of the $<$ as $\leftarrow$) to make more reactant. When $K > Q$, the reaction will proceed to the right (think of the $>$ as $\rightarrow$) to make more product.

THE EQUILIBRIUM CONSTANT (K)

If we consider the equilibrium constant itself. Recall that when we are looking at an equilibrium constant, if $K > 1$, that means there is a greater amount concentration of products than of reactants at equilibrium. If $K < 1$, that means there is a greater amount concentration of reactants than of products at equilibrium. Let us consider a generic reaction.

$$b\,\mathrm{B} + d\,\mathrm{D} \rightleftharpoons e\,\mathrm{E} + f\,\mathrm{F}$$

K, here written as K_c to indicate that the amount of each is measured in terms of amount concentration, takes the form of Equation 18.7:[7]

$$K_c = \frac{[\mathrm{E}]^e[\mathrm{F}]^f}{[\mathrm{B}]^b[\mathrm{D}]^d} \tag{18.7}$$

If it is a gas-phase reaction and we measure the amounts of the chemicals in terms of the partial pressure of each gas (p_X), then K is called K_p (the subscript p means pressure):

$$K_p = \frac{p_E^e p_F^f}{p_B^b p_D^d} \tag{18.8}$$

Since either K_c or K_p could be determined for a gas-phase reaction, it is useful to have a way to relate the two. The two equilibrium constants, K_c or K_p, are related by Equation 18.9:

$$K_p = K_c(RT)^{\Delta n} \tag{18.9}$$

R is the gas constant ($0.082\,06\,\dfrac{\mathrm{L\,atm}}{\mathrm{mol\,K}}$).

T is the temperature (K).

Δn is the difference in moles of gas ($\mathrm{mol_{gas\,product}} - \mathrm{mol_{gas\,reactants}}$).

These concentrations are the amount of each chemical *measured at equilibrium*. This contrasts with Q, where we measure the amount of each chemical at the beginning of a reaction, or any other point. Therefore, K should be thought of as a special case of Q. And so, we can find K without knowing anything about the energy of a reaction. All we need to know is the balanced reaction and a means of measuring the concentrations of the chemicals. Consider the reaction between hydrogen and iodine to make hydrogen iodide:

$$\mathrm{H_2(g)} + \mathrm{I_2(g)} \rightleftharpoons 2\,\mathrm{HI(g)}$$

For this reaction, the equilibrium constant (K_c) expression would have HI in the numerator and raised to the second, power while H_2 and I_2 would both be in the denominator:

$$K_c = \frac{[\mathrm{HI}]^2}{[\mathrm{H_2}][\mathrm{I_2}]}$$

To find the value of this equilibrium constant, we would need to measure the concentrations of each chemical at equilibrium. At 425 °C (like $\Delta_r G°$, the value of K is specific to both a reaction and a particular temperature) it was found that $[\mathrm{H_2}] = 0.0037$ mol/L, $[\mathrm{I_2}] = 0.0037$ mol/L, and $[\mathrm{HI}] = 0.0276$ mol/L. So, the value of K_c is:

$$K_c = \frac{[0.0276]^2}{[0.0037][0.0037]} = 56, \text{ at } 425\,°\mathrm{C}$$

That value tells us that this reaction favors products and the amount of products is greater than the amount of reactants, as can be seen in the concentrations.

Some points that are worth keeping in mind about equilibrium constants:

1. When a reaction is inverted, the value of K for the inverted reaction is the inverse of the original K value. For example:

$$H_2(g) + I_2(g) \rightleftharpoons 2\ HI(g)\ K_c = 56, \text{ at } 425\ °C$$

$$2\ HI(g) \rightleftharpoons H_2(g) + I_2(g)\ K_c = 1/56 = 0.018, \text{ at } 425\ °C$$

2. When writing an equilibrium constant expression, we ignore any solids or liquids. The reason for this harks to the strict definition of the reaction quotient. We are using the concentrations of each chemical as a stand-in for activity. For a pure compound in the solid or liquid state, the activity is 1. So, as you write Q and K_c expressions, it is best practice to write out all the terms in the reaction and then go back through and remove any term where the phase notation is (s) or (l). If we consider the dissociation of aqueous hydrogen fluoride:

$$HF(aq) + H_2O(l) \rightleftharpoons F^-(aq) + H_3O^+(aq)$$

So, when we write K_c we would have:

$$K_c = \frac{[F^-(aq)][H_3O^+(aq)]}{[HF(aq)][H_2O(l)]}$$

Now removing the $H_2O(l)$ term the final K_c expression would be:

$$K_c = \frac{[F^-(aq)][H_3O^+(aq)]}{[HF(aq)]}$$

PROBLEM 18.4
For each reaction, write an expression for Q_c.

a. $2\ H_2(g) + O_2(g) \rightleftharpoons 2\ H_2O(g)$ $\hspace{2cm}$ $Q_c =$

b. $Co(H_2O)_6{}^{2+}(aq) + 4\ Cl^-(aq) \rightleftharpoons CoCl_4{}^{2-}(aq) + 6\ H_2O(l)$ $\hspace{0.5cm}$ $Q_c =$

PROBLEM 18.5
Provided an expression for Q_c, what is the chemical reaction?

$$\rightleftharpoons \hspace{3cm} Q_c = \frac{[H_2S(g)]^2 \left[O_2(g)\right]^3}{\left[SO_2(g)\right]^2 \left[H_2O(g)\right]^2}$$

PROBLEM 18.6
For the following reactions, calculate Q_c using the provided amount concentration values. K_c is provided for each reaction. Indicate whether the reaction is at equilibrium or not. If the reaction is not at equilibrium, will more reactants be produced (shift left) or will more products be produced (shift right)?

a. $H_2(g) + I_2(g) \rightleftharpoons 2\ HI(g)$ $\hspace{2cm}$ $K_c(745\ K) = 50.0$

$\hspace{0.5cm}$ $[H_2] = 0.001\ 25\ mol/L$, $[I_2] = 0.001\ 25\ mol/L$, $[HI] = 0.0375\ mol/L$

b. $O_2(g) + N_2(g) \rightleftharpoons 2\ NO(g)$ $\hspace{2cm}$ $K_c(745\ K) = 1.7 \times 10^{-3}$

$\hspace{0.5cm}$ $[O_2] = 0.25\ mol/L$, $[N_2] = 0.25\ mol/L$, $[NO] = 0.0103\ mol/L$

PROBLEM 18.7
For the following reactions, the amount concentration of each chemical at equilibrium was measured. Calculate K_c.

a. $PCl_5(g) \rightleftharpoons PCl_3(g) + Cl_2(g)$

$\hspace{0.5cm}$ At equilibrium at 250 °C, $[PCl_5] = 4.2 \times 10^{-5}\ mol/L$, $[PCl_3] = 1.3 \times 10^{-2}\ mol/L$, $[Cl_2] = 3.9 \times 10^{-3}$ mol/L.

b. $2\ BrF_5(g) \rightleftharpoons Br_2(g) + 5\ F_2(g)$

$\hspace{0.5cm}$ At equilibrium at 1500 K, $[BrF_5] = 0.0064\ mol/L$, $[Br_2] = 0.0018\ mol/L$, $[F_2] = 0.0090\ mol/L$.

INITIAL, CHANGE, EQUILIBRIUM
Finding the Equilibrium Constant

A useful framework for approaching equilibrium calculations is to use an initial, change, equilibrium (ICE) table. Two common types of problems exist. The first is finding K_c. It is often that when we find the value for K_c, we do not know all the equilibrium amount concentration values of the different chemicals involved. We must find the equilibrium amount concentration values from the information provided.

Consider the following reaction:

$$CO(g) + Cl_2(g) \rightleftarrows COCl_2(g)$$

In any problem that involves an equilibrium, we first need to write an equilibrium expression:

$$K_c = \frac{\left[COCl_2(g)\right]}{\left[CO(g)\right]\left[Cl_2(g)\right]}$$

Next, we need to consider the information provided. Here, a mixture of CO and Cl_2 is placed into a reaction flask and sealed. Initially we have [CO] = 0.0102 mol/L and [Cl_2] = 0.0.006 09 mol/L. When the reaction reaches equilibrium at 600 K, [Cl_2] = 0.00301 mol/L. What are the [CO] and [Cl_2] at equilibrium and what is K_c? We now employ an ICE table in which we have I, C, and E rows where we specify the terms of the equation and plug in what we know.

	CO(g)	Cl$_2$(g)	$\rightleftarrows$	COCl$_2$(g)
Initial			$\rightleftarrows$	
Change			$\rightleftarrows$	
Equilibrium			$\rightleftharpoons$	

We are told, "Initially we have [CO] = 0.0102 mol/L and [Cl_2] = 0.006 09 mol/L." So, we plug those concentrations (ICE tables are only ever for concentrations or pressures) into the I row. Since $COCl_2$ is not mentioned, we assume that its initial amount concentration is 0.

	CO(g)	Cl$_2$(g)	$\rightleftarrows$	COCl$_2$(g)
Initial	0.0102 mol/L	0.006 09 mol/L	$\rightleftarrows$	0 mol/L
Change			$\rightleftarrows$	
Equilibrium			$\rightleftharpoons$	

Next, we need to fill in the C row. Here we will input nx, where n is the stoichiometric coefficient for each chemical and x is the amount by which each chemical is changing. Since we start with only reactants and no product, the reactants will each be $-x$ and the product will then be $+x$.

	CO(g)	Cl$_2$(g)	$\rightleftarrows$	COCl$_2$(g)
I	0.0102 mol/L	0.006 09 mol/L	$\rightleftarrows$	0 mol/L
C	$-x$	$-x$	$\rightleftarrows$	$+x$
E			$\rightleftharpoons$	

To complete the table, we then add the I and C rows to get the E row.

	CO(g)	Cl$_2$(g)	$\rightleftarrows$	COCl$_2$(g)
I	0.0102 mol/L	0.006 09 mol/L	$\rightleftarrows$	0 mol/L
C	$-x$	$-x$	$\rightleftarrows$	$+x$
E	$0.0102 - x$ mol/L	$0.006\ 09 - x$ mol/L	$\rightleftharpoons$	x mol/L

At this stage, we need to know some other information. Here the problem tells us, "When the reaction reaches equilibrium at 600 K, [Cl_2] = 0.003 01 mol/L." Since our table above shows that [Cl_2]

at equilibrium is 0.006 09 – x mol/L and the scenario indicates that the amount concentration at equilibrium is 0.003 01 mol/L, we can solve for x.

$$0.006\ 09 - x = 0.003\ 01$$

$$x = 0.003\ 08$$

So, we can find the equilibrium concentrations of [CO] and [COCl$_2$]:

$$[CO] = 0.0102\ mol/L - 0.003\ 08\ mol/L = 0.0071\ mol/L$$

$$[COCl_2] = x\ mol/L = 0.003\ 08\ mol/L$$

Now we plug our equilibrium expressions into K_c and find:

$$K_c = \frac{0.003\ 08}{[0.0071][0.003\ 01]} = 140$$

PROBLEM 18.8
Consider the reaction of nitrosyl chloride, NOCl, below. Initially, the amount concentration of NOCl is 2.00 mol/L. When equilibrium is established at 462 °C, 0.66 mol/L NO is present. Calculate the equilibrium constant K_c for this reaction.

$$2\ NOCl(g) \rightleftharpoons 2\ NO(g) + Cl_2(g)$$

Finding Equilibrium Concentrations
One of the most important uses of K_c and ICE tables is to find the equilibrium amount concentration values that a reaction will produce. Consider the reaction of butane and isobutane:

$$butane(g) \rightleftharpoons isobutane(g)$$

At room temperature $K_c = 2.5$. If the initial concentrations of butane and isobutane are each 0.100 mol/L, what will the amount concentration values be once the reaction has equilibrium?

Our approach is the same. First, write an expression for K_c and then complete an ICE table. For C, to know whether it is $+x$ or $-x$, we need to use Q. Here, $Q = 1$, which is less than K and so the reaction will shift right and produce more products ($+x$) and decrease reactants ($-x$).

$$K_c = \frac{[isobutane]}{[butane]}$$

	butane(g)	$\rightleftharpoons$	isobutane(g)
I	0.100 mol/L	$\rightleftharpoons$	0.100 mol/L
C	$-x$ mol/L	$\rightleftharpoons$	$+x$ mol/L
E	$0.100 - x$ mol/L	$\rightleftharpoons$	$0.100 + x$ mol/L

Now we plug all our knowns into our K_c expression:

$$2.5 = \frac{(0.100 + x)}{(0.100 - x)}$$

And solve for x algebraically. This will involve, in some instances, using the quadratic formula. Here solving for x, we get $x = 0.043$. Plugging that into E in our table, we find that at equilibrium [butane] = 0.057 mol/L and [isobutane] = 0.143 mol/L.

PROBLEM 18.9
Consider the equilibrium constant K_c for the reaction:

$$Br_2(g) + F_2(g) \rightleftharpoons 2\ BrF(g) \quad K_c = 55.3$$

Calculate what the equilibrium amount concentration values (in mol/L) of all three gases will be if the initial amount concentration of bromine and fluorine were both 0.220 mol/L.

PROBLEM 18.10
Acetic acid (vinegar) dissociates in water according to the following equation:

$$CH_3CO_2H(aq) + H_2O(l) \rightleftharpoons CH_3CO_2^-(aq) + H_3O^+(aq)$$

If the initial concentration of CH_3CO_2H is 0.200 mol/L and the equilibrium constant $K_c = 1.8 \times 10^{-5}$, what are the amount concentration values of acetic acid, acetate, and oxidanium (H_3O^+) at equilibrium? What is the $-\log[H_3O^+]$ (this value is the pH of the solution)?

PROBLEM 18.11
To make carbonated water, gaseous CO_2 is dissolved in water according to the following equation:

$$CO_2(g) \rightleftharpoons CO_2(aq)$$

At 25 °C, the equilibrium constant K_c is 0.837. If the amount concentration of $CO_2(aq)$ is initially 0.12 mol/L, what will be the amount concentration of both types of carbon dioxide at equilibrium?

PROBLEM 18.12
Sodium sulfate dissolves in water according to the following equation:

$$Na_2SO_4(s) \rightleftharpoons 2\,Na^+(aq) + SO_4^{2-}(aq)$$

If the equilibrium constant K_c (at 25 °C) is 0.007 74, what will be the amount concentration of sodium ions and sulfate ions at equilibrium?

REACTION CONTROL: Perturbing Equilibria

Recalling our discussion of chemical kinetics, a reaction at equilibrium will show rapid change in amount concentration values initially until it reaches equilibrium (Figure 18.1). Once a reaction has achieved equilibrium, that equilibrium is a dynamic state where the overall amount concentration of chemicals is not changing, but there is an ongoing interconversion of reactants to products and products to reactants.

The equilibrium state of a reaction is, however, only going to hold so long as the system is closed, and the temperature and pressure are constant. If we were to alter any of these conditions, the equilibrium would be perturbed. When perturbed, equilibria respond in predictable ways. This is generally summarized with Le Chatelier's often cited, ambiguous[8] statement:

> If a stress is applied to a system in chemical equilibrium, the system shifts in the direction that tends to relieve that stress.[9]

In this section, we will consider specific stresses a chemist can employ and how each one leads to a predictable outcome. Note that a catalyst will not perturb an equilibrium and only affects how fast the reaction reaches equilibrium. Stresses that chemists can employ to affect the equilibrium of a reaction are (i) adding or removing chemicals, (ii) increasing or decreasing the volume, (iii) increasing or decreasing the pressure of a gas-phase reaction, and (iv) changing the temperature. In situations i–iii, the way to understand how the equilibrium will shift is to understand how Q

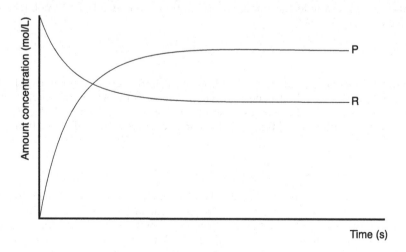

Figure 18.1 Reaction time course plot for the reaction of R → 2 P.

will be affected and then relate K and Q to see how the new conditions will shift. The effect of temperature (iv) on an equilibrium will not be considered in terms of Q in the discussion below.

Adding or Removing Chemicals to a Reaction at Constant Volume and Temperature

By adding or removing a chemical at constant volume and temperature, the amount concentration, in units of amount per volume (mol/L), of that chemical will be affected. This will, in turn, affect the value of Q and cause the reaction to shift. Consider the following scenarios.

Scenario 1: Adding reactant. Consider the following generic reaction that is at equilibrium with [A] = 1.0 mol/L and [B] = 1.0 mol/L:

$$A(aq) \rightleftharpoons B(aq) \qquad K_c = 1.0$$

At equilibrium, $K = Q$, and no significant shifts in amount concentration will happen. If a chemist were to add more A to the reaction, at constant volume, so that the concentration of A is now 2.0 mol/L, the reaction is no longer at equilibrium. Plugging in the current values of [A] and [B] we find that $Q = 0.5$:

$$Q = \frac{[B]}{[A]} = \frac{1.0}{2.0} = 0.5$$

After having added more A, the relationship between K and Q is that $K > Q$. This means that the reaction will shift to the right, converting reactants to products, until K again equals Q. In this example, the amount concentration of B will rise to 1.5 mol/L and the amount concentration of A will decrease to 1.5 mol/L. While the utility of doing this may seem obscure in this hypothetical example, if B were a chemical that we really wanted, adding more A would produce more B.

Scenario 2: Removing reactant. Let's go back to the initial equilibrium, with [A] = 1.0 mol/L and [B] = 1.0 mol/L. If a chemist were to remove some A from the reaction, at constant volume, so that the amount concentration of A is now 0.5 mol/L, the reaction is no longer at equilibrium. Plugging in the current values of [A] and [B] we find that $Q = 2.0$:

$$Q = \frac{[B]}{[A]} = \frac{1.0}{0.5} = 2.0$$

After having removed A, the relationship between K and Q is that $K < Q$. This means that the reaction will shift to the left, converting products to reactants, until K again equals Q. In this example, the amount concentration of B will fall to 0.75 mol/L and A will increase to 0.75 mol/L.

Scenario 3: Removing product. Let's go back to the initial equilibrium, with [A] = 1.0 mol/L and [B] = 1.0 mol/L. If a chemist were to remove some B from the reaction, at constant volume, so that the amount concentration of B is now 0.5 mol/L, the reaction is no longer at equilibrium. Plugging in the current values of [A] and [B] we find that $Q = 0.5$:

$$Q = \frac{[B]}{[A]} = \frac{0.5}{1.0} = 0.5$$

After having removed B, the relationship between K and Q is that $K > Q$. This means that the reaction will shift to the right, converting reactants to products, until K again equals Q. In this example, the amount concentration of B will rise to 0.75 mol/L and the amount concentration of A will decrease to 0.75 mol/L. While the utility of doing this may seem obscure in this hypothetical example, if B were a chemical that we really wanted, removing some B would produce more B. If there is a way to remove B, converting it into a solid or a liquid, then the reaction will keep going to produce B.

Scenario 4: Adding product. For the last time, go back to the initial equilibrium, with [A] = 1.0 mol/L and [B] = 1.0 mol/L. If a chemist were to add some B to the reaction, at constant volume, so that the amount concentration of B is now 2.0 mol/L, the reaction is no longer at equilibrium. Plugging in the current values of [A] and [B] we find that Q = 2.0:

$$Q = \frac{[B]}{[A]} = \frac{2.0}{1.0} = 2.0$$

After having added B, the relationship between K and Q is that $K < Q$. This means that the reaction will shift to the left, converting products to reactants, until K again equals Q. In this example, the amount concentration of B will fall to 1.5 mol/L and the concentration of A will increase to 1.5 mol/L.

Overall, then, we can see that an equilibrium will shift to the right and convert more reactant into product by increasing the amount concentration of reactants or decreasing the amount concentration of products, at constant volume and temperature. Conversely, the reaction will shift to the left and convert more product into reactant by decreasing the amount concentration of reactants or increasing the amount concentration of products, at constant volume and temperature.

Adding or Removing Chemicals to a Reaction at Constant Pressure and Temperature

For a gas-phase reaction at constant pressure (think of a reaction happening in something like a balloon), adding or removing a chemical will change the volume. A change in volume will affect the amount concentration of all chemicals in addition to the effect of adding or removing a chemical. This change can still be determined by using Q, but the analysis can be quite complex. While it is worth highlighting the complexity of adding or removing chemicals to a gas-phase reaction at constant pressure, such problems will not be considered further in this text.

Changing Pressure/Volume of a Gas-Phase Reaction at Constant Temperature

For a gas-phase reaction, one way of shifting the equilibrium is to alter the pressure. This is done by changing the volume (adding an inert gas to increase the pressure has no effect, as an inert gas does not show up in the expression for Q and the amount concentration of each individual chemical remains constant). It is worth highlighting the ideal gas law (Equation 18.10; see Chapter 11) so that we can understand how pressure and volume are related:

$$p = \frac{n}{V}RT \tag{18.10}$$

p is the pressure (atm).
n is the moles of gas (mol).
V is the volume (L).
R is the gas constant (0.082 06 $\frac{L\,atm}{mol\,K}$).
T is the temperature (K).

Considering Equation 18.10, we can see that pressure and volume are inversely related. As the volume of the container decreases, the pressure increases, and as the volume of the container increases, the pressure decreases. Consider this in the context of $NO_2(g)$ in equilibrium with $N_2O_4(g)$:

$$2\,NO_2(g) \rightleftharpoons N_2O_4(g) \; K_c(115\,°C) = 1.0$$

Now, if we think of this in terms of amount concentration and our expression for Q_c, let's expand the normal bracket notation to consider the amount of substance (n) and the volume (V) in the expression. We can see that the numerator changes as $1/V$ and the denominator changes as $1/V^2$, which means that the volume change will have a larger impact on the denominator.

$$Q_c = \frac{[N_2O_4]}{[NO_2]^2} = \frac{\left(\dfrac{n_{N_2O_4}}{V}\right)}{\left(\dfrac{n_{NO_2}}{V}\right)^2}$$

At equilibrium at 115 °C, there is 1.0 mol/L NO_2 and 1.0 mol/L N_2O_4 in a 1.0 L container, which would correspond to 1.0 mol NO_2 and 1.0 mol N_2O_4. If the container volume was doubled to 2.0 L, then the amount concentration of NO_2 and N_2O_4 would both decrease to 0.5 mol/L (1.0 mol/2.0 L) and the value of Q_c would become 2.0.

$$Q_c = \frac{[N_2O_4]}{[NO_2]^2} = \frac{\left(\dfrac{1\,mol}{2.0\,L}\right)}{\left(\dfrac{1.0\,mol}{2.0\,L}\right)^2} = 2.0$$

The value of Q_c is now greater than K ($K < Q_c$) and the reaction would shift to the left and convert products to reactants. If we reconsidered our analysis in terms of the pressure of each gas, we would find the same outcome ($p \propto \dfrac{n}{V}$): the pressure of both product and reactant would decrease, but the denominator is squared and so the denominator has a bigger impact on the value of Q. This is true, however, only because there is a different amount (mol) of gas between reactant and product. If you consider the reaction of hydrogen and fluorine to make hydrogen fluoride, there would be no effect from changing the volume (pressure) because the amount (mol) of gas is constant across the reaction.

$$H_2(g) + F_2(g) \rightleftharpoons 2\,HF(g)$$

In summary, then, gas-phase reactions change in response to changing the volume of the container, and by extension the pressure. We can always find the right result of the shift by finding the value of Q after a change has taken place, but the general result is (i) an increase in pressure and decreases in volume favor the side with fewer moles of gas and (ii) a decrease in pressure and an increase in volume favors the side with more moles of gas.

Changing Volume of a Solution-Phase Reaction at Constant Temperature

For solution-phase reactions, an increase in volume through dilution, i.e., the addition of solvent, shifts the equilibrium, as we saw for gases above, toward the side with more moles of dissolved (aqueous) particles. A decrease in volume, i.e., the removal of solvent, shifts the equilibrium toward the side with fewer moles of dissolved (aqueous) particles.

Changing Temperature

Finally, a reaction at equilibrium can be perturbed by using temperature. For a change in temperature, this does not influence the value of Q, rather the value of the equilibrium constant is directly affected. The relationship between K and T is shown by the van't Hoff equation (Equation 18.11).[10] This is a rearrangement of Equation 18.3.

$$\ln K = -\frac{\Delta_r H^\circ}{RT} + \frac{\Delta_r S^\circ}{R} \tag{18.11}$$

K is the equilibrium constant (unitless).
$\Delta_r H^\circ$ is the reaction enthalpy (kJ/mol).
R is the gas constant (0.008 314 kJ/(mol K)).
T is the temperature (K).
$\Delta_r S^\circ$ is the reaction entropy (kJ/mol).

The van't Hoff equation relates the natural log of the equilibrium constant to the inverse temperature and the relationship depends on the sign of enthalpy ($\Delta_r H^\circ < 0$, exothermic; or $\Delta_r H^\circ > 0$, endothermic). Figure 18.2 shows how the dependence of K on inverse temperature depends on the value of the enthalpy of reaction. Note that since the plot is $\ln K$ versus $1/T$, higher temperatures are on the left and lower temperatures are on the right.

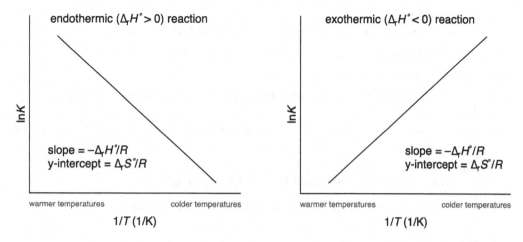

Figure 18.2 Van't Hoff plots of a generic endothermic reaction (left) and exothermic reaction (right).

As an example, consider the (⤳) Haber–Bosch reaction to produce ammonia from nitrogen and hydrogen. This reaction is vital to modern civilization to ensure enough nitrogen fertilizer is available to grow the crops we need to feed the world's population.

$$N_2(g) + 3\,H_2(g) \rightleftharpoons 2\,NH_3(g) \qquad \Delta_rH° = -92.4 \text{ kJ/mol}, K(298.15) > 1$$

At room temperature, this reaction is product favored but does not happen at a detectable rate. As we saw in Chapter 14, increasing the temperature and adding a catalyst makes the reaction occur faster and so industrially this reaction is run at 400 °C with a group 8 metal as a catalyst. At 400 °C, however, the reaction becomes reactant favored; the equilibrium shifts to the left as the temperature is increased because it is an exothermic reaction. The chemical engineering challenge inherent in the Haber–Bosch process is an interesting one to consider for those of you who are interested in engineering and in terms of designing a green chemical synthesis of ammonia.

PROBLEM 18.13
For the reaction $2\,NOBr(g) \rightleftharpoons 2\,NO(g) + Br_2(g)$, what will happen to the equilibrium if more Br_2 is added at constant volume and temperature?

PROBLEM 18.14
For the reaction $H_2O(l) \rightleftharpoons H_2O(g)$, provide an explanation for why an open water bottle will entirely evaporate but a sealed bottle will not.

PROBLEM 18.15
Carbon monoxide is a deadly gas because it strongly binds to the iron in hemoglobin. For people exposed to carbon monoxide, treatment is to place them in a hyperbaric (high pressure) oxygen chamber (the chamber is at constant volume and temperature). Looking at the equilibrium below, explain why this helps (Hb = hemoglobin).

$$Hb(CO)_4(aq) + 4\,O_2(aq) \rightleftharpoons Hb(O_2)_4(aq) + 4\,CO(aq), K_c < 1$$

PROBLEM 18.16
Plants produce sugar through photosynthesis.

$$6\,CO_2(g) + 6\,H_2O(l) \rightleftharpoons C_6H_{12}O_6(aq) + 6\,O_2(g)$$

Provide an explanation why, for a plant that wants to maximize sugar production, the following helps: the plant takes in CO_2 and releases O_2 (assume volume is constant). The plant takes the aqueous glucose and turns it into solid starch.

PROBLEM 18.17

For the following equilibrium: $NH_4HS(s) \rightleftharpoons NH_3(g) + H_2S(g)$ $\Delta_r H^\circ > 0$

a. How will the equilibrium be affected by increasing the temperature?

b. By increasing the pressure?

c. By adding ammonium hydrosulfide ($NH_4SH(s)$) to the container, at constant volume and temperature?

d. By removing ammonia ($NH_3(g)$) from the container, at constant volume and temperature?

PROBLEM 18.18

Consider the following equilibrium: $ClF_5(g) \rightleftharpoons ClF_3(g) + F_2(g)$.

How will the equilibrium be affected if the reaction container is doubled in volume?

PROBLEM 18.19

Consider the following equilibrium: $2\ HBr(g) \rightleftharpoons Br_2(g) + H_2(g)$ $\Delta_r H^\circ = 103.7$ kJ/mol.

a. How will the equilibrium be affected if the reaction container is halved in volume?

b. If the temperature decreases?

PROBLEM 18.20

Consider the reaction: *cis*-but-2-ene $\rightleftharpoons$ *trans*-but-2-ene. A measurement of K_c at various temperatures gave the following data:

Temperature (K)	Equilibrium Constant K_c
500	1.65
600	1.47
700	1.36

Is the reaction from *cis*-but-2-ene to *trans*-but-2-ene exothermic or endothermic? Explain.

PROBLEM 18.21

Consider the following equilibrium: $H_2O(l) \rightleftharpoons H_2O(g)$ $\Delta_r H^\circ = 40.7$ kJ/mol.

a. How will the equilibrium be affected if the pressure is reduced?

b. How will the equilibrium be affected if the temperature is increased?

PROBLEM 18.22

Consider the following equilibrium: $PbCl_2(s) \rightleftharpoons Pb^{2+}(aq) + 2\ Cl^-(aq)$.

a. How will the equilibrium be affected if more water is added?

b. How will the equilibrium be affected if NaCl is added?

↝ **PROBLEM 18.23**

Consider the direct steam reforming reaction equilibrium.

$$CH_4(g) + 2\ H_2O(g) \rightleftharpoons CO_2(g) + 4\ H_2(g) \quad \Delta_r H^\circ = 165 \text{ kJ/mol}$$

a. How will the equilibrium be affected if the container volume is doubled?

b. How will the equilibrium be affected if the temperature is decreased?

c. How will the equilibrium be affected if hydrogen gas is added to the reaction at constant temperature and pressure?

d. How would the equilibrium be affected if pressure increased?

e. This reaction is, currently, the major source of hydrogen production. Hydrogen will likely serve many important roles in a decarbonized future.

i. Why is direct steam reforming reaction problematic for a decarbonized future?

ii. A green chemistry production of hydrogen decomposes water into hydrogen and oxygen:

$$2\ H_2O(l) \rightleftarrows O_2(g) + 2\ H_2(g)$$

Determine the enthalpy of reaction for water decomposition and provide an explanation as to why industry prefers direct steam reforming over water decomposition.

NOTES

1. Berthollet, C.L. *Essai de Statique Chimique*. Imprimerie de Demonville et sœurs (Paris), 1803.

2. van't Hoff, J.H. *Études de Dynamique Chimique*. Frederik Muller & Co. (Amsterdam), 1884.

3. Note the amount of water at equilibrium is not considered. This will be explored further below.

4. Lightly adapted from: Jogalekar, A. The Only Two Equations You Should Know. *The Curious Wavefunction*. 7 February 2018. http://wavefunction.fieldofscience.com/2018/02/the-only-two-equations-that-you-should.html (Accessed 01/27/21).

5. This relationship is called the law of mass action and was first proposed by Wilhelm Ostwald in his master's thesis in 1877.

6. Lewis, G.N. Outlines of a New System of Thermodynamic Chemistry. *Proc. Amer. Acad. Arts.*, **1907**, *43* (7), 259–293. DOI: 10.2307/20022322.

7. (i) This was first proposed by Waage and Guldburg based on their work in kinetics, which is equivalent to the modern expression only where the forward and reverse reactions are both single-step, elementary reactions: Waage, P.; Guldberg, C.M. Studier over Affiniteten. *Forh. Vidensk.-Selsk. Kristiania.* **1864**, 35–41. (Translation by H.I. Abrash: *J. Chem. Educ.*, **1986**, *63* (12), 1044–1047. DOI: 10.1021/ed063p1044.)

(ii) Waage, P.; Guldberg, C.M. Forsøg til Bestemmelse af Lovene for Affiniteten. *Forh. Vidensk.-Selsk. Kristiania.* **1864**, 92–94.

8. Ehrenfest, P. Das Prinzip von Le Chatelier-Braun und die Reziprozitätssätze der Thermodynamik. *Z. Phys. Chem.*, **1911**, *77U* (1), 227–244. DOI: 10.1515/zpch-1911-7714.

9. (i) First enumerated: Le Chatelier, H. Sur un énoncé general des lois des équilibres chimiques. *C. R. Acad. Sci.*, **1884**, *99*, 786–789. https://gallica.bnf.fr/ark:/12148/bpt6k30546 (Accessed 02/15/23).

(ii) Given theoretical justification: (a) Braun, F. Untersuchungen über die Löslichkeit fester Körper und die den Vorgang der Lösung begleitenden Volum- und Energieänderungen. *Z. Phys. Chem.*, **1887**, *1U* (1), 259–272. DOI: 10.1515/zpch-1887-0131. (b) Braun, F. Ueber einen allgemeinen qualitativen Satz für Zustandsänderungen nebst einegen sich anschliessenden Bermerkungen, insbesondere über nich eindeutige Systeme. *Ann. Phys.* (Leipzig), **1888**, *269* (2), 337–353. DOI: 10.1002/andp.18882690210.

10. van't Hoff, J.H. *Études de Dynamique Chimique*. Frederik Muller & Co. (Amsterdam), 1884.

19 Electron Transfer

Throughout the past several chapters, we developed tools – enthalpy, entropy, Gibbs energy, equilibrium constants – for quantifying whether a reaction is reactant favored or product favored. We have also considered how a reaction might be pushed one way or another through coupling reactions, changing amount concentration, changing volume, changing pressure, or changing temperature. During these discussions, however, the identities of the chemicals have been given only passing attention, and the deeper questions of why have not been addressed. This chapter will be the first that seeks to understand why, chemically, a reaction is likely or unlikely to take place. One of the central concepts that we will use to help understand why a reaction is likely or unlikely to take place is electronegativity (χ).[1] Electronegativity is the power of an atom to attract electrons to itself.[2] One of the most widely used scales is the Pauling electronegativity scale (Figure 19.1).

A chemical reaction involves the movement of electrons. That movement could involve the transfer of electrons (the focus of this chapter) or a change in how electrons are shared (the focus of Chapter 20). When electrons are transferred this is called a redox reaction, wherein redox is a portmanteau of *redu*ction[3] and *ox*idation.[4] Reduction, also called electronation, is the complete transfer of one or more electrons to an atom, ion, or molecule.[5] Oxidation, also called de-electronation, is the complete transfer of one or more electrons from an atom, ion, or molecule.[6] The direction of transfer, that is where electrons are most likely to go of their own volition, can often be explained simply by electronegativity: atoms with higher electronegativities tend to get reduced and atoms with lower electronegativities tend to get oxidized.

OXIDATION NUMBERS

The movement of electrons is what defines a chemical reaction, and the transfer of electrons defines redox. The question then arises: How do we keep track of electrons?[7] The method of book-keeping electrons employed in redox reactions is what is called the oxidation number[8] or oxidation state[9] of an atom. The premise of assigning oxidation numbers is to assume that all bonds are perfectly ionic, and each bonding pair of electrons is assigned to the more electronegative atom (unless electronegativity values are equal, in which case the electrons are split evenly). The oxidation number is then the number of valence electrons minus the number of electrons around that atom once bonding pairs are appropriately divvied up. As an example of assigning oxidation numbers graphically, consider ethyl fluoride (Figure 19.2).

In Figure 19.2, the boxes have been added to show the assignment of electrons to the more electronegative atom and in the case of the C–C bond, bifurcating the two electrons equally. If we then assign oxidation numbers, we have +1 for the hydrogen atoms (one valence less zero assigned). For carbon 1, we have four valence electrons less seven assigned, giving an oxidation number of –3. For carbon 2, we have four valence electrons less five assigned, giving an oxidation number of –1. And for fluorine, we have seven valence electrons less eight assigned, giving an oxidation number of –1.

While oxidation numbers can be determined with a Lewis structure, often we only have a chemical formula. To assign oxidation numbers to atoms in a chemical formula, the following rules are used:

1. For any atom in its elemental form, the oxidation number is always zero.

   For example, H in H_2 has an oxidation number of 0. Each P atom in P_4 has an oxidation number of 0. Every Cu atom in copper metal has an oxidation number of 0.

2. For any monatomic ion, the oxidation number always equals the charge.

   For example, in the compound K_2S, the ions are K^+ and S^{2-}. The oxidation number of potassium is +1 and the oxidation number of sulfide is –2.

3. For any polyatomic ion or molecular compound, assign oxidation numbers in the following order.

   a. Fluorine always has an oxidation number of –1.

   b. Oxygen usually has an oxidation number of –2.

   The exceptions are when oxygen is bonded to fluorine or when oxygen is part of peroxide (oxygen has an oxidation number of –1 in peroxides).

DOI: 10.1201/9781003479338-19

1	2	3	4	5	6	7	8	9	10	11	12	13	14	15	16	17	18
1 H 2.20																	2 He
3 Li 0.98	4 Be 1.57											5 B 2.04	6 C 2.55	7 N 3.04	8 O 3.44	9 F 3.98	10 Ne
11 Na 0.93	12 Mg 1.31											13 Al 1.61	14 Si 1.90	15 P 2.19	16 S 2.58	17 Cl 3.16	18 Ar
19 K 0.82	20 Ca 1.00	21 Sc 1.36	22 Ti 1.54	23 V 1.63	24 Cr 1.66	25 Mn 1.55	26 Fe 1.83	27 Co 1.88	28 Ni 1.91	29 Cu 1.90	30 Zn 1.65	31 Ga 1.81	32 Ge 2.01	33 As 2.18	34 Se 2.55	35 Br 2.96	36 Kr 3.0
37 Rb 0.82	38 Sr 0.95	39 Y 1.22	40 Zr 1.33	41 Nb 1.6	42 Mo 2.16	43 Tc 1.9	44 Ru 2.2	45 Rh 2.28	46 Pd 2.20	47 Ag 1.93	48 Cd 1.69	49 In 1.78	50 Sn 1.96	51 Sb 2.05	52 Te 2.30	53 I 2.66	54 Xe 2.6
55 Cs 0.79	56 Ba 0.89	*	72 Hf 1.3	73 Ta 1.5	74 W 2.36	75 Re 1.9	76 Os 2.0	77 Ir 2.20	78 Pt 2.28	79 Au 2.54	80 Hg 2.00	81 Tl 2.04	82 Pb 2.33	83 Bi 2.02	84 Po 2.0	85 At 2.2	86 Rn
87 Fr 0.7	88 Ra 0.9	**	104 Rf	105 Db	106 Sg	107 Bh	108 Hs	109 Mt	110 Ds	111 Rg	112 Cn	113 Nh	114 Fl	115 Mc	116 Lv	117 Ts	118 Og

*	57 La 1.1	58 Ce 1.12	59 Pr 1.13	60 Nd 1.14	61 Pm 1.2	62 Sm 1.17	63 Eu 1.1	64 Gd 1.20	65 Tb 1.2	66 Dy 1.22	67 Ho 1.23	68 Er 1.24	69 Tm 1.25	70 Yb 1.1	71 Lu 1.27
**	89 Ac 1.1	90 Th 1.0	91 Pa 1.5	92 U 1.38	93 Np 1.36	94 Pu 1.28	95 Am 1.30	96 Cm 1.20	97 Bk	98 Cf	99 Es	100 Fm	101 Md	102 No 1.1	103 Lr 1.27

Figure 19.1 Pauling electronegativity values.

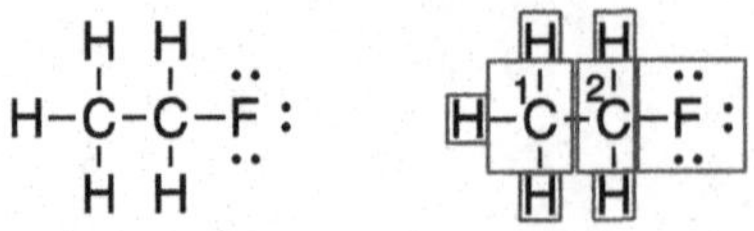

Figure 19.2 Ethyl fluoride (left) and ethyl fluoride with bonding electrons apportioned to the more electronegative atom (right).

  c. The oxidation number of hydrogen is +1 when combined with non-metals and –1 when combined with metals.

   In NaH, sodium has an oxidation number of +1 and hydrogen has an oxidation number of –1.

   In HF, hydrogen has an oxidation number of +1 and fluorine has an oxidation number of –1.

  d. Halogen atoms Cl, Br, and I usually have an oxidation number of –1, except when bonded to another element of higher electronegativity.

   In HCl, hydrogen has an oxidation number of +1 and chlorine has an oxidation number of –1.

   In ClF, chlorine has an oxidation number of +1 and fluorine has an oxidation number of –1.

4. The sum of all oxidation numbers must equal 0 for a neutral molecule or they must equal the charge of a polyatomic ion.

   In ClF_3, the fluorine atoms each have an oxidation number of –1 and the chlorine atom has an oxidation number of +3: $1Cl + 3(–1) = 0$, $Cl = +3$.

   For CO_3^{2-}, oxygen atoms each have an oxidation number of –2 and the carbon atom has an oxidation number of +4: $1C + 3(–2) = –2$, $C = +4$.

PROBLEM 19.1
Assign the oxidation number for each atom or ion.

  a. $V(s)$

  b. $Mg^{2+}(aq)$

  c. $S_8(s)$

  d. $O_2(g)$

  e. $Cl^-(aq)$

PROBLEM 19.2
Assign the oxidation number for each element in the following binary compounds.

  a. $CO(g)$

  b. $CO_2(g)$

  c. $CH_4(g)$

  d. $CuCl_2(s)$

  e. $SF_6(g)$

PROBLEM 19.3
Assign the oxidation number for each element in the following polyatomic ions.

  a. $SO_3^{2-}(aq)$

  b. $SO_4^{2-}(aq)$

  c. $Hg_2^{2+}(aq)$

  d. $BrO_4^-(aq)$

  e. $HF_2^-(aq)$

PROBLEM 19.4
Assign the oxidation number for each element in the following compounds.

a. $NaNO_3(s)$

b. $H_3PO_4(l)$

c. $Cu(CN)_2(s)$

d. $LiOH(s)$

REDUCTION–OXIDATION

When considering a reaction, a good first step is to assign the oxidation number of each element before and after the reaction. Let's consider the reaction of methane and fluorine gas to produce tetrafluoromethane and hydrogen fluoride.

$$CH_4(g) + 4 F_2(g) \rightleftharpoons CF_4(g) + 4 HF(g)$$

Reactant	Product
C = –4	C = +4
H = +1	H = +1
F = 0	F = –1

Looking at the oxidation numbers for this reaction, the carbon atom goes from –4 to +4. The carbon atom oxidation number is becoming more positive or less negative as it loses negatively charged electrons. Losing electrons is called oxidation, and so we can say that carbon is oxidized in this reaction. In contrast, the fluorine atom goes from 0 to –1, which means that the fluorine is becoming less positive or more negative as it gains negatively charged electrons. Gaining electrons is called reduction, and so we can say that fluorine is reduced in this reaction.

There are two mnemonics one can use to remember what oxidation and reduction mean. The first is LEO GER or "lose electrons oxidation and gain electrons reduction." The second is OIL RIG or "oxidation is loss (of electrons), and reduction is gain (of electrons)."

PROBLEM 19.5
For each reaction, identify which element is oxidized and which element is reduced.

a. $CH_4(g) + 2 O_2(g) \rightleftharpoons CO_2(g) + 2 H_2O(g)$

b. $2 Sb(s) + 5 Cl_2(g) \rightleftharpoons 2 SbCl_5(s)$

c. $2 Al(s) + 6 HCl(aq) \rightleftharpoons 2 AlCl_3(aq) + 3 H_2(g)$

d. $2 H_2O(l) \rightleftharpoons 2 H_2(g) + O_2(g)$

HALF-REACTIONS

To see more clearly how electrons are transferred in a redox reaction, an overall reaction can be broken down into two simultaneous reduction and oxidation half-reactions. A half-reaction shows the reactant, products, and the number of electrons lost (oxidation) or gained (reduction). Let's reconsider the reaction of methane and fluorine:

$$CH_4(g) + 4 F_2(g) \rightleftharpoons CF_4(g) + 4 HF(g)$$

Carbon is oxidized in this reaction. We can show the loss of electrons as an oxidation half-reaction. In this half-reaction, we start with CH_4 (the substance that contains the element being oxidized). We then draw a single arrow (→) and write the products. Here we are writing the carbon as C^{4+}, the hydrogen as H^+, and then eight electrons.

Oxidation half-reaction: $CH_4(g) \rightarrow C^{4+} + 4 H^+ + 8 e^-$

Fluorine is reduced in this reaction. We can show the gain of electrons as a reduction half-reaction. In this half-reaction, we start with $4 F_2$ (the substance that contains the element being reduced) and the eight electrons fluorine gets. We then draw a single arrow (→) and write the products. Here we are writing the fluoride as F^-.

Reduction half-reaction: $4 F_2(g) + 8 e^- \rightarrow 8 F^-$

If we were given the two half-reactions, we could determine the overall reaction by adding the two half-reactions together, canceling the electrons, and combining oppositely charged ions.

Oxidation half-reaction: $CH_4(g) \rightarrow C^{4+} + 4\,H^+ + 8\,e^-$

+Reduction half-reaction: $4\,F_2(g) + 8\,e^- \rightarrow 8\,F^-$

Overall reaction: $CH_4(g) + 4\,F_2(g) + 8\,e^- \rightleftarrows C^{4+} + 4\,H^+ + 8\,e^- + 8\,F^-$

Simplified: $CH_4(g) + 4\,F_2(g) \rightleftarrows CF_4(g) + 4\,HF(g)$

PROBLEM 19.6
For each reaction in Problem 19.5, write an appropriate reduction and oxidation half-reaction.

REDOX REACTION STOICHIOMETRY

There are several reasons why determining oxidation numbers and writing half-reactions is important. Initially, we will look at determining the stoichiometry of a redox reaction, which employs both skills in a series of 7 (or 8) steps. As we look at how this system works, let's consider the reaction of aqueous gold(3+) ions with tin under acidic conditions, which means that $H^+(aq)$ may be present in the overall balanced equation.

$$Au^{3+}(aq) + Sn(s) \rightleftarrows Au(s) + SnO_2(s) \text{ in acidic conditions}$$

First (step 1), we must determine the redox reaction stoichiometry and assign the oxidation numbers to every element.

Reactant	Product
Au = +3	Au = 0
Sn = 0	Sn = +4
	O = –2

Now that we can identify that gold is reduced and tin is oxidized, we write, in step 2, the oxidation and reduction half-reactions, starting with only the elements or compounds in the reaction as written.

Oxidation half-reaction: $Sn(s) \rightarrow SnO_2(s)$

Reduction half-reaction: $Au^{3+}(aq) \rightarrow Au(s)$

Once we have the redox half-reactions written, we then in step 3 balance all atoms except for oxygen and hydrogen. For this reaction, there are no other atoms to balance and so we can go on to step 4, where we balance the oxygen atoms by adding water molecules ($H_2O(l)$).

Oxidation half-reaction: $2\,H_2O(l) + Sn(s) \rightarrow SnO_2(s)$

Reduction half-reaction: $Au^{3+}(aq) \rightarrow Au(s)$

In terms of atoms, we finally (step 5) balance any missing hydrogen atoms by adding aqueous hydrons ($H^+(aq)$).

Oxidation half-reaction: $2\,H_2O(l) + Sn(s) \rightarrow SnO_2(s) + 4\,H^+(aq)$

Reduction half-reaction: $Au^{3+}(aq) \rightarrow Au(s)$

To complete each half-reaction (step 6), we balance the charge by adding electrons (e^-).

Oxidation half-reaction: $2\,H_2O(l) + Sn(s) \rightarrow SnO_2(s) + 4\,H^+(aq) + 4\,e^-$

Reduction half-reaction: $Au^{3+}(aq) + 3\,e^- \rightarrow Au(s)$

Now if we consider the two half-reactions, the number of electrons lost and gained is not equal. In step 7, we multiply each half-reaction by an appropriate factor so that the number of reactions gained and lost is equal.

Oxidation half-reaction: $6 H_2O(l) + 3 Sn(s) \rightarrow 3 SnO_2(s) + 12 H^+(aq) + 12 e^-$

Reduction half-reaction: $4 Au^{3+}(aq) + 12 e^- \rightarrow 4 Au(s)$

And we can then add the reactions together to find the overall reaction stoichiometry.

Overall reaction in acidic conditions: $4 Au^{3+}(aq) + 6 H_2O(l) + 3 Sn(s) \rightarrow 3 SnO_2(s) + 12 H^+(aq) + 4 Au(s)$

For this reaction, then, the above overall reaction would be the answer. We can see the 12 H^+(aq) ions that indicate this reaction is under acidic conditions. A reaction may, in the end, show no H^+(aq) ions, which is also referred to as acidic conditions. It is only if hydroxide (OH^-) is present that the reaction would be named as occurring under basic conditions. If we were given the same reaction, but basic conditions were specified:

$Au^{3+}(aq) + Sn(s) \rightleftarrows Au(s) + SnO_2(s)$ *in basic conditions*

We would follow the same seven steps and arrive at the same overall reaction.

Overall reaction: $4 Au^{3+}(aq) + 6 H_2O(l) + 3 Sn(s) \rightleftarrows 3 SnO_2(s) + 12 H^+(aq) + 4 Au(s)$

Then, step 8, to make it a basic reaction, we add OH^-(aq) to both sides in equal number to the number of H^+(aq) ions. This will neutralize the H^+(aq), making water, which may mean that we then cancel some number of water molecules between the reactants and products.

Overall reaction+OH^-(aq): $4 Au^{3+}(aq) + \cancel{6 H_2O(l)} + 3 Sn(s) + 12 OH^-(aq) \rightleftarrows 3 SnO_2(s) + \cancel{12} 6 H_2O(l) + 4 Au(s)$

Overall reaction in basic conditions: $4 Au^{3+}(aq) + 3 Sn(s) + 12 OH^-(aq) \rightleftarrows 3 SnO_2(s) + 6 H_2O(l) + 4 Au(s)$

PROBLEM 19.7

For each of the following, write the reduction and oxidation half-reactions and then provide the overall balanced equation.

a. $CH_2O(g) + O_2(g) \rightleftarrows CO_2(g) + H_2O(l)$ *in acidic solution*

b. $H_2O_2(aq) + Co^{2+}(aq) \rightleftarrows H_2O(l) + Co^{3+}(aq)$ *in acidic solution*

c. $Zn(s) + HgO(s) \rightleftarrows Zn(OH)_2(s) + Hg(l)$ *in basic solution*

d. $Br^-(aq) + MnO_4^-(aq) \rightleftarrows Br_2(g) + MnO_2(s)$ *in basic solution*

e. $As_2O_3(s) + NO_3^-(aq) \rightleftarrows H_3AsO_4(aq) + NO(g)$ *in acidic solution*

f. $CH_3OH(aq) + Cr_2O_7^{2-}(aq) \rightleftarrows CH_2O(aq) + Cr^{3+}(aq)$ *in acidic solution*

ELECTROCHEMICAL CELLS

As stated above, there are several reasons why determining oxidation numbers and writing half-reactions is important. One reason is electrochemistry, which is using chemical change to provide electricity or using electricity to affect chemical change. That is, the two half-reactions that will occur simultaneously inside a reaction vessel can be separated into two half-reaction cells, also called two electrodes, that are connected by a wire and a salt bridge (Figure 19.3).

In an electrochemical cell, it is convention to show the anode, the electrode where oxidation occurs, on the left side and to show the cathode, the electrode where reduction occurs, on the right side.[10] Electrons flow from the anode to the cathode and ions flow into and from the salt bridge to offset the flow of electrons (chloride flows to the anode and potassium ions to the cathode). If a voltmeter is connected, the electrochemical cell potential ($E°_{cell}$) can be measured. A positive voltage indicates that it is favorable for electrons to flow from the anode to the cathode and this is called a galvanic cell, which is the basis of battery technology. If a negative voltage is measured, this indicates that it is unfavorable for electrons to flow from the anode to the cathode and they will not flow without added work. A cell with a negative potential is called an electrolytic cell.

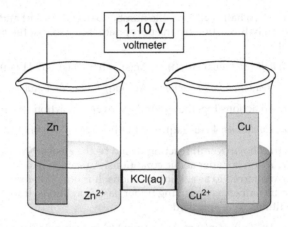

Figure 19.3 An electrochemical, galvanic cell consisting of a zinc anode and copper cathode showing a positive voltage. This was historically called a Daniell cell and was an early basis for the definition of the volt as a unit (a Daniell cell now has a cell potential of 1.10 V with the modern definition of voltage).

For the cell in Figure 19.3, the anode (oxidation) half-reaction is the loss of two electrons by zinc, and the cathode (reduction) half-reaction is the gain of two electrons by copper(2+) ions.

Anode half-reaction: $Zn(s) \rightarrow Zn^{2+}(aq) + 2\ e^-$

Cathode half-reaction: $Cu^{2+}(aq) + 2\ e^- \rightarrow Cu(s)$

Overall reaction: $Zn(s) + Cu^{2+}(aq) \rightarrow Zn^{2+}(aq) + Cu(s)$

While these half-reactions and the overall reaction are useful, it is common to represent the components of electrochemical cells using a shorthand called cell notation.[11] In general, the cell notation is represented by the anode components; two dashed vertical bars, ¦¦, which show the liquid junction between the anode and the salt bridge; the salt bridge components; two dashed vertical bars; and then the cathode components.

$$\text{anode}\ \text{¦¦}\ \text{KCl(aq, saturated)}\ \text{¦¦}\ \text{cathode}$$

It is often the case that the salt bridge components are omitted, and the cell notation is shortened to anode¦¦cathode. For components that are in different phases, they are separated with a solid vertical bar, and for different components in the same phase, they are separated by a comma.[12] For the cell in Figure 19.3, the cell notation would be:

$$Zn(s)\ |\ Zn^{2+}(aq)\ \text{¦¦}\ Cu^{2+}(aq)\ |\ Cu(s)$$

If a cell was constructed from the single-displacement reaction of gold(3+) and tin, we found the stoichiometry for this reaction in the balancing redox reactions example above:

Oxidation half-reaction: $6\ H_2O(l) + 3\ Sn(s) \rightarrow 3\ SnO_2(s) + 12\ H^+(aq) + 12\ e^-$

Reduction half-reaction: $4\ Au^{3+}(aq) + 12\ e^- \rightarrow 4\ Au(s)$

The cell notation would be $Sn(s)|H_2O(l),\ H^+(aq)|SnO_2(s)\text{¦¦}Au^{3+}(aq)|Au(s)$.

PROBLEM 19.8
For each of the following reactions, write the half-reaction that occurs at the cathode and the half-reaction that occurs at the anode.

a. $Zn(s) + Pb(NO_3)_2(aq) \rightleftarrows Zn(NO_3)_2(aq) + Pb(s)$

b. $2\ AgNO_3(aq) + Sn(s) \rightleftarrows 2\ Ag(s) + Sn(NO_3)_2(aq)$

c. $CH_2O(g) + O_2(g) \rightleftarrows CO_2(g) + H_2O(l)$

PROBLEM 19.9
For each reaction in Problem 19.8, write the reaction in cell notation. For reactions where there is not a pure metal, assume that a Pt or C electrode is used.

PROBLEM 19.10
For each cell notation provided, write a balanced equation for the reaction.

a. $Fe(s)|Fe(NO_3)_2(aq)||Cu(NO_3)_2(aq)|Cu(s)$

b. $Pt(s)|Fe^{2+}(aq), Fe^{3+}(aq)||H^+, H_2O_2(aq), H_2O(l)|Pt(s)$

c. $Sn(s)|F^-(aq), SnF_6^{2-}(aq)||ClO_4^-(aq), ClO_3^-(aq), H_2O(l)|Pt(s)$

The cell shorthand is useful because the potential of an electrochemical cell is not dependent upon the balanced reaction but solely on the components that make up each electrode. Generally, the cell potential ($E°_{cell}$) for any cell can be calculated as the difference between the standard potential ($E°$)[13] of the cathode and the anode:

$$E°_{cell} = E°_{cathode} - E°_{anode} \qquad (19.1)$$

$E°_{cell}$ is the cell potential at standard state (V).
$E°_{cathode}$ is the standard reduction potential of the cathode half-cell (V).
$E°_{anode}$ is the standard reduction potential of the anode half-cell (V).

A standard reduction potential is found by using an electrode of interest as a cathode and finding $E°_{cell}$ with the standard hydrogen electrode (SHE) as the anode. The standard hydrogen electrode is a theoretical electrode for the reduction of hydrons, $H^+(aq)$ with the activity of hydrons at unity in an ideal solution, to hydrogen gas ($2 H^+(aq) + 2 e^- \rightarrow H_2(g)$) with a reduction potential assigned as 0.00 V. Practical reference electrodes are the normal hydrogen electrode (NHE), which uses a 1.0 mol/L hydron amount concentration, and the reversible hydrogen electrode (RHE) where the voltage depends on pH. A small sample of standard reduction potentials is shown in Table 19.1 and a more complete list is in Appendix 8. Note that hydrogen does not conduct electricity. So, if you consider the cathode cell notation (Table 19.1) you will see |Pt(s) at the end because platinum is used as a non-reactive electrode to conduct the electricity; C(s, graphite) is also used as a nonreactive electrode.

If you consider the values in Table 19.1, copper(2+) has a higher (more positive) reduction potential (0.34 V) than hydrogen (0.00 V). This means that copper has a greater tendency to be reduced than hydrogen, and copper(2+) ions will oxidize hydrogen gas. In contrast, hydrons will not oxidize copper metal. Now if we look at zinc, it has a lower (more negative) reduction potential (–0.76 V) than hydrogen (0.00 V). This means that hydrons will oxidize zinc metal, but zinc metal will not oxidize hydrogen gas. In general, cells that produce a positive voltage are product favored, and cells with negative voltages are reactant favored.

As an example of using cell voltages, let's determine the voltage of the tin–gold(3+) cell we have worked with so far: $Sn(s)|H_2O(l), H^+(aq)|SnO_2(s)||Au^{3+}(aq)|Au(s)$. The standard reduction voltage for tin(IV) oxide is –0.12 V and for gold(3+) it is 1.50 V. The cell voltage ($E°_{cell}$) is 1.62 V, and the positive voltage indicates a product-favored process. We can explain this in terms of electronegativity: gold is more electronegative than tin and so the electrons are favorably transferred from tin to gold(3+).

PROBLEM 19.11
Consider these half-reactions.

Half-Reaction	$E°$ (V)
$Au^{3+}(aq) + 3 e^- \rightarrow Au(s)$	1.52
$Pt^{2+}(aq) + 2 e^- \rightarrow Pt(s)$	1.12
$Co^{2+}(aq) + 2 e^- \rightarrow Co(s)$	–0.28
$Mn^{2+}(aq) + 2 e^- \rightarrow Mn(s)$	–1.18

Table 19.1 Select Standard Reduction Potential ($E°$) Values

Half-Cell Reaction	Cathode Cell Notation	$E°$ (V)
$Cu^{2+}(aq) + 2 e^- \rightarrow Cu(s)$	$Cu^{2+}(aq)|Cu(s)$	0.34
$2 H^+(aq) + 2 e^- \rightarrow H_2(g)$	$H^+(a_{H^+} = 1)|H_2(g)|Pt(s)$	0.00
$Zn^{2+}(aq) + 2 e^- \rightarrow Zn(s)$	$Zn^{2+}(aq)|Zn(s)$	–0.76

a. Which is the weakest oxidizing agent?

b. Which is the strongest oxidizing agent?

c. Which is the strongest reducing agent?

d. Which is the weakest reducing agent?

e. Can you explain parts a through d in terms of their electronegativity values?

f. Will Co(s) reduce Pt^{2+}(aq) to Pt(s)?

g. Will Pt(s) reduce Co^{2+}(aq) to Co(s)?

h. Which ions can be reduced by Co(s)?

PROBLEM 19.12

Calculate $E°_{cell}$ for each of these reactions. Indicate whether each is a product-favored reaction or a reactant-favored reaction.

a. I_2(s) + Mg(s) $\rightleftarrows$ Mg^{2+}(aq) + 2 I^-(aq)

b. Ag(s) + Fe^{3+}(aq) $\rightleftarrows$ Ag^+(aq) + Fe^{2+}(aq)

c. Sn^{2+}(aq) + 2 Ag^+(aq) $\rightleftarrows$ Sn^{4+}(aq) + 2 Ag(s)

d. 2 Zn(s) + O_2(g) + 2 H_2O(l) $\rightleftarrows$ 2 $Zn(OH)_2$(s)

e. Fe(s)|$Fe(NO_3)_2$(aq)⸬$Cu(NO_3)_2$(aq)|Cu(s)

f. Pt(s)|Fe^{2+}(aq), Fe^{3+}(aq)⸬H^+(aq), H_2O_2(aq), H_2O(l)|Pt(s)

g. Zn(s)|$Zn(NO_3)_2$(aq)⸬$Pb(NO_3)_2$(aq)|Pb(s)

h. Ag(s)|$AgNO_3$(aq)⸬ $Sn(NO_3)_2$(aq)|Sn(s)

i. Pt(s)|$(COOH)_2$(aq)|CO_2(g)⸬O_2(g)|H^+(aq), H_2O(l)|Pt(s)

PROBLEM 19.13

The low-voltage battery in a car (in contrast to the high-voltage, lithium-ion traction battery) is made up of six individual electrochemical cells. Each cell uses electrodes with lead: $PbSO_4$(s)|SO_4^{2-}(aq)|Pb(s) and PbO_2(s)|SO_4^{2-}(aq), H^+(aq), H_2O(l)|$PbSO_4$(s).

a. Using Appendix 8, determine which electrode will be the cathode and which electrode will be the anode to produce a galvanic cell.

b. Write a half-reaction for the reaction occurring at the anode and the cathode.

c. What is the overall, balanced reaction that is occurring in this cell?

PROBLEM 19.14

Using Appendix 8, explain the following observations.

a. Co^{3+} is not stable in aqueous solution.

b. Fe^{2+} is not stable in air.

PROBLEM 19.15

Many metals dissolve in nitric acid (producing different nitrogen oxide gas products in the process).

a. Write a balanced equation for the reaction of iron metal with nitric acid and calculate $E°_{cell}$.

b. Gold does not dissolve in nitric acid. Provide an explanation.

c. Gold will slowly dissolve in nitric acid with aqueous hydrogen chloride, a combination called aqua regia. Using the list of $E°$ values, can you explain why?

CELL POTENTIALS, EQUILIBRIA, AND $\Delta_r G°$ VALUES

In the discussion of cell potential ($E°_{cell}$) above, we covered how positive cell voltages correspond to reactions that proceed to produce products and how negative cell voltages correspond to reactions that do not proceed to make products. We have seen two other means of quantifying whether a reaction is reactant favored or product favored: Gibbs energy and equilibrium constants. To relate the cell potential to Gibbs energy and to an equilibrium constant, we need to know the number of electrons transferred in a reaction (z) with which we can relate $E°_{cell}$ and $\Delta_r G°$ (Equation 19.2 and Equation 19.3). In terms of units, one joule is equal to one volt multiplied by one coulomb.

$$\Delta_r G° = - zFE°_{cell} \qquad (19.2)[14]$$

$\Delta_r G°$ is the Gibbs energy of reaction (J/mol).
z is the number of electrons transferred.
F is the Faraday constant (96 485 C/mol).[15]
$E°_{cell}$ is the standard cell potential (V).

$$E°_{cell} = -\frac{\Delta_r G°}{zF} \qquad (19.3)$$

$E°_{cell}$ is the standard cell potential (V).
$\Delta_r G°$ is the Gibbs energy of reaction (J/mol).
z is the number of electrons transferred.
F is the Faraday constant (96 485 C/mol).

For the tin–gold cell that we have been considering so far, the number of electrons transferred is 12 and the standard cell potential is 1.62 V. With this information we can find the Gibbs energy of reaction.

$$\Delta_r G° = -(12 \text{ electron})(96\,485 \text{ C/mol})(1.62 \text{ V})$$
$$= -1\,880\,000 \text{ J/mol}$$
$$= -1880 \text{ kJ/mol}$$

This reaction has a large negative Gibbs energy of reaction, which corresponds to a very product-favored reaction. It is worth noting that what seems like a small voltage (1.62 V) corresponds to a very large Gibbs energy.

To relate the cell potential to the equilibrium constant (K) we need to know the amount of electrons transferred and the temperature (Equations 19.4 and 19.5).

$$K = e^{\left(zFE°_{cell}/RT\right)} \qquad (19.4)$$

K is the ratio of product amount concentration to reactant amount concentration (unitless).
z is the number of electrons transferred.
F is the Faraday constant (96 485 C/mol).
$E°_{cell}$ is the standard cell potential (V).
R is the ideal gas constant (8.314 J/(mol K)).
T is the temperature (K).

$$E°_{cell} = \frac{RT}{zF} \ln K \qquad (19.5)$$

$E°_{cell}$ is the standard cell potential (V).
z is the number of electrons transferred.
F is the Faraday constant (96 485 C/mol).
R is the ideal gas constant (8.314 J/(mol K)).
T is the temperature (K).
K is the ratio of product amount concentration to reactant amount concentration (unitless).

For the tin–gold cell that we have been considering so far, the number of electrons transferred is 12 and the standard cell potential is 1.62 V. With this information we can find the equilibrium constant at 298 K.

$$K = e^{\left((12)(96\,485\,C/mol)(1.62\,V)\big/(8.314\,J/(mol\,K))(298\,K)\right)}$$

$$= e^{757}$$

This reaction has a *very* large equilibrium constant, e^{757} is so large that neither a spreadsheet nor a calculator can find a numerical value. This corresponds to a *very* product-favored reaction. It is worth noting that what seems like a small voltage (1.62 V) corresponds to a very large equilibrium constant.

PROBLEM 19.16
Consider the reaction of aluminium and copper(2+) ions.

$$2\,Al(s) + 3\,Cu^{2+}(aq) \rightleftarrows 2\,Al^{3+}(aq) + 3\,Cu(s)$$

a. Determine $E°_{cell}$ for this reaction.

b. Using your answer in part a, determine $\Delta_r G°$ at 25 °C.

c. What is K for this reaction?

d. Is this reaction product favored or reactant favored?

PROBLEM 19.17
Hydrazine, N_2H_4, reacts with oxygen in the air to produce nitrogen gas and water.

$$N_2H_4(l) + O_2(g) \rightleftarrows N_2(g) + 2\,H_2O(l)$$

a. Calculate $\Delta_r G°$ for this reaction ($\Delta_f G°(N_2H_4(l)) = 149.3$ kJ/mol).

b. If you wanted to make a basic hydrazine–air fuel cell to power an electric vehicle, what would be the value of $E°_{cell}$?

PROBLEM 19.18
Determine the equilibrium constant K for the reaction between Cd(s) and Cu^{2+}(aq) at 25 °C.

PROBLEM 19.19
Calculate the equilibrium constant K for the reaction between Br_2(l) and Cl^-(aq) at 25 °C.

⤳ ELECTROCHEMISTRY AND REACTION CONTROL

An aspect of electrochemistry, aside from making voltaic cells, that will not be covered here in detail but is of central importance in terms of sustainability in chemistry, is synthetic electrochemistry. Synthetic electrochemistry is using electricity to make chemicals. This is not a new idea; Michael Faraday first demonstrated the synthesis of hydrocarbons from acetic acid in 1834.[16] The advantage of electrochemical synthesis is that electric potential is being used to directly add or remove an electron and initiate change, which can complement existing chemical methods or provide synthetic possibilities that are not otherwise possible.[17] This is done with mild conditions, reducing energy costs associated with increasing a reaction's temperature, and without using added chemical matter, increasing the atom economy.[18] The challenge has been the lack of standardization, and the plethora of unknowns in devising a new electrochemical synthesis, but the synthetic and sustainability benefits are leading to a renewed interest in electrochemistry and a drive to overcome these challenges.[19]

CELL POTENTIAL AT NONSTANDARD CONDITIONS (E_{cell})

So far, we have considered only standard conditions (100 kPa, 298 K, and ions with unit activity). But electrochemical cells are not always at standard conditions. In addition, we know that the potential of a battery (electric batteries are made up of one or more galvanic cells) is not constant, but that it decreases over time and as the battery loses charge. This can be quantified according to the Nernst equation (Equation 19.6),[20] where E_{cell}, the potential of the cell at a nonstandard state, is related to $E°_{cell}$ and modified by the reaction quotient (Q):

$$E_{cell} = E°_{cell} - \frac{RT}{zF}\ln Q \qquad (19.6)$$

E_{cell} is the cell potential at a nonstandard state (V).
$E°_{cell}$ is the cell potential at standard state (V).
R is the ideal gas constant (8.314 J/(mol K)).
T is the temperature (K).
z is the number of electrons transferred.
F is the Faraday constant (96 485 C/mol).

Q is the reaction quotient $\left(\dfrac{[\text{anode (product) aqueous species}]}{[\text{cathode (reactant) aqueous species}]} \right)$.

For our last consideration of the tin–gold cell (Sn(s)|H$_2$O(l), H$^+$(aq)|SnO$_2$(s)⫶Au^{3+}(aq)|Au(s)), let's consider the cell voltage (E_{cell}) for a reaction that starts at pH = 7 ([H$^+$] = 1.0 × 10^{-7} mol/L) and 1.0 mol/L Au^{3+} at 298 K. The standard state voltage is 1.62 V. We can see that the initial cell potential under these conditions is somewhat greater than the standard cell potential conditions.

$$E_{cell} = 1.62 \text{ V} - \frac{(8.314 \text{ J}/(\text{mol K}))(298 \text{ K})}{(12)(96\,485 \text{ C}/\text{mol})}$$

$$\ln\left(\frac{0.000\,000\,1 \text{ mol/L}}{1.0 \text{ mol/L}} \right) = 1.65 \text{ V}$$

The Nernst equation also helps to explain how electricity can be generated by concentration cells, where the anode and cathode consist of the same components but have different dissolved ion concentrations. Let's consider a sodium concentration cell at 298 K. Note this type of concentration cell is how your brain generates electrical impulses.

$$\text{Pt(s)|Na}^+\text{(aq, 0.01 mol/L)⫶Na}^+\text{(aq, 1.0 mol/L)|Pt(s)}$$

In these cells, the standard reduction potential ($E°_{cell}$) is 0.00 V (the cathode and the anode have the same $E°$ value), and the number of electrons (n) is the same as the ion charge. The cell potential (E_{cell}) for this sodium concentration cell would be:

$$E_{cell} = 1.62 \text{ V} - \frac{(8.314 \text{ J}/(\text{mol K}))(298 \text{ K})}{(1)(96485 \text{ C}/\text{mol})} \ln\left(\frac{0.01 \text{ mol/L}}{1.0 \text{ mol/L}} \right) = 0.1 \text{ V}$$

The cell potential of a concentration cell is not necessarily very high, but such small voltages can, as with the functioning of neurons in the brain, have significant ramifications.

PROBLEM 19.20
Consider the galvanic cell made of zinc and cadmium half-cells.

$$\text{Zn(s)} + \text{Cd}^{2+}\text{(aq)} \rightleftharpoons \text{Zn}^{2+}\text{(aq)} + \text{Cd(s)}$$

a. Calculate $E°_{cell}$ for this cell.

b. If [Cd^{2+}] = 0.068 mol/L and [Zn^{2+}] = 1.00 mol/L, what is E_{cell} at 25.0 °C?

c. If E_{cell} is 0.390 V and [Cd^{2+}] = 2.00 mol/L, what is the amount concentration (mol/L) of Zn^{2+}(aq) at 25.0 °C?

PROBLEM 19.21
The standard potential of a Daniell cell, Zn(s)|Zn^{2+}(aq)⫶Cu^{2+}(aq)|Cu(s), is 1.10 V when Zn^{2+}(aq) = Cu^{2+}(aq) = 1.0 mol/L. As the cell operates, the cell potential changes due to amount concentration changes of Zn^{2+}(aq) and Cu^{2+}(aq). Calculate the ratio (at 25.0 °C) of Zn^{2+}/Cu^{2+} when E_{cell} = 0.05 V. What happens, over time, to the [Zn^{2+}(aq)]? What happens to [Cu^{2+}(aq)]?

⤳ **ELECTROCHEMISTRY AND SUSTAINABILITY**
Although electrochemistry has many advantages when it comes to electrochemical synthesis, industrial and commercial electrochemistry is not without its challenges. For example, all cars use a 12 V lead–acid battery. Lead–acid batteries are cheap, easy to mass produce, and durable,

but they use lead, a toxic metal, and sulfuric acid, a hazardous acid. Lithium-ion batteries, which power mobile devices and electric vehicles, use lithium (20 mg per kilogram of Earth's crust), cobalt (25 mg per kilogram of Earth's crust), and nickel (84 mg per kilogram of Earth's crust).[21] Cobalt and nickel are also, in high concentrations, both injurious to human and environmental health. Important work is ongoing, and more is needed to develop batteries that can compete with lithium-ion batteries, reduce our dependence on rare and hazardous metals, and use Earth-abundant metals like iron (56,300 mg per kilogram of Earth's crust).

NOTES

1. Pauling, L. The Nature of the Chemical Bond. IV. The Energy of Single Bonds and the Relative Electronegativity of Atoms. *J. Am. Chem. Soc.*, **1932**, *54* (9), 3570–3582. DOI: 10.1021/ja01348a011.

2. "Electronegativity." IUPAC. *Compendium of Chemical Terminology*, 2nd ed. Compiled by A.D. McNaught and A. Wilkinson. Blackwell Scientific Publications, Oxford, 1997. Online version (2019–) created by S.J. Chalk. DOI: 10.1351/goldbook.E01990.

3. Reduction was named in antithesis to oxidation and meant "to reduce the amount of oxygen."

4. Lavoisier, A.L. *Elements of Chemistry*. Creech: Edinburgh, 1790. Pp. 159–172.

5. "Reduction." IUPAC. *Compendium of Chemical Terminology*, 2nd ed. Compiled by A.D. McNaught and A. Wilkinson. Blackwell Scientific Publications, Oxford, 1997. Online version (2019–) created by S.J. Chalk. DOI: 10.1351/goldbook.R05222.

6. "Oxidation." IUPAC. *Compendium of Chemical Terminology*, 2nd ed. Compiled by A.D. McNaught and A. Wilkinson. Blackwell Scientific Publications, Oxford, 1997. Online version (2019–) created by S.J. Chalk. DOI: 10.1351/goldbook.O04362.

7. First proposed by Lavoisier: Lavoisier, A.L. *Elements of Chemistry*. Creech: Edinburgh, 1790, pp. 159–172. Latimer, W.M. *The Oxidation States of the Elements and their Potentials in Aqueous Solution*. Prentice-Hall: New York, NY, 1938.

8. In English, oxidation number and oxidation state are functionally synonyms. Oxidation number etymologically derives from the now deprecated Stock Number: Stock, A. Einige Nomenklaturfragen der anorganischen Chemie. **1919**, *32* (98), 373–374. DOI: 10.1002/ange.19190329802.

9. "Oxidation state." IUPAC. *Compendium of Chemical Terminology*, 2nd ed. Compiled by A.D. McNaught and A. Wilkinson. Blackwell Scientific Publications, Oxford, 1997. Online version (2019–) created by S.J. Chalk. DOI: 10.1351/goldbook.O04365.

10. Michael Faraday developed the terminology electrode, anode, and cathode: Faraday, M. Experimental Researches in Electricity. Seventh series. *Phil. Trans. R. Soc.*, **1834**, *124* (0), 77–122. DOI: 10.1098/rstl.1834.0008.

11. (i) This may not be the original reference but is an early reference that clearly states that a vertical line represents a phase boundary, specifically with a potential difference across the boundary: Haber, F.; Hlemensiewcz, Z. Über elektrische Phasengrenzkräfte. *Z. Phys. Chem.*, **1909**, *67U* (1), 385–431. DOI: 10.1515/zpch-1909-6720.

    (ii) An early and clear presentation of the use of double vertical lines to represent a cell junction, without a potential difference: Lewis, G.N.; Randall, M. *Thermodynamics and the Free Energy of Chemical Substances*. McGraw-Hill Book Company Inc. (New York), 1923.

12. 2019 IUPAC recommendations, which refine the 1953 Stockholm Report made by Christiansen and Pourbaix: Pingarrón, J.M.; Labuda, J.; Barek, J.; Brett, C.; Camões, M.; Fojta, M.; Hibbert, D. Terminology of Electrochemical Methods of Analysis (IUPAC Recommendations 2019). *Pure and Applied Chemistry*, **2020**, *92* (4), 641–694. DOI: 10.1515/pac-2018-0109.

13. (i) One of the first methods to measure cell potentials ($E°_{cell}$): Poggendorff, J.C. Methode zur quantitativen Bestimmung der elektromotorischen Kraft inconstanter galvanischer Ketten. *Ann. Phys.*, **1841**, *130* (10), 161–191. DOI: 10.1002/andp.18411301002.

(ii) An early investigation of the potentials between different metal-electrodes: Poggendorff, J.C. Untersuchung über die elektromotorischen Kräfte der galvanischen Ströme. *Ann. Phys.*, **1847**, *146* (1), 60–73. DOI: 10.1002/andp.18471460104.

(iii) The first standard electrode potentials against a dropping-mercury electrode: Ostwald, W. Studien zur Kontaktelektrizität. *Z. Phys. Chem.*, **1887**, *1U* (1), 583–610. DOI: 10.1515/zpch-1887-0162.

(iv) The sign convention of electrode potentials underwent a long period of contentious debate, but the convention proposed by Gibbs is the agreed-upon standard: Gibbs, W.J. The Equilibrium of Heterogeneous Substances. *Trans. Conn. Acad.*, **1875–1876**, *3*, 108–248 and **1877–1878**, *3*, 343–524.

(iv) One of the earliest studies of electrode potentials against a hydrogen electrode, which also cites that the cell voltage is the difference of the two half-cells: Smale, F.J. Studien über Grasketten. *Z. Phys. Chem.*, **1894**, *14U* (1), 577–621. DOI: 101515/zpch-1894-1444.

14. From electrostatics, where energy (J) is equal to one volt (V) multiplied by one coulomb (C). This equation is cited as well by Lewis and Randall in 1923: Lewis, G.N.; Randall, M. *Thermodynamics and the Free Energy of Chemical Substances.* McGraw-Hill Book Company Inc. (New York), 1923.

15. (i) The Faraday constant is found by taking the molar mass (kg/mol) and dividing by the electrochemical equivalent of that substance (kg/C). Faraday discusses electrochemical equivalents: Faraday, M. Experimental Researches in Electricity. Seventh series. *Phil. Trans. R. Soc.*, **1834**, *124* (0), 77–122. DOI: 10.1098/rstl.1834.0008.

(ii) One of the first systematic determinations of the electrochemical equivalent of silver: Strutt, J.W.; Sidgwick, E.M. XVII. On The Electro-Chemical Equivalent of Silver, and on the Absolute Electromotive Force of Clark Cells. *Phil. Trans. R. Soc.*, **1884**, *175*, 411–460. DOI: 10.1098/rstl.1884.0018.

16. Faraday, M. Siebente Reihe von Experimental-Untersuchungen über Elektricität. *Ann. Phys.*, **1834**, *109* (31–34), 481–520. DOI: 10.1002/andp.18341093102.

17. Zhu, C.; Ang, N.W.J.; Meyer, T.H.; Qiu, Y.; Ackermann, L. Organic Electrochemistry: Molecular Syntheses with Potential. *ACS Cent. Sci.*, **2021**, *7* (3), 415–431. DOI: 10.1021/acscentsci.0c01532.

18. Horn, E.J.; Rosen, B.R.; Baran, P.S. Synthetic Organic Electrochemistry: An Enabling and Innately Sustainable Method. *ACS Cent. Sci.*, **2016**, *2* (5), 302–308. DOI: 10.1021/acscentsci.6b00091.

19. Yan, M.; Kawamata, Y.; Baran, P.S. Synthetic Organic Electrochemistry: Calling All Engineers. *Angew. Chem. Int. Ed.*, **2018**, *57* (16), 4149–4155. DOI: 10.1002/anie.201707584.

20. Nernst, W. Die elektromotorische Wirksamkeit der Jonen. *Z. Phys. Chem.*, **1889**, *4U* (1), 129–181. DOI: 10.1515/zpch-1889-0412.

21. Abundance values from: Rumble, J.R.; Lide, D.R.; Bruno, T.J. 2019. *CRC Handbook of Chemistry and Physics: A Ready-Reference Book of Chemical and Physical Data*, 100th ed. Boca Raton, FL: CRC Press, 2019–2020.

20 Electron Sharing

This chapter is the second that seeks to explain why, chemically, a reaction is likely or unlikely to take place. As a reminder, a chemical reaction involves the movement of electrons. That movement could involve the transfer of electrons (see Chapter 19) or a change in how electrons are shared, which is the focus of this chapter. To start our discussion, let us consider the reaction of hydrogen chloride with water:

$$HCl(g) + H_2O(l) \rightleftarrows Cl^-(aq) + H_3O^+(aq)$$

If we assign oxidation numbers to each element, we find that hydrogen has an oxidation number of +1 before and after the reaction, chlorine has an oxidation number of –1 before and after the reaction, and oxygen has an oxidation number of –2 before and after the reaction. No electrons have been transferred and nothing has been oxidized nor has anything been reduced. Yet, a reaction has taken place because the electrons are shared differently in the products than they are in the reactants. We can see this more clearly if we consider the Lewis structures of the chemical species (Figure 20.1).

In the reactants, chlorine has three lone pairs, H and Cl share two electrons as a single bond, and oxygen has two lone pairs. In the products, Cl is not sharing any electrons and now has four lone pairs and oxygen only has one lone pair and a new bond to hydrogen. Altogether, we can see the changes and we can talk about them, but there is not a single, quantifiable metric, like oxidation numbers, that we can use to measure the change. Chemists have, therefore, developed the terms "acid" and "base," which are based on definitions from three acid–base theories: Arrhenius, Brønsted–Lowry, and Lewis.

ACID–BASE DEFINITIONS

In the Arrhenius definition,[1] acids are compounds that produce hydrons, $H^+(aq)$, in water, and bases are compounds that produce hydroxide, $HO^-(aq)$, in water. It is worth a brief discussion about $H^+(aq)$ in solution. $H^+(aq)$ does not exist as a freely occurring species, rather $H^+(aq)$ exists as an equilibrium mixture of oxidanium ($H_3O^+(aq)$), the Zundel cation ($H_5O_2^+(aq)$),[2] and the Eigen cation ($H_9O_4^+(aq)$)[3] Figure 20.2. In mathematical formulae, $H^+(aq)$ will be shown per convention; however, note that $H^+(aq)$ implies any of the structures in Figure 20.2. In any depictions of Lewis structures in this text, oxidanium will always be shown – the Zundel and Eigen cations would also work – and not free hydrons.

While the Arrhenius definition of acids and bases is useful in water, not all acid–base reactions happen in water, and not all bases contain or produce hydroxide. The Brønsted–Lowry and Lewis acid–base theories are both more inclusive of the range of acid–base chemistry possible. In Brønsted–Lowry acid–base theory,[4] an acid is a molecule that is an H^+, hydron, donor and a base is a molecule that is an H^+, hydron, acceptor. In Brønsted–Lowry theory, acids and bases are defined in relation to one another. Acids donate a hydron to a base, and this produces a conjugate base while the base accepts the hydron and becomes a conjugate acid (Figure 20.3).

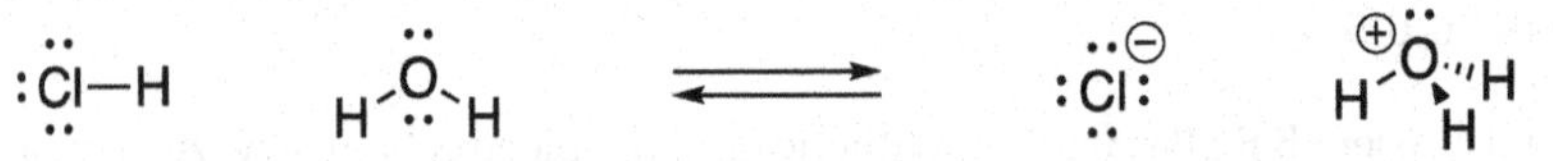

Figure 20.1 Lewis structures showing the reaction of hydrogen chloride with water to produce chloride and oxidanium (hydronium).

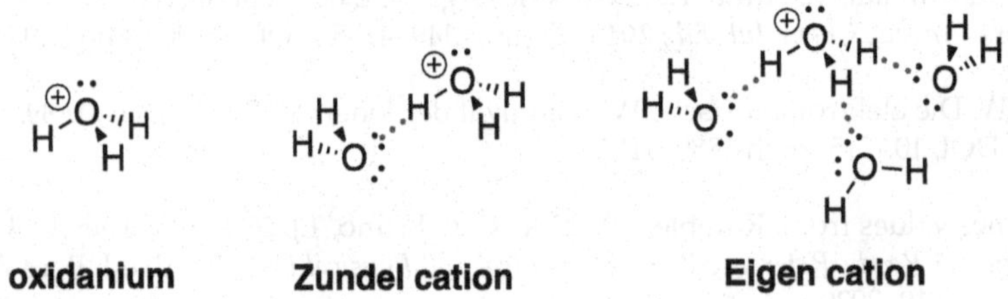

oxidanium   **Zundel cation**   **Eigen cation**

Figure 20.2 Structure of oxidanium, Zundel, and Eigen cations with the red, dotted lines showing hydrogen bonds.

DOI: 10.1201/9781003479338-20

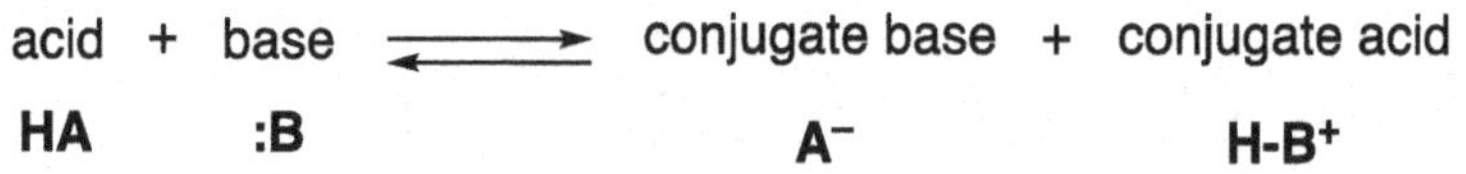

acid + base ⇌ conjugate base + conjugate acid

HA   :B   A⁻   H-B⁺

Figure 20.3 Terminology and symbology of a generic Brønsted–Lowry acid–base reaction.

acid   base   conjugate base   conjugate acid

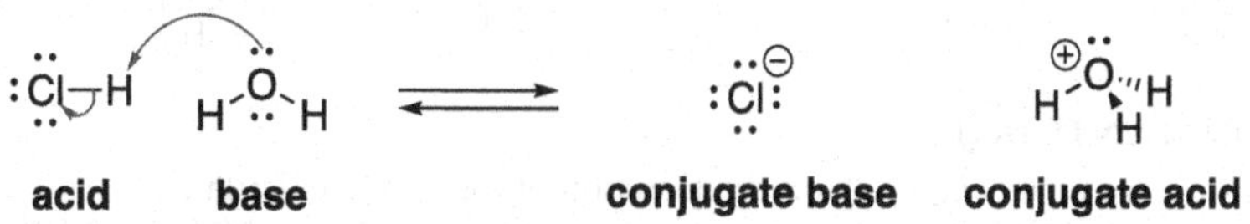

Figure 20.4 Electron-pushing arrows for a Brønsted–Lowry acid–base reaction with each species identified.

It is common to use the shorthand notations **HA** for an acid and **:B** for a base. In a specific example of HCl reacting with water (Figure 20.4), the base (water) takes the hydron from HCl. This is indicated by the arrow from the oxygen atom lone pair to the H atom coupled with the arrow from the Cl–H bond to the chlorine atom, which results in the conjugate base chloride (Cl⁻) and the conjugate acid oxidanium (H_3O^+).

The Lewis acid–base theory is the most inclusive framework for defining acids and bases. In the Lewis acid–base theory,[5] acids are defined as electron-pair acceptors, and bases are defined as electron-pair donors. In Figure 20.4, HCl is the acid because H has a new bond, accepting two electrons from O, and Cl has a new lone pair. Water is the base because oxygen has one fewer lone pair that was donated to H. This assignment is in complete agreement with the Brønsted–Lowry definition, the only change is the perspective (looking at H⁺ for Brønsted–Lowry rather than the electrons for Lewis).

PROBLEM 20.1
For each of the following reactions, identify the acid and the base. For any reactions involving H⁺ transfer, also identify the conjugate acid and conjugate base.

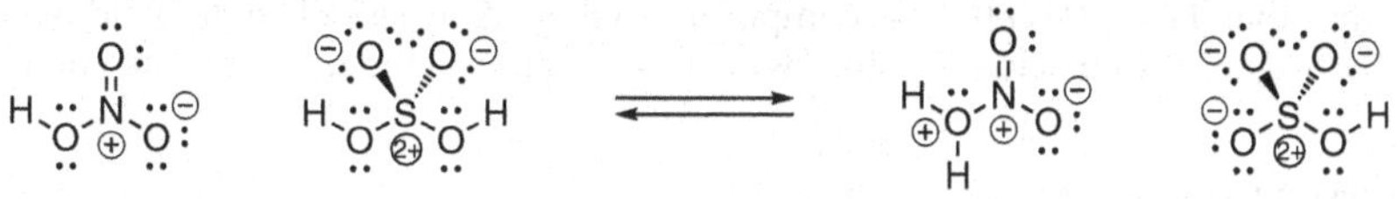

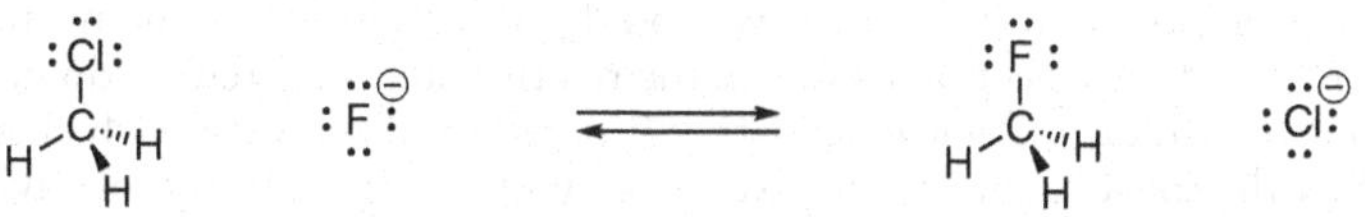

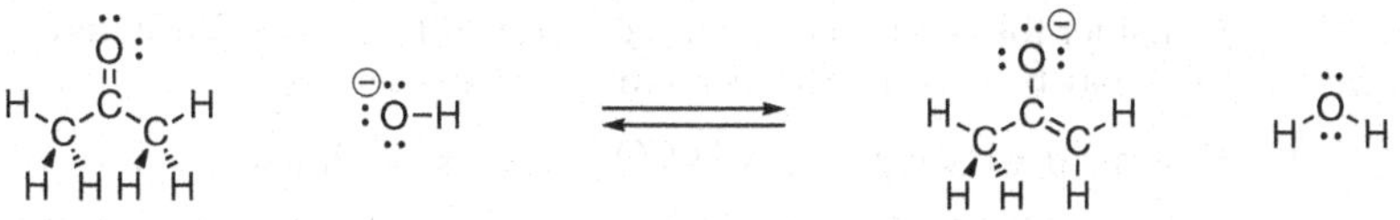

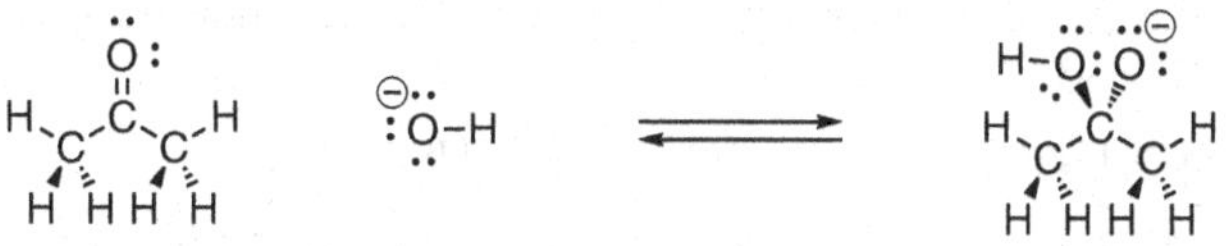

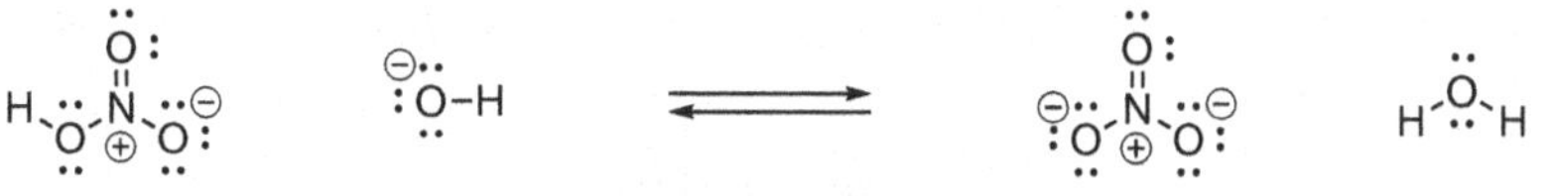

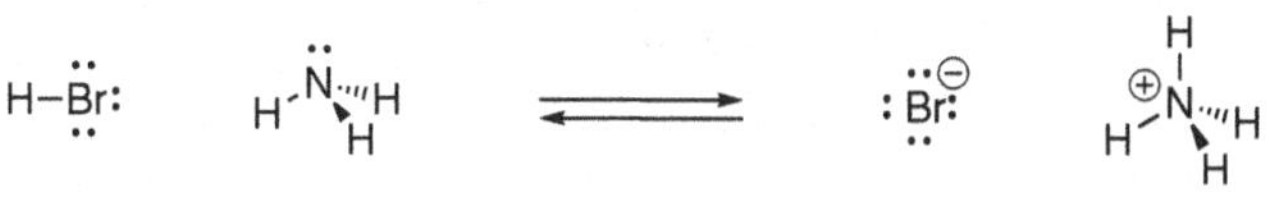

g.

h. H−C≡C−H

ACIDITY CONSTANTS (K_A)

In redox reactions, we utilize standard reduction potentials ($E°$) to calculate $E°_{cell}$ for a reaction. This would allow us to predict whether a reaction was reactant favored ($E°_{cell} < 0$ V) or product favored ($E°_{cell} > 0$ V). We cannot utilize standard reduction potentials for acid–base reactions because no electrons are transferred. Instead, acidity constant (K_a) values are used.[6] Consider the reaction of the generic acid HA:

$$HA(aq) \rightleftarrows H^+(aq) + A^-(aq)$$

The equilibrium constant (K_c) for this reaction (Equation 20.1) is termed K_a because the equilibrium constant corresponds to the dissociation of an acid in water:

$$K_a = K_{c,acid} = \frac{[A^-(aq)]\,[H^+(aq)]}{[HA(aq)]} \tag{20.1}$$

K_a is the acidity constant or the ratio of conjugate base and hydron to acid amount concentration.
$[A^-(aq)]$ is the amount concentration of conjugate base (mol/L).
$[H^+(aq)]$ is the amount concentration of aqueous hydron (mol/L).
$[HA(aq)]$ is the amount concentration of undissociated acid (mol/L).

What do the acidity constants (see Table 20.1) tell us about an acid? For $K_a < 1$, these are acids where A$^-$ is a stronger electron-pair donor than H_2O. We call these compounds, where A$^-$ is more likely to bond and share its electrons with hydrons more than water, weak acids. Another way of saying this is that A$^-$ is a strong conjugate base.[7] For $K_a > 1$, these are acids where A$^-$ is a weaker electron-pair donor than H_2O. We call these compounds, where A$^-$ is less likely to bond (share) its electrons with hydrogen, strong acids. Another way of saying this is that A$^-$ is a weak conjugate base.

APPLYING ACIDITY CONSTANT VALUES

When using standard reduction potential ($E°$) values, we could predict whether a chemical species is likely to be reduced or oxidized in a reaction based on the relative standard reduction potential values. In a similar fashion, we can utilize acidity constant (K_a) values to predict which chemical species is most likely to play the role of acid and, by extension, which is most likely to play the role of base. Note that if one of the chemicals has no hydrogen atoms, it cannot be a Brønsted–Lowry acid and must be a base. Consider the reaction of acetic acid and sulfuric acid. Since they are both named acid, our first instinct may be to think this is an impossible problem.

$$CH_3CO_2H(aq) + H_2SO_4(aq) \rightleftarrows CH_3CO_2H_2^+(aq) + HSO_4^-(aq)$$

Now looking at Table 20.1, the acidity constants for each acid are $K_a(CH_3CO_2H) = 1.8 \times 10^{-5}$ and $K_a(H_2SO_4) = 1.6 \times 10^5$. Since that acidity constant for H_2SO_4 is greater, H_2SO_4 is the stronger acid and so it will donate its H$^+$ to acetic acid.

$$CH_3CO_2H(aq) + H_2SO_4(aq) \rightleftarrows CH_3CO_2H_2^+(aq) + HSO_4^-(aq)$$

In addition, the acidity constant values can then be used to calculate the equilibrium constant (K_c) value as the ratio of the acidity constant of the reactant acid to the acidity constant of the conjugate acid:

$$K_c = \frac{K_a(\text{reactant acid})}{K_a(\text{conjugate acid})} \tag{20.2}$$

Table 20.1   Select Acidity Constant (K_a) Values

Aqueous Acid	Formula (HA)	Conjugate Base (A^-)	K_a	pK_a
Hydrogen iodide	HI	I^-	1.0×10^{10}	-10.00
Hydrogen bromide	HBr	Br^-	1.0×10^9	-9.00
Perchloric acid	$HClO_4$	ClO_4^-	1.0×10^8	-8.00
Hydrogen chloride	HCl	Cl^-	1.0×10^7	-7.00
Acetic acidium	$CH_3CO_2H_2^+$	CH_3CO_2H	1.6×10^6	-6.20
Sulfuric acid	H_2SO_4	HSO_4^-	1.6×10^5	-5.20
Nitric acid	HNO_3	NO_3^-	25	-1.40
Oxidanium	H_3O^+	H_2O	1.0	0.00
Trifluoroacetic acid	CF_3CO_2H	$CF_3CO_2^-$	0.63	0.20
Sulfurous acid	H_2SO_3	HSO_3^-	1.7×10^{-2}	1.77
Hydrogensulfate	HSO_4^-	SO_4^{2-}	1.1×10^{-2}	1.96
Phosphoric acid	H_3PO_4	$H_2PO_4^-$	7.9×10^{-3}	2.10
Nitrous acid	HNO_2	NO_2^-	7.1×10^{-4}	3.15
Hydrogen fluoride	HF	F^-	6.3×10^{-4}	3.20
Formic acid	HCO_2H	HCO_2^-	3.0×10^{-4}	3.52
Hydrogen selenide	H_2Se	HSe^-	1.3×10^{-4}	3.88
Benzoic acid	$C_6H_5CO_2H$	$C_6H_5CO_2^-$	6.3×10^{-4}	4.20
Acetic acid	CH_3CO_2H	$CH_3CO_2^-$	1.8×10^{-5}	4.74
Carbonic acid	H_2CO_3	HCO_3^-	4.3×10^{-7}	6.37
Hydrogen sulfide	H_2S	HS^-	1.0×10^{-7}	7.00
Dihydrogenphosphate	$H_2PO_4^-$	HPO_4^{2-}	6.3×10^{-8}	7.20
Ammonium	NH_4^+	NH_3	5.6×10^{-10}	9.25
Hydrogen cyanide	HCN	CN^-	3.3×10^{-10}	9.48
Phenol	C_6H_5OH	$C_6H_5O^-$	1.0×10^{-10}	10.00
Hydrogen carbonate	HCO_3^-	CO_3^{2-}	4.7×10^{-11}	10.33
Hydrogenphosphate	HPO_4^{2-}	PO_4^{3-}	4.6×10^{-13}	12.34
Water	H_2O	OH^-	1.0×10^{-14}	14.00
Methanol	CH_3OH	CH_3O^-	1.0×10^{-16}	16.00
Sulfanide (bisulfide)	HS^-	S^{2-}	1.0×10^{-17}	17.00
Acetylene	C_2H_2	C_2H^-	1.0×10^{-25}	25.00
Hydrogen	H_2	H^-	1.0×10^{-35}	35.00
Ammonia	NH_3	NH_2^-	1.0×10^{-38}	38.00
Methane	CH_4	CH_3^-	1.0×10^{-50}	50.00

In the example of sulfuric acid and acetic acid, the K_a value for the reactant acid, sulfuric acid, is 1.6×10^5 and for the product acid or conjugate acid, the acetic acidium ion, is 1.6×10^6. Altogether, this gives a value for the equilibrium constant of 0.10, which is slightly reactant favored:

$$K_c = \frac{K_a(\text{reactant acid})}{K_a(\text{conjugate acid})} = \frac{1.6 \times 10^5}{1.6 \times 10^6} = 0.10 \left(\text{reactant favored}\right)$$

Because pK_a values are commonly given for acids – they are typically easier to remember and discuss than acidity constants – the value of K_c can also be determined directly using pK_a values.

$$K_c = 10^{(\text{pKa(conjugate acid)} - \text{pKa(acid)})} \tag{20.3}$$

Using the values for the sulfuric acid–acetic acid example, we find the same value of 0.1 for the equilibrium constant:

$$K_c = 10^{(\text{pKa(conjugate acid)} - \text{pKa(acid)})} = 10^{(-6.2 - -5.2)} = 10^{-1.0} = 0.1 \text{ (reactant favored)}$$

PROBLEM 20.2
Predict the products of each reaction. Once you have the balanced chemical equation, calculate K_c for each reaction and identify whether it is reactant favored or product favored.

a. $H_2O(l) + NH_3(aq) \rightleftharpoons$

b. $HF(aq) + NaH_2PO_4(aq) \rightleftharpoons$

c. $HCN(aq) + NaSH(aq) \rightleftharpoons$

d. $KHCO_3(aq) + HF(aq) \rightleftharpoons$

e. $H_3O^+(aq) + RbF(aq) \rightleftharpoons$

f. $H_2O(l) + H_2O(l) \rightleftharpoons$

⤳ ACID–BASE CATALYSIS AND GREEN CHEMISTRY

An important use of acids and bases is as catalysts (Chapter 14) in (bio)synthetic chemistry. The added acid or base can add or remove a hydron, respectively, which alters the energy of the reaction and allows for transformations that would not be feasible without the added acid or base. Common examples include hydrolysis (breaking of bonds with water) and dehydration (removing water and forming new bonds). In the laboratory, this is frequently accomplished with mineral acids (HCl, H_2SO_4, H_3PO_4) or Lewis acids (commonly transition metal cations or compounds containing a group-13 atom). Strong acids, like sulfuric acid, present a risk as hazardous chemicals. Weaker acids like phosphoric acid can be used to reduce risk, or solid acids have been developed and deployed that are recyclable, significantly reducing waste, improving atom economy, and reducing the chemical hazard of using a corrosive liquid.[8]

In cells, enzyme active sites contain charged amino acid side chains (arginine, histidine, lysine, aspartic acid, and glutamic acid), or Lewis basic side chains (serine, threonine, and cysteine), with water to effect these transformations. A wide number of enzymes use acid–base catalysis, but a common example is hydrolase enzymes, which can break (lyse) bonds using water. Enzymes themselves, including hydrolases, are also used in laboratory synthesis because of their high selectivity, the mild conditions in which they operate, and the rapidity with which they work.[9]

INSIGHTS FROM ACIDITY CONSTANT VALUES

So far, we have considered acidity constant values as only empirical measures of the willingness of a conjugate base (A^-) to share its electrons, with H_2O used as our point of reference. We are now going to try to understand the trends in K_a values, with the goal of gaining insight into what makes something more or less likely to share its electrons.

First, let's consider acids where the hydrogen atom is connected to an atom from period 2 (Table 20.2). There is an incredible range of acidity: there is a difference of roughly 45 orders of magnitude between the weakest acid, CH_4, and the strongest acid, HF. This trend in acidity mirrors

Table 20.2 Chemical Formulae and Lewis Structures of Acids and Conjugate Bases Where the Hydrogen Atom Connected to an Atom from Period 2

Acid (HA)	Lewis Structure	Conjugate Base (A^-)	Lewis Structure	Central Atom (χ_P)	K_a
CH_4		CH_3^-		2.55	1.0×10^{-50}
NH_3		NH_2^-		3.04	1.0×10^{-38}
H_2O		OH^-		3.44	1.0×10^{-14}
HF		F^-		3.98	6.3×10^{-4}

the trend in Pauling electronegativity values (χ_P). That is, the least electronegative atom (carbon with an electronegativity value of 2.5) has the lowest acidity and the most electronegative atom (fluorine with an electronegativity value of 4.0) has the greatest acidity. In general, atoms with low electronegativity (C and N) have a greater willingness to share their electrons, they are stronger conjugate bases, and therefore CH_4 and NH_3 are weaker acids. In contrast, atoms with a high electronegativity (O and F) have a lesser willingness to share their electrons, they are weaker conjugate bases, and therefore H_2O and HF are stronger acids. In general, then, we can conclude that acidity increases from left to right across a period.

Now, consider Table 20.3, which consists of acids where the hydrogen atom is connected to an atom from group 17. In Table 20.3, there is not quite the same dramatic range of acidity values as there was in Table 20.2. There is a difference of only 15 orders of magnitude between the weakest acid, HF, and the strongest acid, HI. This trend in acidity mirrors the trend in atomic or ionic radius. That is, the acid with the smallest atom (fluorine with a van der Waals radius of 146 pm) has the lowest acidity and the largest atom (iodine with a van der Waals radius of 204 pm) has the greatest acidity. In general, small ions like fluoride (133 pm) have a greater willingness to share their electrons, they are stronger conjugate bases, and therefore HF is a weaker acid. In contrast, large ions like iodide (220 pm) have a lesser willingness to share their electrons, they are weaker conjugate bases, and therefore HI is a stronger acid. In general, then, we can conclude that acidity increases from top to bottom in a group.

Now, consider Table 20.4, which consists of acids where the hydrogen atom is connected to an atom from group 17. Here what is changing is the number of oxygen atoms attached to the central chlorine atom. Oxygen is a highly electronegative atom that withdraws electrons inductively, by pulling electrons through bonds, away from the atoms that it is attached to and toward itself. Therefore, we term oxygen, and other highly electronegative atoms like nitrogen, bromine, chlorine, and fluorine, an electron-withdrawing group (EWG). We can see the effect these oxygen atoms have if we consider the calculated charge, what is called a natural population analysis (NPA) charge, of the chlorine atom.[10] In hypochlorite (ClO^-), the chlorine has an NPA charge of −0.062 and this increases with each additional oxygen, and perchlorate (ClO_4^-) has an NPA charge of +2.468. The increasing positive charge of the chlorine atom attracts the oxygen atom electrons, which makes each oxygen atom less likely to share its electrons with H^+. In general, the fewer EWGs (ClO^-) the more likely a chemical species is to share its electrons, the stronger the conjugate base, and therefore HClO is a weaker acid. The more EWGs (ClO_4^-) the less likely a chemical species is to share its electrons, the weaker the conjugate base, and therefore $HClO_4$ is a stronger acid. In general, then, we can conclude that acidity increases with each additional EWGs. Conversely, acidity decreases with each additional electron-donating group (EDG) like carbon and silicon.

Now, consider Table 20.5, which consists of acids that differ in terms of the number of hydrogen atoms and by extension the charge state. The trend in acidity follows the trend in charge. That is, the most negatively charged/least positively charged acid (HPO_4^{2-}) has the lowest acidity, and the least negatively charged/most positively charged acid (H_3PO_4) has the greatest acidity. In general, more negatively charged species have a greater willingness to share their electrons, they are stronger conjugate

Table 20.3 Chemical Formulae and Lewis Structures of Acids and Conjugate Bases Where the Hydrogen Atom Connected to an Atom from Group 17

Acid (HA)	Lewis Structure	Covalent Radius of Halogen (pm)[a]	Conjugate Base (A^-)	Lewis Structure	Ionic Radius of Halide (pm)[a]	K_a
HF	H–F̤	57	F^-	:F̤:⊖	133	6.3×10^{-4}
HCl	H–C̤l	102	Br^-	:C̤l:⊖	181	1.0×10^7
HBr	H–B̤r	120	Cl^-	:B̤r:⊖	196	1.0×10^9
HI	H–Ï:	139	I^-	:Ï:⊖	220	1.0×10^{10}

[a] Values are for coordination number = 6: Shannon, R.D. Revised Effective Ionic Radii and Systemic Studies of Interatomic Distances in Halides and Chalcogenides. *Acta Cryst.*, **1976**, A32, 751–767. DOI: 10.1107/S0567739476001551.

Table 20.4 Chemical Formulae and Lewis Acids and Conjugate Bases with Varying Numbers of Oxygen Atoms Attached to the Central Atom

Acid (HA)	Lewis Structure	Conjugate Base (A⁻)	Lewis Structure	NPA Charge of Chlorine in A⁻	K_a
HClO		ClO⁻		−0.062	3.0×10^{-8}
HClO$_2$		ClO$_2^-$		+0.849	1.2×10^{-2}
HClO$_3$		ClO$_3^-$		+1.741	1.0×10^{1}
HClO$_4$		ClO$_4^-$		+2.468	1.0×10^{8}

bases, and therefore HPO_4^{2-} is a weaker acid. In contrast, acids that are neutral or positive have a lesser willingness to share their electrons, they are weaker conjugate bases, and therefore H_3PO_4 is a stronger acid. In general, then, we can conclude that acidity decreases with increasing negativedecreasing positive charge. Conversely, acidity increases with decreasing negative charge/increasing positive.

Now, consider Table 20.6, which looks at methanol (CH_3OH), a localized acid–conjugate base pair, and at phenol (C_6H_5OH), a delocalized acid–conjugate base pair. Methanol is a weaker acid and the conjugate base methoxide (CH_3O^-) is stronger because the oxygen lone pair is localized on the oxygen atom. In contrast, phenol is a stronger acid and the conjugate base phenoxide ($C_6H_5O^-$) is weaker because the lone pair is delocalized, as shown by the contributing structures (Table 20.6). In general, delocalization of electrons means those electrons are less likely to be shared and the conjugate base is more stable/weaker, which makes the acid stronger. If the atom that has the acidic hydrogen is next to a double or triple bond, the lone pair will be delocalized.

Finally, consider Table 20.7, which looks at the impact of electron geometry on acidity. Ethane (C_2H_6) consists of tetrahedral carbon atoms and is the weakest acid. Ethene (C_2H_4) contains trigonal planar carbon atoms and is a stronger acid. Acetylene (C_2H_2) has linear carbon atoms and is the strongest acid. We can infer that the electron geometry has a role in acidity. This is because the lone pair is in different energy orbitals depending on the electron geometry of the carbon atom. In

Table 20.5 Chemical Formulae and Lewis Structures of Acids and Conjugate Bases with Different Charge States

Acid (HA)	Lewis Structure	Conjugate Base (A⁻)	Lewis Structure	K_a
H$_3$PO$_4$		H$_2$PO$_4^-$		7.9×10^{-3}
H$_2$PO$_4^-$		HPO$_4^{2-}$		6.3×10^{-8}
HPO$_4^{2-}$		PO$_4^{3-}$		4.6×10^{-13}

Table 20.6   Formula and Lewis Structure of a Localized Acid and a Delocalized Acid and Their Conjugate Bases

Acid (HA)	Lewis Structure	Conjugate Base (A⁻)	Lewis Structure	Charge Type	K_a
CH_3OH		CH_3O^-		Localized	1.0×10^{-16}
C_6H_5OH		$C_6H_5O^-$		Delocalized	4.6×10^{-10}

Table 20.7   Formulae and Lewis Structures of Carbon Acids with Different Carbon-Atom Electron Geometries

Acid (HA)	Lewis Structure	Conjugate Base (A⁻)	Lewis Structure	Electronic Geometry of Carbon Atom	K_a
C_2H_6		$C_2H_5^-$		Tetrahedral	1.0×10^{-50}
C_2H_4		$C_2H_3^-$		Trigonal planar	1.0×10^{-44}
C_2H_2		C_2H^-		Linear	1.0×10^{-25}

ethane, the lone pair is in a higher energy sp^3 orbital (p orbitals are higher energy than s orbitals), while in acetylene the lone pair is in a lower energy sp orbital. In general, for the same atom, a linear electron geometry means a weaker conjugate base and a stronger acid than a trigonal planar electron geometry; a trigonal planar electron geometry corresponds to a weaker conjugate base and stronger acid than a tetrahedral geometry.

PROBLEM 20.3
For each reaction, predict the products and calculate K_c. Provide a rationalization for the position of the equilibrium (reactant favored or product favored).

a. $H_2Se(aq) + I^-(aq) \rightleftarrows$

b. $H_3O^+(aq) + HO^-(aq) \rightleftarrows$

c. $HCl(aq) + F^-(aq) \rightleftarrows$

d. $CF_3CO_2H(aq) + CH_3CO_2^-(aq) \rightleftarrows$

PROBLEM 20.4

For each reaction, identify the acid and the base (for reactions involving H⁺ transfer, identify the conjugate acid and the conjugate base), and then predict whether the reaction is most likely product favored or reactant favored. Explain your reasoning.

a.

b.

c.

d.

PROBLEM 20.5

There are lots of chloride (Cl^-) ions floating around in your body and there are lots of thiols (molecules that contain R–SH; note R is a common symbol used to indicate the other, unimportant to the problem, part of a larger structure). Why do we not have to worry about these combining to produce aqueous hydrogen chloride (HCl)? Estimate K_c for this reaction and provide an explanation.

$$R-SH + Cl^- \rightleftharpoons R-S^- + HCl$$

PROBLEM 20.6

Soaps are compounds that contain a large carbon-rich "tail" and a polar or ionic "head." Here are two soaps, one with a carboxylate head and one with a sulfate head.

sodium stearate (carboxylate soap)

sodium stearyl sulfate (sulfate soap)

Soap scum is the precipitation of soap with Mg^{2+} or Ca^{2+} ions. Why do carboxylate soaps cause soap scum while sulfate soaps almost never do?

PROBLEM 20.7

A common structure in organic chemistry and biochemistry is the hexagonal benzene ring (with alternating double and single bonds). If you take organic chemistry in college, you will spend a lot of time with benzene and studying its chemistry, and one of the reactions you'll learn is nitration (adding $-NO_2$ groups).

$K_a = 1.1 \times 10^{-10}$	$K_a = 7.1 \times 10^{-8}$	$K_a = 1.3 \times 10^{-5}$	$K_a = 4.2 \times 10^{-1}$
phenol	4-nitrophenol	2,4-dinitrophenol	picric acid (2,4,6-trinitrophenol)

Considering the nitrated phenols (benzene with an –OH), what explains this trend in acidity?

PROBLEM 20.8

Cysteine is one of the 20 common amino acids. Selenocysteine is a less common amino acid that is identical except that it has a selenium atom rather than a sulfur atom.

cysteine   selenocysteine

Despite their similar structure, there are some differences. Explain why selenocysteine is normally found without an H^+ attached to the selenium, whereas cysteine is normally found with an H^+ connected to the sulfur atom.

PROBLEM 20.9

During cellular respiration in the mitochondria, electrons are transported (through the electron transport chain). This transfer of electrons is used to create an amount concentration gradient (think Nernst) of H^+ ions. The cell then uses this electrochemical potential to create ATP. Here is one small step of the electron transport chain:

ubiquinol   semi-ubiquinone

The semi-ubiquinone then donates H^+ to the inner membrane space of the mitochondria. Why is semi-ubiquinone more likely to donate H^+ than is ubiquinol?

PROBLEM 20.10
Consider the reaction between acetic acid (CH_3CO_2H) and ammonia (NH_3):

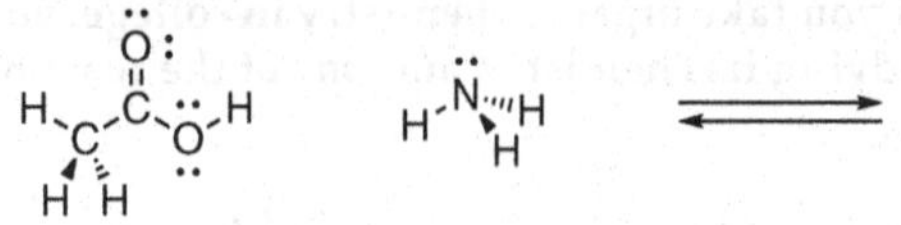

a. Use the K_a values to determine which reactant is the acid and which reactant is the base.

b. Write the products for the above reaction.

c. Calculate K_c for this reaction. Is this a product-favored or reactant-favored process?

d. Provide a rationalization (in terms of acidity trends) for the position of the equilibrium (i.e., why is it product favored or reactant favored).

PROBLEM 20.11
Amino acids contain both a carboxylic acid (R-COOH) and amine (R-NH₂). Consider the equilibrium that exists for glycine in solution:

Use your understanding of acids and bases to explain whether this is a reactant-favored or product-favored equilibrium and explain your answer.

pH

Under the Arrhenius definitions, acids are compounds that produce hydrons, $H^+(aq)$, in water, and bases are compounds that produce hydroxide, $HO^-(aq)$, in water. Within this framework, the amount of $H^+(aq)$ in water is a central value to know and quantify. This led to the development of pH, which is defined as the negative logarithm of $H^+(aq)$ activity:[11]

$$pH = -\log(a_{H+(aq)}) \qquad (20.4)$$

$a_{H+(aq)}$ is the activity of $H^+(aq)$ in solution.

Recall that activity values are the actual or active amount of a chemical in solution. The amount concentration (mol/L) of a chemical, especially at low concentrations, is close to the activity of that species and so we can approximately calculate the pH of a solution with Equation 20.5:

$$pH = -\log[H^+(aq)] \quad (20.5)$$

$[H^+(aq)]$ is the amount concentration (mol/L) of $H^+(aq)$ in solution.

Let us consider the pH scale. To do that we need to consider the central reaction under consideration, the autoionization equilibrium of water:

$$H_2O(l) \rightleftharpoons H^+(aq) + OH^-(aq) \qquad K_w(25.0 \text{ °C}) = 1.0 \times 10^{-14}$$

The equilibrium constant (given the importance of this K_c value is called K_w) for this reaction at 25.0 °C is 1.0×10^{-14}. The expression for this equilibrium constant is $K_w = [H^+(aq)][OH^-(aq)] = 1.0 \times 10^{-14}$. This means that at 25.0 °C the neutral pH, when the amount concentration of $H^+(aq)$ equals the amount concentration of $OH^-(aq)$, the amount concentration of each is 1.0×10^{-7} mol/L and the pH is 7.0. Since the equilibrium constant changes with temperature, the pH of a neutral solution, where the amount concentration of $H^+(aq)$ equals the amount concentration of $OH^-(aq)$, is 7.47 at 0.0 °C and 6.14 at 100.0 °C. For the rest of this discussion, we will consider only the pH scale at 25.0 °C.

On the pH scale, an acidic, less than 7.0, pH means the amount concentration of $H^+(aq)$ is greater than the amount concentration of $OH^-(aq)$. For example, vinegar is typically pH 2.4, which is less than 7.0 and therefore an acidic solution. We can use this value to determine the amount concentration of hydrons to be 0.004 mol/L:

$$2.4 = -\log[H^+(aq)]$$
$$10^{-2.4} = [H^+(aq)]$$
$$0.004 \text{ mol/L} = [H^+(aq)]$$

And using the expression for K_w at 25.0 °C, we find that the amount concentration of hydroxide ions is 3×10^{-12} mol/L:

$$[H^+(aq)][OH^-(aq)] = 1.0 \times 10^{-14}$$

$$[0.004][OH^-(aq)] = 1.0 \times 10^{-14}$$

$$[OH^-(aq)] = 3 \times 10^{-12} \text{ mol/L}$$

On the pH scale, a basic, more than 7.0, pH means the amount concentration of $H^+(aq)$ is less than the amount concentration of $OH^-(aq)$. For example, ammonia solution, a common household cleaner, is typically around pH 11.0, which is more than 7.0 and therefore a basic solution. We can use this value to determine the amount concentration of hydrons to be 1×10^{-11} mol/L:

$$11.0 = -\log[H^+(aq)]$$
$$1 \times 10^{-11} \text{ mol/L} = [H^+(aq)]$$

And using the expression for K_w at 25.0 °C, we find that the amount concentration of hydroxide ions is 0.001 mol/L:

$$[H^+(aq)][OH^-(aq)] = 1.0 \times 10^{-14}$$

$$[1 \times 10^{-11}][OH^-(aq)] = 1.0 \times 10^{-14}$$

$$[OH^-(aq)] = 0.001 \text{ mol/L}$$

Given the importance of pH, much work has gone into accurately measuring pH values. This is done electrochemically using selective-ion electrodes. For hydrogen-ion electrodes, pH is easily determined as the potential difference between a reference cell (H^+(aq, 1×10^{-7} mol/L)$\|$KCl(aq)$|$AgCl(s)$|$Ag(s)) and a test cell (H^+(aq, unknown mol/L)$\|$KCl(aq)$|$AgCl(s)$|$Ag(s)):

$$Ag(s)|AgCl(s)|KCl(aq)\|H^+(aq, 1 \times 10^{-7} \text{ mol/L})\|H^+(aq, \text{unknown mol/L})|KCl(aq)|AgCl(s)|Ag(s)$$

PROBLEM 20.12
Aqueous hydrogen fluoride dissociates in water according to the following equation:

$$HF(aq) + H_2O(l) \rightleftarrows F^-(aq) + H_3O^+(aq)$$

If the initial amount concentration of HF is 0.200 mol/L, what are the concentrations of aqueous hydrogen fluoride, fluoride, and oxidanium (H_3O^+) at equilibrium? What is the pH of the solution? Provide an explanation – in terms of acidity trends – for why this is a reactant-favored process.

HENDERSON–HASSELBALCH EQUATION AND ACID SPECIATION
Given the importance of pH and the relative ease of determining pH with hydrogen-ion electrodes, pH plays a central role in studying the equilibrium of acid–base reactions. To see the importance of this, let's consider Equation 20.1:

$$K_a = \frac{[A^-][H^+(aq)]}{[HA]}$$

Now let us take the negative logarithm of both sides of the acidity constant equation:

$$-\log K_a = -\log[H^+(aq)] - \log\frac{[A^-]}{[HA]}$$

In this new equation, we have $-\log K_a$, which is pK_a, and $-\log[H^+(aq)]$, which is pH. With a small rearrangement, we have the Henderson–Hasselbalch equation:[12]

$$pH = pK_a + \log\left(\frac{[A^-]}{[HA]}\right) \tag{20.6}$$

[HA] is the amount concentration of acid (mol/L).
[A$^-$] is the amount concentration of conjugate base (mol/L).

The Henderson–Hasselbalch equation allows us to relate the pH of a solution to the pK_a and the ratio of conjugate base to acid. With this equation, we can easily determine the pK_a of an acid, or with a known pK_a value we can determine the ratio of conjugate base to acid.

PROBLEM 20.13
The pH of a solution of unknown acid is 2.12, the amount concentration of HA is 0.1983 mol/L, and the amount concentration of A⁻ is 0.0017 mol/L. What is the pK_a of the acid?

PROBLEM 20.14
Acetic acid is in a solution at pH 5.74. What is the ratio of [A⁻]/[HA]?

PROBLEM 20.15
Acetic acid is in a solution at pH 3.74. What is the ratio of [A⁻]/[HA]?

Determining the ratio of conjugate base to acid provides a way to understand the speciation – the distribution of a chemical between different forms – of an acid. Review Figure 20.3, which considers the speciation of ammonium, NH_4^+(aq), and ammonia, NH_3(aq), as a function of pH. At pH values below 9.25, the pK_a of ammonium, the predominant chemical species in solution is NH_4^+, and above 9.25 the predominant chemical species in solution is NH_3.

Ammonium is a monoprotic acid; there is only one H⁺ that can be removed between pH 0 and pH 14. The speciation of monoprotic acids all look like Figure 20.5 with the only change being the crossover point (the pK_a). In contrast, phosphoric acid (H_3PO_4) is an example of a polyprotic acid. There are three H⁺ ions that can be removed between pH 0 and pH 14. The speciation of phosphoric acid is shown in Figure 20.6. Notice there are three crossover points, which correspond to pK_{a1} at 2.10, pK_{a2} at 7.20, and pK_{a3} at 12.34.

BUFFERS
Last, we will consider an important application of acids, bases, and speciation: buffers. A buffer is a combination of acid and conjugate base in a solution whose pH is near the pK_a. For example, our blood has a combination of dihydrogenphosphate ($H_2PO_4^-$(aq)) and hydrogenphosphate (HPO_4^{2-}(aq)). Why might this be the case? Consider the equilibrium dihydrogenphosphate in water.

$$H_2PO_4^-(aq) + H_2O(l) \rightleftharpoons HPO_4^{2-}(aq) + H_3O^+(aq)$$

If acid is added to this solution, the equilibrium will shift to the left but so long as the ratio of A⁻ to HA stays between 0.1 and 10 the pH will not change by more than 1. The range $pK_a \pm 1$ is the buffer range of a given acid/conjugate base. Let's consider the dihydrogenphosphate–blood buffer in

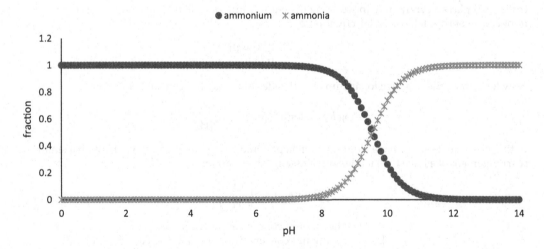

Figure 20.5 Speciation of ammonium (NH_4^+(aq)) and ammonia (NH_3(aq)) as a function of pH.

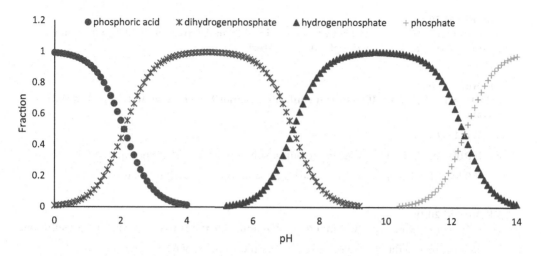

Figure 20.6 Speciation of phosphoric acid (H_3PO_4(aq)), dihydrogenphosphate ($H_2PO_4^-$(aq)), hydrogenphosphate (HPO_4^{2-}(aq)), and phosphate (PO_4^{3-}(aq)) as a function of pH.

more detail. If there is 0.468 mmol/L $H_2PO_4^-$(aq) and 0.732 mmol/L HPO_4^{2-}(aq), that corresponds to a blood pH of 7.39:

$$pH = 7.20 + \log\left(\frac{0.732 \text{ mmol/L}}{0.468 \text{ mmol/L}}\right) = 7.39$$

Now if 0.100 mM of acid is added, then the amount concentration of $H_2PO_4^-$(aq) would be 0.568 mmol/L (0.468 mmol/L + 0.100 mmol/L) and the amount concentration of HPO_4^{2-}(aq) would be 0.632 mmol/L (0.732 mmol/L - 0.100 mmol/L). The new pH of the solution would be 7.25:

$$pH = 7.20 + \log\left(\frac{0.632 \text{ mmol/L}}{0.568 \text{ mmol/L}}\right) = 7.25$$

While the phosphate buffer that we have looked at so far is only one buffer combination, and only one of the several buffers present in our blood, there is a wide variety of buffer combinations possible. A list of common buffers is shown in Table 20.8.

PROBLEM 20.16
If a researcher wanted to make a buffer with the following pH values, what acid/conjugate base pair(s) would be best?

a. pH 4.0

b. pH 6.5

c. pH 8.5

d. pH 2.5

Table 20.8 Common Buffer Combinations

Buffer	Acid (HA)	Conjugate Base (A⁻)	pK_a
Citric acid	$C_3H_5O(CO_2H)_3$	$C_3H_5O(CO_2H)_2CO_2^-$	3.13
Acetic acid	CH_3CO_2H	$CH_3CO_2^-$	4.74
Dihydrogencitrate	$C_3H_5O(CO_2H)_2CO_2^-$	$C_3H_5OCO_2H(CO_2)_3^{2-}$	4.76
Hydrogencitrate	$C_3H_5OCO_2H(CO_2)_3^{2-}$	$C_3H_5O(CO_2)_3^{3-}$	6.40
Dihydrogenphosphate	$H_2PO_4^-$	HPO_4^{2-}	7.20
Boric acid	$B(OH)_3$	$B(OH)_4^-$	9.24
Hydrogencarbonate	HCO_3^-	CO_3^{2-}	10.33

PROBLEM 20.17

A pyruvic acid ($K_a = 3.2 \times 10^{-3}$) buffer contains 0.50 mol/L pyruvic acid and 0.60 mol/L sodium pyruvate. What is the pH of the buffer?

PROBLEM 20.18

A lactic acid ($K_a = 1.4 \times 10^{-4}$) buffer contains 0.15 mol/L each of lactic acid and sodium lactate.

a. What is the initial pH of the solution?

b. If 0.050 mol/L HCl is added to the buffer solution, what will the new pH be?

c. If 0.10 mol/L NaOH is added to the buffer solution, what will the new pH be?

PROBLEM 20.19

A buffer is prepared with 0.50 mol NaH_2PO_4 and 0.30 mol Na_2HPO_4 in 0.500 L of solution.

a. Determine whether the solution can buffer the addition of 6.2 g KOH.

b. Determine whether the solution can buffer the addition of 46.0 mL of 6.0 mol/L HCl.

NOTES

1. Arrhenius, S. Thesis: Recherches sur la conductibilité galvanique des électrolytes. 1884.

2. Zundel, G.; Metzger, H. Energiebänder der tunnelnden Überschuß-Protonen in flüssigen Säuren. Eine IR-spectroskopische Untersuchung der Natur der Gruppierungen $H_5O_2^+$. Z. *Phys. Chem.* **1968,** *58* (5–6), 225-245. DOI: 10.1524/zpch.1968.58.5_6.225.

3. Wicke, E.; Eigen, M.; Ackerman, T. Über den Zustand des Protons (Hydroniumions) in wäßriger Lösung. Z. *Phys. Chem. (N F).* **1954,** *1* (5–6), 340–364. DOI: 10.1524/zpch.1954.1.5_6.340.

4. (i) Brönsted, J. N. Einige Bemerkungen über den Begriff der Säuren und Basen. *Recl. Trav. Chim. Pays-Bas.* **1923,** *42* (8), 718–728. DOI: 10.1002/recl.19230420815.

   (ii) Lowry, T.M. The Uniqueness of Hydrogen. *J. Soc. Chem. Ind. (London).* **1923,** *42* (3), 43–47. DOI: 10.1002/jctb.5000420302.

5. Lewis, G. N. *Valence and the Structure of Atoms and Molecules.* The Chemical Catalog Co., Inc.: New York, NY, 1923.

6. Also called the acid dissociation constant. "Acidity constant." IUPAC. *Compendium of Chemical Terminology,* 2nd ed. Compiled by A.D. McNaught and A. Wilkinson. Blackwell Scientific Publications, Oxford, 1997. Online version (2019–) created by S.J. Chalk. DOI: 10.1351/goldbook.A00080.

7. Conjugate comes from the Latin meaning "to yoke together," and so the term conjugate here means that it is the base formed from HA giving up H^+ rather than some random, unrelated base.

8. Gong, S.; Liu, L.; Zhang, J.; Cui, Q. Stable and eco-friendly solid acids as alternative to sulfuric acid in the liquid phase nitration of toluene. *Process Saf. Environ. Prot.,* **2014,** *92* (6), 577–582. DOI: 10.1016/j.psep.2013.03.005.

9. *Organic Synthesis Using Biocatalysis.* Animesh Goswami and Jon D. Stewart, Eds. Elsevier: Oxford, 2015. DOI: 10.1016/C2012-0-07124-4.

10. Thank you to Dr. Brian Esselman for running these calculations. There is no definitive way to measure or to calculate the real charge, but we can approximate the real charge with computational analysis. One method is natural population analysis (NPA; Reed, A.E.; Weinstock, R.B.; Weinhold, F. Natural Population Analysis. *J. Chem. Phys.* **1985**, *83* (2), 735-746. DOI: 10.1063/1.449486.), which is a result of a natural bond orbital (*NBO 7.0.* E.D. Glendening, J.K. Badenhoop, A.E. Reed, J.E. Carpenter, J.A. Bohmann, C. M. Morales, P. Karafiloglou, C. R. Landis, and F. Weinhold, Theoretical Chemistry Institute, University of Wisconsin, Madison, 2018) calculation.

11. Sørensen, S.P.L. Über die Messung und die Bedeutung der Wasserstoffionenkonzentration bei enzymatischen Prozessen. *Biochem. Z.*, **1909**, *21*, 131–304.

12. (i) Henderson, L.J. Concerning the Relationship Between the Strength of Acids and Their Capacity to Preserve Neutrality. *Am. J. Physiol.*, **1908**, *21* (2), 173–179. DOI: 10.1152/ajplegacy.1908.21.2.173.(ii) Hasselbalch, K.A. Die Berechnung der Wasserstoffzahl des Blutes aus der freien und gebundenen Kohlensäure desselben, und die Sauerstoffbindung des Blutes al Funktion der Wasserstoffzahl. *Biochem. Z.*, **1917**, *78*, 112–144.

Appendices

APPENDIX 1 – PHYSICAL CONSTANTS AND CONVERSION FACTORS
Physical Constants

Quantity	Symbol	Value
Atomic mass constant	m_u	$1.660\ 539\ 066\ 60 \times 10^{-27}$ kg
Avogadro constant	N_A, L	$6.022\ 140\ 76 \times 10^{23}$ 1/mol
Boltzmann constant	k_B	$1.380\ 649 \times 10^{-23}$ J/K
Electric constant (vacuum permittivity)	E_0	$8.854\ 187\ 8128 \times 10^{-12}$ F/m
Electron mass	m_e	$9.109\ 383\ 7015 \times 10^{-31}$ kg
Elementary charge	e	$1.602\ 176\ 634 \times 10^{-19}$ C
Faraday constant	$F, N_A e$	$96\ 485.332\ 12$ C/mol
Gas constant	R	$8.314\ 462\ 618$ J/(mol K)
		$0.082\ 057\ 366$ (L atm)/(mol K)
Molar mass constant	$M_u, N_A m_u$	$0.999\ 999\ 999\ 65 \times 10^{-3}$ kg/mol
Neutron mass	m_n	$1.674\ 927\ 498\ 04 \times 10^{-27}$ kg
Planck constant	h	$6.626\ 070\ 15 \times 10^{-34}$ J s
Proton mass	m_p	$1.672\ 621\ 923\ 69 \times 10^{-27}$ kg
Rydberg constant	R_∞	$10\ 973\ 731.568\ 160$ 1/m
Speed of light in a vacuum	c	$299\ 792\ 458$ m/s
Dalton, unified atomic mass unit	Da, u	$1.660\ 539\ 066\ 60 \times 10^{-27}$ kg

Conversion Factors

Conversion	Factor
Liter-atmosphere to joule	$\dfrac{101.325\,\text{J}}{\text{L atm}}$
Density of water (at 298.15 K)	$\dfrac{0.9970\,\text{g}}{\text{mL}}$
Atmosphere to pascal	$\dfrac{101\ 325\,\text{Pa}}{\text{atm}}$
Bar to pascal	$\dfrac{100\ 000\,\text{Pa}}{\text{bar}}$
Millimeters of mercury to pascal	$\dfrac{133.322\,\text{Pa}}{\text{mmHg}}$
Calorie to joule	$\dfrac{4.184\,\text{J}}{\text{cal}}$
Electronvolt to kJ/mol	$\dfrac{96.485\ 332\ 1\,\text{kJ/mol}}{\text{eV}}$

Source: Tiesinga, E.; Mohr, P.J.; Newell, D.B.; Taylor, B.N. CODATA Recommended Values of the Fundamental Physical Constants: 2018, *J. Phys. Chem. Ref. Data.* 2021, *50* (3), 033105-1 – 033105-61. DOI: 10.1063/5.0064853.

APPENDIX 2 – TABLE OF RECOMMENDED RELATIVE ATOMIC MASS (STANDARD ATOMIC WEIGHT) VALUES, $A_r°(E)$, WHICH HAVE BEEN ABRIDGED TO FIVE SIGNIFICANT FIGURES

Atomic Number (Z)	Name	Symbol	$A_r°(E)$	Atomic Number (Z)	Name	Symbol	$A_r°(E)$
1	hydrogen	H	1.0080	47	silver	Ag	107.87
2	helium	He	4.0026	48	cadmium	Cd	112.41
3	lithium	Li	6.94	49	indium	In	114.82
4	beryllium	Be	9.0122	50	tin	Sn	118.71
5	boron	B	10.81	51	antimony	Sb	121.76
6	carbon	C	12.011	52	tellurium	Te	127.60
7	nitrogen	N	14.007	53	iodine	I	126.90
8	oxygen	O	15.999	54	xenon	Xe	131.29
9	fluorine	F	18.998	55	caesium	Cs	132.91
10	neon	Ne	20.180	56	barium	Ba	137.33
11	sodium	Na	22.990	57	lanthanum	La	138.91
12	magnesium	Mg	24.305	58	cerium	Ce	140.12
13	aluminium	Al	26.982	59	praseodymium	Pr	140.91
14	silicon	Si	28.085	60	neodymium	Nd	144.24
15	phosphorus	P	30.974	61	promethium	Pm	—
16	sulfur	S	32.06	62	samarium	Sm	150.36
17	chlorine	Cl	35.45	63	europium	Eu	151.96
18	argon	Ar	39.95	64	gadolinium	Gd	157.25
19	potassium	K	39.098	65	terbium	Tb	158.93
20	calcium	Ca	40.078	66	dysprosium	Dy	162.50
21	scandium	Sc	44.956	67	holmium	Ho	164.93
22	titanium	Ti	47.867	68	erbium	Er	167.36
23	vanadium	V	50.942	69	thulium	Tm	168.93
24	chromium	Cr	51.996	70	ytterbium	Yb	173.05
25	manganese	Mn	54.938	71	lutetium	Lu	174.97
26	iron	Fe	55.845	72	hafnium	Hf	178.49
27	cobalt	Co	58.933	73	tantalum	Ta	180.95
28	nickel	Ni	58.693	74	tungsten	W	183.84
29	copper	Cu	63.546	75	rhenium	Re	186.21
30	zinc	Zn	65.38	76	osmium	Os	190.23
31	gallium	Ga	69.723	77	iridium	Ir	192.22
32	germanium	Ge	72.630	78	platinum	Pt	195.08
33	arsenic	As	74.922	79	gold	Au	196.97
34	selenium	Se	78.971	80	mercury	Hg	200.59
35	bromine	Br	79.904	81	thallium	Tl	204.38
36	krypton	Kr	83.798	82	lead	Pb	207.2
37	rubidium	Rb	85.468	83	bismuth	Bi	208.98
38	strontium	Sr	87.62	84	polonium	Po	—
39	yttrium	Y	88.906	85	astatine	At	—
40	zirconium	Zr	91.224	86	radon	Rn	—
41	niobium	Nb	92.906	87	francium	Fr	—
42	molybdenum	Mo	95.95	88	radium	Ra	—
43	technetium	Tc	—	89	actinium	Ac	—
44	ruthenium	Ru	101.07	90	thorium	Th	232.04
45	rhodium	Rh	102.91	91	protactinium	Pa	231.04
46	palladium	Pd	106.42	92	uranium	U	238.03

Atomic Number (Z)	Name	Symbol	$A_r°(E)$	Atomic Number (Z)	Name	Symbol	$A_r°(E)$
93	neptunium	Np	—	106	seaborgium	Sg	—
94	plutonium	Pu	—	107	bohrium	Bh	—
95	americium	Am	—	108	hassium	Hs	—
96	curium	Cm	—	109	meitnerium	Mt	—
97	berkelium	Bk	—	110	darmstadtium	Ds	—
98	californium	Cf	—	111	roentgenium	Rg	—
99	einsteinium	Es	—	112	copernicium	Cn	—
100	fermium	Fm	—	113	nihonium	Nh	—
101	mendelevium	Md	—	114	flerovium	Fl	—
102	nobelium	No	—	115	moscovium	Mc	—
103	lawrencium	Lr	—	116	livermorium	Lv	—
104	rutherfordium	Rf	—	117	tennessine	Ts	—
105	dubnium	Db	—	118	oganesson	Og	—

Source: Prohaska, T., et al. Standard Atomic Weights of the Elements 2021 (IUPAC Technical Report). *Pure Appl. Chem.*, 2022, 94 (5), 573–600. DOI: 10.1515/pac-2019-0603.

APPENDIX 3 – ELECTRON CONFIGURATION AND ORBITAL ENERGY DIAGRAMS
A. Electron Configuration of Each Element

Element	Electron Configuration	Noble Gas Configuration
H	$1s^1$	$1s^1$
He	$1s^2$	$1s^2$
Li	$1s^22s^1$	$[He]2s^1$
Be	$1s^22s^2$	$[He]2s^2$
B	$1s^22s^22p^1$	$[He]2s^22p^1$
C	$1s^22s^22p^2$	$[He]2s^22p^2$
N	$1s^22s^22p^3$	$[He]2s^22p^3$
O	$1s^22s^22p^4$	$[He]2s^22p^4$
F	$1s^22s^22p^5$	$[He]2s^22p^5$
Ne	$1s^22s^22p^6$	$[He]2s^22p^6$
Na	$1s^22s^22p^63s^1$	$[Ne]3s^1$
Mg	$1s^22s^22p^63s^2$	$[Ne]3s^2$
Al	$1s^22s^22p^63s^23p^1$	$[Ne]3s^23p^1$
Si	$1s^22s^22p^63s^23p^2$	$[Ne]3s^23p^2$
P	$1s^22s^22p^63s^23p^3$	$[Ne]3s^23p^3$
S	$1s^22s^22p^63s^23p^4$	$[Ne]3s^23p^4$
Cl	$1s^22s^22p^63s^23p^5$	$[Ne]3s^23p^5$
Ar	$1s^22s^22p^63s^23p^6$	$[Ne]3s^23p^6$
K	$1s^22s^22p^63s^23p^64s^1$	$[Ar]4s^1$
Ca	$1s^22s^22p^63s^23p^64s^2$	$[Ar]4s^2$
Sc	$1s^22s^22p^63s^23p^63d^14s^2$	$[Ar]3d^14s^2$
Ti	$1s^22s^22p^63s^23p^63d^24s^2$	$[Ar]3d^24s^2$
V	$1s^22s^22p^63s^23p^63d^34s^2$	$[Ar]3d^34s^2$
Cr	$1s^22s^22p^63s^23p^63d^54s^1$	$[Ar]3d^54s^1$
Mn	$1s^22s^22p^63s^23p^63d^54s^2$	$[Ar]3d^54s^2$
Fe	$1s^22s^22p^63s^23p^63d^64s^2$	$[Ar]3d^64s^2$
Co	$1s^22s^22p^63s^23p^63d^74s^2$	$[Ar]3d^74s^2$
Ni	$1s^22s^22p^63s^23p^63d^84s^2$	$[Ar]3d^84s^2$

Element	Electron Configuration	Noble Gas Configuration
Cu	$1s^22s^22p^63s^23p^63d^{10}4s^1$	$[Ar]3d^{10}4s^1$
Zn	$1s^22s^22p^63s^23p^63d^{10}4s^2$	$[Ar]3d^{10}4s^2$
Ga	$1s^22s^22p^63s^23p^63d^{10}4s^24p^1$	$[Ar]3d^{10}4s^24p^1$
Ge	$1s^22s^22p^63s^23p^63d^{10}4s^24p^2$	$[Ar]3d^{10}4s^24p^2$
As	$1s^22s^22p^63s^23p^63d^{10}4s^24p^3$	$[Ar]3d^{10}4s^24p^3$
Se	$1s^22s^22p^63s^23p^63d^{10}4s^24p^4$	$[Ar]3d^{10}4s^24p^4$
Br	$1s^22s^22p^63s^23p^63d^{10}4s^24p^5$	$[Ar]3d^{10}4s^24p^5$
Kr	$1s^22s^22p^63s^23p^63d^{10}4s^24p^6$	$[Ar]3d^{10}4s^24p^6$
Rb	$1s^22s^22p^63s^23p^63d^{10}4s^24p^65s^1$	$[Kr]5s^1$
Sr	$1s^22s^22p^63s^23p^63d^{10}4s^24p^65s^2$	$[Kr]5s^2$
Y	$1s^22s^22p^63s^23p^63d^{10}4s^24p^64d^15s^2$	$[Kr]4d^15s^2$
Zr	$1s^22s^22p^63s^23p^63d^{10}4s^24p^64d^25s^2$	$[Kr]4d^25s^2$
Nb	$1s^22s^22p^63s^23p^63d^{10}4s^24p^64d^45s^1$	$[Kr]4d^45s^1$
Mo	$1s^22s^22p^63s^23p^63d^{10}4s^24p^64d^55s^1$	$[Kr]4d^55s^1$
Tc	$1s^22s^22p^63s^23p^63d^{10}4s^24p^64d^55s^2$	$[Kr]4d^55s^2$
Ru	$1s^22s^22p^63s^23p^63d^{10}4s^24p^64d^75s^1$	$[Kr]4d^75s^1$
Rh	$1s^22s^22p^63s^23p^63d^{10}4s^24p^64d^85s^1$	$[Kr]4d^85s^1$
Pd	$1s^22s^22p^63s^23p^63d^{10}4s^24p^64d^{10}$	$[Kr]4d^{10}$
Ag	$1s^22s^22p^63s^23p^63d^{10}4s^24p^64d^{10}5s^1$	$[Kr]4d^{10}5s^1$
Cd	$1s^22s^22p^63s^23p^63d^{10}4s^24p^64d^{10}5s^2$	$[Kr]4d^{10}5s^2$
In	$1s^22s^22p^63s^23p^63d^{10}4s^24p^64d^{10}5s^25p^1$	$[Kr]4d^{10}5s^25p^1$
Sn	$1s^22s^22p^63s^23p^63d^{10}4s^24p^64d^{10}5s^25p^2$	$[Kr]4d^{10}5s^25p^2$
Sb	$1s^22s^22p^63s^23p^63d^{10}4s^24p^64d^{10}5s^25p^3$	$[Kr]4d^{10}5s^25p^3$
Te	$1s^22s^22p^63s^23p^63d^{10}4s^24p^64d^{10}5s^25p^4$	$[Kr]4d^{10}5s^25p^4$
I	$1s^22s^22p^63s^23p^63d^{10}4s^24p^64d^{10}5s^25p^5$	$[Kr]4d^{10}5s^25p^5$
Xe	$1s^22s^22p^63s^23p^63d^{10}4s^24p^64d^{10}5s^25p^6$	$[Kr]4d^{10}5s^25p^6$
Cs	$1s^22s^22p^63s^23p^63d^{10}4s^24p^64d^{10}5s^25p^66s^1$	$[Xe]6s^1$
Ba	$1s^22s^22p^63s^23p^63d^{10}4s^24p^64d^{10}5s^25p^66s^2$	$[Xe]6s^2$
La	$1s^22s^22p^63s^23p^63d^{10}4s^24p^64d^{10}5s^25p^65d^16s^2$	$[Xe]5d^16s^2$
Ce	$1s^22s^22p^63s^23p^63d^{10}4s^24p^64d^{10}5s^25p^64f^15d^16s^2$	$[Xe]4f^15d^16s^2$
Pr	$1s^22s^22p^63s^23p^63d^{10}4s^24p^64d^{10}5s^25p^64f^36s^2$	$[Xe]4f^36s^2$
Nd	$1s^22s^22p^63s^23p^63d^{10}4s^24p^64d^{10}5s^25p^64f^46s^2$	$[Xe]4f^46s^2$
Pm	$1s^22s^22p^63s^23p^63d^{10}4s^24p^64d^{10}5s^25p^64f^56s^2$	$[Xe]4f^56s^2$
Sm	$1s^22s^22p^63s^23p^63d^{10}4s^24p^64d^{10}5s^25p^64f^66s^2$	$[Xe]4f^66s^2$
Eu	$1s^22s^22p^63s^23p^63d^{10}4s^24p^64d^{10}5s^25p^64f^76s^2$	$[Xe]4f^76s^2$
Gd	$1s^22s^22p^63s^23p^63d^{10}4s^24p^64d^{10}5s^25p^64f^75d^16s^2$	$[Xe]4f^75d^16s^2$
Tb	$1s^22s^22p^63s^23p^63d^{10}4s^24p^64d^{10}5s^25p^64f^96s^2$	$[Xe]4f^96s^2$
Dy	$1s^22s^22p^63s^23p^63d^{10}4s^24p^64d^{10}5s^25p^64f^{10}6s^2$	$[Xe]4f^{10}6s^2$
Ho	$1s^22s^22p^63s^23p^63d^{10}4s^24p^64d^{10}5s^25p^64f^{11}6s^2$	$[Xe]4f^{11}6s^2$
Er	$1s^22s^22p^63s^23p^63d^{10}4s^24p^64d^{10}5s^25p^64f^{12}6s^2$	$[Xe]4f^{12}6s^2$
Tm	$1s^22s^22p^63s^23p^63d^{10}4s^24p^64d^{10}5s^25p^64f^{13}6s^2$	$[Xe]4f^{13}6s^2$
Yb	$1s^22s^22p^63s^23p^63d^{10}4s^24p^64d^{10}5s^25p^64f^{14}6s^2$	$[Xe]4f^{14}6s^2$
Lu	$1s^22s^22p^63s^23p^63d^{10}4s^24p^64d^{10}5s^25p^64f^{14}5d^16s^2$	$[Xe]4f^{14}5d^16s^2$
Hf	$1s^22s^22p^63s^23p^63d^{10}4s^24p^64d^{10}5s^25p^64f^{14}5d^26s^2$	$[Xe]4f^{14}5d^26s^2$
Ta	$1s^22s^22p^63s^23p^63d^{10}4s^24p^64d^{10}5s^25p^64f^{14}5d^36s^2$	$[Xe]4f^{14}5d^36s^2$
W	$1s^22s^22p^63s^23p^63d^{10}4s^24p^64d^{10}5s^25p^64f^{14}5d^46s^2$	$[Xe]4f^{14}5d^46s^2$
Re	$1s^22s^22p^63s^23p^63d^{10}4s^24p^64d^{10}5s^25p^64f^{14}5d^56s^2$	$[Xe]4f^{14}5d^56s^2$

Element	Electron Configuration	Noble Gas Configuration
Os	$1s^2 2s^2 2p^6 3s^2 3p^6 3d^{10} 4s^2 4p^6 4d^{10} 5s^2 5p^6 4f^{14} 5d^6 6s^2$	$[Xe]4f^{14}5d^6 6s^2$
Ir	$1s^2 2s^2 2p^6 3s^2 3p^6 3d^{10} 4s^2 4p^6 4d^{10} 5s^2 5p^6 4f^{14} 5d^7 6s^2$	$[Xe]4f^{14}5d^7 6s^2$
Pt	$1s^2 2s^2 2p^6 3s^2 3p^6 3d^{10} 4s^2 4p^6 4d^{10} 5s^2 5p^6 4f^{14} 5d^9 6s^1$	$[Xe]4f^{14}5d^9 6s^1$
Au	$1s^2 2s^2 2p^6 3s^2 3p^6 3d^{10} 4s^2 4p^6 4d^{10} 5s^2 5p^6 4f^{14} 5d^{10} 6s^1$	$[Xe]4f^{14}5d^{10} 6s^1$
Hg	$1s^2 2s^2 2p^6 3s^2 3p^6 3d^{10} 4s^2 4p^6 4d^{10} 5s^2 5p^6 4f^{14} 5d^{10} 6s^2$	$[Xe]4f^{14}5d^{10} 6s^2$
Tl	$1s^2 2s^2 2p^6 3s^2 3p^6 3d^{10} 4s^2 4p^6 4d^{10} 5s^2 5p^6 4f^{14} 5d^{10} 6s^2 6p^1$	$[Xe]4f^{14}5d^{10} 6s^2 6p^1$
Pb	$1s^2 2s^2 2p^6 3s^2 3p^6 3d^{10} 4s^2 4p^6 4d^{10} 5s^2 5p^6 4f^{14} 5d^{10} 6s^2 6p^2$	$[Xe]4f^{14}5d^{10} 6s^2 6p^2$
Bi	$1s^2 2s^2 2p^6 3s^2 3p^6 3d^{10} 4s^2 4p^6 4d^{10} 5s^2 5p^6 4f^{14} 5d^{10} 6s^2 6p^3$	$[Xe]4f^{14}5d^{10} 6s^2 6p^3$
Po	$1s^2 2s^2 2p^6 3s^2 3p^6 3d^{10} 4s^2 4p^6 4d^{10} 5s^2 5p^6 4f^{14} 5d^{10} 6s^2 6p^4$	$[Xe]4f^{14}5d^{10} 6s^2 6p^4$
At	$1s^2 2s^2 2p^6 3s^2 3p^6 3d^{10} 4s^2 4p^6 4d^{10} 5s^2 5p^6 4f^{14} 5d^{10} 6s^2 6p^5$	$[Xe]4f^{14}5d^{10} 6s^2 6p^5$
Rn	$1s^2 2s^2 2p^6 3s^2 3p^6 3d^{10} 4s^2 4p^6 4d^{10} 5s^2 5p^6 4f^{14} 5d^{10} 6s^2 6p^6$	$[Xe]4f^{14}5d^{10} 6s^2 6p^6$
Fr	$1s^2 2s^2 2p^6 3s^2 3p^6 3d^{10} 4s^2 4p^6 4d^{10} 5s^2 5p^6 4f^{14} 5d^{10} 6s^2 6p^6 7s^1$	$[Rn]7s^1$
Ra	$1s^2 2s^2 2p^6 3s^2 3p^6 3d^{10} 4s^2 4p^6 4d^{10} 5s^2 5p^6 4f^{14} 5d^{10} 6s^2 6p^6 7s^2$	$[Rn]7s^2$
Ac	$1s^2 2s^2 2p^6 3s^2 3p^6 3d^{10} 4s^2 4p^6 4d^{10} 5s^2 5p^6 4f^{14} 5d^{10} 6s^2 6p^6 6d^1 7s^2$	$[Rn]6d^1 7s^2$
Th	$1s^2 2s^2 2p^6 3s^2 3p^6 3d^{10} 4s^2 4p^6 4d^{10} 5s^2 5p^6 4f^{14} 5d^{10} 6s^2 6p^6 6d^2 7s^2$	$[Rn]6d^2 7s^2$
Pa	$1s^2 2s^2 2p^6 3s^2 3p^6 3d^{10} 4s^2 4p^6 4d^{10} 5s^2 5p^6 4f^{14} 5d^{10} 6s^2 6p^6 5f^2 6d^1 7s^2$	$[Rn]5f^2 6d^1 7s^2$
U	$1s^2 2s^2 2p^6 3s^2 3p^6 3d^{10} 4s^2 4p^6 4d^{10} 5s^2 5p^6 4f^{14} 5d^{10} 6s^2 6p^6 5f^3 6d^1 7s^2$	$[Rn]5f^3 6d^1 7s^2$
Np	$1s^2 2s^2 2p^6 3s^2 3p^6 3d^{10} 4s^2 4p^6 4d^{10} 5s^2 5p^6 4f^{14} 5d^{10} 6s^2 6p^6 5f^4 6d^1 7s^2$	$[Rn]5f^4 6d^1 7s^2$
Pu	$1s^2 2s^2 2p^6 3s^2 3p^6 3d^{10} 4s^2 4p^6 4d^{10} 5s^2 5p^6 4f^{14} 5d^{10} 6s^2 6p^6 5f^6 7s^2$	$[Rn]5f^6 7s^2$
Am	$1s^2 2s^2 2p^6 3s^2 3p^6 3d^{10} 4s^2 4p^6 4d^{10} 5s^2 5p^6 4f^{14} 5d^{10} 6s^2 6p^6 5f^7 7s^2$	$[Rn]5f^7 7s^2$
Cm	$1s^2 2s^2 2p^6 3s^2 3p^6 3d^{10} 4s^2 4p^6 4d^{10} 5s^2 5p^6 4f^{14} 5d^{10} 6s^2 6p^6 5f^7 6d^1 7s^2$	$[Rn]5f^7 6d^1 7s^2$
Bk	$1s^2 2s^2 2p^6 3s^2 3p^6 3d^{10} 4s^2 4p^6 4d^{10} 5s^2 5p^6 4f^{14} 5d^{10} 6s^2 6p^6 5f^9 7s^2$	$[Rn]5f^9 7s^2$
Cf	$1s^2 2s^2 2p^6 3s^2 3p^6 3d^{10} 4s^2 4p^6 4d^{10} 5s^2 5p^6 4f^{14} 5d^{10} 6s^2 6p^6 5f^{10} 7s^2$	$[Rn]5f^{10} 7s^2$
Es	$1s^2 2s^2 2p^6 3s^2 3p^6 3d^{10} 4s^2 4p^6 4d^{10} 5s^2 5p^6 4f^{14} 5d^{10} 6s^2 6p^6 5f^{11} 7s^2$	$[Rn]5f^{11} 7s^2$
Fm	$1s^2 2s^2 2p^6 3s^2 3p^6 3d^{10} 4s^2 4p^6 4d^{10} 5s^2 5p^6 4f^{14} 5d^{10} 6s^2 6p^6 5f^{12} 7s^2$	$[Rn]5f^{12} 7s^2$
Md	$1s^2 2s^2 2p^6 3s^2 3p^6 3d^{10} 4s^2 4p^6 4d^{10} 5s^2 5p^6 4f^{14} 5d^{10} 6s^2 6p^6 5f^{13} 7s^2$	$[Rn]5f^{13} 7s^2$
No	$1s^2 2s^2 2p^6 3s^2 3p^6 3d^{10} 4s^2 4p^6 4d^{10} 5s^2 5p^6 4f^{14} 5d^{10} 6s^2 6p^6 5f^{14} 7s^2$	$[Rn]5f^{14} 7s^2$
Lr	$1s^2 2s^2 2p^6 3s^2 3p^6 3d^{10} 4s^2 4p^6 4d^{10} 5s^2 5p^6 4f^{14} 5d^{10} 6s^2 6p^6 5f^{14} 6d^1 7s^2$	$[Rn]5f^{14}6d^1 7s^2$
Rf	$1s^2 2s^2 2p^6 3s^2 3p^6 3d^{10} 4s^2 4p^6 4d^{10} 5s^2 5p^6 4f^{14} 5d^{10} 6s^2 6p^6 5f^{14} 6d^2 7s^2$	$[Rn]5f^{14}6d^2 7s^2$
Db	$1s^2 2s^2 2p^6 3s^2 3p^6 3d^{10} 4s^2 4p^6 4d^{10} 5s^2 5p^6 4f^{14} 5d^{10} 6s^2 6p^6 5f^{14} 6d^3 7s^2$	$[Rn]5f^{14}6d^3 7s^2$
Sg	$1s^2 2s^2 2p^6 3s^2 3p^6 3d^{10} 4s^2 4p^6 4d^{10} 5s^2 5p^6 4f^{14} 5d^{10} 6s^2 6p^6 5f^{14} 6d^4 7s^2$	$[Rn]5f^{14}6d^4 7s^2$
Bh	$1s^2 2s^2 2p^6 3s^2 3p^6 3d^{10} 4s^2 4p^6 4d^{10} 5s^2 5p^6 4f^{14} 5d^{10} 6s^2 6p^6 5f^{14} 6d^5 7s^2$	$[Rn]5f^{14}6d^5 7s^2$
Hs	$1s^2 2s^2 2p^6 3s^2 3p^6 3d^{10} 4s^2 4p^6 4d^{10} 5s^2 5p^6 4f^{14} 5d^{10} 6s^2 6p^6 5f^{14} 6d^6 7s^2$	$[Rn]5f^{14}6d^6 7s^2$
Mt	$1s^2 2s^2 2p^6 3s^2 3p^6 3d^{10} 4s^2 4p^6 4d^{10} 5s^2 5p^6 4f^{14} 5d^{10} 6s^2 6p^6 5f^{14} 6d^7 7s^2$	$[Rn]5f^{14}6d^7 7s^2$
Ds	$1s^2 2s^2 2p^6 3s^2 3p^6 3d^{10} 4s^2 4p^6 4d^{10} 5s^2 5p^6 4f^{14} 5d^{10} 6s^2 6p^6 5f^{14} 6d^8 7s^2$	$[Rn]5f^{14}6d^8 7s^2$
Rg	$1s^2 2s^2 2p^6 3s^2 3p^6 3d^{10} 4s^2 4p^6 4d^{10} 5s^2 5p^6 4f^{14} 5d^{10} 6s^2 6p^6 5f^{14} 6d^9 7s^2$	$[Rn]5f^{14}6d^9 7s^2$
Cn	$1s^2 2s^2 2p^6 3s^2 3p^6 3d^{10} 4s^2 4p^6 4d^{10} 5s^2 5p^6 4f^{14} 5d^{10} 6s^2 6p^6 5f^{14} 6d^{10} 7s^2$	$[Rn]5f^{14}6d^{10} 7s^2$
Nh	$1s^2 2s^2 2p^6 3s^2 3p^6 3d^{10} 4s^2 4p^6 4d^{10} 5s^2 5p^6 4f^{14} 5d^{10} 6s^2 6p^6 5f^{14} 6d^{10} 7s^2 7p^1$	$[Rn]5f^{14}6d^{10} 7s^2 7p^1$
Fl	$1s^2 2s^2 2p^6 3s^2 3p^6 3d^{10} 4s^2 4p^6 4d^{10} 5s^2 5p^6 4f^{14} 5d^{10} 6s^2 6p^6 5f^{14} 6d^{10} 7s^2 7p^2$	$[Rn]5f^{14}6d^{10} 7s^2 7p^2$
Mc	$1s^2 2s^2 2p^6 3s^2 3p^6 3d^{10} 4s^2 4p^6 4d^{10} 5s^2 5p^6 4f^{14} 5d^{10} 6s^2 6p^6 5f^{14} 6d^{10} 7s^2 7p^3$	$[Rn]5f^{14}6d^{10} 7s^2 7p^3$
Lv	$1s^2 2s^2 2p^6 3s^2 3p^6 3d^{10} 4s^2 4p^6 4d^{10} 5s^2 5p^6 4f^{14} 5d^{10} 6s^2 6p^6 5f^{14} 6d^{10} 7s^2 7p^4$	$[Rn]5f^{14}6d^{10} 7s^2 7p^4$
Ts	$1s^2 2s^2 2p^6 3s^2 3p^6 3d^{10} 4s^2 4p^6 4d^{10} 5s^2 5p^6 4f^{14} 5d^{10} 6s^2 6p^6 5f^{14} 6d^{10} 7s^2 7p^5$	$[Rn]5f^{14}6d^{10} 7s^2 7p^5$
Og	$1s^2 2s^2 2p^6 3s^2 3p^6 3d^{10} 4s^2 4p^6 4d^{10} 5s^2 5p^6 4f^{14} 5d^{10} 6s^2 6p^6 5f^{14} 6d^{10} 7s^2 7p^6$	$[Rn]5f^{14}6d^{10} 7s^2 7p^6$

Source: All reported configurations conform with those reported by NIST Atomic Spectra Database: Kramida, A.; Ralchenko, Y.; Reader, J.; NIST ASD Team. *NIST Atomic Spectra Database* (ver. 5.10), [Online]. Available: https://physics.nist.gov/asd. [2023 April 28]. National Institute of Standards and Technology, Gaithersburg, MD. DOI: https://doi.org/10.18434/T4W30F.

Note: For meitnerium (Mt) and beyond, the configurations are theoretical following expected trends.

B. Ground State Orbital Energy Diagram of Elements
Hydrogen (Z = 1) through Krypton (Z = 36)

Orbital images and energy values calculated at the CCSD(T)/cc-pVTZ level of theory and basis set. WebMO uses a red/blue color scheme for filled orbitals (those that have electrons) and a yellow/green color scheme for unfilled orbitals (those without electrons). Note that the interaction of the electron spin and the orbital angular momentum leads to slightly (<1 eV) different energy values for filled or partially filled orbitals that are degenerate when unfilled: p, d, and f. The importance and effects of spin–orbit coupling are beyond the scope of this document and the values in the diagrams shown here do not include the effects of spin–orbit coupling.

Hydrogen
Electron configuration
H: 1s^1

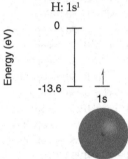

Helium
Electron configuration
He: 1s^2

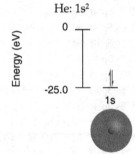

Lithium
Electron configuration
Li: 1s^{2}2s^1

Noble gas configuration
Li: [He]2s^1

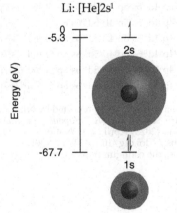

Beryllium
Electron configuration
Be: $1s^2 2s^2$

Noble gas configuration
Be: $[He]2s^2$

Boron
Electron configuration
B: $1s^2 2s^2 2p^1$

Noble gas configuration
B: $[He]2s^2 2p^1$

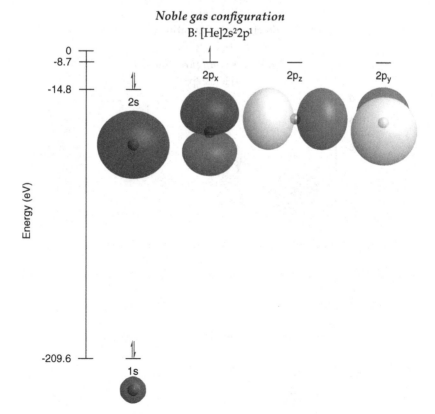

Carbon
Electron configuration
C: $1s^2 2s^2 2p^2$
Noble gas configuration
C: $[He]2s^2 2p^2$

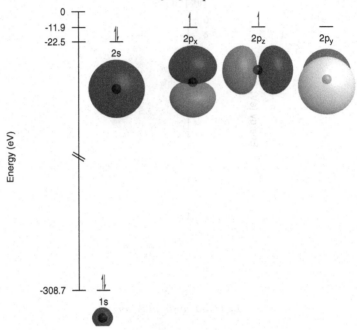

Nitrogen
Electron configuration
N: $1s^2 2s^2 2p^3$
Noble gas configuration
N: $[He]2s^2 2p^3$

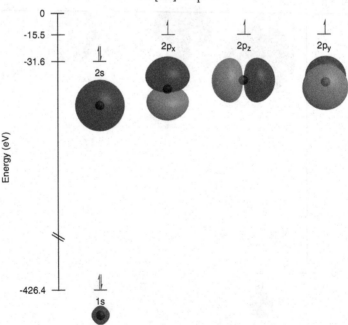

Oxygen
Electron configuration
O: $1s^2 2s^2 2p^4$

Noble gas configuration
O: $[He]2s^2 2p^4$

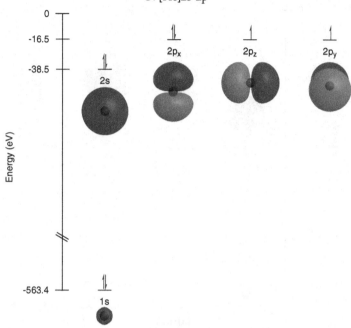

Fluorine
Electron configuration
F: $1s^2 2s^2 2p^5$

Noble gas configuration
F: $[He]2s^2 2p^5$

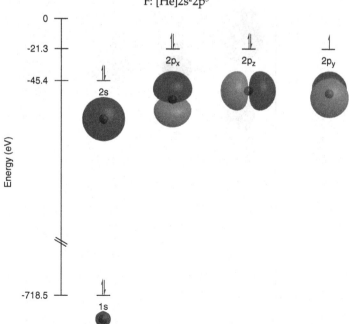

Neon
Electron configuration
Ne: $1s^2 2s^2 2p^6$

Noble gas configuration
Ne: $[He]2s^2 2p^6$

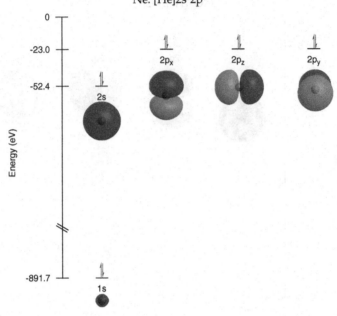

Sodium
Electron configuration
Na: $1s^2 2s^2 2p^6 3s^1$
Noble gas configuration
Na: $[Ne]3s^1$

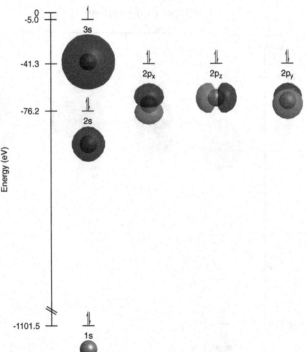

Magnesium
Electron configuration
Mg: $1s^2 2s^2 2p^6 3s^2$

Noble gas configuration
Mg: [Ne]$3s^2$

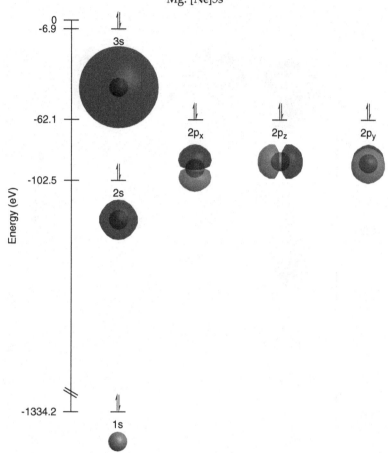

Aluminium
Electron configuration
Al: $1s^2 2s^2 2p^6 3s^2 3p^1$

Noble gas configuration
Al: $[Ne]3s^2 3p^1$

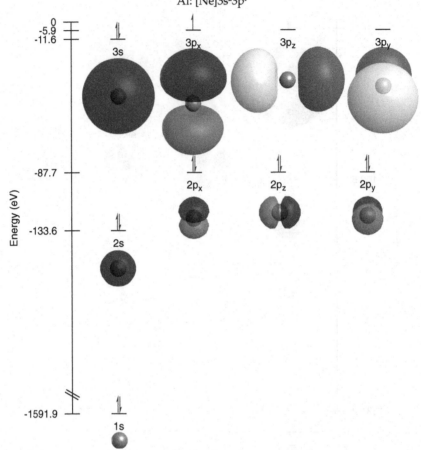

Silicon

Electron configuration
Si: $1s^2 2s^2 2p^6 3s^2 3p^2$

Noble gas configuration
Si: $[Ne]3s^2 3p^2$

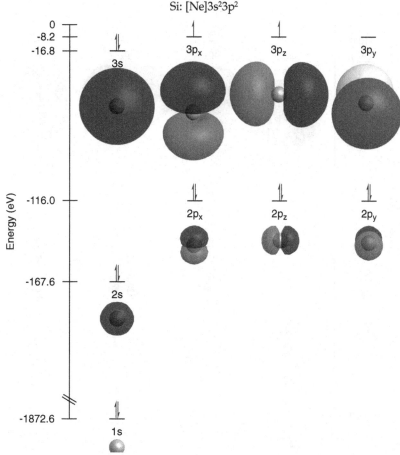

Phosphorus
Electron configuration
P: $1s^2 2s^2 2p^6 3s^2 3p^3$

Noble gas configuration
P: $[Ne]3s^2 3p^3$

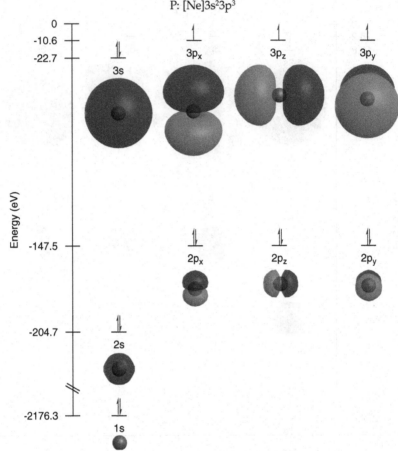

Sulfur

Electron configuration
S: $1s^2 2s^2 2p^6 3s^2 3p^4$

Noble gas configuration
S: $[Ne]3s^2 3p^4$

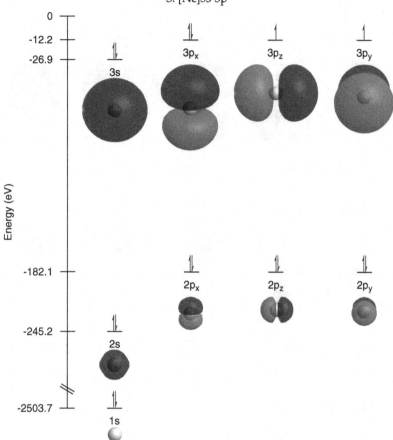

Chlorine
Electron configuration
Cl: $1s^2 2s^2 2p^6 3s^2 3p^5$

Noble gas configuration
Cl: $[Ne]3s^2 3p^5$

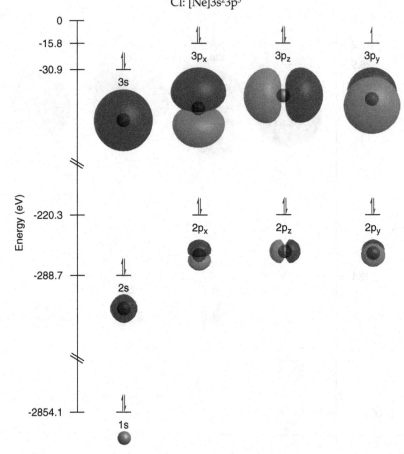

Argon
Electron configuration
Ar: $1s^2 2s^2 2p^6 3s^2 3p^6$

Noble gas configuration
Ar: $[Ne]3s^2 3p^6$

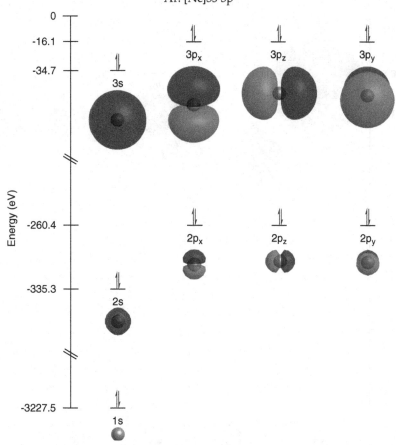

Potassium
(Orbital energy diagram not included because potassium is not included in the basis set.)

Electron configuration	Noble gas configuration
K: $1s^2 2s^2 2p^6 3s^2 3p^6 4s^1$	K: $[Ar]4s^1$

Calcium
Electron configuration
Ca: $1s^2 2s^2 2p^6 3s^2 3p^6 4s^2$

Noble gas configuration
Ca: $[Ar]4s^2$

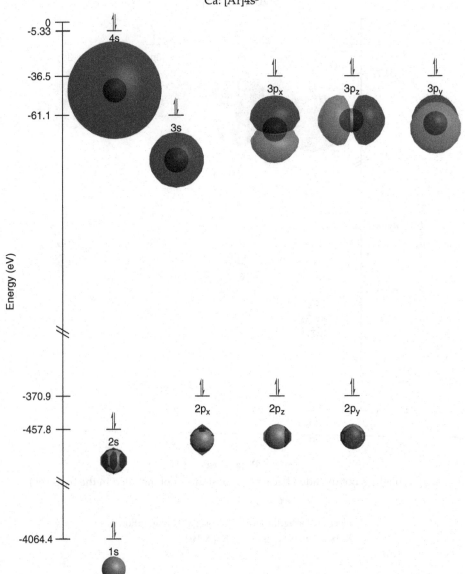

Scandium

Electron configuration

Sc: $1s^2 2s^2 2p^6 3s^2 3p^6 3d^1 4s^2$

Noble gas configuration

Sc: $[Ar]3d^1 4s^2$

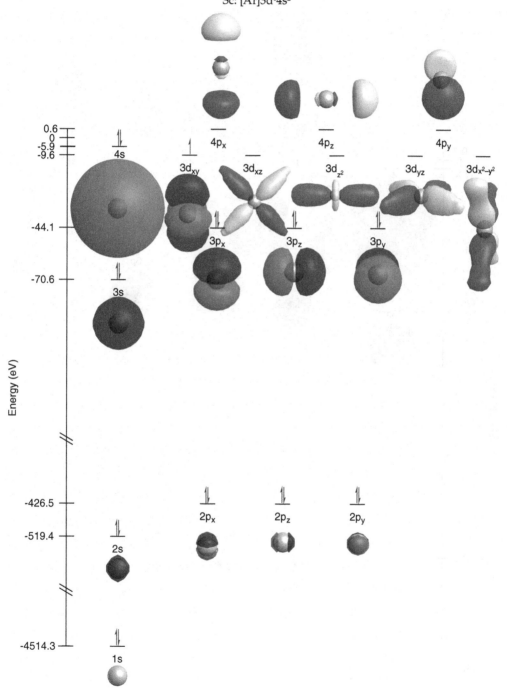

Titanium
Electron configuration
Ti: $1s^2 2s^2 2p^6 3s^2 3p^6 3d^2 4s^2$

Noble gas configuration
Ti: $[Ar]3d^2 4s^2$

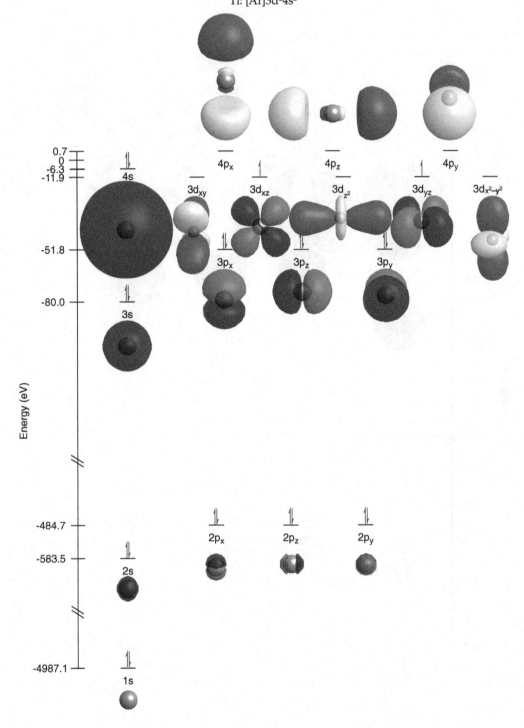

Vanadium

Electron configuration
V: $1s^2 2s^2 2p^6 3s^2 3p^6 3d^3 4s^2$

Noble gas configuration
V: [Ar]$3d^3 4s^2$

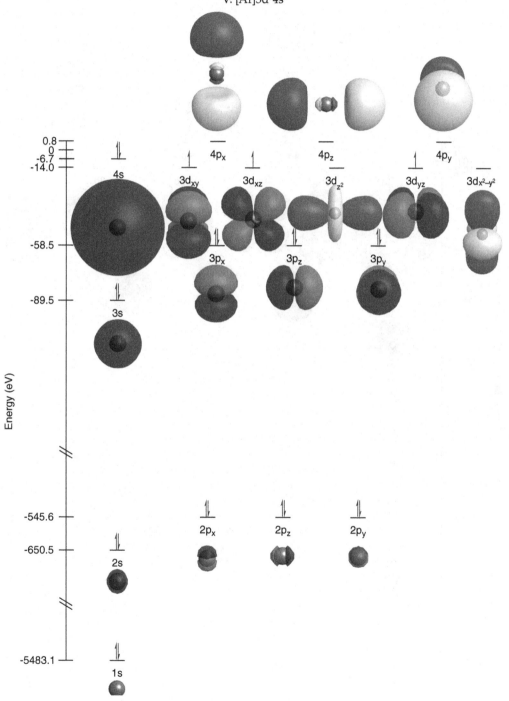

Chromium

Electron configuration
Cr: $1s^2 2s^2 2p^6 3s^2 3p^6 3d^5 4s^1$

Noble gas configuration
Cr: [Ar]$3d^5 4s^1$

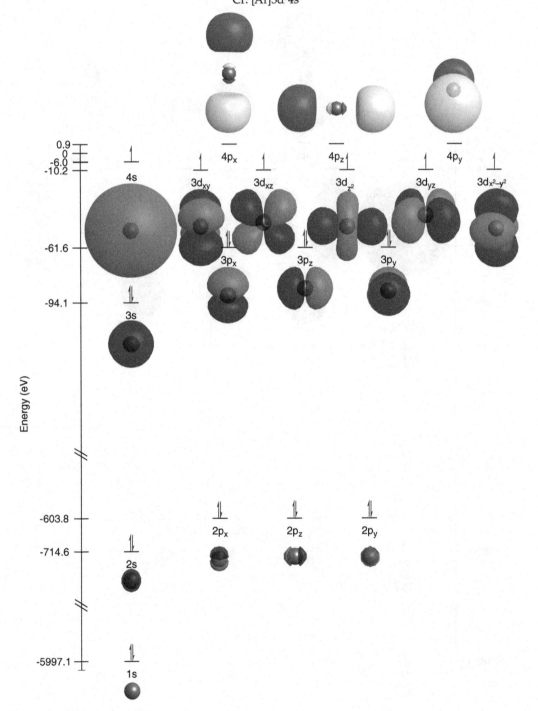

Manganese

Electron configuration

Mn: $1s^2 2s^2 2p^6 3s^2 3p^6 3d^5 4s^2$

Noble gas configuration

Mn: $[Ar]3d^5 4s^2$

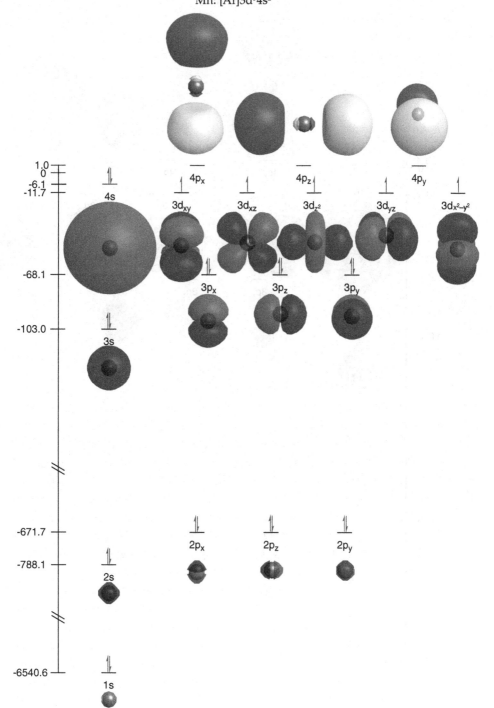

Iron

Electron configuration
Fe: $1s^2 2s^2 2p^6 3s^2 3p^6 3d^6 4s^2$

Noble gas configuration
Fe: $[Ar]3d^6 4s^2$

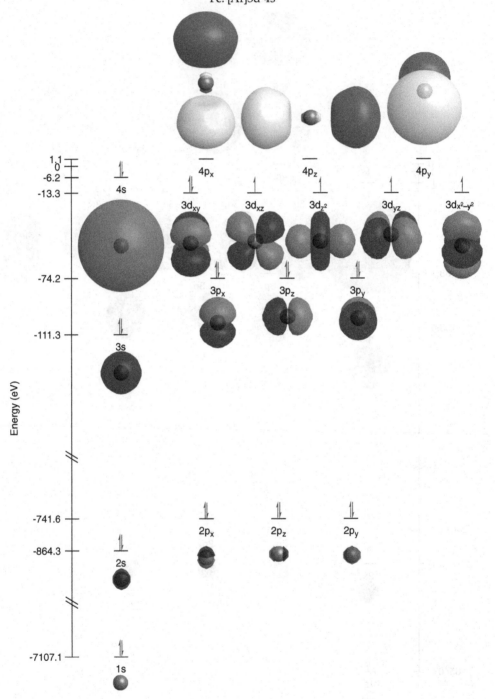

Cobalt

Electron configuration
Co: $1s^2 2s^2 2p^6 3s^2 3p^6 3d^7 4s^2$

Noble gas configuration
Co: $[Ar]3d^7 4s^2$

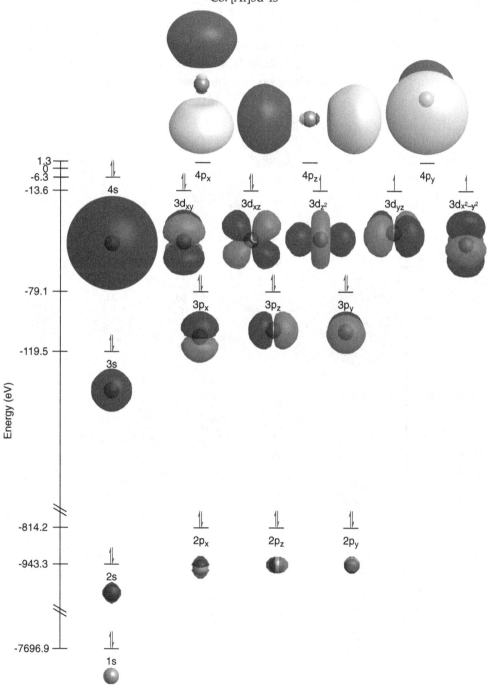

Nickel

(Orbital energy diagram not included due to issues with ground state nickel calculations.)

Electron configuration

Ni: $1s^2 2s^2 2p^6 3s^2 3p^6 3d^8 4s^2$

Noble gas configuration

Ni: $[Ar]3d^8 4s^2$

Copper

Electron configuration

Cu: $1s^2 2s^2 2p^6 3s^2 3p^6 3d^{10} 4s^1$

Noble gas configuration

Cu: $[Ar]3d^{10} 4s^1$

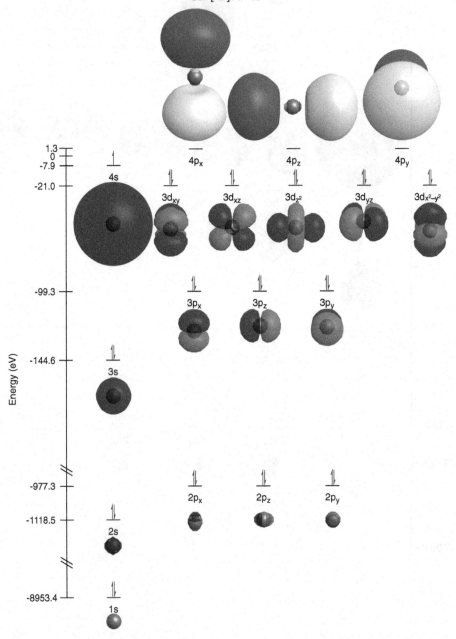

Zinc

Electron configuration

Zn: $1s^2 2s^2 2p^6 3s^2 3p^6 3d^{10} 4s^2$

Noble gas configuration

Zn: $[Ar]3d^{10}4s^2$

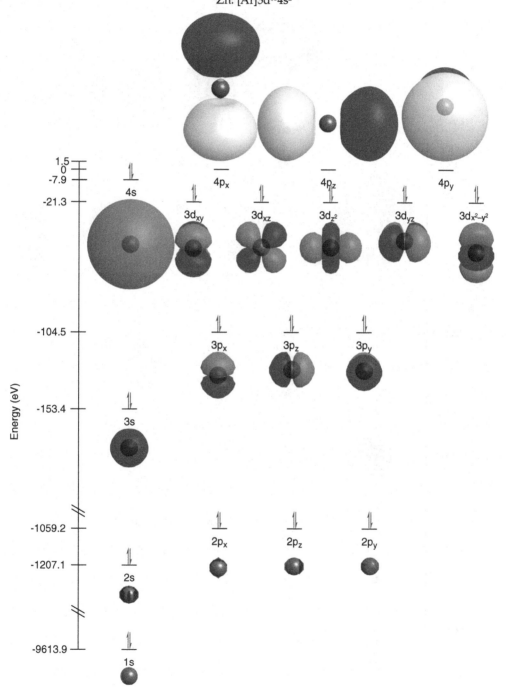

Gallium

Electron configuration

Ga: $1s^2 2s^2 2p^6 3s^2 3p^6 3d^{10} 4s^2 4p^1$

Noble gas configuration

Ga: $[Ar]3d^{10}4s^2 4p^1$

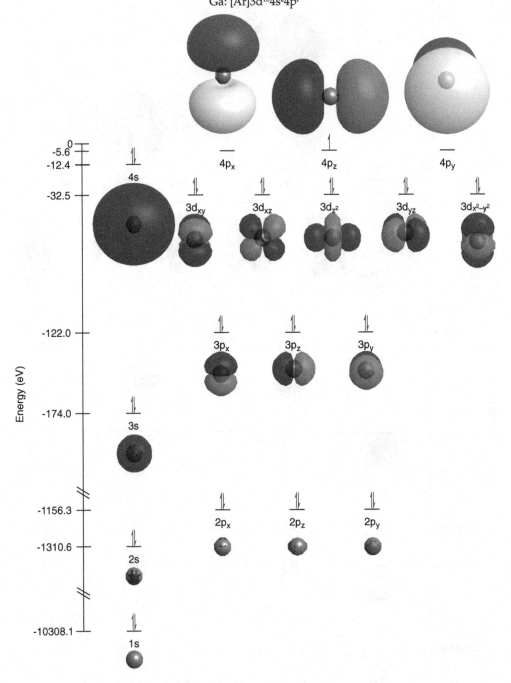

Germanium
Electron configuration
Ge: $1s^2 2s^2 2p^6 3s^2 3p^6 3d^{10} 4s^2 4p^2$

Noble gas configuration
Ge: $[Ar]3d^{10}4s^2 4p^2$

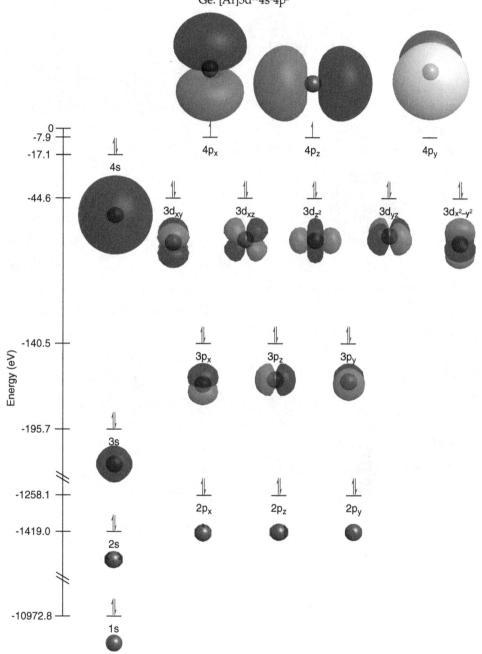

Arsenic

Electron configuration
Ge: $1s^2 2s^2 2p^6 3s^2 3p^6 3d^{10} 4s^2 4p^3$

Noble gas configuration
Ge: $[Ar]3d^{10}4s^2 4p^3$

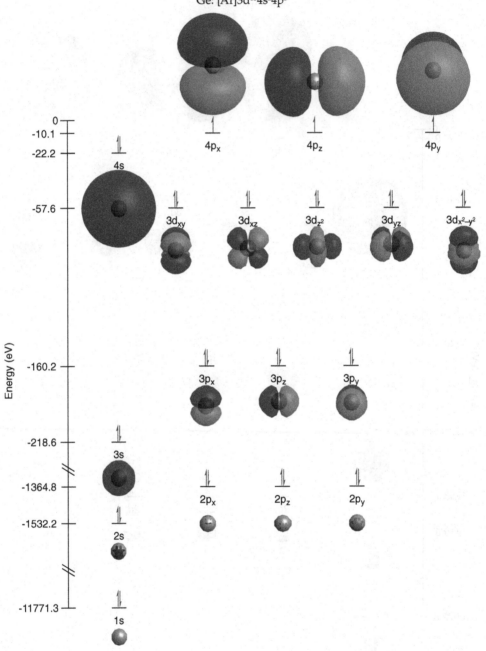

Selenium

Electron configuration

Se: $1s^2 2s^2 2p^6 3s^2 3p^6 3d^{10} 4s^2 4p^4$

Noble gas configuration

Se: $[Ar]3d^{10}4s^2 4p^4$

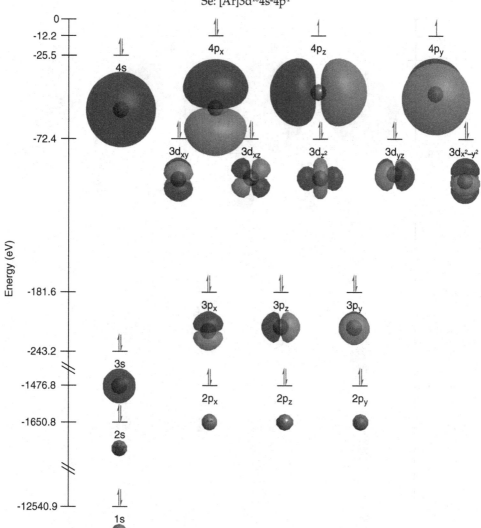

Bromine
Electron configuration
Br: $1s^2 2s^2 2p^6 3s^2 3p^6 3d^{10} 4s^2 4p^5$

Noble gas configuration
Br: $[Ar]3d^{10}4s^2 4p^5$

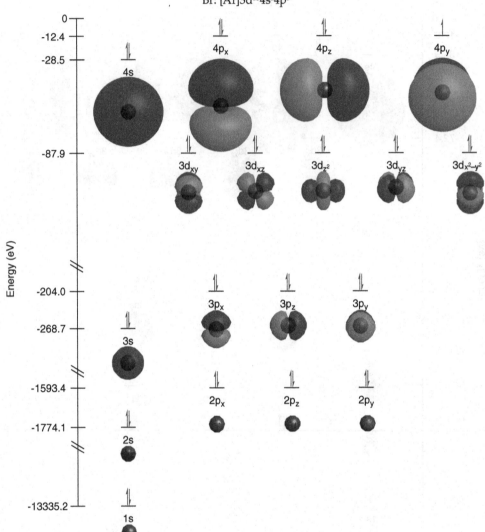

Krypton

Electron configuration

Kr: $1s^2 2s^2 2p^6 3s^2 3p^6 3d^{10} 4s^2 4p^6$

Noble gas configuration

Kr: $[Ar]3d^{10}4s^2 4p^6$

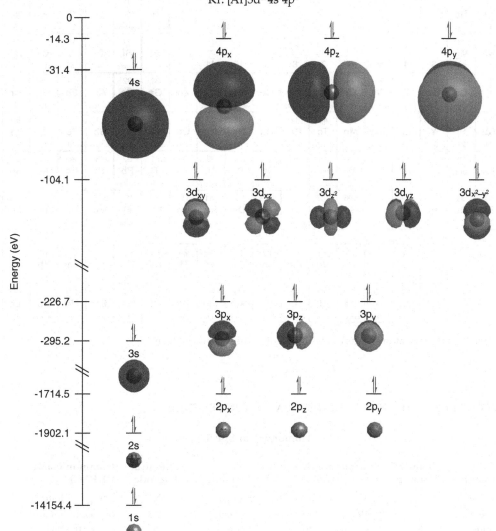

APPENDIX 4 – COMMON CHARGE FOR EACH ELEMENT AS A MONATOMIC ION

1	2	3	4	5	6	7	8	9	10	11	12	13	14	15	16	17	18
1 **H** 1+/1–																	2 **He** 0
3 **Li** 1+	4 **Be** 2+											5 **B** 3+	6 **C** variable	7 **N** 3–	8 **O** 2–	9 **F** 1–	10 **Ne** 0
11 **Na** 1+	12 **Mg** 2+											13 **Al** 3+	14 **Si** variable	15 **P** 3–	16 **S** 2–	17 **Cl** 1–	18 **Ar** 0
19 **K** 1+	20 **Ca** 2+	21 **Sc** 3+	22 **Ti** variable	23 **V** variable	24 **Cr** variable	25 **Mn** variable	26 **Fe** variable	27 **Co** variable	28 **Ni** variable	29 **Cu** variable	30 **Zn** 2+	31 **Ga** 3+	32 **Ge** variable	33 **As** variable	34 **Se** 2–	35 **Br** 1–	36 **Kr** 0
37 **Rb** 1+	38 **Sr** 2+	39 **Y** 3+	40 **Zr** 4+	41 **Nb** variable	42 **Mo** variable	43 **Tc** variable	44 **Ru** variable	45 **Rh** variable	46 **Pd** variable	47 **Ag** 1+	48 **Cd** 2+	49 **In** variable	50 **Sn** variable	51 **Sb** variable	52 **Te** variable	53 **I** 1–	54 **Xe** 0
55 **Cs** 1+	56 **Ba** 2+	*	72 **Hf** 4+	73 **Ta** variable	74 **W** variable	75 **Re** variable	76 **Os** variable	77 **Ir** variable	78 **Pt** variable	79 **Au** variable	80 **Hg** variable	81 **Tl** variable	82 **Pb** variable	83 **Bi** variable	84 **Po** variable	85 **At** variable	86 **Rn** 0
87 **Fr** 1+	88 **Ra** 2+	**	104 **Rf** unknown	105 **Db** unknown	106 **Sg** unknown	107 **Bh** unknown	108 **Hs** unknown	109 **Mt** unknown	110 **Ds** unknown	111 **Rg** unknown	112 **Cn** unknown	113 **Nh** unknown	114 **Fl** unknown	115 **Mc** unknown	116 **Lv** unknown	117 **Ts** unknown	118 **Og** unknown

*	57 **La** 3+	58 **Ce** 3+	59 **Pr** 3+	60 **Nd** 3+	61 **Pm** 3+	62 **Sm** 3+	63 **Eu** 3+	64 **Gd** 3+	65 **Tb** 3+	66 **Dy** 3+	67 **Ho** 3+	68 **Er** 3+	69 **Tm** 3+	70 **Yb** 3+	71 **Lu** 3+
**	89 **Ac** 3+	90 **Th** variable	91 **Pa** variable	92 **U** variable	93 **Np** variable	94 **Pu** variable	95 **Am** 3+	96 **Cm** 3+	97 **Bk** 3+	98 **Cf** 3+	99 **Es** 3+	100 **Fm** 3+	101 **Md** 3+	102 **No** variable	103 **Lr** 3+

0 means that monatomic ions of this element do not naturally form.

APPENDIX 5 – CHEMISTRY NOMENCLATURE REFERENCE

Homonuclear Species

Element	Chemical Formula	Common Name (IUPAC)	Element	Chemical Formula	Common Name (IUPAC)
bromine	Br_2	bromine (dibromine)	nitrogen	N_2	nitrogen (dinitrogen)
chlorine	Cl_2	chlorine (dichlorine)	oxygen	O_2	oxygen (dioxygen)
fluorine	F_2	Fluorine (difluorine)	oxygen	O_3	ozone (trioxygen)
hydrogen	H_2	hydrogen (dihydrogen)	phosphorus	P_4	phosphorus (tetraphosphorus)
iodine	I_2	iodine (diiodine)	sulfur	S_8	sulfur (*cyclo*-octasulfur)

Polyatomic Ions

a. Cations

Common Name (IUPAC)	Chemical Formula	Common Name (IUPAC)	Chemical Formula
ammonium (azanium)	NH_4^+	hydronium* (oxidanium)	H_3O^+
guanidinium (carbamimidoylazanium)	$C(NH_2)_3^+$	mercury(I) (dimercury(2+))	Hg_2^{2+}

*Hydronium is a commonly used name for H_3O^+, but its use has been officially deprecated and the names oxidanium or oxonium should be used preferentially.

b. Anions

Common Name (IUPAC)	Chemical Formula	Common Name (IUPAC)	Chemical Formula
acetate	$CH_3CO_2^-$	hydroxide (oxidanide)	OH^-
amide (azanide)	NH_2^-	hypobromite	BrO^-
azide (trinitride(1–))	N_3^-	hypochlorite	ClO^-
bromate	BrO_3^-	hypoiodite	IO^-
bromite	BrO_2^-	iodate	IO_3^-
bicarbonate (hydrogencarbonate)	HCO_3^-	iodite	IO_2^-
bifluoride (difluoridohydrogenate(1–))	HF_2^-	nitrate	NO_3^-
bisulfate (hydrogensulfate)	HSO_4^-	nitrite	NO_2^-
bisulfide (sulfanide)	HS^-	perbromate	BrO_4^-
bisulfite (hydrogensulfite)	HSO_3^-	perchlorate	ClO_4^-
carbonate	CO_3^{2-}	periodate	IO_4^-
chromate	CrO_4^{2-}	permanganate	MnO_4^-
chlorate	ClO_3^-	peroxide (dioxide(2–))	O_2^{2-}
chlorite	ClO_2^-	phosphate	PO_4^{3-}
cyanate	OCN^-	pyrophosphate (diphosphate)	$P_2O_7^{4-}$
cyanide	CN^-	sulfate	SO_4^{2-}
dichromate (heptaoxidodichromate(2–))	$Cr_2O_7^{2-}$	sulfite	SO_3^{2-}
dihydrogenphosphate	$H_2PO_4^-$	tetrafluoridoborate	BF_4^-
formate	HCO_2^-	thiocyanate	SCN^-
hydrogenphosphate	HPO_4^{2-}	thiosulfate	$S_2O_3^{2-}$

APPENDIX 6 – DISSOCIATION VALUES

Selected A-H Bond Dissociation Energy Values (kJ/mol)

H–H
436

B–H	H–C	H–N	H–O	H–F
326	439	450.	497	570.
Al–H	Si–H	P–H	H–S	H–Cl
285	383	351	381	431
Ga–H	Ge–H	As–H	Se–H	H–Br
276	349	319	335	366
				H–I
				298

A-A Bond Dissociation Energy Values (kJ/mol)

C–C	N–N	O–O	F–F
377	277	213	159
C=C	N=N	O=O	Cl–Cl
728	456	498	243
C≡C	N≡N		Br–Br
965	945		194
			I–I
			152

A-X Bond Dissociation Energy Values (kJ/mol)

Al–F	C–F	N–F	O–F
675	460.	287	215
Al–Cl	C–Cl	N–Cl	O–Cl
502	350.	253	233
Al–Br	C–Br	N–Br	O–Br
429	294	285	210.
Al–I	C–I	N–I	O–I
370.	232	234	213

A-O bond dissociation energy values (kJ/mol)

C–O	N–O	O–O	F–O
385	265	213	215
C=O	N=O	O=O	Cl–O
732	481	498	233
C≡O			Br–O
1075			210.
C=O (in CO$_2$)			I–O
532			213

A-N Bond Dissociation Energy Values (kJ/mol)

C–N	N–N	O–N	F–N
331	277	265	287
C=N	N=N	O=N	Cl–N
644	456	481	253
C≡N	N≡N		Br–N
937	945		285
			I–N
			234

Sources: (i) Blanksby, S.J.; Ellison, G.B. Bond Dissociation Energies of Organic Molecules. *Acc. Chem. Res.*, **2003**, *36* (4), 255–263. DOI: 10.1021/ar020230d. (ii) Berkowitz, J.; Ellison, G.B.; Gutman, D. Three Methods to Measure RH Bond Energies. *J. Phys. Chem.*, **1994**, *98* (11), 2744–2765. DOI: 10.1021/j100062a009. (iii) Benson, S.W. III - Bond Energies. *J. Chem. Educ.*, **1965**, *42* (9), 502–518. DOI: 10.1021/ed042p502. (iv) Darwent, B. de B. Bond Dissociation Energies in Simple Molecules. Washington D.C.: National Bureau of Standards, 1970. (v) Wiberg, K.B. Bond Dissociation Energies of H$_2$NX Compounds: Comparison with CH$_3$X, HOX, and FX Compounds. *J. Phys. Chem.*, **1992**, *96* (14), 5800–5803. DOI: 10.1021/j100193a028. (vi) Pople, J.A.; Curtiss, L.A. A Theoretical Study of the Energy of Hypofluorous Acid, HOF. *J. Chem. Phys,* **1989**, *90* (5), 2833. DOI: 10.1063/1.455934. (vii) Kerr, J.A. Bond Dissociation Energies by Kinetic Methods. *Chem. Rev.*, **1966**, *66* (5), 465–500. (viii) From: Rumble, J.R.; Lide, D.R.; Bruno, T.J. *CRC Handbook of Chemistry and Physics: A Ready-Reference Book of Chemical and Physical Data*, 100th ed. Boca Raton, FL: CRC Press, 2019–2020.

APPENDIX 7 – SELECTED THERMOCHEMICAL DATA

Silver	$\Delta_f H°\ (\frac{kJ}{mol})$	$\Delta_f G°\ (\frac{kJ}{mol})$	$S°\ (\frac{J}{mol\ K})$
$Ag(s)$	0.0	0.0	42.6
$Ag(g)$	284.9	246.0	173.0
$Ag^+(aq)$	105.6	77.1	72.7
$AgBr(s)$	−100.4	−96.9	107.1
$Ag_2CO_3(s)$	−505.8	−436.8	167.4
$AgCl(s)$	−127.0	−109.8	96.3
$AgI(s)$	−61.8	−66.2	115.5
$AgNO_3(s)$	−124.4	−33.4	140.9
$Ag_2O(s)$	−31.1	−11.2	121.3

Aluminium	$\Delta_f H°\ (\frac{kJ}{mol})$	$\Delta_f G°\ (\frac{kJ}{mol})$	$S°\ (\frac{J}{mol\ K})$
$Al(s)$	0.0	0.0	28.3
$Al(g)$	330.0	289.4	164.6
$Al^{3+}(aq)$	−531.0	−485.0	−321.7
$AlCl_3(s)$	−704.2	−628.8	109.3
$AlN(s)$	−318.0	−287.0	20.2
$Al_2O_3(s)$	−1675.7	−1582.3	50.9
$AlPO_4(s)$	−1733.8	−1617.9	90.8

Argon	$\Delta_f H°\ (\frac{kJ}{mol})$	$\Delta_f G°\ (\frac{kJ}{mol})$	$S°\ (\frac{J}{mol\ K})$
$Ar(g)$	0.0	0.0	154.8
$Ar(aq)$	−12.1	16.4	59.4

Arsenic	$\Delta_f H°\ (\frac{kJ}{mol})$	$\Delta_f G°\ (\frac{kJ}{mol})$	$S°\ (\frac{J}{mol\ K})$
$As(s)$	0.0	0.0	35.1
$As_2(g)$	222.2	171.9	239.4
$As(g)$	302.5	261.0	174.2
$AsH_3(g)$	66.4	68.9	222.8
$As_2O_5(s)$	−924.9	−782.3	105.4
$As_2S_3(s)$	−169.0	−168.6	163.6

Gold	$\Delta_f H°\ (\frac{kJ}{mol})$	$\Delta_f G°\ (\frac{kJ}{mol})$	$S°\ (\frac{J}{mol\ K})$
$Au(s)$	0.0	0.0	47.4
$Au^{3+}(aq)$	409.2	433.5	−229.3

Boron	$\Delta_f H°\ (\frac{kJ}{mol})$	$\Delta_f G°\ (\frac{kJ}{mol})$	$S°\ (\frac{J}{mol\ K})$
$B(s)$	0.0	0.0	5.9
$B(g)$	565.0	521.0	153.4
$BBr_3(l)$	−239.7	−238.5	229.7
$BBr_3(g)$	−205.6	−232.5	324.2
$BCl_3(l)$	−427.2	−387.4	206.3
$BCl_3(g)$	−403.8	−388.7	290.1
$BF_3(g)$	−1136.0	−1119.4	254.4
$BF_4^-(aq)$	−1574.9	−1486.9	180.0
$BH_3(g)$	89.2	93.3	188.2

BH$_4^-$(aq)	48.2	114.4	110.5
B$_2$H$_6$(g)	36.4	87.6	232.1
BI$_3$(g)	71.1	20.7	349.2
B(OH)$_3$(s)	−1094.3	−968.9	90.0
B$_2$O$_3$(s)	−1273.5	−1194.3	54.0

Barium	$\Delta_f H°\ (\dfrac{kJ}{mol})$	$\Delta_f G°\ (\dfrac{kJ}{mol})$	$S°\ (\dfrac{J}{mol\ K})$
Ba(s)	0.0	0.0	62.5
Ba(g)	180.0	146.0	170.2
Ba^{2+}(aq)	-537.6	-560.8	9.6
BaCO$_3$(s)	-1213.0	-1134.4	112.1
BaCl$_2$(s)	-855.0	-806.7	123.7
BaO(s)	-548.0	-520.3	72.1
BaSO$_4$(s)	-1473.2	-1362.2	132.2

Beryllium	$\Delta_f H°\ (\dfrac{kJ}{mol})$	$\Delta_f G°\ (\dfrac{kJ}{mol})$	$S°\ (\dfrac{J}{mol\ K})$
Be(s)	0.0	0.0	9.5
Be(g)	324.0	286.6	136.3
Be^{2+}(aq)	-382.8	-379.7	-129.7
BeCl$_2$(s)	-490.4	-445.6	75.8
BeF$_2$(s)	−1026.8	−979.4	53.4
BeI$_2$(s)	−192.6	−210.0	130.0
BeO(s)	−609.4	−580.1	13.8
Be(OH)$_2$(s)	−902.5	−815.0	45.5

Bismuth	$\Delta_f H°\ (\dfrac{kJ}{mol})$	$\Delta_f G°\ (\dfrac{kJ}{mol})$	$S°\ (\dfrac{J}{mol\ K})$
Bi(s)	0.0	0.0	56.7
Bi(g)	207.1	168.2	187.0
BiCl$_3$(s)	−379.1	−315.0	177.0
Bi$_2$O$_3$(s)	−573.9	−493.7	151.5

Bromine	$\Delta_f H°\ (\dfrac{kJ}{mol})$	$\Delta_f G°\ (\dfrac{kJ}{mol})$	$S°\ (\dfrac{J}{mol\ K})$
Br$_2$(l)	0.0	0.0	152.2
Br$_2$(aq)	−2.6	3.9	130.5
Br$_2$(g)	30.9	3.1	245.5
Br(g)	111.9	82.4	175.0
Br$^-$(aq)	−121.6	−104.0	82.4
BrCl(g)	14.6	−1.0	240.1
BrF$_3$(g)	−255.6	−229.4	292.5
BrF$_3$(l)	−300.8	−240.5	178.2
BrF$_5$(g)	−428.9	−350.6	320.2
BrF$_5$(l)	−458.6	−351.8	225.1
BrNO(g)	82.2	82.4	273.7

Carbon	$\Delta_f H°\ (\dfrac{kJ}{mol})$	$\Delta_f G°\ (\dfrac{kJ}{mol})$	$S°\ (\dfrac{J}{mol\ K})$
C(s, graphite)	0	0	5.7
C(s, diamond)	1.9	2.9	2.4
C$_2$(g)	831.9	775.9	199.4

$CH_4(g)$	−74.6	−50.5	186.3
$CH_3(g)$	145.7	147.9	194.2
$CH_3OH(l)$	−239.2	−166.6	126.8
$CH_3OH(aq)$	−245.9	−175.3	133.1
$CH_3OH(g)$	−201.0	−162.3	239.9
$CH_2O(g)$	−108.6	−102.5	218.8
$HCO_2^-(aq)$	−425.6	−351.0	92.0
$HCO_3^-(aq)$	−692.0	−578.1	92.5
$HCO_2H(aq)$	−425.4	−372.3	163.0
$HCO_2H(l)$	−425.0	−361.4	129.0
$CCl_2O(g)$	−219.1	−204.9	283.5
$CN^-(aq)$	150.6	172.4	94.1
$CO(g)$	−110.5	−137.2	197.7
$CO_2(g)$	−393.5	−394.4	213.8
$CS_2(l)$	89.0	64.6	151.3
$CS_2(g)$	116.7	67.1	237.8
$CO_3^{2-}(aq)$	−677.1	−527.8	−56.9
$C_2H_2(g)$	227.4	209.9	200.9
$C_2H_4(g)$	52.4	68.4	219.3
$C_2H_6(g)$	−84.0	−32.0	229.2
$C_2H_5Cl(g)$	−112.1	-60.4	276.0
$C_2H_5Cl(l)$	−136.8	−59.3	190.8
$C_2H_5OH(l)$	−277.6	−174.8	160.7
$C_2H_5OH(g)$	−234.8	−167.9	281.6
$C_2H_5OH(aq)$	−288.3	−181.6	148.5
$CH_3CO_2H(l)$	−484.3	−389.9	159.8
$CH_3CO_2H(g)$	−432.2	−374.2	283.5
$CH_3CO_2^-(aq)$	−486.0	−369.3	86.6
$C_3H_8(g)$	−103.8	−23.4	270.3
$C_6H_6(l)$	49.1	124.5	173.4

Calcium	$\Delta_f H° \, (\dfrac{kJ}{mol})$	$\Delta_f G° \, (\dfrac{kJ}{mol})$	$S° \, (\dfrac{J}{mol\,K})$
Ca(s)	0.0	0.0	41.6
Ca(g)	177.8	144.0	154.9
$Ca^{2+}(aq)$	−542.8	−553.6	−53.1
$CaBr_2(s)$	−682.8	−663.6	130.0
$CaC_2(s)$	−59.8	−64.9	70.0
$CaCO_3(s, calcite)$	−1207.6	−1129.1	91.7
$CaCl_2(s)$	−795.4	−748.8	108.4
$CaF_2(s)$	−1228.0	−1175.6	68.5
$CaH_2(s)$	−181.5	−142.5	41.4
CaO(s)	−634.9	−603.3	38.1
$Ca(OH)_2(s)$	−985.2	−897.5	83.4
$Ca_3(PO_4)_2(s)$	−4120.8	−3884.7	236.0
$CaSO_4(s)$	−1434.5	−1322.0	106.5

Chlorine	$\Delta_f H° \, (\dfrac{kJ}{mol})$	$\Delta_f G° \, (\dfrac{kJ}{mol})$	$S° \, (\dfrac{J}{mol\,K})$
$Cl_2(g)$	0.0	0.0	223.1
Cl(g)	121.3	105.3	165.2
$Cl^-(aq)$	−167.2	−131.2	56.5

ClF(g)	–50.3	–51.8	217.9
ClF$_3$(g)	–163.2	–123.0	281.6
ClNO(g)	51.7	66.1	261.7

Cobalt	$\Delta_f H°\ (\dfrac{kJ}{mol})$	$\Delta_f G°\ (\dfrac{kJ}{mol})$	$S°\ (\dfrac{J}{mol\ K})$
Co(s)	0.0	0.0	30.0
Co(g)	424.7	380.3	179.5
Co^{2+}(aq)	–58.2	–54.4	–113.0
Co^{3+}(aq)	92.0	134.0	–305.0
CoCl$_2$(s)	–312.5	–269.8	109.2

Chromium	$\Delta_f H°\ (\dfrac{kJ}{mol})$	$\Delta_f G°\ (\dfrac{kJ}{mol})$	$S°\ (\dfrac{J}{mol\ K})$
Cr(s)	0.0	0.0	23.8
Cr(g)	396.6	351.8	174.5
CrCl$_2$(s)	–395.4	–356.0	115.3
CrCl$_3$(s)	–556.5	–486.1	123.0
CrO$_4^{2-}$(aq)	–881.2	–727.8	50.2
Cr$_2$O$_3$(s)	–1139.7	–1058.1	81.2
Cr$_2$O$_7^{2-}$(aq)	–1490.3	–1301.1	261.9

Copper	$\Delta_f H°\ (\dfrac{kJ}{mol})$	$\Delta_f G°\ (\dfrac{kJ}{mol})$	$S°\ (\dfrac{J}{mol\ K})$
Cu(s)	0.0	0.0	33.2
Cu(g)	337.4	297.7	166.4
Cu$^+$(aq)	71.7	50.0	40.6
Cu^{2+}(aq)	64.8	65.5	–99.6
CuCl(s)	–137.2	–119.9	86.2
CuCl$_2$(s)	–220.1	–175.7	108.1
Cu$_2$O(s)	–168.6	–146.0	93.1
CuO(s)	–157.3	–129.7	42.6
CuSO$_4$(s)	–771.4	–662.2	109.2

Fluorine	$\Delta_f H°\ (\dfrac{kJ}{mol})$	$\Delta_f G°\ (\dfrac{kJ}{mol})$	$S°\ (\dfrac{J}{mol\ K})$
F$_2$(g)	0.0	0.0	202.8
F(g)	79.4	62.3	158.8
F$^-$(aq)	–332.6	–278.8	–13.8
F$_2$O(g)	24.5	41.8	247.5
F$_2$O$_2$(g)	19.2	58.2	277.2
FNO(g)	–66.5	–51.0	248.1

Iron	$\Delta_f H°\ (\dfrac{kJ}{mol})$	$\Delta_f G°\ (\dfrac{kJ}{mol})$	$S°\ (\dfrac{J}{mol\ K})$
Fe(s)	0.0	0.0	27.3
Fe(g)	416.3	370.7	180.5
Fe^{2+}(aq)	–89.1	–78.9	–137.7
Fe^{3+}(aq)	–48.5	–4.7	–315.9
FeBr$_2$(s)	–249.8	–238.1	140.6
FeCO$_3$(s)	–740.6	–666.7	92.9
FeCl$_2$(s)	–341.8	–302.3	118.0
FeCl$_3$(s)	–399.5	–334.0	142.3

	$\Delta_f H° (\frac{kJ}{mol})$	$\Delta_f G° (\frac{kJ}{mol})$	$S° (\frac{J}{mol\ K})$
$Fe_2O_3(s)$	−824.2	−742.2	87.4
$FeSO_4(s)$	−928.4	−820.8	107.5

Hydrogen	$\Delta_f H° (\frac{kJ}{mol})$	$\Delta_f G° (\frac{kJ}{mol})$	$S° (\frac{J}{mol\ K})$
$H_2(g)$	0.0	0.0	130.7
$H(g)$	218.0	203.3	114.7
$H^+(aq)$	0.0	0.0	0.0
$HBr(g)$	−36.3	−53.4	198.7
$HCl(g)$	−92.3	−95.3	186.9
$HF(g)$	−273.3	−275.4	173.8
$HI(g)$	26.5	1.7	206.6
$HNO_2(g)$	−79.5	−46.0	254.1
$HNO_3(l)$	−174.1	−80.7	155.6
$HNO_3(g)$	−133.9	−73.5	266.9
$H_2O(l)$	−285.8	−237.1	70.0
$H_2O(g)$	−241.8	−228.6	188.8
$HO(g)$	39.0	34.2	183.7
$^1H^2HO(l)$	−289.9	−241.9	79.3
$^2H_2O(l)$	−294.6	−243.4	75.9
$H_2O_2(l)$	−187.8	−120.4	109.6
$H_2O_2(aq)$	−191.2	−134.0	143.9
$H_2O_2(g)$	−136.3	−105.6	232.7
$H_3PO_4(l)$	−1271.7	−1123.6	150.8
$H_2S(g)$	−20.6	−33.4	205.8
$H_2SO_4(l)$	−814.0	−690.0	156.9
$H_2SO_4(aq)$	−909.3	−744.5	20.1
$HSO_4^-(aq)$	−887.3	−755.9	131.8

Helium	$\Delta_f H° (\frac{kJ}{mol})$	$\Delta_f G° (\frac{kJ}{mol})$	$S° (\frac{J}{mol\ K})$
$He(g)$	0.0	0.0	126.2

Mercury	$\Delta_f H° (\frac{kJ}{mol})$	$\Delta_f G° (\frac{kJ}{mol})$	$S° (\frac{J}{mol\ K})$
$Hg(l)$	0.0	0.0	75.9
$Hg(g)$	61.4	31.8	175.0
$Hg_2^{2+}(aq)$	172.4	153.5	84.5
$Hg^{2+}(aq)$	171.1	164.4	−32.2
$Hg_2Br_2(s)$	−206.9	−181.1	218.0
$HgBr_2(s)$	−170.7	−153.1	172.0
$Hg_2Cl_2(s)$	−265.4	−210.7	191.6
$HgCl_2(s)$	−224.3	−178.6	146.0
$HgO(s)$	−90.8	−58.5	70.3
$HgS(s)$	−58.2	−50.6	82.4

Iodine	$\Delta_f H° (\frac{kJ}{mol})$	$\Delta_f G° (\frac{kJ}{mol})$	$S° (\frac{J}{mol\ K})$
$I_2(s)$	0.0	0.0	116.1
$I_2(g)$	62.4	19.3	260.7
$I(g)$	106.8	70.2	180.8
$I^-(aq)$	−55.2	−51.6	111.3
$IBr(g)$	40.8	3.7	258.8

ICl(g)	17.8	–5.5	247.6
ICl(l)	–23.9	–13.6	135.1

Potassium	$\Delta_f H° \left(\dfrac{kJ}{mol}\right)$	$\Delta_f G° \left(\dfrac{kJ}{mol}\right)$	$S° \left(\dfrac{J}{mol\ K}\right)$
K(s)	0.0	0.0	64.7
K(g)	89.0	60.5	160.3
K_2(g)	123.7	87.5	249.7
K^+(aq)	–252.4	–283.3	102.5
K_2CO_3(s)	–1151.0	–1063.5	155.5
KBr(s)	–393.8	–380.7	95.9
KCl(s)	–436.5	–408.5	82.6
KF(s)	–567.3	–537.8	66.6
KI(s)	–327.9	–324.9	106.3
$KMnO_4$(s)	–837.2	–737.6	171.7
KNO_3(s)	–494.6	–394.9	133.1
KOH(s)	–424.6	–379.4	81.2

Krypton	$\Delta_f H° \left(\dfrac{kJ}{mol}\right)$	$\Delta_f G° \left(\dfrac{kJ}{mol}\right)$	$S° \left(\dfrac{J}{mol\ K}\right)$
Kr(g)	0.0	0.0	164.1

Lithium	$\Delta_f H° \left(\dfrac{kJ}{mol}\right)$	$\Delta_f G° \left(\dfrac{kJ}{mol}\right)$	$S° \left(\dfrac{J}{mol\ K}\right)$
Li(s)	0.0	0.0	29.1
Li(g)	159.3	126.6	138.8
Li_2(g)	215.9	174.4	197.0
Li^+(aq)	–278.5	–293.3	13.4
$LiAlH_4$(s)	–116.3	–44.7	78.7
LiBr(s)	–351.2	–342.0	74.3
Li_2CO_3(s)	–1215.9	–1132.1	90.4
LiCl(s)	–408.6	–384.4	59.3
LiF(s)	–616.0	–587.7	35.7
LiH(s)	–90.5	–68.3	20.0
LiI(s)	–270.4	–270.3	86.8
$LiNO_3$(s)	–483.1	–381.1	90.0
Li_2O(s)	–597.9	–561.2	37.6
LiOH(s)	–487.5	–441.5	42.8

Magnesium	$\Delta_f H° \left(\dfrac{kJ}{mol}\right)$	$\Delta_f G° \left(\dfrac{kJ}{mol}\right)$	$S° \left(\dfrac{J}{mol\ K}\right)$
Mg(s)	0.0	0.0	32.7
Mg(g)	147.1	112.5	148.6
Mg^{2+}(aq)	–466.9	–454.8	–138.1
$MgBr_2$(s)	–524.3	–503.8	117.2
$MgCO_3$(s)	–1095.8	–1012.1	65.7
$MgCl_2$(s)	–641.3	–591.8	89.6
MgF_2(s)	–1124.2	–1071.1	57.2
MgH_2(s)	–75.3	–35.9	31.1
$Mg(NO_3)_2$(s)	–790.7	–589.4	164.0
MgO(s)	–601.6	–569.3	27.0
$Mg(OH)_2$(s)	–924.5	–833.5	63.2
$MgSO_4$(s)	–1284.9	–1170.6	91.6

Manganese	$\Delta_f H°\ (\frac{kJ}{mol})$	$\Delta_f G°\ (\frac{kJ}{mol})$	$S°\ (\frac{J}{mol\ K})$
Mn(s)	0	0	32.0
Mn(g)	280.7	238.5	173.7
$MnCl_2$(s)	−481.3	−440.5	118.2
MnO(s)	−385.2	−362.9	59.7
Mn_2O_3(s)	−959.0	−881.1	110.5
MnO_2(s)	−520.0	−465.1	53.1
MnO_4^-(aq)	−541.4	−447.2	191.2
MnS(s)	−214.2	−218.4	78.2

Sodium	$\Delta_f H°\ (\frac{kJ}{mol})$	$\Delta_f G°\ (\frac{kJ}{mol})$	$S°\ (\frac{J}{mol\ K})$
Na(s)	0.0	0.0	51.3
Na(g)	107.5	77.0	153.7
Na_2(g)	142.1	103.9	230.2
Na^+(aq)	−240.1	−261.9	59.0
NaBr(s)	−361.1	−349.0	86.8
NaCN(s)	−87.5	−76.4	115.6
Na_2CO_3(s)	−1130.7	−1044.4	135.0
$NaCH_3CO_2$(s)	−708.8	−607.2	123.0
NaCl(s)	−411.2	−384.1	72.1
NaF(s)	−576.6	−546.3	51.1
NaH(s)	−56.3	−33.5	40.0
NaI(s)	−287.8	−286.1	98.5
NaN_3(s)	21.7	93.8	96.9
$NaNO_3$(s)	−467.9	−367.0	116.5
Na_2O(s)	−414.2	−375.5	75.1
NaOH(s)	−425.8	−379.7	64.4
Na_2SO_4(s)	−1387.1	−1270.2	149.6

Neon	$\Delta_f H°\ (\frac{kJ}{mol})$	$\Delta_f G°\ (\frac{kJ}{mol})$	$S°\ (\frac{J}{mol\ K})$
Ne(g)	0.0	0.0	146.3

Nitrogen	$\Delta_f H°\ (\frac{kJ}{mol})$	$\Delta_f G°\ (\frac{kJ}{mol})$	$S°\ (\frac{J}{mol\ K})$
N_2(g)	0	0	191.6
N(g)	472.7	455.5	153.3
N_3^-(aq)	275.1	348.2	107.9
NF_3(g)	−132.1	−90.6	260.8
NH_3(g)	−45.9	−16.4	192.8
NH_4^+(aq)	−132.5	−79.3	113.4
NH_4Br(s)	−270.8	−175.2	113.0
NH_4Cl(s)	−314.4	−202.9	94.6
NH_4F(s)	−464.0	−348.7	72.0
NH_4I(s)	−201.4	−112.5	117.0
NO(g)	91.3	87.6	210.8
NO_2^-(aq)	−104.6	−32.2	123.0
NO_2(g)	33.2	51.3	240.1
NO_3^-(aq)	−207.4	−111.3	146.4
N_2H_4(l)	50.6	149.3	121.2

	$\Delta_f H° (\frac{kJ}{mol})$	$\Delta_f G° (\frac{kJ}{mol})$	$S° (\frac{J}{mol\ K})$
$N_2O(g)$	81.6	103.7	220.0
$N_2O_4(g)$	11.1	99.8	304.4
$N_2O_4(l)$	–19.5	97.5	209.2

Nickel	$\Delta_f H° (\frac{kJ}{mol})$	$\Delta_f G° (\frac{kJ}{mol})$	$S° (\frac{J}{mol\ K})$
$Ni(s)$	0.0	0.0	29.9
$Ni(g)$	429.7	384.5	182.2
$NiCl_2(s)$	–305.3	–259.0	97.7
$NiF_2(s)$	–651.4	–604.1	73.6
$NiS(s)$	–82.0	–79.5	53.0

Oxygen	$\Delta_f H° (\frac{kJ}{mol})$	$\Delta_f G° (\frac{kJ}{mol})$	$S° (\frac{J}{mol\ K})$
$O_2(g)$	0.0	0.0	205.2
$O_2(aq)$	–11.7	110.9	16.4
$O(g)$	249.2	231.7	161.1
$O_3(g)$	142.7	163.2	238.9
$OH^-(aq)$	–230.0	–157.2	–10.8

Osmium	$\Delta_f H° (\frac{kJ}{mol})$	$\Delta_f G° (\frac{kJ}{mol})$	$S° (\frac{J}{mol\ K})$
$Os(s)$	0.0	0.0	32.6
$Os(g)$	791.0	745.0	192.6
$OsO_4(g)$	–337.2	–292.8	293.8
$OsO_4(s)$	–394.0	–304.9	143.9

Phosphorus	$\Delta_f H° (\frac{kJ}{mol})$	$\Delta_f G° (\frac{kJ}{mol})$	$S° (\frac{J}{mol\ K})$
$P(s, white)$	0.0	0.0	41.1
$P(s, red)$	–17.6	–12.2	22.8
$P_2(g)$	144.0	103.5	218.1
$P(g)$	316.5	280.1	163.2
$PBr_3(l)$	–184.5	–175.7	240.2
$PCl_3(g)$	–287.0	–267.8	311.8
$PCl_3(l)$	–319.7	–272.3	217.1
$PCl_5(g)$	–374.9	–305.0	364.6
$PF_3(g)$	–958.4	–936.9	273.1
$PF_5(g)$	–1594.4	–1520.7	300.8
$PH_3(g)$	5.4	13.5	210.2
$PO_4^{3-}(aq)$	–1277.4	–1018.7	–220.5

Lead	$\Delta_f H° (\frac{kJ}{mol})$	$\Delta_f G° (\frac{kJ}{mol})$	$S° (\frac{J}{mol\ K})$
$Pb(s)$	0.0	0.0	64.8
$Pb(g)$	195.2	162.2	175.4
$Pb^{2+}(aq)$	–1.7	–24.4	10.5
$PbBr_2(s)$	–278.7	–261.9	161.5
$PbCO_3(s)$	–699.1	–625.5	131.0
$PbCl_2(s)$	–359.4	–314.1	136.0
$PbO(s, litharge)$	–219.0	–188.9	66.5
$PbO(s, massicot)$	–217.3	–187.9	68.7
$PbO_2(s)$	–277.4	–217.3	68.6

PbS(s)	−100.4	−98.7	91.2
PbSO$_4$(s)	−920.0	−813.0	148.5

Sulfur	$\Delta_f H° \, (\dfrac{kJ}{mol})$	$\Delta_f G° \, (\dfrac{kJ}{mol})$	$S° \, (\dfrac{J}{mol\ K})$
S(s, rhombic)	0.0	0.0	32.1
S$_2$(g)	128.6	79.7	228.2
S(g)	277.2	236.7	167.8
SCN$^-$(aq)	76.4	92.7	144.3
SF$_4$(g)	−763.2	−722.0	299.6
SF$_6$(g)	−1220.5	−1116.5	291.5
SO(g)	6.3	−19.9	222.0
SO$_2$(g)	−296.8	−300.1	248.2
SO$_3$(g)	−395.7	−371.1	256.8
SO$_3$(l)	−441.0	−373.8	113.8
SO$_3$(s)	−454.5	−374.2	70.7
SO$_4^{2-}$(aq)	−909.3	−744.5	20.1
HS$^-$(aq)	−17.6	12.1	62.8

Antimony	$\Delta_f H° \, (\dfrac{kJ}{mol})$	$\Delta_f G° \, (\dfrac{kJ}{mol})$	$S° \, (\dfrac{J}{mol\ K})$
Sb(s)	0.0	0.0	45.7
Sb$_2$(g)	235.6	187.0	254.9
Sb(g)	262.3	222.1	180.3
SbCl$_3$(s)	−382.2	−323.7	184.1
SbH$_3$(g)	145.1	147.8	232.8
Sb$_2$O$_5$(s)	−971.9	−829.2	125.1

Silicon	$\Delta_f H° \, (\dfrac{kJ}{mol})$	$\Delta_f G° \, (\dfrac{kJ}{mol})$	$S° \, (\dfrac{J}{mol\ K})$
Si(s)	0.0	0.0	18.8
Si(g)	450.0	405.5	168.0
SiBr$_4$(l)	−457.3	−443.8	277.8
SiC(s, cubic)	−65.3	−62.8	16.6
SiC(s, hexagonal)	−62.8	−60.2	16.5
SiCl$_4$(g)	−657.0	−617.0	330.7
SiF$_4$(g)	−1615.0	−1572.8	282.8
SiH$_4$(g)	34.3	56.9	204.6
SiO$_2$(s)	−910.7	−856.3	41.5

Tin	$\Delta_f H° \, (\dfrac{kJ}{mol})$	$\Delta_f G° \, (\dfrac{kJ}{mol})$	$S° \, (\dfrac{J}{mol\ K})$
Sn(s, white)	0.0	0.0	51.2
Sn(s, gray)	−2.1	0.1	44.1
Sn(g)	301.2	266.2	168.5
Sn^{2+}(aq)	−8.8	−27.2	−17.0
SnBr$_4$(s)	−377.4	−350.2	264.4
SnCl$_4$(l)	−511.3	−440.1	258.6
SnH$_4$(g)	162.8	188.3	227.7
SnO(s)	−280.7	−251.9	57.2
SnO$_2$(s)	−577.6	−515.8	49.0

Titanium	$\Delta_f H° (\frac{kJ}{mol})$	$\Delta_f G° (\frac{kJ}{mol})$	$S° (\frac{J}{mol\ K})$
Ti(s)	0.0	0.0	30.7
Ti(g)	473.0	428.4	180.3
TiBr$_4$(s)	−616.7	−589.5	243.5
TiBr$_4$(g)	−549.4	−568.2	398.4
TiCl$_4$(l)	−804.2	−737.2	252.3
TiCl$_4$(g)	−763.2	−726.3	353.2
TiI$_4$(s)	−375.7	−371.5	249.4
TiO$_2$(s)	−944.0	−888.8	50.6

Uranium	$\Delta_f H° (\frac{kJ}{mol})$	$\Delta_f G° (\frac{kJ}{mol})$	$S° (\frac{J}{mol\ K})$
U(s)	0.0	0.0	50.2
U(g)	533.0	488.4	199.8
U^{3+}(aq)	−489.1	−476.2	−188.0
UCl$_3$(s)	−866.5	−799.1	159.0
UCl$_4$(s)	−1019.2	−930.0	197.1
UCl$_4$(g)	−809.6	−786.6	419.0
UCl$_6$(s)	−1092.0	−962.0	285.0
UCl$_6$(g)	−1013.0	−928.0	431.0
UF$_4$(s)	−1914.2	−1823.3	151.7
UF$_4$(g)	−1598.7	−1572.7	368.0
UF$_6$(s)	−2197.0	−2068.5	227.6
UF$_6$(g)	−2147.4	−2063.7	377.9
UH$_3$(s)	−127.2	−72.8	63.7
UO$_2$(s)	−1085.0	−1031.0	77.0
UO$_3$(s)	−1223.8	−1145.7	96.1

Xenon	$\Delta_f H° (\frac{kJ}{mol})$	$\Delta_f G° (\frac{kJ}{mol})$	$S° (\frac{J}{mol\ K})$
Xe(g)	0.0	0.0	169.7

Zinc	$\Delta_f H° (\frac{kJ}{mol})$	$\Delta_f G° (\frac{kJ}{mol})$	$S° (\frac{J}{mol\ K})$
Zn(s)	0.0	0.0	41.6
Zn(g)	130.4	94.8	161.0
Zn^{2+}(aq)	−153.9	−147.1	−112.1
ZnBr$_2$(s)	−328.7	−312.1	138.5
ZnCO$_3$(s)	−812.8	−731.5	82.4
ZnCl$_2$(s)	−415.1	−369.4	111.5
ZnI$_2$(s)	−208.0	−209.0	161.1
ZnO(s)	−350.5	−320.5	43.7
Zn(OH)$_2$(s)	−641.9	−553.5	81.2
ZnS(s)	−206.0	−201.3	57.7
ZnSO$_4$(s)	−982.8	−871.5	110.5

Source: Rumble, J.R.; Lide, D.R.; Bruno, T.J. *CRC Handbook of Chemistry and Physics: A Ready-Reference Book of Chemical and Physical Data*, 100th ed. Boca Raton, FL: CRC Press, 2019–2020.

APPENDIX 8 – HALF-CELL REACTIONS AND THEIR STANDARD REDUCTION POTENTIAL VALUES ($E°$) AT 25 °C

Acidic Half-Cell Reaction	$E°$ (V)
$F_2(g) + 2 e^- + 2 H^+(aq) \rightarrow 2 HF(aq)$	3.05
$F_2(g) + 2 e^- \rightarrow 2 F^-(aq)$	2.87
$XeF_2(aq) + 2 H^+(aq) + 2 e^- \rightarrow Xe(g) + 2 HF(aq)$	2.32
$O_3(g) + 2 H^+(aq) + 2 e^- \rightarrow O_2(g) + H_2O(l)$	2.08
$Co^{3+}(aq) + e^- \rightarrow Co^{2+}(aq)$	1.92
$Au^+(aq) + e^- \rightarrow Au(s)$	1.83
$H_2O_2(aq) + 2 H^+(aq) + 2 e^- \rightarrow 2 H_2O(l)$	1.78
$Ce^{4+}(aq) + e^- \rightarrow Ce^{3+}(aq)$	1.72
$PbO_2(s) + SO_4^{2-}(aq) + 4 H^+(aq) + 2 e^- \rightarrow PbSO_4(s) + 2 H_2O(l)$	1.69
$NiO_2(s) + 4 H^+(aq) + 2 e^- \rightarrow Ni^{2+}(aq) + 2 H_2O(l)$	1.68
$2 HClO(aq) + 2 H^+(aq) + 2 e^- \rightarrow Cl_2(g) + 2 H_2O(l)$	1.61
$MnO_4^-(aq) + 8 H^+(aq) + 5 e^- \rightarrow Mn^{2+}(aq) + 4 H_2O(l)$	1.51
$Au^{3+}(aq) + 3 e^- \rightarrow Au(s)$	1.50
$2 BrO_3^-(aq) + 12 H^+(aq) + 10 e^- \rightarrow Br_2(l) + 6 H_2O(l)$	1.48
$2 ClO_3^-(aq) + 12 H^+(aq) + 10 e^- \rightarrow Cl_2(g) + 6 H_2O(l)$	1.47
$Cr_2O_7^{2-}(aq) + 14 H^+(aq) + 6 e^- \rightarrow 2 Cr^{3+}(aq) + 7 H_2O(l)$	1.36
$Cl_2(g) + 2 e^- \rightarrow 2 Cl^-(aq)$	1.36
$MnO_2(s) + 4 H^+(aq) + 2 e^- \rightarrow Mn^{2+}(aq) + 2 H_2O(l)$	1.22
$O_2(g) + 4 H^+(aq) + 4 e^- \rightarrow 2 H_2O(l)$	1.23
$2 IO_3^-(aq) + 12 H^+(aq) + 10 e^- \rightarrow I_2(s) + 6 H_2O(l)$	1.20
$ClO_4^-(aq) + 2 H^+(aq) + 2 e^- \rightarrow ClO_3^-(aq) + H_2O(l)$	1.19
$Pt^{2+}(aq) + 2 e^- \rightarrow Pt(s)$	1.18
$Br_2(aq) + 2 e^- \rightarrow 2 Br^-(aq)$	1.09
$Br_2(l) + 2 e^- \rightarrow 2 Br^-(aq)$	1.07
$OsO_4(s) + 4 H^+(aq) + 4 e^- \rightarrow OsO_2(s) + 2 H_2O(l)$	1.02
$AuCl_4^-(aq) + 3 e^- \rightarrow Au(s) + 4 Cl^-(aq)$	1.00
$NO_3^-(aq) + 4 H^+(aq) + 3 e^- \rightarrow NO(g) + 2 H_2O(l)$	0.96
$Pd^{2+}(aq) + 2 e^- \rightarrow Pd(s)$	0.95
$NO_3^-(aq) + 3 H^+(aq) + 2 e^- \rightarrow HNO_2(aq) + H_2O(l)$	0.93
$2 Hg^{2+}(aq) + 2 e^- \rightarrow Hg_2^{2+}(aq)$	0.92
$Hg^{2+}(aq) + 2 e^- \rightarrow Hg(l)$	0.85
$Ag^+(aq) + e^- \rightarrow Ag(s)$	0.80
$Hg_2^{2+}(aq) + 2 e^- \rightarrow 2 Hg(l)$	0.80
$Fe^{3+}(aq) + e^- \rightarrow Fe^{2+}(aq)$	0.77
$PtCl_4^{2-}(aq) + 2 e^- \rightarrow Pt(s) + 4 Cl^-(aq)$	0.76
$Fe_2O_3(s) + 6 H^+(aq) + 2 e^- \rightarrow 2 Fe^{2+}(aq) + 3 H_2O(l)$	0.73
$O_2(g) + 2 H^+(aq) + 2 e^- \rightarrow H_2O_2(aq)$	0.70
$PtCl_6^{2-}(aq) + 2 e^- \rightarrow PtCl_4^{2-}(aq) + 2 Cl^-(aq)$	0.68
$TeO_2(s) + 4 H^+(aq) + 4 e^- \rightarrow Te(s) + 2 H_2O(l)$	0.60
$I_2(s) + 2 e^- \rightarrow 2 I^-(aq)$	0.54
$Cu^+(aq) + e^- \rightarrow Cu(s)$	0.52
$CO(g) + 2 H^+(aq) + 2 e^- \rightarrow C(s) + H_2O(l)$	0.52
$CH_3OH(aq) + 2 H^+(aq) + 2 e^- \rightarrow CH_4(g) + H_2O(l)$	0.50
$Cu^{2+}(aq) + 2 e^- \rightarrow Cu(s)$	0.34
$Bi^{3+}(aq) + 3 e^- \rightarrow Bi(s)$	0.31
$Hg_2Cl_2(s) + e^- \rightarrow 2 Hg(l) + 2 Cl^-(aq)$	0.27
$AgCl(s) + e^- \rightarrow Ag(s) + Cl^-(aq)$	0.22

$SO_4^{2-}(aq) + 4\,H^+(aq) + 2\,e^- \rightarrow H_2SO_3(aq) + H_2O(l)$	0.17
$Cu^{2+}(aq) + e^- \rightarrow Cu^+(aq)$	0.15
$Sn^{4+}(aq) + 2\,e^- \rightarrow Sn^{2+}(aq)$	0.15
$S(s) + 2\,H^+(aq) + 2\,e^- \rightarrow H_2S(aq)$	0.14
$C(s) + 4\,H^+(aq) + 4\,e^- \rightarrow CH_4(g)$	0.13
$H_2CO(aq) + 2\,H^+(aq) + 2\,e^- \rightarrow CH_3OH(aq)$	0.13
$AgBr(s) + e^- \rightarrow Ag(s) + Br^-(aq)$	0.07
$2\,H^+(aq) + 2\,e^- \rightarrow H_2(g)$	0.00
$HCO_2H(aq) + 2\,H^+(aq) + 2\,e^- \rightarrow CH_2O(aq) + H_2O(l)$	−0.03
$Fe^{3+}(aq) + 3\,e^- \rightarrow Fe(s)$	−0.04
$CO_2(g) + 2\,H^+(aq) + 2\,e^- \rightarrow CO(g) + H_2O(l)$	−0.11
$SnO_2(s) + 4\,H^+(aq) + 4\,e^- \rightarrow Sn(s) + 2\,H_2O(l)$	−0.12
$Pb^{2+}(aq) + 2\,e^- \rightarrow Pb(s)$	−0.13
$Sn^{2+}(aq) + 2\,e^- \rightarrow Sn(s)$	−0.14
$AgI(s) + e^- \rightarrow Ag(s) + I^-(aq)$	−0.15
$CO_2(g) + 2\,H^+(aq) + 2\,e^- \rightarrow HCOOH(aq)$	−0.20
$Ni^{2+}(aq) + 2\,e^- \rightarrow Ni(s)$	−0.25
$Co^{2+}(aq) + 2\,e^- \rightarrow Co(s)$	−0.28
$In^{3+}(aq) + 3\,e^- \rightarrow In(s)$	−0.34
$PbSO_4(s) + 2\,e^- \rightarrow Pb(s) + SO_4^{2-}(aq)$	−0.36
$Cd^{2+}(aq) + 2\,e^- \rightarrow Cd(s)$	−0.40
$Cr^{3+}(aq) + e^- \rightarrow Cr^{2+}(aq)$	−0.41
$Fe^{2+}(aq) + 2\,e^- \rightarrow Fe(s)$	−0.44
$2\,CO_2(g) + 2\,H^+(aq) + 2\,e^- \rightarrow (COOH)_2(aq)$	−0.48
$TiO_2(s) + 4\,H^+(aq) + 2\,e^- \rightarrow Ti^{2+}(aq) + 2\,H_2O(l)$	−0.50
$Ga^{3+}(aq) + 3\,e^- \rightarrow Ga(s)$	−0.55
$Cr^{3+}(aq) + 3\,e^- \rightarrow Cr(s)$	−0.74
$Zn^{2+}(aq) + 2\,e^- \rightarrow Zn(s)$	−0.76
$B(OH)_3(aq) + 3\,H^+(aq) + 3\,e^- \rightarrow B(s) + 3\,H_2O(l)$	−0.87
$Cr^{2+}(aq) + 2\,e^- \rightarrow Cr(s)$	−0.91
$V^{2+}(aq) + 2\,e^- \rightarrow V(s)$	−1.18
$Mn^{2+}(aq) + 2\,e^- \rightarrow Mn(s)$	−1.19
$U^{3+}(aq) + 3\,e^- \rightarrow U(s)$	−1.66
$Al^{3+}(aq) + 3\,e^- \rightarrow Al(s)$	−1.68
$H_2(g) + 2\,e^- \rightarrow 2\,H^-(aq)$	−2.23
$Mg^{2+}(aq) + 2\,e^- \rightarrow Mg(s)$	−2.37
$Na^+(aq) + e^- \rightarrow Na(s)$	−2.71
$Ca^{2+}(aq) + 2\,e^- \rightarrow Ca(s)$	−2.87
$K^+(aq) + e^- \rightarrow K(s)$	−2.93
$Li^+(aq) + e^- \rightarrow Li(s)$	−3.05

Basic Half-Cell Reaction

Basic Half-Cell Reaction	$E°$ (V)
$OOH^-(aq) + H_2O(l) + 2\,e^- \rightarrow 3\,OH^-(aq)$	0.88
$ClO^-(aq) + H_2O(l) + 2\,e^- \rightarrow Cl^-(aq) + 2\,OH^-(aq)$	0.81
$ClO_3^-(aq) + 3\,H_2O(l) + 6\,e^- \rightarrow Cl^-(aq) + 6\,OH^-(aq)$	0.62
$MnO_4^-(aq) + 2\,H_2O(l) + 3\,e^- \rightarrow MnO_2(s) + 4\,OH^-(aq)$	0.60
$O_2(g) + 2\,H_2O(l) + 4\,e^- \rightarrow 4\,OH^-(aq)$	0.40
$ClO_4^-(aq) + H_2O(l) + 2\,e^- \rightarrow ClO_3^-(aq) + 2\,OH^-(aq)$	0.36
$HgO(s) + H_2O(l) + 2\,e^- \rightarrow Hg(l) + 2\,OH^-(aq)$	0.10
$NO_3^-(aq) + H_2O(l) + 2\,e^- \rightarrow NO_2^-(aq) + 2\,OH^-(aq)$	0.01
$CrO_4^{2-}(aq) + 4\,H_2O(l) + 3\,e^- \rightarrow Cr(OH)_3(s) + 5\,OH^-(aq)$	−0.13

$Cu_2O(s) + H_2O(l) + 2 e^- \rightarrow 2 Cu(s) + 2 OH^-(aq)$	-0.36
$NiO_2(aq) + 2 H_2O(l) + 2 e^- \rightarrow Ni(OH)_2(s) + 2 OH^-(aq)$	-0.49
$Fe_2O_3(s) + 3 H_2O(l) + 2 e^- \rightarrow 2 Fe(OH)_2(s) + 2 OH^-(aq)$	-0.86
$2 H_2O(l) + 2 e^- \rightarrow H_2(g) + 2 OH^-(aq)$	-0.87
$Fe(OH)_2(s) + 2 e^- \rightarrow Fe(s) + 2 OH^-(aq)$	-0.89
$SO_4^{2-}(aq) + H_2O(l) + 2 e^- \rightarrow SO_3^{2-}(aq) + 2 OH^-(aq)$	-0.94
$N_2(g) + 4 H_2O(l) + 4 e^- \rightarrow N_2H_4(aq) + 4 OH^-(aq)$	-1.16
$Zn(OH)_4^{2-}(aq) + 2 e^- \rightarrow Zn(s) + 4 OH^-(aq)$	-1.20
$Zn(OH)_2(s) + 2 e^- \rightarrow Zn(s) + 2 OH^-(aq)$	-1.25
$Cr(OH)_3(s) + 3 e^- \rightarrow Cr(s) + 3 OH^-(aq)$	-1.48
$SiO_3^{2-}(aq) + 3 H_2O(l) + 4 e^- \rightarrow Si(s) + 6 OH^-(aq)$	-1.70
$H_2BO_3^-(aq) + H_2O(l) + 3 e^- \rightarrow B(s) + 6 OH^-(aq)$	-1.79
$Al(OH)_3(s) + 3 e^- \rightarrow Al(s) + 3 OH^-(aq)$	-2.30
$Mg(OH)_2(s) + 2 e^- \rightarrow Mg(s) + 2 OH^-(aq)$	-2.69

Source: Rumble, J.R.; Lide, D.R.; Bruno, T.J. *CRC Handbook of Chemistry and Physics: A Ready-Reference Book of Chemical and Physical Data*, 100th ed. Boca Raton, FL: CRC Press, 2019–2020.

APPENDIX 9 – ACIDITY CONSTANT (K_a) VALUES

Aqueous Acid	Chemical Formula (HA)	Conjugate Base (A^-)	K_a	pK_a
Hydrogen iodide	HI	I^-	1.0×10^{10}	-10.00
Hydrogen bromide	HBr	Br^-	1.0×10^9	-9.00
Perchloric acid	$HClO_4$	ClO_4^-	1.0×10^8	-8.00
Hydrogen chloride	HCl	Cl^-	1.0×10^7	-7.00
Acetic acidium	$CH_3CO_2H_2^+$	CH_3CO_2H	1.6×10^6	-6.20
Sulfuric acid	H_2SO_4	HSO_4^-	1.6×10^5	-5.20
Nitric acid	HNO_3	NO_3^-	25	-1.40
Oxidanium	H_3O^+	H_2O	1.0	0.00
Trifluoroacetic acid	CF_3CO_2H	$CF_3CO_2^-$	0.63	0.20
Sulfurous acid	H_2SO_3	HSO_3^-	1.7×10^{-2}	1.77
Hydrogensulfate	HSO_4^-	SO_4^{2-}	1.1×10^{-2}	1.96
Phosphoric acid	H_3PO_4	$H_2PO_4^-$	7.9×10^{-3}	2.10
Nitrous acid	HNO_2	NO_2^-	7.1×10^{-4}	3.15
Hydrogen fluoride	HF	F^-	6.3×10^{-4}	3.20
Formic acid	HCO_2H	HCO_2^-	3.0×10^{-4}	3.52
Hydrogen selenide	H_2Se	HSe^-	1.3×10^{-4}	3.88
Benzoic acid	$C_6H_5CO_2H$	$C_6H_5CO_2^-$	6.3×10^{-4}	4.20
Acetic acid	CH_3CO_2H	$CH_3CO_2^-$	1.8×10^{-5}	4.74
Carbonic acid	H_2CO_3	HCO_3^-	4.3×10^{-7}	6.37
Hydrogen sulfide	H_2S	HS^-	1.0×10^{-7}	7.00
Dihydrogenphosphate	$H_2PO_4^-$	HPO_4^{2-}	6.3×10^{-8}	7.20
Ammonium	NH_4^+	NH_3	5.6×10^{-10}	9.25
Hydrogen cyanide	HCN	CN^-	3.3×10^{-10}	9.48
Phenol	C_6H_5OH	$C_6H_5O^-$	1.0×10^{-10}	10.00
Hydrogencarbonate	HCO_3^-	CO_3^{2-}	4.7×10^{-11}	10.33
Hydrogenphosphate	HPO_4^{2-}	PO_4^{3-}	4.6×10^{-13}	12.34
Water	H_2O	OH^-	1.0×10^{-14}	14.00

Aqueous Acid	Chemical Formula (HA)	Conjugate Base (A⁻)	K_a	pK_a
Methanol	CH_3OH	CH_3O^-	1.0×10^{-16}	16.00
Sulfanide (bisulfide)	HS^-	S^{2-}	1.0×10^{-17}	17.00
Acetylene	C_2H_2	C_2H^-	1.0×10^{-25}	25.00
Hydrogen	H_2	H^-	1.0×10^{-35}	35.00
Ammonia	NH_3	NH_2^-	1.0×10^{-38}	38.00
Methane	CH_4	CH_3^-	1.0×10^{-50}	50.00

Answers

CHAPTER 2 ANSWERS

PROBLEM 2.1
Calculate the kinetic energy (E_k) in each of the following examples.

a. A car (1628 kg) travels at 105 km/h.

$$105 \frac{km}{h} \left(\frac{1000\,m}{km}\right)\left(\frac{h}{60\,min}\right)\left(\frac{min}{60\,s}\right) = 29.2 \text{ m/s}$$

$$E_k = \frac{1}{2}mv^2 = \frac{1}{2}(1628\text{ kg})(29.2\text{ m/s})^2 = 6.92 \times 10^5 \text{ J}$$

b. A human (89 kg) walks at 4.8 km/h.

$$4.8 \frac{km}{h} \left(\frac{1000\,m}{km}\right)\left(\frac{h}{60\,min}\right)\left(\frac{min}{60\,s}\right) = 1.3 \text{ m/s}$$

$$E_k = \frac{1}{2}mv^2 = \frac{1}{2}(89\text{ kg})(1.3\text{ m/s})^2 = 79 \text{ J}$$

c. A proton (1.62×10^{-27} kg) moves at 3.00×10^6 m/s.

$$E_k = \frac{1}{2}mv^2 = \frac{1}{2}(1.62 \times 10^{-27}\text{ kg})(3.00 \times 10^6\text{ m/s})^2 = 7.29 \times 10^{-15} \text{ J}$$

PROBLEM 2.2
For each of the following, calculate the thermal energy transfer (J) and identify whether each is endothermic or exothermic.

a. 155 kg water (the average mass of water in a bathtub) is heated from 25.0 °C to 38.5 °C, where $c_p(H_2O(l)) = 4.184$ J/(g K)

$$\Delta T = 38.5\text{ °C} - 25.0\text{ °C} = 13.5\text{ °C} = 13.5\text{ K}$$

$$q = mc_p\Delta T = (155\text{ kg}(\frac{1000\,g}{kg}))(4.184\text{ J/(g K)})(13.5\text{ K}) = 8.76 \times 10^6 \text{ J (endothermic)}$$

b. 18.0 g of water evaporates ($\Delta H = 43.9$ kJ/mol).

$$n = 18.0\text{ g}(\frac{1\,kg}{1000\,g})(\frac{1\,mol}{0.0180\,kg}) = 1.00 \text{ mol}$$

$$q = n\Delta H = 1.00\text{ mol}(43.9\text{ kJ/mol}) = 43.9\text{ kJ}(1000\text{ J/kJ}) = 43\,900 \text{ J (endothermic)}$$

c. 3660 g aluminium is heated from 298.2 K to 311.7 K, where $c_p(Al(s)) = 0.89$ J/(g K)

$$q = mc_p\Delta T = (3660\text{ g})(0.89\text{ J/(g K)})(311.7\text{ K} - 298.2\text{ K}) = 44\,000 \text{ J (endothermic)}$$

d. 1.0 g hydrogen gas is cooled from 100.0 °C to –77.2 °C, where $c_p(H_2(g)) = 1.030$ J/(g K)

$$\Delta T = -77.2\text{ °C} - 100.0\text{ °C} = -177.2\text{ °C} = -177.2\text{ K}$$

$$q = mc_p\Delta T = (1.0\text{ g})(1.030\text{ J/(g K)})(-177.2\text{ K}) = -180 \text{ J (exothermic)}$$

e. 1.5 g methane gas combusts ($\Delta H = -890.4$ kJ/mol)

$$n = 1.5\text{ g}(\frac{1\,kg}{1000\,g})(\frac{1\,mol}{0.016\,04\,kg}) = 0.094 \text{ mol}$$

$$q = n\Delta H = 0.094\text{ mol}(-890.4\text{ kJ/mol}) = -83\text{ kJ}(1000\text{ J/kJ}) = -83\,000 \text{ J (exothermic)}$$

PROBLEM 2.3

For each of the following, calculate the work (J) and identify whether each is endoworkic or exoworkic.

a. A piston compresses from 1.5 L to 0.5 L with an external pressure of 1.05 atm.

$$w = -p\Delta V = -(1.05 \text{ atm})(0.5 \text{ L} - 1.5 \text{ L}) = 1.1 \text{ L atm}(\frac{101.325 \text{ J}}{1 \text{ L atm}}) = 110 \text{ J (endoworkic)}$$

b. A sample of 18.00 mL liquid water evaporates and becomes 31 L of steam with an external pressure of 0.985 atm.

$$18.00 \text{ mL}(\frac{1 \text{ L}}{1000 \text{ mL}}) = 0.018 \text{ L}$$

$$w = -p\Delta V = -(0.985 \text{ atm})(31 \text{ L} - 0.018 \text{ L}) = -31 \text{ L atm}(\frac{101.325 \text{ J}}{1 \text{ L atm}}) = -3100 \text{ J (exoworkic)}$$

c. If $UF_6(s)$ has a density of 5.09 g/mL and $UF_6(g)$ at 51.7 °C has a molar volume of 26.7 L/mol, determine the work (at 1 atm) associated with 200 g of $UF_6(s)$ converting to $UF_6(g)$

$$200 \text{ g}(\frac{\text{mL}}{5.09 \text{ g}})(\frac{1 \text{ L}}{1000 \text{ mL}}) = 0.04 \text{ L}$$

$$200 \text{ g}(\frac{1 \text{ kg}}{1000 \text{ g}})(\frac{1 \text{ mol}}{0.352 \, 02 \text{ g}})(\frac{26.7 \text{ L}}{1 \text{ mol}}) = 20 \text{ L}$$

$$w = -p\Delta V = -(1 \text{ atm})(20 \text{ L} - 0.04 \text{ L}) = -20 \text{ L atm}(\frac{101.325 \text{ J}}{1 \text{ L atm}}) = -2000 \text{ J (exoworkic)}$$

PROBLEM 2.4

Answer the following problems where a sample of gas is contained in a cylinder-and-piston arrangement.

a. Assume that the cylinder and piston are perfect insulators and they do not allow thermal energy to enter or leave the enclosed volume. What is the value for q (<0, 0, or >0) for this change? What is the value for w (<0, 0, or >0) for this change? What is the value for ΔU (<0, 0, or >0) for this change?

If the piston and cylinder are perfect insulators and do not allow thermal energy to enter or leave, then the value for q (heat) must be 0. Since the enclosed volume is becoming smaller, then $\Delta V < 0$ and $w = -p\Delta V$ must be a positive value (endoworkic). Therefore, the change in internal energy (ΔU) is equal to the sum of heat, and work ($q + w$) must also be positive (0 + a positive value) and the change is endergonic.

b. Problem 2.13. Assume that the cylinder and piston are conductors and during this state change the cylinder becomes warm to the touch. What is the value for q (<0, 0, or >0) for this change? What is the value for w (<0, 0, or >0) for this change? What is the value for ΔU (<0, 0, or >0) for this change?

If the piston and cylinder are conductors that become hot to the touch, then the value for q (heat) must be <0 (heat is leaving the system and flowing to the surroundings, i.e., exothermic). Since the enclosed volume is becoming smaller, then $\Delta V < 0$ and $w = -p\Delta V$ must be a positive value (endoworkic). The change in internal energy (ΔU) cannot be predicted in this case, at least without more information. Depending on the magnitudes of heat and work, ΔU could be < 0 (if $q > w$), 0 (if $q = w$), or > 0 (if $q < w$).

PROBLEM 2.5

For each of the following scenarios, calculate the change in internal energy (ΔU). Specify whether the system is undergoing an endergonic or an exergonic process.

a. 18.0 g of water evaporates and goes from 18.0 mL to 31 L with an external pressure of 1.0 atm.

$$n = 18.0 \text{ g}(\frac{1 \text{ kg}}{1000 \text{ g}})(\frac{1 \text{ mol}}{0.0180 \text{ kg}}) = 1.00 \text{ mol}$$

$$18.00 \text{ mL}(\frac{1 \text{ L}}{1000 \text{ mL}}) = 0.018 \text{ L}$$

$$\Delta U = n\Delta H - p\Delta V = (1.00 \text{ mol})(43.9 \text{ kJ/mol}) - [(0.985 \text{ atm})(31 \text{ L} - 0.018 \text{ L})(\frac{101.325 \text{ J}}{1 \text{ L atm}})(\frac{1 \text{ kJ}}{1000 \text{ J}})]$$
$$\Delta U = 40.8 \text{ kJ (endergonic process)}$$

b. 10.0 g of sodium chloride dissolves in water ($\Delta H = 3.9$ kJ/mol) with negligible volume change at 1.00 atm.

$$n = 10.0 \text{ g}(\frac{1 \text{ kg}}{1000 \text{ g}})(\frac{1 \text{ mol}}{0.058 \, 44 \text{ g}}) = 0.171 \text{ mol}$$

$$\Delta U = n\Delta H - p\Delta V = (0.171 \text{ mol})(3.9 \text{ kJ/mol}) - (1.00 \text{ atm})(0 \text{ L})$$
$$\Delta U = 0.67 \text{ kJ (endergonic process)}$$

c. 14.0 g nitrogen gas is cooled from 100.0 °C to –77.2 °C, where $c_p(N_2(g)) = 1.040 \, \frac{J}{g \, K}$, and changes from 15.3 L to 8.1 L at a pressure of 1.0 atm.

$$\Delta T = -77.2 \text{ °C} - 100.0 \text{ °C} = -177.2 \text{ °C} = -177.2 \text{ K}$$

$$\Delta U = mc_p\Delta T - p\Delta V = (14.0 \text{ g})(1.040 \, \frac{J}{g \, K})(-177.2 \text{ K}) - [(1.0 \text{ atm})(8.1 \text{ L} - 15.3 \text{ L})(\frac{101.325 \text{ J}}{1 \text{ L atm}})]$$
$$\Delta U = -1850 \text{ J (exergonic)}$$

d. 500.0 g of copper metal is heated from room temperature (25.0 °C) to being red hot (460 °C), where $c_p(Cu(s)) = 0.386 \, \frac{J}{g \, K}$, with negligible change in volume at 1.10 atm

$$\Delta T = 460 \text{ °C} - 25.0 \text{ °C} = 440 \text{ °C} = 440 \text{ K (with correct sig figs)}$$

$$\Delta U = mc_p\Delta T - p\Delta V = (500.0 \text{ g})(0.386 \, \frac{J}{g \, K})(435 \text{ K}) - (1.10 \text{ atm})(0 \text{ L})$$
$$\Delta U = 84 \, 000 \text{ J (endergonic)}$$

Note: The unrounded 435 K is shown in the calculation for ΔU. Best practice is to wait until the end to round a final answer. However, if an individual step is being shown, then that step's answer must be shown with correct significant figures.

CHAPTER 3 ANSWERS

PROBLEM 3.1
Early experimentation (1894) to determine the density of nitrogen gas found that nitrogen gas produced from chemical reactions was reproducibly 0.4% lighter than "atmospheric" nitrogen gas. Considering this difference, answer the following questions:

a. Why is the law of conservation of mass important in this example?

If mass were not conserved, then the difference between the density of "atmospheric" nitrogen gas and synthetic nitrogen gas could be attributed to their different methods of production. However, the law of conservation of mass means that these differences in density are important and encourage future investigation.

b. Why is precision (the number of significant figures in a measurement) important in this example?

Precision is of central importance in this example. The density of nitrogen from air was found to be 1.2583 g/L and synthetically produced nitrogen was 1.2529 g/L. The percent difference in density between the two sources of nitrogen gas is 0.4%. This small difference could only be elucidated if highly precise measurements were taken, which allow the determination of the density values to at least four significant figures.

c. Provide a testable hypothesis for what you think might be able to explain this difference in the density of chemically produced nitrogen gas and "atmospheric" nitrogen gas.

One hypothesis is that nitrogen gas from air may be a mixture of gases (nitrogen and a gas with more mass). To test this, efforts could be undertaken to separate the gaseous mixtures (Lord Rayleigh found argon gas in this way).

Another hypothesis is that the synthesis of nitrogen may not be clean, and the nitrogen gas may be a mixture of gases (nitrogen and a gas with less mass). To test this, efforts could be undertaken to separate the gaseous mixtures.

Finally, another, though by no means the last possible hypothesis, is that Lord Rayleigh's results were those of a single researcher and two experimental setups (one for atmospheric gas and one for synthetic gas). There could be a systemic error in his setup that leads to the difference in density values. If another research or another experimental setup was tried, this could validate, or invalidate, Rayleigh's results.

PROBLEM 3.2
Provide an explanation for how atomic theory supports and explains:

a. The law of conservation of mass

According to atomic theory, atoms cannot be created from nothing, and they cannot be destroyed. Since atoms are the fundamental units of mass, this means that mass can also not be created from nothing and cannot be destroyed, which means that in a normal chemical process mass is conserved.

b. The law of constant proportions

A compound consists of a given combination of atoms (H_2O, for example, in water). Since all water is a combination of two atoms of hydrogen and one atom of oxygen, then the proportion of hydrogen to oxygen must always be fixed.

c. The law of multiple proportions

Atoms combine to form compounds and whole numbers of atoms must combine, e.g., CO and CO_2. The whole numbers observed by Dalton are the result of the indivisibility of atoms.

PROBLEM 3.3
Use the mole map (Figure 3.7) to answer the following questions.

a. What is the mass (in kg) of 5.01×10^{25} carbon atoms?

Using the values from Figure 3.4 for $A_r°(E)$:

$$5.01 \times 10^{25} \text{ atoms C}\left(\frac{\text{mol}}{6.022 \times 10^{23} \text{ atom}}\right)\left(\frac{0.012 \text{ kg}}{\text{mol}}\right) = 0.998 \text{ kg C}$$

Using the values from Appendix 2 for $A_r°(E)$:

$$5.01 \times 10^{25} \text{ atoms C}\left(\frac{\text{mol}}{6.022 \times 10^{23} \text{ atom}}\right)\left(\frac{0.012\,011 \text{ kg}}{\text{mol}}\right)\left(\frac{\text{kg}}{1000 \text{ g}}\right) = 0.999 \text{ kg C}$$

b. What volume (L at STP) does 1.23 g helium gas occupy?

Using the values from Figure 3.4 for $A_r°(E)$:

$$1.23 \text{ g He}\left(\frac{\text{kg}}{1000 \text{ g}}\right)\left(\frac{\text{mol}}{0.004\,00 \text{ kg}}\right)\left(\frac{22.7 \text{ L}}{\text{mol}}\right) = 6.98 \text{ L He}$$

Using the values from Appendix 2 for $A_r°(E)$:

$$1.23 \text{ g He}\left(\frac{\text{kg}}{1000 \text{ g}}\right)\left(\frac{\text{mol}}{0.004\,0026 \text{ kg}}\right)\left(\frac{22.7 \text{ L}}{\text{mol}}\right) = 6.98 \text{ L He}$$

c. If a solution is 0.900 mol/L NaCl, how many grams of NaCl are in 500.0 mL?

Using the values from Figure 3.4 for $A_r°(E)$:

$$500.0 \text{ mL NaCl solution} \left(\frac{L}{1000 \text{ mL}}\right)\left(\frac{0.900 \text{ mol}}{L \text{ solution}}\right)\left(\frac{0.0585 \text{ kg}}{\text{mol}}\right)\left(\frac{1000 \text{ g}}{\text{kg}}\right) = 26.3 \text{ g NaCl}$$

Using the values from Appendix 2 for $A_r°(E)$:

$$500.0 \text{ mL NaCl solution} \left(\frac{L}{1000 \text{ mL}}\right)\left(\frac{0.900 \text{ mol}}{L \text{ solution}}\right)\left(\frac{0.058 \text{ 44 kg}}{\text{mol}}\right)\left(\frac{1000 \text{ g}}{\text{kg}}\right) = 26.30 \text{ g NaCl}$$

PROBLEM 3.4
If the CODATA recommended value for the charge-to-mass quotient for H^+ is 9.578 833 1560 × 10^7 C/kg, what are the two possible inferences one could make about cathode rays?

Because the value is the ratio of charge to mass, there are two possible inferences as to why the value for cathode rays is larger than the value for H^+. The first inference is that if the mass is the same (same denominator), then the magnitude of the charge (the numerator) must be larger. The second inference is that if the magnitude of charge is the same (same numerator), then the mass (the denominator) must be smaller.

PROBLEM 3.5
Following Avogadro's initial hypothesis, that the volume of gas is proportional to the number of atoms, the Avogadro constant was determined predominantly through gas studies. Millikan's value for the elementary charge ($1.592 × 10^{-19}$ C) provided a way of determining the Avogadro constant from the Faraday constant (96 485.332 12 C/mol). What is the value of the Avogadro constant using Millikan's $1.592 × 10^{-19}$ C value for the elementary charge?

$$\frac{96\ 485.332\ 12\dfrac{C}{\text{mol}}}{1.592 \times 10^{-19}\ C} = 6.061 \times 10^{23}\ \frac{1}{\text{mol}}$$

PROBLEM 3.6
Complete the following table.

Atomic Symbol	Isotope Designation	Atomic Number (Z)	Mass Number (A)	Charge Number (z)	Proton Number (Z)	Neutron Number (N)	Electron Number
^{24}Mg	magnesium-24	12	24	0	12	12	12
$^{39}K^+$	potassium-39(1+)	19	39	1+	19	20	18
$^{60}Fe^{2+}$	iron-60(2+)	26	60	2+	26	34	24
^{35}Cl	chlorine-35	17	35	0	17	18	17
$^{35}Cl^-$	chloride-35(1−)	17	35	1−	17	18	18
$^{56}Fe^{3+}$	iron-56(3+)	26	56	3+	26	30	23
$^{15}N^{3-}$	nitride-15(3−)	7	15	3−	7	8	10
$^{16}O^{2-}$	oxide-16(2−)	8	16	2−	8	8	10
$^{24}Na^+$	sodium-24(1+)	11	13	1+	11	13	10
$^{27}Al^{3+}$	aluminium-27(3+)	13	27	3+	13	14	10
^{62}Ni	nickel-62	28	62	0	28	34	28

CHAPTER 4 ANSWERS

PROBLEM 4.1

Boron has two stable nuclides, ^{10}B (m_a = 10.012 94 u) and ^{11}B (m_a = 11.009 31 u). Calculate the binding energies per nucleon (MeV/nucleon) of these two nuclei and compare their stabilities.

$$\Delta m(^{10}B) = 10.012\ 94\ u - [5(1.007\ 83\ u) + 5(1.008\ 66\ u)] = -0.069\ 51\ u$$

$$E = -0.069\ 51\ u\ c^2(\frac{931.494\ 102\ MeV}{u\ c^2}) = -64.75\ MeV$$

$$\frac{-64.75\ MeV}{10\ nucleon} = -6.475\ MeV/nucleon$$

$$\Delta m(^{11}B) = 11.009\ 31\ u - [5(1.007\ 83\ u) + 6(1.008\ 66\ u)] = -0.081\ 80\ u$$

$$E = -0.081\ 80\ u\ c^2(\frac{931.494\ 102\ MeV}{u\ c^2}) = -76.20\ MeV$$

$$\frac{-76.20\ MeV}{11\ nucleon} = -6.920\ MeV/nucleon$$

Boron-11 is more stable (has a larger binding energy) than boron-10. This is due to the greater stability of the even numbers of protons and/or neutrons over the odd numbers of protons and/or neutrons.

PROBLEM 4.2

Consider the first seven isotopes of hydrogen and their atomic masses in Table 4.1 and the five isotones (equal number of neutrons) of $N = 7$ and their atomic masses in Table 4.2.

Explain, using Table 4.1 and Table 4.2, why it is not possible to determine the mass of a neutron or the mass of a proton from these data.

If the change in mass (Δm) is considered (see added columns in Table 4.1 and Table 4.2), we can see that the change in mass is not constant. If the change in mass were constant, then the mass of a proton or neutron could potentially be determined from the atomic mass values. The change in mass is not constant, however, due to binding energy. As nucleons come together to form a nucleus, the nuclide has a mass less than the sum of its constituent particles. This mass loss is the origin of the binding energy that holds nucleons together: the more mass lost (the greater the mass defect) the greater the binding energy. As such, the mass of a proton and the mass of a neutron can only be determined from a free particle and not in a nucleus.

Table 4.1 Isotopes ($Z = 1$) of Hydrogen and Their Atomic Masses (u)

Isotope	Atomic Symbol	Atomic Mass (u)	Δm (u)
Protium	^1H	1.0078	—
Deuterium	^2H	2.0141	1.0063
Tritium	^3H	3.0161	1.0020
Hydrogen-4	^4H	4.0264	1.0103
Hydrogen-5	^5H	5.0353	1.0089
Hydrogen-6	^6H	6.0450	1.0097
Hydrogen-7	^7H	7.0527	1.0077

Table 4.2 Isotones ($N = 7$) and their atomic masses (u)

Isotope	Atomic symbol	Atomic Mass (u)	Δm (u)
Beryllium-11	^{11}Be	11.0217	–
Boron-12	^{12}B	12.0144	0.9927
Carbon-13	^{13}C	13.0034	0.9890
Nitrogen-14	^{14}N	14.0031	0.9997
Oxygen-15	^{15}O	15.0031	1.0000

PROBLEM 4.3
Consider the graph of binding energy per nucleon (Figure 4.2).

a. What does the fact that binding energy is negative mean? Could a nucleus have positive binding energy? Explain.

Binding energy is negative because it is the energy released (or given off) when a nucleus forms. The fact that energy is released tells us that the nucleus is more stable than the free nucleons are. A nucleus could never have a positive bonding energy because this would mean that the free nucleons were more stable than the nucleus and the nucleus would therefore fall apart into the free nucleons.

b. Why does binding energy increase in magnitude from $A = 1$ to $A = 62$?

Binding energy increases in magnitude (becomes more negative) from $A = 1$ to $A = 62$ because up until $A = 62$ the greater the volume of nucleons, the more stable the nucleus. This is because, within a larger volume, there are more favorable nucleon–nucleon interactions because of the nuclear force.

c. Why does binding energy decrease in magnitude from $A = 62$ to $A = 238$?

While the number of nucleon–nucleon interactions increases, the proton–proton repulsion also increases substantially. This means that for nuclides where $A > 62$ the proton–proton repulsion starts to diminish the binding energy of the nucleus.

PROBLEM 4.4
Consider Table 4.3 of nuclear binding energy values (MeV/nucleon).

Table 4.3 Binding Energy per Nucleon Values (MeV/nucleon) for Select Nuclides

Nuclide	Mass of Particles (u)	Atomic Mass (u)	Δm (u)	Binding Energy (MeV/ nucleon)
^{2}H	2.016	2.014	−0.002	−1.11
^{3}He	3.024	3.016	−0.008	−2.57
^{4}He	4.033	4.003	−0.030	−7.07
^{6}Li	6.049	6.015	−0.034	−5.33
^{7}Li	7.058	7.016	−0.042	−5.60
^{9}Be	9.075	9.012	−0.062	−6.46
^{12}C	12.099	12.000	−0.099	−7.68
^{13}C	13.108	13.003	−0.104	−7.47
^{14}C	14.116	14.003	−0.113	−7.52
^{14}N	14.115	14.003	−0.112	−7.47
^{15}N	15.124	15.000	−0.124	−7.70

a. In general, what happens to binding energy per nucleon as atomic mass increases?

In general, the binding energy per nucleon increases due to the increasing number of stabilizing nucleon–nucleon interactions.

b. What produces greater stability (higher binding energy per nucleon): having an even or an odd number of neutrons? Provide specific evidence to support your claim.

A more stable nuclide is the result of having an even number of neutrons. For example, carbon-12 (six protons and six neutrons) has a higher binding energy per nucleon than carbon-13 (six protons and seven neutrons). We can see the same pattern if we compare the binding energy of helium-3 (one neutron, lower binding energy per nucleon) versus helium-4 (two neutrons, higher energy per nucleon), lithium-6 (three neutrons, lower binding energy per nucleon) versus lithium-7 (four neutrons, higher binding energy per nucleon), and nitrogen-14 (seven neutrons, lower binding energy per nucleon) versus nitrogen-15 (eight neutrons, high binding energy per nucleon).

c. What produces greater stability (higher binding energy per nucleon): having an even or an odd number of protons? Provide specific evidence to support your claim.

This is harder to identify in the dataset provided, but it is more stable to have an even number of protons. Consider the following two pieces of evidence: carbon-14 (six protons and eight neutrons) and nitrogen-14 (seven protons and seven neutrons) are isobars (same number of nucleons), and carbon-14 has a greater binding energy per nucleon. This example, however, has both a changing number of protons and a changing number of neutrons, which is less convincing. More convincing data are the isotones carbon-13 (six protons and seven neutrons) and nitrogen-14 (seven protons and seven neutrons). They have the same binding energy per nucleon, even though carbon-13 has one fewer nucleon and based on nucleon number alone should be less stable.

d. Consider the periodic table in Figure 4.4. Can you use your answer in part c to explain the pattern observed?

Even numbers of protons are more stable than odd numbers of protons. As such, there is an even/odd striped pattern to the periodic table with even numbers of protons giving rise to more stable nuclides and odd numbers of protons having fewer stable nuclides.

PROBLEM 4.5
Look at the periodic table in Figure 4.4. Do you notice anything significant about the stability of heavy elements? Is there a clear cut-off that you may want to hold in mind as we move forward?

Yes, there is a sharp cut-off at lead ($Z = 82$). Above lead, no nuclides are stable.

PROBLEM 4.6
Beryllium-8 ($Z = 4$) is a unique example of a nuclide that undergoes α decay even though it is significantly smaller in size than lead ($Z = 82$). Provide an explanation for this anomalous α decay.

Beryllium-8 undergoes α decay because it produces two exceptionally stable helium-4 nuclei:

$$^8\text{Be} \rightarrow {}^4\text{He} + {}^4\text{He}$$

If you consider the binding energy diagram in Figure 4.2, you can see that the two helium-4 nuclei products are substantially more stable than the initial beryllium-8.

PROBLEM 4.7
For each of the following, predict the α decay product nuclide or starting nuclide.

a. $^{210}\text{Po} \rightarrow {}^{206}\text{Pb} + \alpha$

b. $^{240}\text{Pu} \rightarrow {}^{236}\text{U} + \alpha$

c. $^{220}\text{Rn} \rightarrow {}^{216}\text{Po} + \alpha$

d. $^{237}\text{Np} \rightarrow {}^{233}\text{Pa} + \alpha$

PROBLEM 4.8
Predict the β decay product nuclide or starting nuclide.

a. $^{18}\text{F} \rightarrow {}^{18}\text{O} + \beta^+ + \nu_e$

b. $^3\text{H} \rightarrow {}^3\text{He} + \beta^- + \bar{\nu}_e$

c. $^{11}\text{C} \rightarrow {}^{11}\text{B} + \beta^+ + \nu_e$

d. $^{233}\text{Th} \rightarrow {}^{233}\text{Pa} + \beta^- + \bar{\nu}_e$

e. $^{49}\text{V} + e^- \rightarrow {}^{49}\text{Ti} + \nu_e$

f. $^{14}\text{C} \rightarrow {}^{14}\text{N} + \beta^- + \bar{\nu}_e$

g. $^{133}Xe \rightarrow {}^{133}Cs + \beta^- + \bar{\nu}_e$

h. $^{99}Tc \rightarrow {}^{99}Ru + \beta^- + \bar{\nu}_e$

PROBLEM 4.9

Yttrium is a monoisotopic element, that is, it possesses only one stable nuclide.

a. Explain why yttrium-89 is the only stable nuclide of yttrium.

Yttrium-89 has an odd number of protons ($Z = 39$) and if we consider the nuclear shells, we can see that if there were one fewer neutron there would be two unpaired nucleons of equal energy that would prefer to combine and pair up (through β^+ decay). Any more neutrons would start a new neutron shell, which would leave two unpaired nucleons with a much higher energy neutron. This would undergo β^- decay, which would lower the energy of the system and pair up the unpaired nucleons.

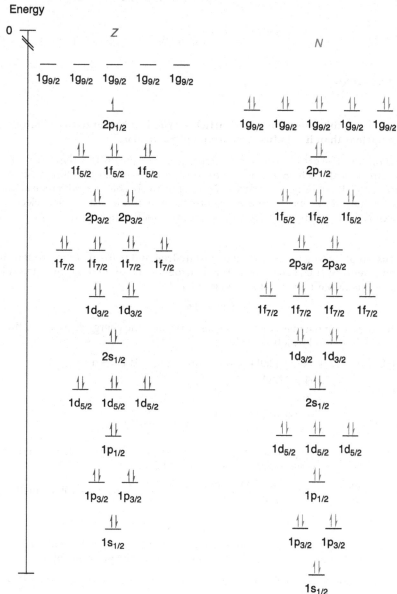

b. What type of decay is yttrium-88 most likely to undergo? Write a nuclear equation to show this decay.

See the answer to part a; yttrium-88 is most likely to undergo β^+ decay:

$$^{88}Y \rightarrow {}^{88}Sr + \beta^+ + v_e$$

c. What type of decay is yttrium-90 most likely to undergo? Write a nuclear equation to show this decay.

See the answer to part a; yttrium-90 is most likely to undergo β^- decay:

$$^{90}Y \rightarrow {}^{90}Zr + \beta^- + \bar{v}_e$$

PROBLEM 4.10
Identify the decay products for each step of the thorium series. The thorium decay series starts with thorium-232 and undergoes the following decay steps to produce lead-208: α, β^-, β^-, α, α, α, α, β^-, β^-, α.

$$^{232}Th \xrightarrow{\alpha} {}^{228}Ra \xrightarrow{\beta^-} {}^{228}Ac \xrightarrow{\beta^-} {}^{228}Th \xrightarrow{\alpha} {}^{224}Ra \xrightarrow{\alpha} {}^{220}Rn \xrightarrow{\alpha} {}^{216}Po \xrightarrow{\alpha} {}^{212}Pb \xrightarrow{\beta^-} {}^{212}Bi \xrightarrow{\beta^-} {}^{212}Po \xrightarrow{\alpha} {}^{208}Pb$$

CHAPTER 5 ANSWERS

PROBLEM 5.1
Fusion requires the combination of nuclear particles. Why is it harder to fuse a proton or α particle with a nucleus than it is to fuse a neutron with a nucleus?

Nuclei (Z+) and protons (1+) are both positively charged, which means that they are repelled from one another electrostatically, which means that fusion of a nucleus and a proton must overcome electrostatic repulsion. In addition, α particles (2+) are positively charged, which means they also suffer from electrostatic repulsion. In contrast, neutrons are neutral, so there is no electrostatic repulsion between a neutron and a nucleus, which means that fusion is easier.

PROBLEM 5.2
Consider the fusion of hydrogen-2 (2.014 10 u) and hydrogen-3 (3.016 05 u) to produce helium-4 (4.002 60 u) and a neutron (1.008 66 u). Note that hydrogen-2 is called deuterium (atomic symbol D) and hydrogen-3 is called tritium (atomic symbol T):[1]

$$T(D, n)^4He$$

a. Determine the change in mass for this process (Δm) and the energy released by this fusion process (in MeV and in kJ/mol).

$$\Delta m = 4.002\ 60\ u + 1.008\ 66\ u - (2.014\ 10\ u + 3.016\ 05\ u) = -0.018\ 89\ u$$

$$E = -0.018\ 89\ u\ c^2(\frac{931.494\ 102\ MeV}{u\ c^2}) = -17.60\ MeV$$

$$-17.60\ MeV(\frac{1 \times 10^6\ eV}{MeV})(\frac{96.485\ 332\ 1\ kJ/mol}{eV}) = -1.698 \times 10^9\ kJ/mol$$

b. If burning methane (CH_4) produces 890 kJ/mol, what mass (kg) of methane needs to be combusted to produce the same energy as the fusion of 1.0 kg tritium?

$$1.0\ kg\ T(\frac{mol}{0.003016505kg})(\frac{1.698 \times 10^9\ kJ}{mol})(\frac{mol}{890\ kJ})(\frac{0.016\ 043\ kg}{mol}) = 1.0 \times 10^{10}\ kg\ methane$$

PROBLEM 5.3
While the process for heavy elements is endergonic, fusion has allowed scientists to synthesize the transuranic (Z > 92) elements and push the boundaries of the periodic table.

The synthesis of curium-242 (242.058 83 u) results from the fusion of plutonium-239 (239.052 16 u) with an α particle (4.002 60 u).

$$^{239}Pu(\alpha, n)^{242}Cm$$

a. Determine the change in mass for this process (Δm) and the energy required to make this fusion process occur (in MeV and in kJ/mol).

Δm = 242.058 83 u +1.008 66 u − (239.052 16 u + 4.002 60 u) = 0.012 73 u

$$E = 0.012\ 73 \text{ u } c^2(\frac{931.494\ 102 \text{ MeV}}{\text{u } c^2}) = 11.86 \text{ MeV}$$

$$11.86 \text{ MeV}(\frac{1\times10^6 \text{ eV}}{\text{MeV}})(\frac{96.485\ 3321 \text{ kJ/mol}}{\text{eV}}) = 1.144 \times 10^9 \text{ kJ/mol}$$

b. What must be the velocity (m/s) of α particles (M = 0.004 002 60 kg/mol) to impart enough kinetic energy to make plutonium–α particle fusion occur (1 J = 1 (kg m²)/s²)? What percentage of the speed of light is this?

$$1.144 \times 10^9 \text{ kJ/mol}(\frac{1000 \text{ J}}{\text{kJ}}) = 1.144 \times 10^{12} \text{ J/mol}$$

$E_{k, molar} = \frac{1}{2}Mv^2$

1.144×10^{12} J/mol = ½(0.004 002 60 kg/mol)v^2

$v = 1.691 \times 10^7$ m/s

%c = (1.691 × 10⁷ m/s)/(299 792 458 m/s) = 5.369% of the speed of light

PROBLEM 5.4
Consider the fission of uranium-235 (235.04393 u) with a neutron (1.00866 u) to produce barium-144 (143.92295 u), krypton-89 (88.91763 u), and three neutrons.

a. Determine the change in mass for this process (Δm) and the energy released by this fusion process (in MeV and in kJ/mol).

Δm = 143.922 95 u + 88.917 63 u + 3(1.008 66 u) − (235.043 93 u + 1.008 66 u) = −0.186 03 u

$$E = -0.186\ 03 \text{ u } c^2(\frac{931.494\ 102 \text{ MeV}}{\text{u } c^2}) = -173.29 \text{ MeV}$$

$$-173.29 \text{ MeV}(\frac{1\times10^6 \text{ eV}}{\text{MeV}})(\frac{96.485\ 3321 \text{ kJ/mol}}{\text{eV}}) = -1.672\ 0 \times 10^{10} \text{ kJ/mol}$$

b. If burning methane (CH_4) produces 890 kJ/mol, what mass (kg) of methane needs to be combusted to produce the same energy as the fission of 1.0 kg uranium-235?

$$1.0 \text{ kg } ^{235}\text{U}(\frac{\text{mol}}{0.235\ 04393 \text{ g}})(\frac{1.6720\times10^{10} \text{ kJ}}{\text{mol}})(\frac{\text{mol}}{890 \text{ kJ}})(\frac{0.016\ 043 \text{ kg}}{\text{mol}}) = 1.3 \times 10^9 \text{ kg methane}$$

CHAPTER 6 ANSWERS

PROBLEM 6.1
Convert each wavelength value into frequency.

a. 590 nm

$$v = \frac{c}{\lambda} = \frac{299\ 792\ 458 \text{ m/s}}{590\times10^{-9} \text{ m}} = 5.1 \times 10^{14} \text{ Hz} = 510\ 000 \text{ GHz}$$

b. 100.0 mm

$$v = \frac{c}{\lambda} = \frac{299\ 792\ 458 \text{ m/s}}{0.1000 \text{ m}} = 2.998 \times 10^9 \text{ Hz} = 2.998 \text{ GHz}$$

PROBLEM 6.2
Convert each frequency value into wavelength.

a. 89.7 MHz (the frequency of WGBH in Boston)

$$\lambda = \frac{c}{\nu} = \frac{299\,792\,458\ \text{m/s}}{89.7 \times 10^6\ \text{Hz}} = 3.34\ \text{m}$$

b. 10.0 Hz

$$\lambda = \frac{c}{\nu} = \frac{299\,792\,458\ \text{m/s}}{10.0\ \text{Hz}} = 3.00 \times 10^8\ \text{m}$$

PROBLEM 6.3
Convert each wavelength value into energy (J).

a. 590 nm

$$E = h\nu = \frac{hc}{\lambda} = \frac{(6.626\,070\,15 \times 10^{-34}\ \text{Js})(299\,792\,458\ \text{m/s})}{590 \times 10^{-9}\ \text{m}} = 3.4 \times 10^{-19}\ \text{J}$$

or

$$E = h\nu = (6.626\,070\,15 \times 10^{-34}\ \text{J s})(5.1 \times 10^{14}\ \text{Hz}) = 3.4 \times 10^{-19}\ \text{J}$$

b. 100.0 mm

$$E = h\nu = \frac{hc}{\lambda} = \frac{(6.626\,070\,15 \times 10^{-34}\ \text{Js})(299\,792\,458\,\text{m/s})}{0.1000\,\text{m}} = 1.986 \times 10^{-24}\ \text{J}$$

or

$$E = h\nu = (6.626\,070\,15 \times 10^{-34}\ \text{J s})(2.998 \times 10^9\ \text{Hz}) = 1.987 \times 10^{-24}\ \text{J}$$

PROBLEM 6.4
Convert each frequency value into energy (J).

a. 89.7 MHz (the frequency of WGBH in Boston)

$$E = h\nu = (6.626\,070\,15 \times 10^{-34}\ \text{J s})(89.7 \times 10^6\ \text{Hz}) = 5.94 \times 10^{-26}\ \text{J}$$

b. 10.0 Hz

$$E = h\nu = (6.626\,070\,15 \times 10^{-34}\ \text{J s})(10.0\ \text{Hz}) = 6.63 \times 10^{-33}\ \text{J}$$

PROBLEM 6.5
On average, one square meter receives 225 kJ of solar energy every three minutes. Assuming all the light has a wavelength of 530 nm, how many photons does this correspond to?

$$\text{The energy of a photon: } E = h\nu = \frac{hc}{\lambda} = \frac{(6.626\,070\,15 \times 10^{-34}\ \text{Js})(299\,792\,458\ \text{m/s})}{530 \times 10^{-9}\ \text{m}} = 3.7 \times 10^{-19}\ \text{J}$$

Total energy 225 kJ = 225 000 J

$$225\,000\,\text{J} \left(\frac{\text{photon}}{3.7 \times 10^{-19}\ \text{J}} \right) = 6.0 \times 10^{23}\ (1\ \text{mol of photons})$$

PROBLEM 6.6
What are n_i and n_f for the other three lines in Figure 6.6?

656 nm (red) is for $n_i = 3$ and $n_f = 2$ (as seen in text).
486 nm (teal) is for $n_i = 4$ and $n_f = 2$.
434 nm (blue) is for $n_i = 5$ and $n_f = 2$.
410 nm (purple) is for $n_i = 6$ and $n_f = 2$.

PROBLEM 6.7

Calculate the wavelength in nanometers for n_i equals two and n_f equals one. Can humans see this light (humans see roughly 400 nm to 800 nm light)?

$$\frac{1}{\lambda} = 1.096\ 775\ 83 \times 10^7\ \frac{1}{m}\left(\frac{1}{2^2} - \frac{1}{1^2}\right) = -8\ 225\ 818.7\frac{1}{m}$$

$\lambda = -1.2157 \times 10^{-7}$ m (or -121.57 nm)

No, humans would not be able to see this light as it falls in the ultraviolet part of the electromagnetic spectrum.

CHAPTER 7 ANSWERS

PROBLEM 7.1

An electron has a velocity of 6.06×10^6 m/s and a mass (m_e) of $9.109\ 383\ 7015 \times 10^{-31}$ kg. What is the wavelength of an electron with this velocity? How does this wavelength compare to the van der Waals radius of a hydrogen atom (1.20×10^{-10} m)?

$$\lambda = \frac{h}{p} = \frac{6.626\ 070\ 15 \times 10^{-34}\ J\,s}{(9.109\ 383\ 7015 \times 10^{-31}\ kg)(6.06 \times 106\ m/s)} = 1.20 \times 10^{-10}\ m.\ \text{At this velocity, the electron}$$

wavelength is comparable to the radius of the hydrogen atom.

PROBLEM 7.2

An average human (89 kg) has a velocity of 1.3 m/s when walking. What is the wavelength of a human with this velocity? How does this wavelength compare to the height of the average human (1.7 m)?

$$\lambda = \frac{h}{p} = \frac{6.626\ 070\ 15 \times 10^{-34}\ J\,s}{(89\ kg)(1.3\ m/s)} = 5.73 \times 10^{-36}\ m.\ \text{At this velocity, the wavelength is insignificant}$$

in comparison to the size of a human.

PROBLEM 7.3

A neutron is found to have a de Broglie wavelength of 181 pm. What is the velocity (m/s) given that the mass of a neutron (m_n) is $1.674\ 927\ 498\ 04 \times 10^{-27}$ kg?

$$\lambda = \frac{h}{mv}\quad 181 \times 10^{-12}\ m = \frac{6.626\ 070\ 15 \times 10^{-34}\ J\,s}{(1.674\ 927\ 498\ 04 \times 10^{-27}\ kg)(v)}$$

$$v = \frac{6.626\ 070\ 15 \times 10^{-34}\ J\,s}{(1.674\ 927\ 498\ 04 \times 10^{-27}\ kg)(181 \times 10^{-12}\ m)} = 2190\ m/s$$

PROBLEM 7.4

Consider the following scenarios.

a. The energy of an electron around an atom is known with a high level of precision (a very low uncertainty or low standard deviation). Can we say where the electron is around the atom?

If we know the energy of an electron (which is related to its mass and momentum) to a high level of precision, this means that σ_p is very small. Given $\sigma_x\sigma_p \geq \dfrac{h}{4\pi}$, this would mean that there would be significant uncertainty in the position (σ_x) and we would not be able to provide any firm location for where the electron is.

b. The position of an electron around an atom is known precisely (almost no uncertainty). Can we say what the energy of that electron is?

If we know the exact position of an electron, this means that σ_x is incredibly small. Given $\sigma_x\sigma_p \geq \dfrac{h}{4\pi}$, this would mean that there would be incredible uncertainty in the momentum (σ_p) and, therefore, the energy.

PROBLEM 7.5

Consider the two scenarios in Problem 7.4. Given that there are inherent trade-offs in what we can know about the electron, which do you think is more important to know to a high level of precision: the position or the momentum/energy?

Various answers are reasonable. For the author, it would make the most sense to know the momentum/energy. Electrons are in constant motion, so knowing the exact position would provide an interesting snapshot, but it would not provide the most useful information over the long term. In contrast, knowing the energy can tell us how much energy it takes to remove, how likely the electron is to engage in bonding or chemistry, or (if the energy/momentum of multiple electrons is known) the relative energy of an electron in the configuration.

PROBLEM 7.6

Using Figure 7.4, determine the maximum number of electrons that can fit in the s, p, and d subshells.

The s subshell has one orbital, which can hold two electrons, and so the s subshell can hold, at most, two electrons.

The p subshell has three orbitals, which can hold two electrons each, and so the p subshell can hold, at most, six electrons.

The d subshell has five orbitals, which can hold two electrons each, and so the d subshell can hold, at most, ten electrons.

PROBLEM 7.7

The angular momentum quantum number (ℓ) for the f subshell is 3. Determine:

a. How many f orbitals there are.

If $\ell = 3$, then m_ℓ could have values of $-3, -2, -1, 0, 1, 2$, and 3. This would correspond to seven f orbitals.

b. The maximum number of electrons that can go into the f subshell.

Since there are seven orbitals, which can hold two electrons each, and so the f subshell can hold, at most, 14 electrons.

PROBLEM 7.8

Write the electron configuration for Be and B. Identify the number of core electrons and the number of valence electrons.

Be: $1s^2 2s^2$. There are two core electrons ($1s^2$) and two valence electrons ($2s^2$).
B: $1s^2 2s^2 2p^1$. There are two core electrons ($1s^2$) and three valence electrons ($2s^2 2p^1$).

PROBLEM 7.9

For each electron configuration identify it as ground state, excited state, or impossible. Identify what neutral atom each configuration (that is not impossible) corresponds to.

a. $1s^2 2s^2 2p^6 3s^2 3p^1$. This is a ground-state configuration, which follows the Aufbau principle (and, so far as we can tell from this notation, Hund's rule and the Pauli exclusion principle). There are 13 electrons, which would correspond to aluminium ($Z = 13$).

b. $1s^2 2s^2 2p_x^2 2p_z^1 2p_y^0$. This is an excited-state configuration that follows the Aufbau principle (and, so far as we can tell from this notation, the Pauli exclusion principle) but violates Hund's rule.

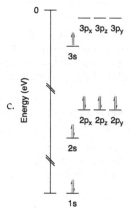

c.

This is an impossible configuration. The Pauli exclusion principle is violated because both 3s-orbital electrons have the same spin quantum number

d. $1s^2 2s^2 2p^6 3s^2 3p^6 3d^{12} 4s^2$. This is an impossible configuration. The Pauli exclusion principle is violated because there are more d-orbital electrons ($3d^{12}$) than can fit in the d subshell (10 maximum).

e. $1s^1 2s^1 2p^6$. This is an excited-state configuration that follows (so far as we can tell from this notation) the Pauli exclusion principle but violates the Aufbau principle.

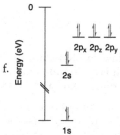

f.

This is a ground-state configuration that follows the Aufbau principle and the Pauli exclusion principle. There are ten electrons, which would correspond to neon ($Z = 13$).

PROBLEM 7.10

How would the periodic table change if there were three possible spin quantum number values for electrons ($-\frac{1}{2}$, 0, and $+\frac{1}{2}$) instead of just two ($-\frac{1}{2}$ and $+\frac{1}{2}$)?

This would allow for three electrons per orbital (rather than two). This would mean that the s subshell could hold three electrons, the p subshell could hold nine electrons, the d subshell could hold 15 electrons, and the f subshell could hold 21 electrons. This would mean that the s block would have three groups (rather than two), the p block would consist of nine groups (rather than six), the d block would contain 15 groups rather than 10, and the f block would comprise 21 groups rather than 14.

PROBLEM 7.11

Using Figures 7.11 and Figure 7.13, answer the following questions.

a. (Why do all group-3 elements commonly form 3+ ions?

All group-3 elements are unstable because the valence s and p subshells are not completely filled. There are three valence electrons more than the preceding noble gas, and so group-3 elements are most likely to lose those three electrons to achieve a stable noble gas configuration. Losing three, negatively charged, electrons would form ions with a 3+ charge.

(ii) In Chapter 3 we discussed that the lanthanoids caused a lot of frustration for Mendeleev. Why would he be led to think that they all belonged to group 3?

Mendeleev's periodic table structure was based on chemical formulae formed when the compounds formed oxides. Elements in group 3 form oxides with the general formula M_2O_3. The lanthanides all commonly form 3+ ions, which means they would form oxides with the general formula M_2O_3. This means that, per Mendeleev's schema, all the lanthanoids should be in group 3 rather than in other groups.

b. Transition metals can form several different ionic charges, but a 2+ ion is a common charge for most transition metals. Why would 2+ ions commonly form for transition metals?

Most transition metals have ns^2 as part of their electron configuration. The two valence electrons in the s subshell have the highest energy and are most likely to be lost first. This means that all transition metals that have ns^2 in their electron configuration are likely to form a 2+ ion.

PROBLEM 7.12
Write the electron configuration for each of the following ions.

a. Fe^{3+}: $1s^2 2s^2 2p^6 3s^2 3d^5$

b. P^{3-}: $1s^2 2s^2 2p^6 3s^2 3p^6$

c. Sn^{2+}: $1s^2 2s^2 2p^6 3s^2 3p^6 3d^{10} 4s^2 4p^6 4d^{10} 5s^2$

d. I^-: $1s^2 2s^2 2p^6 3s^2 3p^6 3d^{10} 4s^2 4p^6 4d^{10} 5s^2 5p^6$

e. Sc^{3+}: $1s^2 2s^2 2p^6 3s^2 3p^6$

PROBLEM 7.13
For each of the following configurations: (i) identify if it is a ground-state, excited-state, or impossible electron configuration and explain your answer; and (ii) if it is a ground-state configuration, identify whether it is a stable ground state or unstable ground state and explain.

a. $[Ne]3s^2 3p^1$ – Ground-state configuration (follows the Aufbau principle, Hund's rule, and the Pauli exclusion principle). This is an unstable, ground-state configuration because the valence s and p subshells are not filled. This atom is likely to lose three electrons and form a 3+ ion. This corresponds to aluminium, which will tend to form Al^{3+}.

b. $[Ar]3d^6 4s^2$ – Ground-state configuration (follows the Aufbau principle, Hund's rule, and the Pauli exclusion principle). This is an unstable, ground-state configuration because the valence s and p subshells are not filled. This atom is likely to lose electrons, but the number is not certain. This configuration corresponds to iron, which can lose two electrons to form a 2+ ion, three to form a 3+ ion, and four to form a 4+ ion.

c.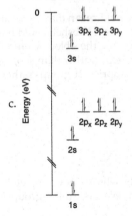

This corresponds to an excited-state configuration. It follows the Aufbau principle and the Pauli exclusion principle, but it does not follow Hund's rule.

d. $[Xe]4f^{14}5d^{10}6s^26p^5$ – Ground-state configuration (follows the Aufbau principle, Hund's rule, and the Pauli exclusion principle). This is an unstable, ground-state configuration because the valence s and p subshells are not filled. This atom is likely to gain one electron and would form a 1– ion. This configuration corresponds to astatine, which can gain one electron to form At^-.

e.

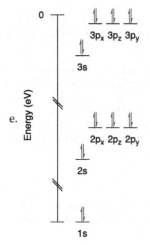

Ground-state configuration (follows the Aufbau principle, Hund's rule, and the Pauli exclusion principle). This is a stable, ground-state configuration because the valence s and p subshells are filled. This configuration corresponds to the noble gas argon, which will neither gain nor lose electrons.

f. $[Kr]4d^85s^1$ – Ground-state configuration (follows the Aufbau principle, Hund's rule, and the Pauli exclusion principle). This is an unstable, ground-state configuration because the valence s and p subshells are not filled. This atom is likely to lose electrons, but the number is not certain. This configuration corresponds to rhodium, which can lose anywhere from one to six electrons to form ions ranging from Rh^+ to Rh^{6+}.

g. $1s^22s^22p^63s^23p^64s^23d^94p^6$. This corresponds to an excited-state configuration. It follows the Pauli exclusion principle, but it does not follow the Aufbau principle.

CHAPTER 8 ANSWERS

PROBLEM 8.1
What will happen to the attractive force ($F_{electrostatic}$) between the nucleus and a valence electron as Z_{eff} increases?

$F_{electrostatic}$ and Z_{eff} are linearly related; therefore, an increase in Z_{eff} will lead to an increase in $F_{electrostatic}$, which will mean a stronger pull on the electron by the nucleus.

PROBLEM 8.2
What will happen to the force ($F_{electrostatic}$) between the nucleus and a valence electron as r increases?

$F_{electrostatic}$ and r are inversely related; therefore, an increase in r will lead to a decrease in $F_{electrostatic}$, which will mean a weaker pull on the electron by the nucleus. Because of the inverse square relationship, $F_{electrostatic}$ will decrease very quickly with distance.

PROBLEM 8.3
Considering lithium to fluorine, what is changing to cause Z_{eff} to increase?

Moving across period 2, the shell number (2) does not change, but the number of protons in the nucleus is increasing. More protons mean a greater positive charge, which is reflected in a greater value for Z_{eff}.

PROBLEM 8.4
Considering hydrogen to caesium (Figure 8.1), the change in Z_{eff} is relatively small. Why does Z_{eff} for lithium (1.26) show only a 26% increase compared to hydrogen, while Z shows a 200% increase?

Lithium has three protons, while hydrogen only has one (which means a 200% increase in Z for lithium over hydrogen). First instinct, then, is that lithium's nucleus should exert 200% more pull on its valence electron, but the actual or effective charge (Z_{eff}) for valence electrons is 1.26. This is because the core electrons ($1s^2$) are between the valence electron ($2s^1$) and the nucleus, and these core electrons shield the nucleus from the valence electron. Because the core electrons shield, the valence electron only feels the effective charge +1.26, which is significantly less than the +3 charge of the nucleus.

PROBLEM 8.5
Given the relationship between shell number and radius (Table 8.1), how does the attractive force ($F_{electrostatic}$) between the nucleus and a valence electron in shell 3 compare to the same nucleus and an electron in shell 2?

A larger shell number corresponds to a greater distance (r is proportional to n). Since r is in the denominator, $F_{electrostatic}$ and r are inversely related, and $F_{electrostatic}$ will be lower for an electron in shell 3 and greater for an electron in shell 2. This means the electron in shell 2 is more strongly attracted to the nucleus and the electron in shell 3 is more weakly attracted to the nucleus.

PROBLEM 8.6
Given the relationship between shell number and radius (Table 8.1), what would you predict for the trend in atomic size for an atom in period 1 compared to an atom in period 2 compared to an atom in period 3? Explain.

For an atom in period 1, the valence electron(s) are in shell 1. For an atom in period 2, their valence electron(s) are in shell 2. For an atom in period 3, the valence electron(s) are in shell 3. Since higher shell numbers correspond to greater distances from the nucleus, atoms in period 1 should be smaller (electrons are closer to the nucleus) than atoms in period 2. Atoms in period 2 should be smaller than atoms in period 3.

PROBLEM 8.7
Consider the elements in group 1, Figure 8.3. Using the electron configuration of each element, provide an explanation for the trend $r_{cov}(H) < r_{cov}(Li) < r_{cov}(Na) < r_{cov}(K) < r_{cov}(Rb) < r_{cov}(Cs) < r_{cov}(Fr)$.

Considering the electron configurations (here just of the first three elements):

Hydrogen: $1s^1$
Lithium: $[He]2s^1$
Sodium: $[Ne]3s^1$

As you move down group 1, the valence electron is in a successively higher shell number. Since higher shell numbers correspond to greater distances from the nucleus, atoms in period 1 should be smaller (electrons are closer to the nucleus) than atoms in period 2. Atoms in period 2 should be smaller than atoms in period 3.

PROBLEM 8.8
Consider the elements in period 4, Figure 8.3. Using the electron configuration of each element and Coulomb's law, provide an explanation for the general decrease in covalent radius from potassium to krypton.

For elements potassium (19) through krypton (36), the highest shell number is 4. The number of shells does not change. What does change is the increasing number of protons and concomitantly Z_{eff} increases, which increases the pull/force the nucleus exerts on the electrons ($F_{electrostatic}$). This means the electrons are pulled in closer to the nucleus and the radius decreases. $F_{electrostatic}$ and Z_{eff} are linearly related; therefore, an increase in Z_{eff} will lead to an increase in $F_{electrostatic}$, which will mean a stronger pull on the electrons by the nucleus and a decrease in the atomic radius.

PROBLEM 8.9
Provide an explanation for why cations are smaller than their corresponding neutral atoms. Think about how Z_{eff} and r might change in going from the neutral atom to the ion.

Compare the electron configuration of lithium ($1s^2 2s^1$) to a lithium cation ($1s^2$). The shell number (n) decreases, which means that r decreases. Z_{eff} will also be higher for the remaining electrons and so $F_{electrostatic}$ will increase. Together, this pulls the electrons closer to the nucleus (making the cation smaller than the corresponding neutral atom).

PROBLEM 8.10
Arrange the following ions in order of radius size (from smallest to largest): Ca^{2+}, Cl^-, S^{2-}, K^+. What is the determining factor (Z_{eff} or shell number) in this order?

They all have the same electron configuration, so the shell number is not the dominant factor. The only significant change is Z_{eff}. More protons will mean higher $F_{electrostatic}$ and so the order is Ca^{2+}, K^+, Cl^-, S^{2-} (see Table 8.2 for actual values).

PROBLEM 8.11
Notice that $E_i(H) < E_i(He)$. Provide an explanation for this.

He should have a higher Z_{eff}. No value is listed in Figure 8.1, but the general trend is that Z_{eff} increases across a period, which means a higher $F_{electrostatic}$ for helium than for hydrogen. This means that more energy would have to be added (greater E_i) to overcome the greater attraction between the helium nucleus and its electrons than between hydrogen and its electron.

PROBLEM 8.12
Provide an explanation for this trend in ionization: $E_i(H) > E_i(Li) > E_i(Na) > E_i(K) > E_i(Rb) > E_i(Cs)$.

Lithium's valence electron is in a higher shell (n) than hydrogen, which means the electron is farther from the nucleus (greater r) than hydrogen's electron. This corresponds to a lower $F_{electrostatic}$ and means that less energy would have to be added (lower E_i) to overcome the weaker attraction between the lithium nucleus and its valence electron than between hydrogen and its electron. This trend continues down group 1 as each successive element has its valence electron in a higher shell number.

PROBLEM 8.13
Moving from lithium to neon, we cover the entire second period of elements. Can you make a generalization for the overall trend for E_i within a period?

In general, E_i increases across a period, which follows the increase in Z_{eff} and the concomitant increase in $F_{electrostatic}$.

PROBLEM 8.14

Does the generalized trend you proposed in Problem 8.13 hold up for the other periods in Figure 8.5?

It does. Each period shows, with some irregularities, an increase in E_i across the period.

PROBLEM 8.15

If we consider E_{ea}, is there really a trend (that is, could you draw a linear line) for elements 3–10 (period 2)?

No. There is not really a trend for E_{ea} across a period. The data are nonlinear across an entire period. There is, however, a linear increase in E_{ea} within the s, p, d, and f blocks.

PROBLEM 8.16

We can, however, analyze E_{ea} values for different groups.

a. The elements of which group have the largest electron affinity values? Why are their electron affinity values so high?

The elements with the biggest E_{ea} values are the halogens (group 17). They are all only one electron away from a noble gas configuration and so they have a very high affinity for electrons as they become significantly more stable upon the addition of one electron.

b. Why do group 1 and group 11 elements have an appreciable electron affinity?

The group 1 and 11 elements have an appreciable E_{ea} because they have a nearly filled s and d subshell, respectively, and if one electron is added that would complete the subshell. Filled subshells are local energy minima (compared to the global energy minimum of a filled shell), and so group 1 and group 11 elements show appreciable electron affinity values.

c. Why do groups 2, 12, and 18 have an electron affinity of zero?

Groups 2 and 12 have filled subshells and group 18 has a filled shell. These are all energy minima and adding an electron would be destabilizing as each element would be shifted away from an energy minimum by the addition of one electron.

PROBLEM 8.17

Consider the electrostatic potential maps of lithium hydride, hydrogen, and hydrogen fluoride.

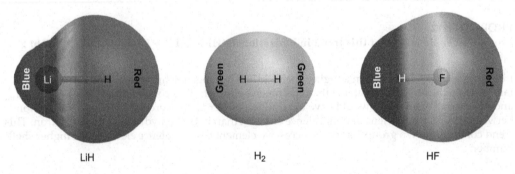

LiH   H₂   HF

Red indicates areas of high electron density, blue indicates areas of low electron density, and green indicates a middling amount of electron density.

Using the electronegativity values in Figure 8.6 and your knowledge of what electronegativity means, can you explain the differences in color for H in each of the above compounds?

The difference in the images above is due to the electronegativity of the atom H is combined with. If we look at the middle image, H is combined with H, and so there is no electronegativity difference, and they equally share the electrons (which is why the surface is green).

If we look at LiH, H (χ_P = 2.20) is *more* electronegative than Li (χ_P = 0.98) and so the H atom attracts the electrons toward itself (giving it a red color and larger size) and away from the Li atom (leaving it electron poor, smaller, and blue in the image above).

If we look at HF, H (χ_P = 2.20) is *less* electronegative than F (χ_P = 3.98) and so the F atom attracts the electrons toward itself (giving it a red color and a larger size) and away from the H atom (leaving it electron poor, smaller, and blue in the image above).

PROBLEM 8.18
Look at the electronegativity values in Figure 8.6.

a. LiF and RbCl are example ionic compounds. What is the difference in electronegativity ($\Delta\chi_P$) for Li and F? For Rb and Cl?

Li, χ_P = 0.98   Rb, χ_P = 0.82

F, χ_P = 3.98   Cl, χ_P = 3.16

$\Delta\chi_P$ = 3.00 $\Delta\chi_P$ = 2.34

b. Now consider CH_4 and N_2O_4, example molecular compounds. What is the difference in electronegativity ($\Delta\chi_P$) for C and H? For N and O?

C, χ_P = 2.55   N, χ_P = 3.04

H, χ_P = 2.20   O, χ_P = 3.44

$\Delta\chi_P$ = 0.35 $\Delta\chi_P$ = 0.40

c. Can you make any generalization about the magnitude of the electronegativity differences ($\Delta\chi_P$) for the ionic compounds versus the molecular compounds?

In ionic compounds, there is a significant difference in electronegativity ($\Delta\chi_P$ > 2.00), which leads to the electrons being attracted so strongly to the more electronegative atom that the more electronegative atom takes the electrons away from the less electronegative atom. In contrast, in molecular compounds, the difference in electronegativity is usually small to moderate ($\Delta\chi_P$ > 2.00), so electrons may be shared unequally but they are still shared.

PROBLEM 8.19
For each pair of elements, determine their electronegativity difference ($\Delta\chi_P$) and indicate which atom should have high electron density and which should have low electron density.

a. C and H $\Delta\chi_P$ __0.35_ high electron density atom_C_ low electron density atom_H_

b. Fe and H $\Delta\chi_P$ __0.37_ high electron density atom_H_ low electron density atom_Fe_

c. Br and Al $\Delta\chi_P$ __1.35_ high electron density atom_Br_ low electron density atom_Al_

d. C and F $\Delta\chi_P$ __1.43_ high electron density atom_F_ low electron density atom_C_

e. Fr and F $\Delta\chi_P$ __3.19_ high electron density atom_F_ low electron density atom_Fr_

PROBLEM 8.20
When sodium fluoride (NaF) forms from sodium and fluorine, the final compound consists of a sodium cation (Na^+) and fluoride anion (F^-). Sodium fluoride will *not* form with a sodide anion (Na^-) and fluorine cation (F^+).

a. Provide an explanation why sodium is always the cation and fluorine is always the anion in sodium fluoride based on the electronegativity values (χ) of sodium and fluorine.

Fluorine is more electronegative (χ_P = 3.98) than sodium (χ_P = 0.93), which means that fluorine has a much greater ability to attract electrons toward itself than sodium. This means that the fluorine will take an electron from sodium, which converts the fluorine into fluoride (F^-) and the sodium into a sodium cation (Na^+). For an electron to go from fluorine to sodium would require that an electron goes from a more electronegative to a more electropositive atom, which goes against the fundamental electrostatic forces at play.

b. A more rigorous consideration looks at the energetics of the processes involved. Using Figure 8.4, provide an explanation based on energy why sodium will always form a cation and fluorine will always form an anion when sodium and fluorine react together to make sodium fluoride.

If we consider the ionization energy values (Figure 8.4), we can estimate that sodium has an ionization energy of 5 eV and fluorine has an ionization energy of 17 eV. And so, it is more than three times harder to remove an electron from fluorine than it is to remove an electron from sodium.

If we consider the electron affinity values (Figure 8.4), we can estimate that sodium has an electron affinity of 0.5 eV and fluorine has an electron affinity of 4 eV. And so, it is eight times more stabilizing for fluorine to get an electron than it is for sodium.

If we look at the overall energetics of each process ($E_i - E_{ea}$), sodium and fluorine atoms combine to make Na^+ and F^- have an overall energy of 1 eV (5 eV – 4 eV). In contrast, sodium and fluorine combine to make Na^- and F^+ have an overall energy of 17 eV (17 eV – 0.5 eV). Energetically, then, Na^+ and F^- are *much* more likely to form than Na^- and F^+.

CHAPTER 9 ANSWERS

PROBLEM 9.1
Consider the lattice enthalpy values of MgO ($\Delta_{lattice}H° = -3929$ kJ/mol) and MgF$_2$ ($\Delta_{lattice}H° = -2922$ kJ/mol).

a. Provide an explanation of why magnesium oxide has a larger lattice enthalpy value than magnesium fluoride.

Both compounds contain Mg^{2+} ions but they differ in the anion. MgO has an O^{2-} anion and MgF$_2$ has an F^- anion. We can attribute the difference in lattice energy to the difference in anion charge. The greater charge of O^{2-} means a larger coulombic attraction and therefore larger lattice energy for MgO than for MgF$_2$, which has less coulombic attraction between the Mg^{2+} and each F^- ion.

b. The boiling point of MgO is 3600 °C and the boiling point of MgF$_2$ is 2260 °C. Provide an explanation for the difference in boiling points.

Boiling an ionic compound requires that the ions separate so that they can gasify. The greater lattice energy for MgO means that there is a greater amount of energy required to separate the Mg^{2+} and O^{2-} ions than it takes to separate the Mg^{2+} and F^- ions. Since it takes more energy to separate the Mg^{2+} and O^{2-} ions, this requires a higher temperature (greater thermal energy).

PROBLEM 9.2
F, Cl, Br, and I are called halogens, which means "salt maker" in Greek. These are so named because halogens typically react to make ionic compounds (salts). Salts that contain Cl^-, Br^-, or I^- are very soluble in water, while salts that contain F^- are usually insoluble in water. In the context of lattice energy, explain why salts with fluoride are typically insoluble in water.

The halide ions all have the same magnitude of charge (1–) but differ in terms of their ionic radii. Fluoride is the smallest halide, which would have the smallest distance between ions (r) for any of the halogens, which would therefore have the strongest coulombic attraction and the greatest lattice energy. Ionic compounds with fluoride, then, would be less soluble than those containing chloride, bromide, or iodide, which are all large ions with weaker coulombic attraction and weaker lattice energy.

PROBLEM 9.3

Ionic liquids are ionic compounds whose melting point is at or below room temperature. Below is an example compound that is an ionic liquid (its melting point is –71 °C). Propose an explanation of why this ionic compound, 1-butyl-3-methylimidazolium tetrafluoroborate, has such a low melting point.

1-butyl-3-methylimdazolium tetrafluoroborate

The polyatomic ions in 1-butyl-3-methylimdazolium tetrafluoroborate are substantially larger in size than the monatomic ions in a compound like NaCl. The substantially larger size and distance between the charged centers of the molecules would lead to a weaker coulombic attraction, a weaker lattice energy, and a lower melting point.

PROBLEM 9.4

You find two bottles (bottle A and bottle B) labeled as chromium chloride. Remembering the nomenclature from chemistry class, you know that chromium chloride could mean chromium(II) chloride ($CrCl_2$) or chromium(III) chloride ($CrCl_3$). You determine the melting point of the chemical in each bottle. Bottle A's compound melts at 1152 °C and bottle B's compound melts at 824 °C. Using this information and your knowledge of ionic compounds, identify which compound ($CrCl_2$ or $CrCl_3$) is in each bottle and briefly explain your reasoning.

Both compounds contain Cl^- ions but they differ in the cation. $CrCl_2$ has a Cr^{2+} cation and $CrCl_3$ has a Cr^{3+} cation. We can attribute the difference in melting point to differences in lattice energy, which stem from a difference in cation charge. The greater charge of Cr^{3+} means a larger coulombic attraction and therefore larger lattice energy and higher melting point for $CrCl_3$ than for $CrCl_2$, which has less coulombic attraction between the Cr^{2+} and each Cl^- ion.

PROBLEM 9.5

Draw Lewis structures for each of the following chemical formulae: CH_4, NH_3, HF, NH_2^-, NH_4^+, HS^-, BH_4^-.

CH_4 The carbon atom has four valence electrons: $1C(4 e^-)= 4 e^-$
     The hydrogen atoms each have one valence electron: $+4H(1 e^-)= 4 e^-$
     The total number of valence electrons is: $8 e^-$ total

     The carbon atom needs eight electrons: $1C(8 e^-) = 8 e^-$
     Each hydrogen atom needs two electrons: $+4H(2 e^-) = 8 e^-$
     The total number of electrons required is: $16 e^-$ needed

$$16 e^- \text{ needed}$$
$$\underline{-8 e^- \text{ total}}$$
$$8 e^- \text{ shared (or four bonds)}$$

$$8 e^- \text{ total}$$
$$\underline{-8 e^- \text{ shared}}$$
$$0 e^- \text{ unshared}$$

Lewis structure

```
    H
    |
H - C - H
    |
    H
```

NH_3 The nitrogen atom has five valence electrons: $1N(5\ e^-) = 5\ e^-$
    The hydrogen atoms each have one valence electron: $+3H(1\ e^-) = 3\ e^-$
    The total number of valence electrons is: $\overline{8\ e^-\ \text{total}}$

    The carbon atom needs eight electrons: $1N(8\ e^-) = 8\ e^-$
    Each hydrogen atom needs two electrons: $+3H(2\ e^-) = 6\ e^-$
    The total number of electrons required is: $\overline{14\ e^-\ \text{needed}}$

$$14\ e^-\ \text{needed}$$
$$\underline{-\ 8\ e^-\ \text{total}}$$
$$6\ e^-\ \text{shared (or three bonds)}$$

$$8\ e^-\ \text{total}$$
$$\underline{-\ 6\ e^-\ \text{shared}}$$
$$2\ e^-\ \text{unshared}$$

Lewis structure

H–N̈–H
|
H

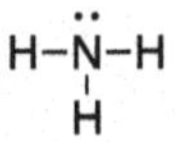

HF The fluorine atom has seven valence electrons: $1F(7\ e^-) = 7\ e^-$
    The hydrogen atom has one valence electron: $+1H(1\ e^-) = 1\ e^-$
    The total number of valence electrons is: $\overline{8\ e^-\ \text{total}}$

    The fluorine atom needs eight electrons: $1F(8\ e^-) = 8\ e^-$
    The hydrogen atom needs two electrons: $+1H(2\ e^-) = 2\ e-$
    The total number of electrons required is: $\overline{10\ e^-\ \text{needed}}$

$$10\ e^-\ \text{needed}$$
$$\underline{-\ 8\ e^-\ \text{total}}$$
$$2\ e^-\ \text{shared (or one bond)}$$

$$8\ e^-\ \text{total}$$
$$\underline{-\ 2\ e^-\ \text{shared}}$$
$$6\ e^-\ \text{unshared}$$

Lewis structure

H–F̈:

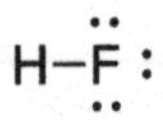

NH_2^- The nitrogen atom has five valence electrons: $1N(5\ e^-) = 5\ e^-$
    Each hydrogen atom has one valence electron: $2H(1\ e^-) = 2\ e^-$
    The negative charged adds one valence electron: $+1(-) = 1\ e^-$
    The total number of valence electrons is: $\overline{8\ e^-\ \text{total}}$

    The nitrogen atom needs eight electrons: $1N(8\ e^-) = 8\ e^-$
    Each hydrogen atom needs two electrons: $+2H(2\ e^-) = 4\ e^-$
    The total number of electrons required is: $\overline{12\ e^-\ \text{needed}}$

$$12\ e^-\ \text{needed}$$
$$\underline{-\ 8\ e^-\ \text{total}}$$
$$4\ e^-\ \text{shared (or two bonds)}$$

$$8\ e^-\ \text{total}$$
$$\underline{-\ 4\ e^-\ \text{shared}}$$
$$4\ e^-\ \text{unshared}$$

Lewis structure

H–N̈–H ⌐ ⁻
    ⋅⋅

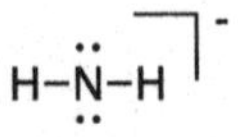

NH_4^+ The nitrogen atom has five valence electrons: $1N(5\ e^-) = 5\ e^-$
     Each hydrogen atom has one valence electron: $+4H(1\ e^-) = 4\ e^-$
     The positive charged removes one valence electron: $1(+)$ $= -1\ e^-$
     The total number of valence electrons is: $8\ e^-$ total

     The nitrogen atom needs eight electrons: $1N(8\ e^-) = 8\ e^-$
     Each hydrogen atom needs two electrons: $+4H(2\ e^-) = 8\ e^-$
     The total number of electrons required is: $16\ e^-$ needed

                                                   $16\ e^-$ needed
                                                   $-\ 8\ e^-$ total
                                                   $8\ e^-$ shared (or four bonds)

                                                   $8\ e^-$ total
                                                   $-\ 8\ e^-$ shared
                                                   $0\ e^-$ unshared

Lewis structure

```
      H   ⌝ +
      |
  H − N − H
      |
      H
```

HS^- The sulfur atom has six valence electrons: $1S(6\ e^-) = 6\ e^-$
     The hydrogen atom has one valence electron: $1H(1\ e^-) = 1\ e^-$
     The negative charged adds one valence electron: $+1(-)$ $= 1\ e^-$
     The total number of valence electrons is: $8\ e^-$ total

     The sulfur atom needs eight electrons: $1S(8\ e^-) = 8\ e^-$
     The hydrogen atom needs two electrons: $+1H(2\ e^-) = 2\ e^-$
     The total number of electrons required is: $10\ e^-$ needed

                                                   $10\ e^-$ needed
                                                   $-\ 8\ e^-$ total
                                                   $2\ e^-$ shared (or one bond)

                                                   $8\ e^-$ total
                                                   $-\ 2\ e^-$ shared
                                                   $6\ e^-$ unshared

Lewis structure

```
     ..  ⌝ -
 H − S :
     ..
```

BH_4^- The boron atom has three valence electrons: $1B(3\ e^-) = 3\ e^-$
     Each hydrogen atom has one valence electron: $4H(1\ e^-) = 4\ e^-$
     The negative charged adds one valence electron: $+1(-)$ $= 1\ e^-$
     The total number of valence electrons is $8\ e^-$ total

     The boron atom needs eight electrons: $1B(8\ e^-) = 8\ e^-$
     Each hydrogen atom needs two electrons: $+4H(2\ e^-) = 8\ e^-$
     The total number of electrons required is: $16\ e^-$ needed

                                                   $16\ e^-$ needed
                                                   $-\ 8\ e^-$ total
                                                   $8\ e^-$ shared (or four bonds)

                                                   $8\ e^-$ total
                                                   $-\ 8\ e^-$ shared
                                                   $0\ e^-$ unshared

Lewis structure

$$
\begin{array}{c}
\text{H} \\
| \\
\text{H}-\overset{|}{\underset{|}{\text{B}}}-\text{H} \\
| \\
\text{H}
\end{array}
\quad \rceil^{\,-}
$$

CH₃ The carbon atom has four valence electrons:

The hydrogen atoms each have one valence electron:

The total number of valence electrons is:

$$
\begin{array}{r}
1C(4\ e^-) = 4\ e^- \\
+3H(1\ e^-) = 3\ e^- \\
\hline
7\ e^-\ \text{total}
\end{array}
$$

The carbon atom needs eight electrons:

Each hydrogen atom needs two electrons:

The total number of electrons required is:

$$
\begin{array}{r}
1C(8\ e^-) = 8\ e^- \\
+3H(2\ e^-) = 6\ e^- \\
\hline
14\ e^-\ \text{needed}
\end{array}
$$

$$
\begin{array}{r}
14\ e^-\ \text{needed} \\
-7\ e^-\ \text{total} \\
\hline
7\ e^-\ \text{shared (or 3.5 bonds)}
\end{array}
$$

$$
\begin{array}{r}
7\ e^-\ \text{total} \\
-7\ e^-\ \text{shared} \\
\hline
0\ e^-\ \text{unshared}
\end{array}
$$

Lewis structure

$$
\begin{array}{c}
\overset{\displaystyle \cdot}{} \\
\text{H}-\overset{\cdot}{\underset{|}{\text{C}}}-\text{H} \\
| \\
\text{H}
\end{array}
$$

PROBLEM 9.6

For each of the following molecules, calculate the formal charge on each atom and indicate any formal charges by appropriately drawing in the formal charge(s).

$$
\begin{array}{c}
\text{H} \\
| \\
\text{H}-\overset{\ominus}{\underset{|}{\text{B}}}-\text{H} \\
| \\
\text{H}
\end{array}
\qquad
\begin{array}{c}
\text{H} \\
| \\
\text{H}-\overset{|}{\underset{|}{\text{C}}}-\text{H} \\
| \\
\text{H}
\end{array}
\qquad
\begin{array}{c}
\text{H} \\
| \\
\text{H}-\overset{\oplus}{\underset{|}{\text{N}}}-\text{H} \\
| \\
\text{H}
\end{array}
$$

PROBLEM 9.7

For the following series of Lewis structures drawn for the formula CH₄O, correctly assign all formal charges to each atom and then choose which Lewis structure is the best depiction of the molecule.

$$
\begin{array}{c}
\text{H} \\
\overset{2\ominus}{} \quad | \quad \overset{2\oplus}{} \\
\text{H}-\text{C}-\text{O}-\text{H} \\
\overset{\cdot\cdot}{} \quad | \\
\text{H}
\end{array}
\qquad
\left(
\begin{array}{c}
\text{H} \\
| \\
\text{H}-\text{C}-\overset{\cdot\cdot}{\underset{\cdot\cdot}{\text{O}}}-\text{H} \\
| \\
\text{H}
\end{array}
\right)
\qquad
\begin{array}{c}
\overset{\ominus}{\overset{\cdot\cdot}{}} \quad \overset{\oplus}{\overset{\cdot\cdot}{}} \\
\text{H}-\text{C}-\text{O}-\text{H} \\
| \quad\ | \\
\text{H} \quad \text{H}
\end{array}
$$

best, fewest formal charges

PROBLEM 9.8

Draw a Lewis structure for each of the following. Include all nonzero formal charges: CO, HCN, CH₂O, SOCl₂.

CO The carbon atom has four valence electrons:

The oxygen atom has six valence electrons:

The total number of valence electrons is:

$$
\begin{array}{r}
1C(4\ e^-) = 4\ e- \\
+1(O)(6\ e^-) = 6\ e^- \\
\hline
10\ e^-\ \text{total}
\end{array}
$$

The carbon atom needs eight electrons

The oxygen atom needs eight electrons:

The total number of valence electrons is:

$$
\begin{array}{r}
1C(8\ e^-) = 8\ e^- \\
+1(O)(8\ e^-) = 8\ e^- \\
\hline
16\ e^-\ \text{needed}
\end{array}
$$

$$16 \ e^- \ \text{needed}$$
$$\underline{- 10 \ e^- \ \text{total}}$$
$$6 \ e^- \ \text{shared (or three}$$

bonds)

$$26 \ e^- \ \text{total}$$
$$\underline{- 6 \ e^- \ \text{shared}}$$
$$20 \ e^- \ \text{unshared}$$

Lewis structure

$$: \overset{\ominus}{C} \equiv \overset{\oplus}{O} :$$

HCN The hydrogen atom has one valence electron: $1H(1 \ e^-) = 1 \ e^-$
The carbon atom has four valence electrons: $1C(4 \ e^-) = 4 \ e^-$
The nitrogen atom has five valence electrons: $\underline{+1N(5 \ e^-) = 5 \ e^-}$
The total number of valence electrons is: $10 \ e^- \ \text{total}$

The hydrogen atom needs two electrons: $1H(2 \ e^-) = 2 \ e^-$
The carbon atom needs eight electrons: $1C(8 \ e^-) = 8 \ e^-$
The nitrogen atom needs eight electrons: $\underline{+1N(8 \ e^-) = 8 \ e^-}$
The total number of valence electrons is: $18 \ e^- \ \text{needed}$

$$18 \ e^- \ \text{needed}$$
$$\underline{- 10 \ e^- \ \text{total}}$$
$$8 \ e^- \ \text{shared (or four bonds)}$$

$$10 \ e^- \ \text{total}$$
$$\underline{- 8 \ e^- \ \text{shared}}$$
$$2 \ e^- \ \text{unshared}$$

Lewis structure

$$\text{H--C} \equiv \text{N} :$$

CH$_2$O Each hydrogen atom has one valence electron: $2H(1 \ e^-) = 2 \ e^-$
The carbon atom has four valence electrons: $1C(4 \ e^-) = 4 \ e^-$
The oxygen atom has six valence electrons: $\underline{+1O(6 \ e^-) = 6 \ e^-}$
The total number of valence electrons is: $12 \ e^- \ \text{total}$

Each hydrogen atom needs two electrons: $2H(2 \ e^-) = 4 \ e^-$
The carbon atom needs eight electrons: $1C(8 \ e^-) = 8 \ e^-$
The oxygen atom needs eight electrons: $\underline{+1O(8 \ e^-) = 8 \ e^-}$
The total number of valence electrons is: $20 \ e^- \ \text{needed}$

$$20 \ e^- \ \text{needed}$$
$$\underline{- 12 \ e^- \ \text{total}}$$
$$8 \ e^- \ \text{shared (or four bonds)}$$

$$12 \ e^- \ \text{total}$$
$$\underline{- 8 \ e^- \ \text{shared}}$$
$$4 \ e^- \ \text{unshared}$$

Lewis structure

$$\overset{\displaystyle ..}{\underset{\displaystyle H \diagdown \underset{\displaystyle H}{C}}{\overset{\displaystyle \|}{O} :}}$$

SOCl$_2$ The sulfur atom has six valence electrons: $1S(6 \ e^-) = 6 \ e^-$
The oxygen atom has six valence electrons: $1(O)(6 \ e^-) = 6 \ e^-$
Each chlorine atom has seven valence electrons: $\underline{+2Cl(7 \ e^-) = 14 \ e^-}$
The total number of valence electrons is: $26 \ e^- \ \text{total}$

The sulfur atom needs eight electrons: 1S(8 e⁻) = 8 e⁻

The oxygen atom needs eight electrons: 1(O)(8 e⁻) = 8 e⁻

Each chlorine atom needs eight electrons: +2Cl(8 e⁻) = 16 e⁻

The total number of valence electrons is: 32 e⁻ needed

$$32 \text{ e}^- \text{ needed}$$
$$\underline{-26 \text{ e}^- \text{ total}}$$
$$6 \text{ e}^- \text{ shared (or three bonds)}$$

$$26 \text{ e}^- \text{ total}$$
$$\underline{-6 \text{ e}^- \text{ shared}}$$
$$20 \text{ e}^- \text{ unshared}$$

Lewis structure

$$:\overset{\cdot\cdot}{\underset{\oplus}{\overset{\ominus}{\text{O}}}}:$$
$$:\overset{\cdot\cdot}{\underset{\cdot\cdot}{\text{Cl}}}-\overset{|}{\underset{\oplus}{\text{S}}}-\overset{\cdot\cdot}{\underset{\cdot\cdot}{\text{Cl}}}:$$

PROBLEM 9.9

For each of the following chemical formulae, draw all contributing structures.

O_3 Each oxygen atom has six valence electrons: $\underline{3(O)6 \text{ e}^- = 18 \text{ e}^-}$

     The total number of valence electrons is: 18 e⁻ total

     Each oxygen atom needs eight electrons: $\underline{3(O) \text{ } 8 \text{ e}^- = 24 \text{ e}^-}$

     The total number of electrons required is: 24 e⁻ needed

$$24 \text{ e}^- \text{ needed}$$
$$\underline{-18 \text{ e}^- \text{ total}}$$
$$6 \text{ e}^- \text{ shared (or three bonds)}$$

$$18 \text{ e}^- \text{ total}$$
$$\underline{-6 \text{ e}^- \text{ shared}}$$
$$12 \text{ e}^- \text{ unshared}$$

Lewis structure

$$:\text{O}=\overset{}{\underset{\oplus}{\text{O}}}-\overset{\cdot\cdot}{\underset{\cdot\cdot}{\text{O}}}\overset{\ominus}{:} \qquad \longleftrightarrow \qquad \overset{\ominus}{:}\overset{\cdot\cdot}{\underset{\cdot\cdot}{\text{O}}}-\overset{}{\underset{\oplus}{\text{O}}}=\text{O}:$$

BF_3 The boron atom has three valence electrons: 1B(3 e⁻) = 3 e⁻

     Each fluorine atom has seven valence electrons: $\underline{+3F(7 \text{ e}^-) = 21 \text{ e}^-}$

     The total number of valence electrons is: 24 e⁻ total

     The boron atom needs eight electrons: 1B(8 e⁻) = 8 e⁻

     Each fluorine atom needs eight electrons: $\underline{+3F(8 \text{ e}^-) = 24 \text{ e}^-}$

     The total number of electrons required is: 32 e⁻ needed

$$32 \text{ e}^- \text{ needed}$$
$$\underline{-24 \text{ e}^- \text{ total}}$$
$$8 \text{ e}^- \text{ shared (or four bonds)}$$

$$24 \text{ e}^- \text{ total}$$
$$\underline{-8 \text{ e}^- \text{ shared}}$$
$$16 \text{ e}^- \text{ unshared}$$

Lewis structure

$$\overset{\oplus}{:}\overset{..}{F}=\overset{..}{\underset{:}{B}}-\overset{..}{\underset{..}{F}}:\quad\longleftrightarrow\quad\overset{\oplus..}{:}\overset{..}{\underset{:}{F}}\overset{..}{=}\overset{..}{\underset{\oplus}{B}}-\overset{..}{\underset{..}{F}}:\quad\longleftrightarrow\quad :\overset{..}{\underset{..}{F}}-\overset{..}{\underset{\ominus}{B}}=\overset{..}{F}:$$

N$_2$O Each nitrogen atom has five valence electrons $2N(5\ e^-) = 10\ e^-$
The oxygen atom has six valence electrons: $+1(O)6\ e^- = \ \ 6\ e^-$
The total number of valence electrons is $\overline{\qquad 16\ e^-\ \text{total}}$

Each nitrogen atom needs eight electrons: $2N(8\ e^-) = 16\ e^-$
The oxygen atom needs eight electrons: $+1(O)8\ e^- = \ \ 8\ e^-$
The total number of electrons required is: $\overline{\qquad 24\ e^-\ \text{needed}}$

$$\begin{array}{r} 24\ e^-\ \text{needed} \\ -16\ e^-\ \text{total} \\ \hline 8\ e^-\ \text{shared (or four bonds)} \end{array}$$

$$\begin{array}{r} 16\ e^-\ \text{total} \\ -8\ e^-\ \text{shared} \\ \hline 8\ e^-\ \text{unshared} \end{array}$$

Lewis structure

$$:N\equiv\overset{\oplus}{N}-\overset{..\ominus}{\underset{..}{O}}:\quad\longleftrightarrow\quad \overset{\ominus..}{:}\overset{..}{N}=\overset{\oplus}{N}=\overset{..}{\underset{..}{O}}:\quad\longleftrightarrow\quad \overset{\circleddash..}{:}\overset{..}{\underset{..}{N}}-\overset{\oplus}{N}\equiv\overset{\oplus}{O}:$$

COFCl The carbon atom has four valence electrons: $1C(4\ e^-) = \ \ 4\ e^-$
The chlorine atom has seven valence electrons: $1Cl(7\ e^-) = 7\ e^-$
The fluorine atom has seven valence electrons: $1F(7\ e^-) = \ \ 7\ e^-$
The oxygen atom has six valence electrons: $+1(O)6\ e^- = 6\ e^-$
The total number of valence electrons is: $\overline{\qquad 24\ e^-\ \text{total}}$

The carbon atom needs eight electrons: $1C(8\ e^-) = \ \ 8\ e^-$
The chlorine atom needs eight electrons: $1Cl(8\ e^-) = 8\ e^-$
The fluorine atom needs eight electrons: $1F(8\ e^-) = \ \ 8\ e^-$
The oxygen atom needs eight electrons: $+1(O)8\ e^- = 8\ e^-$
The total number of electrons required is: $\overline{\qquad 32\ e^-\ \text{needed}}$

$$\begin{array}{r} 32\ e^-\ \text{needed} \\ -24\ e^-\ \text{total} \\ \hline 8\ e^-\ \text{shared (or four bonds)} \end{array}$$

$$\begin{array}{r} 24\ e^-\ \text{total} \\ -8\ e^-\ \text{shared} \\ \hline 16\ e^-\ \text{unshared} \end{array}$$

Lewis structure

$$\overset{\oplus}{:}\overset{..\ominus}{\underset{..}{O}}{:}\ \ \overset{..\ominus}{\underset{..}{F}}=\overset{..}{\underset{..}{C}}-\overset{..}{\underset{..}{Cl}}:\quad\longleftrightarrow\quad :\overset{..}{\underset{..}{F}}-\overset{..}{\underset{..}{C}}-\overset{..}{\underset{..}{Cl}}:\quad\longleftrightarrow\quad :\overset{..}{\underset{..}{F}}-\overset{..}{\underset{..}{C}}=\overset{..}{\underset{\oplus}{Cl}}:$$

PROBLEM 9.10

Draw the Lewis structures for each of the following chemical formulae: SO_3, PF_5, SF_6, PO_4^{3-}, ClO_2^-, and I_3^-. Check the Lewis structures you have drawn in this problem against the results found online, where hypervalent structures are almost exclusively presented.

SO_3 The sulfur atom has six valence electrons:
  Each oxygen atom has six valence electrons:
  The total number of valence electrons is:

$$\begin{array}{ll} 1S(6\ e^-) = & 6\ e^- \\ +3(O)6\ e^- = & 18\ e^- \\ \hline & 24\ e^-\ \text{total} \end{array}$$

  The sulfur atom needs eight electrons:
  Each oxygen atom needs eight electrons:
  The total number of electrons required is:

$$\begin{array}{ll} 1S(8\ e^-) = & 8\ e^- \\ +3(O)8\ e^- = & 24\ e^- \\ \hline & 32\ e^-\ \text{needed} \end{array}$$

$$\begin{array}{l} 32\ e^-\ \text{needed} \\ -24\ e^-\ \text{total} \\ \hline 8\ e^-\ \text{shared (or four bonds)} \end{array}$$

$$\begin{array}{l} 24\ e^-\ \text{total} \\ -8\ e^-\ \text{shared} \\ \hline 16\ e^-\ \text{unshared} \end{array}$$

Lewis structure:

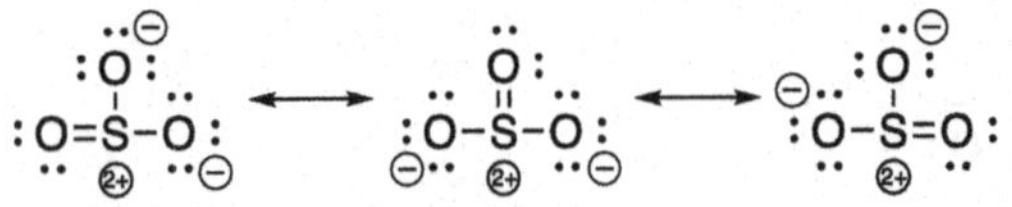

Hypervalent structure:

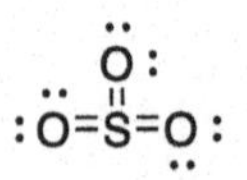

PF_5 The phosphorus atom has five valence electrons:
  Each fluorine atom has seven valence electrons:
  The total number of valence electrons is:

$$\begin{array}{ll} 1P(5\ e^-) = & 5\ e^- \\ +5F(7\ e^-) = & 35\ e^- \\ \hline & 40\ e^-\ \text{total} \end{array}$$

  The phosphorus atom needs eight electrons:
  Each fluorine atom needs eight electrons:
  The total number of electrons required is:

$$\begin{array}{ll} 1P(8\ e^-) = & 8\ e^- \\ +5F(8\ e^-) = & 40\ e^- \\ \hline & 48\ e^-\ \text{needed} \end{array}$$

$$\begin{array}{l} 48\ e^-\ \text{needed} \\ -40\ e^-\ \text{total} \\ \hline 8\ e^-\ \text{shared (or four bonds)} \end{array}$$

$$\begin{array}{l} 40\ e^-\ \text{total} \\ -8\ e^-\ \text{shared} \\ \hline 32\ e^-\ \text{unshared} \end{array}$$

Lewis structure

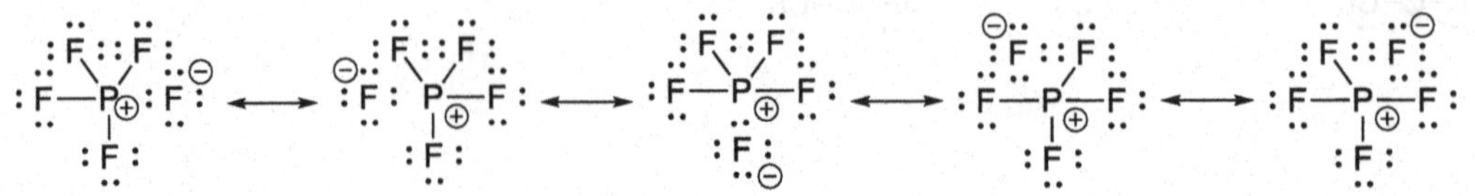

Hypervalent structure:

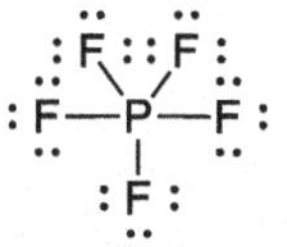

SF$_6$ The sulfur atom has six valence electrons:

Each fluorine atom has seven valence electrons:

The total number of valence electrons is:

$$1S(8\ e^-) = \ \ 6\ e^-$$
$$+6F(7\ e^-) = 42\ e^-$$
$$48\ e^-\ \text{total}$$

The sulfur atom needs eight electrons:

Each fluorine atom needs eight electrons:

The total number of electrons required is:

$$1S(8\ e^-) = \ \ 8\ e^-$$
$$+6F(8\ e^-) = 48\ e^-$$
$$56\ e^-\ \text{needed}$$

$$56\ e^-\ \text{needed}$$
$$-48\ e^-\ \text{total}$$
$$8\ e^-\ \text{shared (or four bonds)}$$

$$48\ e^-\ \text{total}$$
$$-8\ e^-\ \text{shared}$$
$$40\ e^-\ \text{unshared}$$

Lewis structure (note this has many contributing structures):

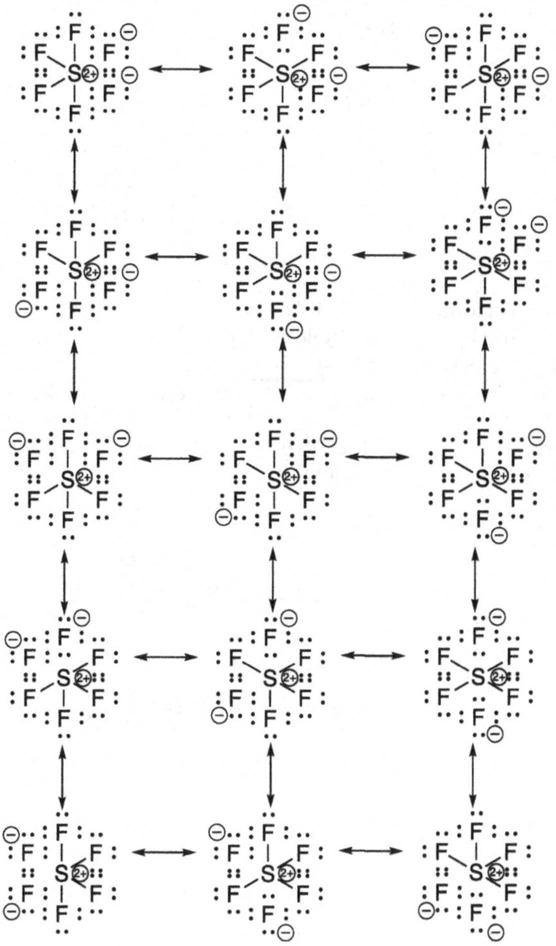

Hypervalent structure:

$$
\begin{array}{c}
\ddots \overset{\cdot\cdot}{\underset{\cdot\cdot}{:F:}}\cdot\cdot \\
:F \diagdown \mid \diagup F: \\
:: \diagup S \diagdown :: \\
:F \diagup \mid \diagdown F: \\
\ddots :\overset{}{\underset{\cdot\cdot}{F}}:\ddots
\end{array}
$$

$PO_4{}^{3-}$ The phosphorus atom has five valence electrons $1P(5\ e^-) =\ \ 5\ e^-$
Each oxygen atom has six valence electrons: $4(O)6\ e^- = 24\ e^-$
Each negative charged adds one valence electron: $+3(-)\quad\ \ = \ \ 3\ e^-$
The total number of valence electrons is: $32\ e^-$ total

The phosphorus atom needs eight electrons: $1P(8\ e^-) =\ \ 8\ e^-$
Each oxygen atom needs eight electrons: $+4(O)8\ e^- = 32\ e^-$
The total number of electrons required is: $40\ e^-$ needed

$$
\begin{array}{r}
40\ e^-\ \text{needed}\\
-32\ e^-\ \text{total}\\
\hline
8\ e^-\ \text{shared (or four bonds)}
\end{array}
\qquad
\begin{array}{r}
32\ e^-\ \text{total}\\
-8\ e^-\ \text{shared}\\
\hline
24\ e^-\ \text{unshared}
\end{array}
$$

Lewis structure:

$$
\overset{\ominus}{}\!\!\cdots\!\!\begin{array}{c}\overset{\cdot\cdot}{:O:}\\\parallel\\:O\!-\!P\!-\!O:\\\mid\\\underset{\cdot\cdot\ominus}{:O:}\end{array}\!\!\cdots\overset{\ominus}{}
\longleftrightarrow
\cdots
\longleftrightarrow
\cdots
\longleftrightarrow
\cdots
$$

Hypervalent structure:

$$
\overset{\ominus}{}\!\!\cdots\!\!\begin{array}{c}\overset{\cdot\cdot\ominus}{:O:}\\\mid\\:O\!-\!\overset{\oplus}{P}\!-\!O:\\\mid\\\underset{\cdot\cdot\ominus}{:O:}\end{array}\!\!\cdots\overset{\ominus}{}
$$

$ClO_2{}^-$ The chlorine atom has seven valence electrons: $1Cl(7\ e^-) =\ \ 7\ e^-$
Each oxygen atom has six valence electrons: $2(O)6\ e- = 12\ e^-$
Each negative charged adds one valence electron: $+1(-)\qquad = \ \ 1\ e^-$
The total number of valence electrons is: $20\ e^-$ total

The chlorine atom needs eight electrons: $1Cl(8\ e^-) =\ \ 8\ e^-$
Each oxygen atom needs eight electrons: $+2(O)8\ e^- = 16\ e^-$
The total number of electrons required is: $24\ e^-$ needed

$$
\begin{array}{r}
24\ e^-\ \text{needed}\\
-20\ e^-\ \text{total}\\
\hline
4\ e^-\ \text{shared (or two bonds)}
\end{array}
$$

$$
\begin{array}{r}
20\ e^-\ \text{total}\\
-4\ e^-\ \text{shared}\\
\hline
16\ e^-\ \text{unshared}
\end{array}
$$

Lewis structure:

$$
\overset{\ominus}{}\!\!\cdots\ \ \overset{\oplus}{\underset{\cdot\cdot}{:O\!-\!Cl\!-\!O:}}\!\!\cdots\overset{\ominus}{}
$$

Hypervalent structure:

$$
\overset{\ominus}{}\!\!\cdots\!\!:O\!-\!\overset{\cdot\cdot}{Cl}\!=\!O:
\longleftrightarrow
:O\!=\!\overset{\cdot\cdot}{Cl}\!-\!\overset{\cdot\cdot}{O}:\!\!\cdots\overset{\ominus}{}
$$

I_3^- Each iodine atom has seven valence electrons: $3I(7\ e^-) = 21\ e^-$
The negative charged adds one valence electron: $+1(-)\quad = 1\ e^-$
The total number of valence electrons is: $\overline{22\ e^-\ \text{total}}$

Each iodine atom needs eight electrons: $3I(8\ e^-) = 24\ e^-$
The total number of electrons required is: $\overline{24\ e^-\ \text{needed}}$

$24\ e^-\ \text{needed}$
$\underline{-\ 22\ e^-\ \text{total}}$
$2\ e^-\ \text{shared (or one bond)}$

$22\ e^-\ \text{total}$
$\underline{-\ 2\ e^-\ \text{shared}}$
$20\ e^-\ \text{unshared}$

Lewis structure: Hypervalent structure:

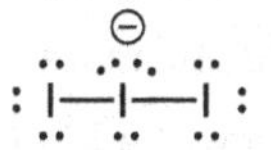

PROBLEM 9.11
Which has a larger ionization energy (E_i): H or H_2? Explain your answer.

Dihydrogen (H_2) has a larger ionization energy than hydrogen (H). Looking at Figure 9.23, the energy of the electrons in dihydrogen is lower (more stable) than the energy of the electron in hydrogen. This means it would take more energy (16.2 eV) to remove an electron from dihydrogen than it would to remove an electron from hydrogen (13.6 eV).

PROBLEM 9.12
The dihelium cation (He_2^+) is a stable polyatomic ion with a dissociation energy (E_d) of 2.47 eV. Using Figure 9.23, explain why He_2^+ is stable in contrast to unstable He_2.

He_2 is unstable because both the σ orbital and the σ^* orbital are filled with electrons. This means there is no energetic stabilization that comes from the two atoms being connected and the bond order would be 0. In contrast, He_2^+ has one less electron and so the bond order is 0.5, which means there is a partial covalent bond connecting the two atoms. Therefore, He_2^+ would be stabilized by being connected – bonded.

PROBLEM 9.13
Unlike σ bonds, π bonds cannot rotate. Compare Figure 9.26 with Figure 9.24. Provide an explanation of why π bonds cannot rotate.

A π bond is made up of facially aligned p orbitals. If the bond were to rotate, the p orbitals would no longer be aligned and the π bond would break:

aligned p-orbitals   misaligned p-orbitals
(π bond)   (rotated, broken π bond)

PROBLEM 9.14
Answer the following questions about dichlorine (Cl_2).

a. Draw a Lewis structure for dichlorine and identify the type of bond that exists between the chlorine atoms.

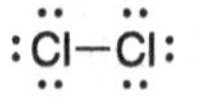

A single, σ bond exists between the two chlorine atoms.

b. Dichlorine (Cl_2) has a dissociation energy (E_d) of 2.52 eV, but when one electron is added (Cl_2^-) the dissociation energy decreases to 1.26 eV.

i. Where (into which molecular orbital) does this added electron go?

Any electrons that are added to dichlorine would add into the σ* orbital.

ii. Why does this decrease the dissociation energy?

The addition of an electron to the σ* orbital would decrease the bonding between the two chlorine atoms (the bond order would decrease from 1.0 to 0.5). The lower bond order and weaker bonding would lead to a lower dissociation energy

c. The bond lengths for these two are 201 pm and 264 pm. Which bond length goes with which molecular entity? Explain.

A weaker bond is a longer bond, and a stronger bond is a shorter bond. Since dichlorine (Cl_2) has a higher dissociation energy (2.52 eV), it would have the shorter bond length of 201 pm. The weaker dissociation energy of Cl_2^- implies a longer bond of 264 pm.

d. What would adding one more electron do to the bond energy and the bond length? Explain

If one more electron was added, forming Cl_2^{2-}, then the σ* orbital would be filled. The bond would break, the dissociation energy would become 0, and the bond length would become infinite as the atoms would move apart from one another.

PROBLEM 9.15

Use MO theory to explain the following bond length data (in picometers [pm]), and dissociation energy (E_d) data (in kJ/mol). Can you provide a trend for how bond order relates to bond length and bond dissociation energy?

O_2^+ bond length = 112 pm; E_d = 6.48 eV → $p_{OO(+)} = \frac{1}{2}(10 - 5) = 2.5$
O_2 bond length = 121 pm; E_d = 5.16 eV → $p_{OO} = \frac{1}{2}(10 - 6) = 2.0$
O_2^- bond length = 128 pm; E_d = 4.55 eV → $p_{OO(-)} = \frac{1}{2}(10 - 7) = 1.5$
O_2^{2-} bond length = 149 pm; E_d = 2.21 eV → $p_{OO(2-)} = \frac{1}{2}(10 - 8) = 1.0$

Bond length trends in the opposite direction of p_{rs} (higher p_{rs} = shorter bond, lower p_{rs} = longer bond).

E_d and p_{rs} trend in the same direction (higher p_{rs} = higher E_d, lower p_{rs} = lower E_d).

CHAPTER 10 ANSWERS

PROBLEM 10.1

Identify the electron geometry and hybridization of each carbon atom.

a. H–N–C–N–H (with O above C, H below each N)

b. H–N=C=O

c. H–C–C–H (with H H above and H H below)

Three areas of electron density Trigonal planar and sp²

Two areas of electron density and sp

Four areas of electron density Tetrahedral and sp³

PROBLEM 10.2

For each of the following Lewis structures, count the number of areas of electron density and redraw the Lewis structure (using Table 10.2) to appropriately show the correct three-dimensional shape. Provide the electron geometry shape name and identify the bond angle(s).

a. Cl–N–Cl (with Cl below N)

N with Cl, Cl, and Cl

Four areas of electron density give a tetrahedral geometry. Bond angles <109.5° (less than ideal because of the lone pair).

b.

Two areas of electron density give a linear geometry. Bond angles 180°.

c.

Four areas of electron density give a tetrahedral geometry. Bond angles <109.5° (less than ideal because of the lone pair).

d.

Five areas of electron density give a trigonal bipyramidal geometry. Bond angles 180° (the iodide atoms try to get as far apart as possible).

e.

Five areas of electron density give a trigonal bipyramidal geometry. Bond angles 180° (between axial fluorine atoms) and 120° for equatorial fluorine atoms); 90° between the equatorial and axial fluorine atoms.

f.

Four areas of electron density give a tetrahedral geometry. Bond angles 109.5°.

PROBLEM 10.3

What is the hybridization of each carbon atom in the following molecule? Redraw the molecule showing the appropriate bond angles and shape of each carbon atom.

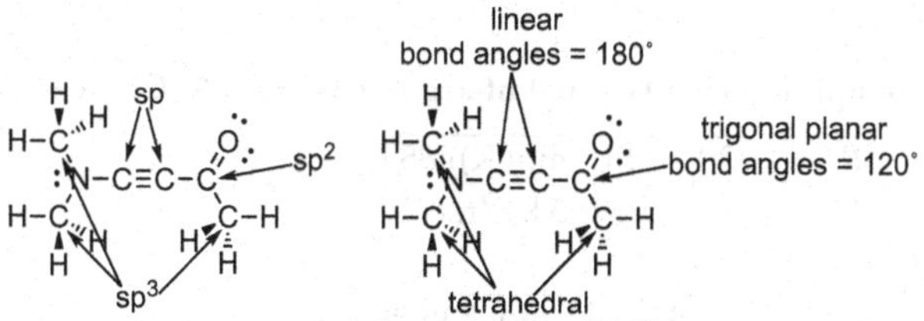

PROBLEM 10.4

For each of the following chemical formulae, draw a Lewis structure that shows the correct three-dimensional shape of the molecule.

a. SF_6 (octahedral)

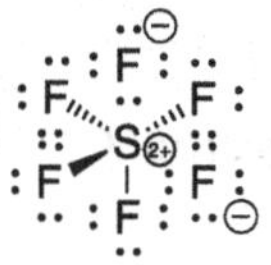

b. COS (linear)

c. BF_3 (trigonal planar)

307

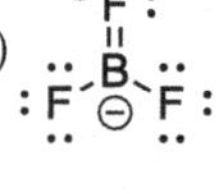

d. ClF_3 (trigonal bipyramidal)

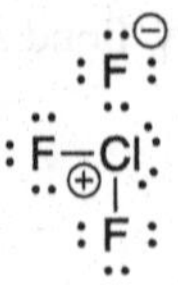

e. PO_4^{3-} (tetrahedral)

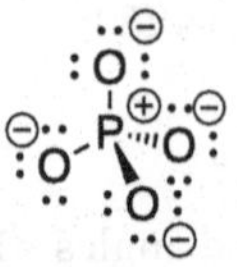

f. SO_2Cl_2 (tetrahedral)

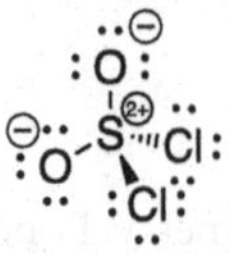

CHAPTER 11 ANSWERS

PROBLEM 11.1
We can approximate the change in mean free path (*l*) in air using:

$$l = 2.22 \times 10^{-10} \frac{m}{K} T$$

l is the mean free path (m).

T is the temperature (K).

Argon gas condenses into liquid argon at –186 °C. What is the mean free path (*l*) at –185 °C? How does this compare to the van der Waals diameter (0.366 nm) of an argon atom?

Temperature (in kelvins): –185 + 273.15 = 88 K

$$I = (2.22 \times 10^{-10} \frac{m}{K})(88 \text{ K}) = 2.0 \times 10^8 \text{ m or } 20 \text{ nm}$$

20. nm/0.366 nm = 55, that is, the mean free path is 55 times greater than the diameter of argon.

PROBLEM 11.2
What is the root-mean-square velocity (m/s) of argon gas at –185 °C? At 25 °C? At 323 °C?

$$v_{rms}(\text{Ar, 88 K}) = \sqrt{\frac{3RT}{M}} = \sqrt{\frac{3(8.314 \text{ J}/(\text{mol K}))(88 \text{ K})}{0.039\,95 \text{ kg}/\text{mol}}} = 230 \text{ m/s}$$

$$v_{rms}(\text{Ar, 298 K}) = \sqrt{\frac{3RT}{M}} = \sqrt{\frac{3(8.314 \text{ J}/(\text{mol K}))(298 \text{ K})}{0.039\,95 \text{ kg}/\text{mol}}} = 431 \text{ m/s}$$

$$v_{rms}(\text{Ar, 596 K}) = \sqrt{\frac{3RT}{M}} = \sqrt{\frac{3(8.314 \text{ J}/(\text{mol K}))(596 \text{ K})}{0.039\,95 \text{ kg}/\text{mol}}} = 610. \text{ m/s}$$

PROBLEM 11.3
What is the root-mean-square velocity (m/s) of dihydrogen gas at 25 °C? Of dioxygen gas at 25 °C? Of sulfur hexafluoride gas at 25 °C?

$$v_{rms}(\text{H}_2, 298 \text{ K}) = \sqrt{\frac{3RT}{M}} = \sqrt{\frac{3(8.314 \text{ J}/(\text{mol K}))(298 \text{ K})}{0.002\,0160 \text{ kg}/\text{mol}}} = 1920 \text{ m/s}$$

$$v_{rms}(O_2, 298 \text{ K}) = \sqrt{\frac{3RT}{M}} = \sqrt{\frac{3(8.314 \text{ J}/(\text{mol K}))(298 \text{ K})}{0.031\,998 \text{ kg}/\text{mol}}} = 482 \text{ m/s}$$

$$v_{rms}(SF_6, 298 \text{ K}) = \sqrt{\frac{3RT}{M}} = \sqrt{\frac{3(8.314 \text{ J}/(\text{mol K}))(298 \text{ K})}{0.146\,05 \text{ kg}/\text{mol}}} = 226 \text{ m/s}$$

PROBLEM 11.4

Two rigid containers of gas are 1.00 L in size, which will have a greater amount of kinetic energy: the container with a pressure of 101 kPa or the container with a pressure of 202 kPa (note that 1000 L = 1 m³)?

The kinetic energy of a gas increases with pressure $(\overline{E_k} = \frac{3}{2}pV)$, so the container at 202 kPa

$(\overline{E_k} = \frac{3}{2}pV = \frac{3}{2}(202\,000 \text{ Pa})(0.001 \text{ m}^3) = 152 \text{ J})$ will have a greater kinetic energy than the container

at 101 kPa $(\overline{E_k} = \frac{3}{2}pV = \frac{3}{2}(101\,000 \text{ Pa})(0.001 \text{ m}^3) = 303 \text{ J})$.

PROBLEM 11.5

Calculate the pressure (atm) exerted by 1.55 g of Xe gas at 25 °C in a 560 mL container.

p is unknown.

$$V = 560 \text{ mL}(\frac{1 \text{ L}}{1000 \text{ mL}}) = 0.56 \text{ L}$$

$$n = 1.55 \text{ g}(\frac{1 \text{ kg}}{1000 \text{ g}})(\frac{1 \text{ mol}}{0.131\,29 \text{ kg}}) = 0.0118 \text{ mol}$$

$$R = 0.082\,06 \frac{\text{L atm}}{\text{mol K}}$$

$$T = 25 + 273.15 = 298 \text{ K}$$

$$pV = nRT$$

$$p = \frac{(0.0118 \text{ mol})(0.082\,06 \text{ (L atm)}/(\text{mol K}))(298 \text{ K})}{0.56 \text{ L}} = 0.52 \text{ atm}$$

PROBLEM 11.6

Once filled with 4.4 g of CO_2 gas, a flask of unknown volume is found to have a pressure of 0.961 atm. If the flask is 27 °C, what is the volume of the flask (L)?

$p = 0.961 \text{ atm}$

V is unknown.

$$n = 4.4 \text{ g}(\frac{1 \text{ kg}}{1000 \text{ g}})(\frac{1 \text{ mol}}{0.044\,009 \text{ kg}}) = 0.010 \text{ mol}$$

$$R = 0.082\,06 \frac{\text{L atm}}{\text{mol K}}$$

$$T = 27 + 273.15 = 300 \text{ K}$$

$$pV = nRT$$

$$V = \frac{(0.010 \text{ mol})(0.082\,06 \text{ L atm}/\text{mol K})(300 \text{ K})}{0.961 \text{ atm}} = 2.6 \text{ L}$$

PROBLEM 11.7
A 4.23 g sample of unknown gas exerts a pressure of 0.965 atm in a 1.00 L container at 445.7 K.

a. Calculate the amount of gas (mol) present in the container.

$p = 0.965$ atm

$V = 1.00$ L

n is unknown.

$R = 0.082\,06 \dfrac{\text{L atm}}{\text{mol K}}$

$T = 445.7$ K

$pV = nRT$

$n = \dfrac{(0.965 \text{ atm})(1.00 \text{ L})}{(0.082\,06 \text{ L atm/mol K})(445.7 \text{ K})} = 0.0264$ mol

b. Calculate the molar mass (kg/mol) given the mass (4.23 g) and your answer to part a.

$m = 4.23 \text{ g}(\dfrac{1 \text{ kg}}{1000 \text{ g}}) = 0.004\,23$ kg

$M = \dfrac{m}{n} = \dfrac{0.004\,23 \text{ kg}}{0.0264 \text{ mol}} = 0.160$ kg/mol

PROBLEM 11.8
Two gas containers at 22.5 °C are connected by a valve. Container A is 9.2 L and the pressure inside is 1.75 atm. Container B is 5.4 L and the pressure inside is 0.82 atm.

a. Calculate the amount of gas (mol) present in each container.

Container A Container B

$p = 1.75$ atm $p = 0.82$ atm

$V = 9.2$ L $V = 5.4$ L

n is unknown. n is unknown.

$R = 0.082\,06 \dfrac{\text{L atm}}{\text{mol K}}$ $R = 0.082\,06 \dfrac{\text{L atm}}{\text{mol K}}$

$T = 22.5 + 273.15 = 295.7$ K $T = 22.5 + 273.15 = 295.7$ K

$pV = nRT$ $pV = nRT$

$n = \dfrac{(1.75 \text{ atm})(9.2 \text{ L})}{(0.082\,06 \text{ L atm/mol K})(295.7 \text{ K})} = 0.66$ mol $\quad n = \dfrac{(0.82 \text{ atm})(5.4 \text{ L})}{(0.082\,06 \text{ L atm/mol K})(295.7 \text{ K})} = 0.18$ mol

b. When the valve is opened, what is the total volume of the container AB? What is the total amount of gas inside the container AB?

The total amount of gas inside the container AB is 0.84 mol (0.66 mol + 0.18 mol).

c. Using your answers in part b., what will be the final pressure (atm) after the valve is opened?

The total volume inside the container (AB) is 14.6 L (9.2 L + 5.4 L).

p is unknown.

$V = 14.6$ L

$n = 0.84$ mol

$R = 0.082\,06 \dfrac{\text{L atm}}{\text{mol K}}$

$T = 295.7$ K

$pV = nRT$

$$p = \frac{(0.84 \text{ mol})(0.082\ 06 \text{ L atm/mol K})(295.7 \text{ K})}{14.6 \text{ L}} = 1.4 \text{ atm}$$

PROBLEM 11.9

When inhaling, a person's chest cavity increases in volume, and when exhaling their chest cavity decreases in volume.

a. Provide an explanation for how increasing our chest cavity volume helps us inspire, that is, get air into our lungs.

According to Boyle's law ($p_1V_1 = p_2V_2$), as the volume of the chest cavity increases, the pressure inside the chest cavity will decrease. Gas from outside the lungs rushes into the lungs to offset the reduction in pressure.

b. Provide an explanation as to why decreasing our chest cavity volume pushes air out of our lungs.

According to Boyle's law ($p_1V_1 = p_2V_2$), as the volume of the chest cavity decreases, the pressure inside the chest cavity will increase. Gas from inside the lungs will be pushed out of the lungs to offset the increase in pressure.

PROBLEM 11.10

Consider the data in Figure 11.4 for 1.0 mol helium gas at 1.0 atm showing the relationship between volume (V) and temperature (T). The trendline for these data and the linear fit equation are shown. Determine the x intercept of these data. What is the significance of this temperature?

Linear fit equation: $y = 0.082x + 22.402$

$0 = 0.0820x + 22.402$

$-22.402 = 0.082x$

$x = -273$ °C

The y intercept corresponds to absolute zero. Note that the implication of this is that the gas would have zero volume at −273 °C, which is not physically possible.

PROBLEM 11.11

Consider the data in Figure 11.5 for 5.0 mol dinitrogen gas in a 10.0 L fire extinguisher showing the relationship between pressure (p) and temperature (T). Fire extinguishers are great tools for helping to escape from fires, or to put out a small fire, but fire extinguishers put into a fire can be quite dangerous. Consider Figure 11.5 and provide an explanation as to why a fire extinguisher dropped into a fire is dangerous.

If a pressurized cylinder (like a fire extinguisher) were put into a fire, then the increase in temperature would lead to an increase in pressure. If the pressure were to get high enough, then the fire extinguisher could (and would) explode.

PROBLEM 11.12

In a 1.0 L container is 1.31 g of dioxygen gas and in a separate 1.0 L container is 0.0826 g of dihydrogen gas. Both containers are at 1.0 atm and 25 °C.

a. What can we say about the number of particles in the two containers?

As per Avogadro's law, if the volume is the same (at constant temperature and pressure), then the number of particles in the two containers is the same.

b. What is the relative mass of dihydrogen gas to dioxygen gas (m_{H2}/m_{O2})?

The relative mass of dihydrogen gas to dioxygen gas (m_{H2}/m_{O2}) is 0.0631.

c. Now, if we were to set that atomic mass of oxygen as 16.000, what is the relative atomic mass of hydrogen? How does this compare with the value in Appendix 2?

If the atomic mass of oxygen is set as 16.000, then the relative atomic mass of hydrogen would be 1.01. Rounded to three significant figures, this is the same as the recommended relative atomic mass of hydrogen in Appendix 2.

PROBLEM 11.13
A sample of air in a 10.0 L container has the following partial pressures of each gas: p_{O2} = 20.9 kPa, p_{N2} = 78.1 kPa, p_{Ar} = 0.97 kPa, p_{H2O} = 1.28 kPa, p_{CO2} = 0.05 kPa. What is the total pressure of the sample?

$$p_{Total} = 20.9 \text{ kPa} + 78.1 \text{ kPa} + 0.97 \text{ kPa} + 1.28 \text{ kPa} + 0.05 \text{ kPa} = 101.3 \text{ kPa}$$

PROBLEM 11.14
In the lungs, the normal mole fraction of CO_2 (y_{CO2}) is 0.046. What is the partial pressure (kPa) of CO_2 in your lungs (total pressure is 101.3 kPa)?

$$p_{CO2} = 101.3 \text{ kPa}(0.046) = 4.7 \text{ kPa}$$

PROBLEM 11.15
Natural uranium is a mixture of fissionable uranium-235 (0.72% abundance, $m_{U\text{-}235}$ = 235.04 u) and non-fissionable uranium-238 (99.3% abundance, $m_{U\text{-}238}$ = 238.05 u). One method of creating enriched uranium (increasing the uranium-235 content) is to take advantage of the differential rate of effusion of uranium hexafluoride (UF_6) gas.

a. What is the ratio of root-mean-square velocity magnitudes for $^{235}UF_6$ and $^{238}UF_6$?

$$\frac{v_{rms}(\text{lighter gas})}{v_{rms}(\text{heavier gas})} = \sqrt{\frac{M_{\text{heavier gas}}}{M_{\text{lighter gas}}}} = \sqrt{\frac{0.352\,04 \text{ kg/mol}}{0.349\,03 \text{ kg/mol}}} = 1.004\,30$$

b. Given the ratio calculated in part a, how effective do you think effusion is as a method of enrichment?

Given the very small difference in molar mass, the rate of effusion of $^{235}UF_6$ is only marginally faster than the rate of $^{238}UF_6$ (1.004 30:1). Given the very small difference, a single effusion is a poor method for enriching uranium. To enrich uranium through effusion, the process must be repeated thousands of times.

CHAPTER 12 ANSWERS

PROBLEM 12.1
What makes a gas ideal? What makes a gas real?

An ideal gas is a zero-dimensional point mass that has no molecular interactions (when it collides with other gas particles, the collisions are perfectly elastic).

A real gas has a nonzero molecular volume and interacts with other gas particles. This is represented by the volume and pressure correction terms in the van der Waals equation.

PROBLEM 12.2
Consider the following gases: Ar, SF_6, H_2, CO_2, and CH_4. In terms of their structure (not looking up the values), rank them in order of smallest to biggest b value. Explain your answer.

Smaller molecular entities have smaller b values. Argon (Ar) is a monatomic gas particle and should therefore be one of the smallest (b = 0.032 L/mol) along with H_2, which is diatomic but made up of two very small atoms (H_2, b = 0.026 51 L/mol). Methane (CH_4, b value = 0.043 01 L/mol) and carbon dioxide (CO_2, b = 0.0427 L/mol) are probably similar in size (given the disparity in size between O and H). Sulfur hexafluoride (SF_6) is the biggest (seven atoms, all larger atoms) gas particle ($b = 0.087\,86$ L/mol).

Reasonable orderings are (SF_6 is the biggest):

$$Ar < H_2 < CH_4 < CO_2 < SF_6$$
$$Ar < H_2 < CO_2 < CH_4 < SF_6$$
$$H_2 < Ar < CH_4 < CO_2 < SF_6$$
$$H_2 < Ar < CO_2 < CH_4 < SF_6$$

PROBLEM 12.3

Considering the intermolecular forces involved, can you provide an explanation for the difference in the boiling point between these two compounds?

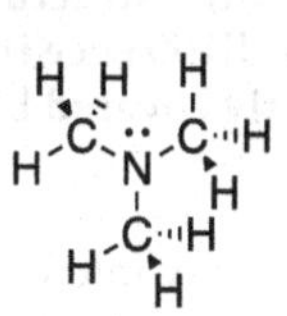

C_3H_9N
$b = 0.1$ L/mol
boiling point = 7 °C

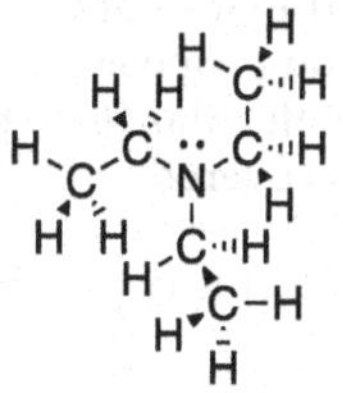

C_6H_7N
$b = 0.2$ L/mol
boiling point = 89 °C

There is a significant difference in b values. The molecule on the right (triethylamine) has a b value twice as large as the b value of the molecule on the left (trimethylamine). All other things being equal, the strength of intermolecular interactions, as represented by a values, is strongly correlated with b values: as b values increase, the a values will also increase. The strength of the intermolecular forces between particles is directly related to the boiling point of a chemical species; therefore, the stronger intermolecular interactions (stronger dispersion due to larger size) of triethylamine lead to a higher boiling point.

PROBLEM 12.4

Sulfur hexafluoride (SF_6) has a b value roughly equivalent to that of propane (C_3H_8). Sulfur hexafluoride's boiling point is 10 K lower than that of propane. Provide an explanation for this effect.

Sulfur hexafluoride and propane have a similar size, which means they should have similar boiling points (all other things being equal). Here we see, however, that sulfur hexafluoride has a lower boiling point. This is because sulfur hexafluoride is highly fluorinated, which means that sulfur hexafluoride will have weaker dispersion interactions. A decrease in dispersion interactions means a decrease in intermolecular forces holding the molecules together and therefore a lower boiling point for sulfur hexafluoride.

PROBLEM 12.5

Consider the following set of data looking at fluorinated ethane compounds. Provide an explanation for the trend in boiling points.

Name	Fluoroethane	1,1,1-Trifluoroethane	1,1,1,2,2,2-Hexafluoroethane
Structure	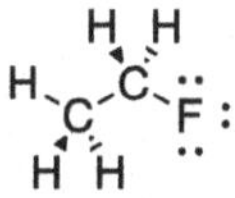	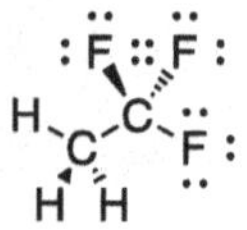	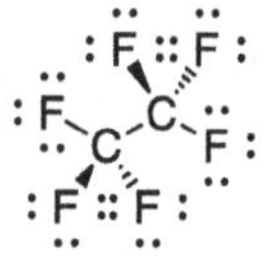
b (L/mol)	0.076 61	0.094 17	0.098 92
Boiling point (K)	236	226	195

Bigger molecules, in general, have stronger van der Waals interactions and therefore higher boiling points in the condensed phase. Here we see, however, that the bigger molecules, which are also more highly fluorinated, have a lower boiling point (and therefore must have lower intermolecular attraction). So, while the molecules are becoming bigger in size, they are also becoming more saturated with the highly electronegative fluorine atom, which limits dispersion interactions. Here, then, the bigger size trends in the opposite direction from the boiling point because of the decrease in dispersion interactions and the decrease in intermolecular forces holding the larger molecules together.

PROBLEM 12.6

Methane (CH_4) and carbon dioxide (CO_2) are similar in terms of b values. Methane boils at 111.6 K, while carbon dioxide sublimates at 194.7 K. Draw the structures, consider the distribution of charge, and propose an explanation for this difference in behavior. Note: This involves a type of intermolecular force that was not explicitly covered but an extension of the noncovalent interactions covered.

Methane is a nonpolar molecule ($\chi_C = 2.55$ and $\chi_H = 2.20$) that only interacts through dispersion interactions.

$$
\begin{array}{c}
H \\
| \\
H-C\cdots H \\
H
\end{array}
$$

Carbon dioxide does have polar bonds ($\chi_O = 3.44$ and $\chi_C = 2.55$) but does not have an electric *dipole* moment because of its structure.

$$: \! O = C = O \! :$$

But carbon dioxide does have two electron-rich ends and an electron-poor middle. This is referred to as an electric *quadrupole* moment:

Here there is a strong quadrupole–quadrupole interaction, which enhances the intermolecular attraction between carbon dioxide molecules and leads to a higher temperature for a phase change.

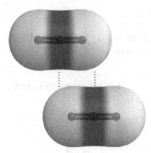

PROBLEM 12.7
Considering the intermolecular forces involved, can you provide an explanation for the difference in the boiling point for the two compounds below?

C_3H_9N
$b = 0.1$ L/mol
boiling point = 7 °C

C_3H_7NO
$b = 0.1$ L/mol
boiling point = 153 °C

There is no significant difference in b values. The molecule on the left has no polar bonds (χ_N = 3.04 and χ_C = 2.55, χ_H = 2.20), while the molecule on the right side does have a polar C=O bond (χ_O = 3.44 and χ_C = 2.55). This means that the molecule on the right interacts through both dispersion and dipole–dipole interactions, a more strongly attracting set of interactions, while the molecule on the left interacts through only dispersion, a weaker interaction. The strength of the intermolecular forces between particles is directly related to the boiling point of a chemical species; therefore, the stronger intermolecular interactions (due to dipole–dipole interactions) of the molecule on the right lead to a higher boiling point.

PROBLEM 12.8
Considering the intermolecular forces involved, can you provide an explanation for the difference in the boiling point for the two isomers below?

C_3H_7NO
$b = 0.1$ L/mol
boiling point = 153 °C

C_3H_7NO
$b = 0.1$ L/mol
boiling point = 206 °C

There is no significant difference in b values. The molecule on the left does not have the ability to form a significant hydrogen bond (no N–H, O–H, or F–H), while the molecule on the right can form a significant hydrogen bond (it contains an N–H). This means that the molecule on the right interacts through both dispersion, dipole–dipole, and hydrogen bond interactions, a more strongly attracting set of interactions, while the molecule on the left interacts through only dispersion and dipole–dipole interactions, a weaker set of interactions. The strength of the intermolecular forces between particles is directly related to the boiling point of a chemical species; therefore, the stronger intermolecular interactions (due to hydrogen bonding) of the molecule on the right leads to a higher boiling point.

PROBLEM 12.9
Consider the following alcohols, their structures, and their miscibility in water (miscible means that they will always form a homogeneous mixture, regardless of amounts). Provide an explanation for the trend you see.

miscible

miscible

miscible

73 g/L

22 g/L

All the alcohols have an –O–H group, which can hydrogen bond with water; however, the -O–H becomes a less significant group given the growing size of the hydrocarbon chain. The decreasing solubility of the alcohols is due to the increasingly nonpolar nature of the molecules, which interact predominantly through dispersion, which means they will interact less and less well with water (which interacts predominantly through hydrogen bonds and dipole–dipole interactions).

PROBLEM 12.10

Surfactants are large molecules that have pieces with dissimilar intermolecular forces. Most commonly, surfactants have a hydrophilic (water loving) and hydrophobic (water fearing) component. Consider tetraethylene glycol monodecyl ether.

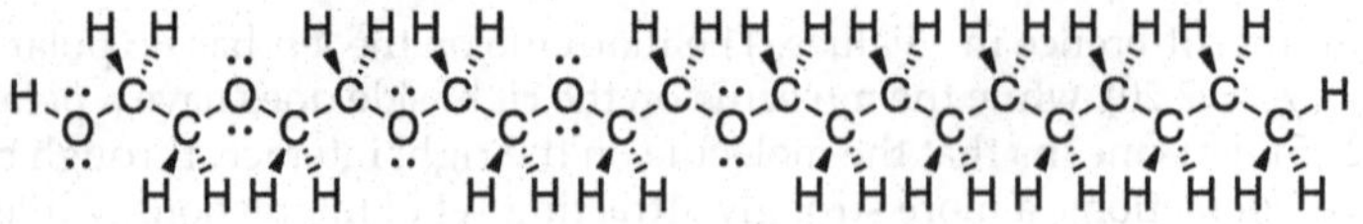

Tetraethylene glycol monodecyl ether

This is made from tetraethylene glycol and decane.

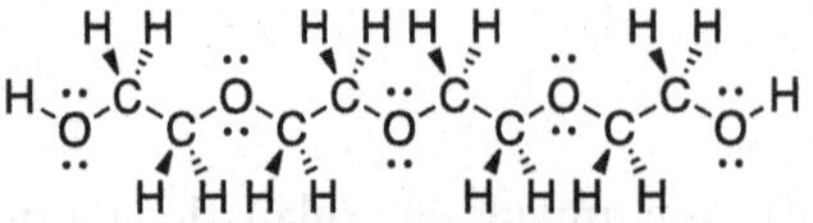

tetraethylene glycol

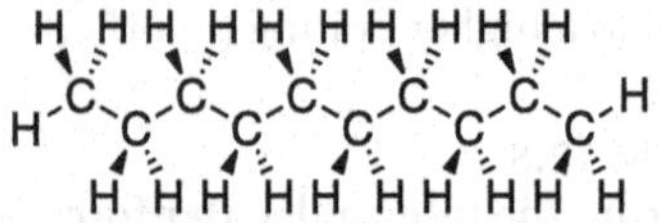

decane

a. Both tetraethylene glycol and decane have very similar b values (0.3 L/mol). The two a values are 51.6 $\dfrac{\text{atm L}^2}{\text{mol}^2}$ and 56.9 $\dfrac{\text{atm L}^2}{\text{mol}^2}$. Assign the a value to the correct structure and explain your answer.

Tetraethylene glycol will have the greater a value because it has polar bonds and the ability to hydrogen bond, while decane can only interact through dispersion interactions.

b. Reconsider the structure of tetraethylene glycol monodecyl ether. Using your answer in part a, identify which part of the molecule is hydrophobic and which part is hydrophilic. Explain your answer.

The tetraethylene glycol portion of the molecule is hydrophilic because it can interact with water through strong H-bond interactions. Decane, by contrast, can only interact through dispersion, so it will not interact strongly with water and will be hydrophobic.

CHAPTER 13 ANSWERS

PROBLEM 13.1

For each of the following, identify the state(s) of matter (solid, liquid, or gas) and whether the material is a singular phase or more than one phase.

a. A pencil

The pencil is a solid with several different phases (with different chemical compositions): the graphite core, the wood, the paint, the metal eraser holder, and the eraser.

b. Partially melted gallium metal

The partially melted gallium is a mixture of two states/phases (liquid and solid).

PROBLEM 13.2

Consider Figure 13.2. Rank the mean kinetic energy ($\overline{E_k}$) of a solid (ice) versus a liquid (water) versus a gas (steam) from highest to lowest. Explain your answer.

From highest to lowest mean kinetic energy, the phases are, in order: gas, liquid, and solid.

A gas has the highest mean kinetic energy because the particles are moving the fastest (see Chapter 11), which is evidenced by the wide spacing between particles.

A solid has the lowest mean kinetic energy because the particles only vibrate (and do not rotate nor translate) and they are closely packed.

A liquid is intermediate between the gas and solid phases because the particles are spread out more than a solid and do vibrate, rotate, and translate, but they translate at a much slower speed than a gas.

PROBLEM 13.3

If we compare carbon dioxide (sublimates at –78 °C at 1 atm) and carbon disulfide (boils at 46 °C at 1 atm), we can see that carbon disulfide must be given more thermal energy to vaporize.

a. What can we say about the strength of the intermolecular forces between carbon dioxide molecules versus the strength of the intermolecular forces between carbon disulfide?

   Because it requires a higher temperature to vaporize, we can conclude that carbon disulfide has stronger intermolecular forces than carbon dioxide.

b. Can you provide an explanation for the differences in intermolecular forces?

   Given the larger size of the sulfur atoms, compared to the oxygen atoms, carbon disulfide has a larger surface area and therefore stronger van der Waals interactions.

PROBLEM 13.4

The enthalpy of fusion ($\Delta_{fus}H°$) for water is 6.01 kJ/mol and the enthalpy of vaporization ($\Delta_{vap}H°$) for water is 40.65 kJ/mol. Provide an explanation of why more heat is required to vaporize water than is required to melt water. Consider the particles (Figure 13.2) and how they change upon vaporization versus melting.

Gases particles are widely separated so that there are no particle–particle interactions. Liquid and solid particles are close together and so there are particle–particle interactions. Given this difference, melting a solid only requires enough energy so that particles can move past one another, but particle–particle interactions still occur. To vaporize a substance, the particle–particle interactions (intermolecular forces) must be entirely disrupted, which requires a substantial amount of energy.

PROBLEM 13.5

The normal (1.00 atm) boiling point of water ($\Delta_{vap}H° = 43.9$ kJ/mol) is 100.0 °C.

a. What is the boiling point (°C) of water on the top of Mount Everest (0.31 atm)?

$$\ln\left(\frac{p_2}{p_1}\right) = \frac{-\Delta_{vap}H°}{R}\left(\frac{1}{T_2} - \frac{1}{T_1}\right)$$

$$\ln\left(\frac{0.31\text{ atm}}{1.00\text{ atm}}\right) = \frac{-(43\,900\text{ J/mol})}{8.314\text{ J/(mol K)}}\left(\frac{1}{T_2} - \frac{1}{373.2\text{ K}}\right)$$

$$T_2 = 344.7\text{ K} = 71.5\text{ °C}$$

b. Will it take more time or less time to cook something on the top of Mount Everest? Explain.

   Given the lower temperature that water boils at on top of Mount Everest, it would take substantially longer to cook food than it does at sea level. The rate at which food cooks is dependent upon temperature and so the lower temperature would dramatically slow the rate at which food cooks.

PROBLEM 13.6

The critical point of methane is −82.3 °C and 45.79 atm. The critical point of ammonia is 132.4 °C and 111.3 atm.

a. Calculate the a and b values for methane and for ammonia.

Methane

$$a = \frac{27}{64}\frac{(RT_c)^2}{p_c} = \frac{27}{64}\frac{((0.082\,06\,(\text{L atm})/(\text{mol K}))(190.9\,\text{K}))^2}{45.79\,\text{atm}} = 2.26\,(\text{atm L}^2)/\text{mol}^2$$

$$b = \frac{RT_c}{8p_c} = \frac{(0.082\,06\,(\text{L atm})/(\text{mol K}))(190.9\,\text{K})}{45.79\,\text{atm}} = 0.0428\,\text{L/mol}$$

Ammonia

$$a = \frac{27}{64}\frac{(RT_c)^2}{p_c} = \frac{27}{64}\frac{((0.082\,06\,(\text{L atm})/(\text{mol K}))(405.6\,\text{K}))^2}{111.3\,\text{atm}} = 4.20\,(\text{atm L}^2)/\text{mol}^2$$

$$b = \frac{RT_c}{8p_c} = \frac{(0.082\,06\,(\text{L atm})/(\text{mol K}))(405.6\,\text{K})}{111.3\,\text{atm}} = 0.0374\,\text{L/mol}$$

b. Which substance (methane or ammonia) has stronger intermolecular forces between particles? Based on their structures, provide an explanation for the difference.

Given the larger a value, ammonia has stronger intermolecular forces. The greater intermolecular attraction between ammonia molecules is due to its ability to hydrogen bond, which methane cannot do.

CHAPTER 14 ANSWERS

PROBLEM 14.1

Determine the initial rates (mol/(L s)) for each of the following datasets.

a. Reaction: $Br_2(g) + CH_4(g) \xrightarrow{\text{light}} HBr(g) + CH_3Br(g)$. See also Table 14.1.

Table 14.1 Time Course Data Showing Disappearance of Bromine Over Time at 25 °C

Time (s)	[Br$_2$] (mol/L)	Initial Rate of Disappearance of Bromine (mol/(L s))
0	1.49×10^{-3}	—
60	1.44×10^{-3}	$\dfrac{(1.44\times10^{-3} - 1.49\times10^{-3})\,\text{mol/L}}{(60-0)\,\text{s}} = -8 \times 10^{-7}$
180	1.35×10^{-3}	$\dfrac{(1.35\times10^{-3} - 1.44\times10^{-3})\,\text{mol/L}}{(180-60)\,\text{s}} = -7.5 \times 10^{-7}$
300	1.26×10^{-3}	$\dfrac{(1.26\text{x}10^{-3} - 1.35\times10^{-3})\,\text{mol/L}}{(300-180)\,\text{s}} = -8 \times 10^{-7}$
420	1.17×10^{-3}	$\dfrac{(1.17\times10^{-3} - 1.26\times10^{-3})\,\text{mol/L}}{(420-300)\,\text{s}} = -7.5 \times 10^{-7}$
600	1.04×10^{-3}	$\dfrac{(1.04\times10^{-3} - 1.17\times10^{-3})\,\text{mol/L}}{(600-420)\,\text{s}} = -7 \times 10^{-7}$

Source: Kistiakowsky, G.B.; Van Artsdalen, E.R. Bromination of Hydrocarbons. I. Photochemical and Thermal Bromination of Methane and Methyl Bromine. Carbon Hydrogen Bond Strength in Methane. *J. Chem. Phys.* 1944, 12 (12), 469–478. DOI: 10.1063/1.1723896.

All of the initial rates do not need to be calculated here, but we can see that there is a roughly constant rate among these initial data points. The average initial rate is -8×10^{-7} mol/(L s).

b. Reaction: $2\,NaN_3(s) \xrightarrow{330°C} 2\,Na(s) + 3\,N_2(g)$. See also Table 14.2.

Table 14.2 Time Course Data Showing Appearance of Nitrogen Over Time at 330 °C

Time (s)	[N$_2$] (mol/L)	Initial Rate of Appearance of Nitrogen (mol/(L s))
0	0	—
1	6.60×10^{-4}	$\dfrac{(6.60 \times 10^{-4} - 0)\,\text{mol/L}}{(1-0)\,\text{s}} = 7 \times 10^{-4}$
2	1.131×10^{-3}	$\dfrac{(1.131 \times 10^{-3} - 6.60 \times 10^{-4})\,\text{mol/L}}{(2-1)\,\text{s}} = 5 \times 10^{-4}$
3	1.467×10^{-3}	$\dfrac{(1.467 \times 10^{-3} - 1.131 \times 10^{-3})\,\text{mol/L}}{(3-2)\,\text{s}} = 4 \times 10^{-4}$

Source: Walker, R.F. Thermal Decomposition of Sodium Azide: Crystal Size Effects, Topochemistry and Gas Analyses. *J. Phys. Chem. Solids.* **1968**, *29* (6), 985–1000. DOI: 10.1016/0022-3697(68)90235-7.

All of the initial rates do not need to be calculated here, but we can see that the initial rate does vary for the data points are being considered. This suggests that the reaction is fast enough that even considering the initial rates during the first three seconds, our approximation of linearity is a much rougher approximation than it was for Problem 14.1a. The average initial rate is 5×10^{-4} mol/(L s).

c. Reaction: $(CH_3)_3CBr(sln) + NaOH(sln) \xrightarrow{water, ethanol} (CH_3)_3COH(sln) + NaBr(sln)$. See also Figure 14.3.

Let's consider the rate between each data point in Figure 14.3:

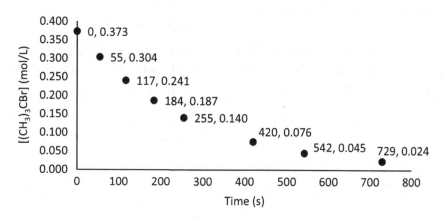

Figure 14.3 Time course data showing the disappearance of *tert*-butyl bromide, $(CH_3)_3CBr$, over time. (Data from Bateman, L.C.; Cooper, K.A.; Hughes, E.D.; Ingold, C.K. 178. Mechanism of Substitution at a Saturated Carbon Atom. Part XIII. Mechanisms Operative in the Hydrolysis of Methyl, Ethyl, Isopropyl, and *tert*.-Butyl Bromides in Aqueous Solutions. *J. Chem. Soc. (Resumed).* **1940**, 925–935. DOI: 10.1039/JR9400000925.)

Time (s)	[(CH₃)₃CBr] (mol/L)	Initial Rate of Disappearance of (CH₃)₃CBr (mol/(L s))
0	0.373	—
55	0.304	$\dfrac{(0.304 - 0.373)\ \text{mol/L}}{(55 - 0)\ \text{s}} = -0.0013$
117	0.241	$\dfrac{(0.241 - 0.304)\ \text{mol/L}}{(117 - 55)\ \text{s}} = -0.001\,02$
184	0.187	$\dfrac{(0.187 - 0.241)\ \text{mol/L}}{(184 - 117)\ \text{s}} = -0.000\,806$
255	0.140	$\dfrac{(0.140 - 0.187)\ \text{mol/L}}{(255 - 184)\ \text{s}} = -0.000\,662$
420	0.076	$\dfrac{(0.076 - 0.140)\ \text{mol/L}}{(420 - 255)\ \text{s}} = -0.00\,039$
542	0.045	$\dfrac{(0.045 - 0.076)\ \text{mol/L}}{(542 - 420)\ \text{s}} = -0.000\,25$
729	0.024	$\dfrac{(0.024 - 0.045)\ \text{mol/L}}{(729 - 542)\ \text{s}} = -0.000\,11$

What we can see is that initial rates are relatively linear between 0 and 184 seconds (roughly –0.001 mol/(L s)). The average initial rate (for 0 to 184 seconds) is –0.0010 mol/(L s). After 184 seconds, the rate begins to change rapidly between each data point.

PROBLEM 14.2
Reconsider Tables 14.1 and 14.2 and Figure 14.3 and determine the initial rate of reaction for each dataset.

The first reaction has a 1:1 ratio of bromine to the overall reaction:

$$Br_2(g) + CH_4(g) \xrightarrow{\text{Light}} HBr(g) + CH_3Br(g)$$

Therefore, to calculate the rate of reaction, we need to multiply the rate of disappearance of bromine by $-\frac{1}{1}$ (negative corrects for the fact that the rate of reaction is always positive and $\frac{1}{1}$ is the reaction/bromine ratio). Rate of reaction $= -\frac{1}{1} - 8 \times 10^{-7}$ mol/(L s) $= 8 \times 10^{-7}$ mol/(L s).

The second reaction has a 3:1 ratio of nitrogen to the overall reaction:

$$2\,NaN_3(s) \xrightarrow{330°\,C} 2\,Na(s) + 3\,N_2(g)$$

Therefore, to calculate the rate of reaction, we need to multiply the rate of appearance of nitrogen by $\frac{1}{3}$ ($\frac{1}{3}$ is the reaction/nitrogen ratio). Rate of reaction $= \frac{1}{3} 5 \times 10^{-4}$ mol/(L s) $= 2 \times 10^{-4}$ mol/(L s).

The third reaction has a 1:1 ratio of *tert*-butyl bromide to the overall reaction:

$$(CH_3)_3CBr(sln) + NaOH(sln) \xrightarrow{\text{water, ethanol}} (CH_3)_3COH(sln) + NaBr(sln)$$

Therefore, to calculate the rate of reaction, we need to multiply the rate of disappearance of *tert*-butyl bromide by $-\frac{1}{1}$ (negative corrects for the fact that the rate of reaction is always positive and $\frac{1}{1}$ is the reaction/*tert*-butyl bromide ratio). Rate of reaction $= -\frac{1}{1} - 0.0010$ mol/(L s) $= 0.0010$ mol/(L s).

PROBLEM 14.3

For the following dataset, determine the rate law for the reaction. Also see Table 14.4.

$$C_{11}H_{13}NO_2(sln) + C_4H_3NO_2(sln) \xrightarrow{\text{Iridium catalyst}} C_{15}H_{16}N_2O_4(sln)$$

Table 14.4 Initial Rate of Reaction Analysis for the Reaction of Quinolone ($C_{11}H_{13}NO_2$) and Maleimide ($C_4H_3NO_2$) with an Iridium Catalyst at 25 °C

Experiment	$[C_{11}H_{13}NO_2]$ (mol/L)	$[C_4H_3NO_2]$ (mol/L)	[Iridium Catalyst] (mol/L)	Rate of Reaction (mol/L)
1	0.0396	0.103	7.00×10^{-4}	1.70×10^{-6}
2	0.0496	0.103	7.00×10^{-4}	2.03×10^{-6}
3	2.00×10^{-5}	0.052	2.80×10^{-4}	1.17×10^{-6}
4	2.00×10^{-5}	0.075	2.80×10^{-4}	1.17×10^{-6}
5	2.00×10^{-5}	0.103	4.60×10^{-4}	1.17×10^{-6}
6	2.00×10^{-5}	0.103	6.20×10^{-4}	1.23×10^{-6}

Order dependence for $[C_{11}H_{13}NO_2]$:

The amount concentration increases by 1.25 ($\frac{0.0496 \text{ mol/L}}{0.0396 \text{ mol/L}} = 1.25$) and the rate increases by 1.19 ($\frac{2.03 \times 10^{-6} \text{ mol/(L s)}}{1.70 \times 10^{-6} \text{ mol/(L s)}} = 1.19$), which suggests a first-order dependence on $[C_{11}H_{13}NO_2]$.

Order dependence for $[C_4H_3NO_2]$:

The concentration increases by 1.44 ($\frac{0.075 \text{ mol/L}}{0.052 \text{ mol/L}} = 1.44$) and the rate does not change ($\frac{1.17 \times 10^{-6} \text{ mol/(L s)}}{1.17 \times 10^{-6} \text{ mol/(L s)}} = 1$), which suggests a zero-order dependence on $[C_4H_3NO_2]$, that is, the change in concentration does not affect the rate.

Order dependence for [iridium catalyst]:

The concentration increases by 1.34 ($\frac{6.20 \times 10^{-4} \text{ mol/L}}{4.60 \times 10^{-4} \text{ mol/L}} = 1.34$) and the rate does not significantly change ($\frac{1.23 \times 10^{-6} \text{ mol/(L s)}}{1.17 \times 10^{-6} \text{ mol/(L s)}} = 1.05$), which suggests a zero-order dependence on [iridium catalyst], that is, the change in concentration does not affect the rate.

Overall, then, the rate law is:

rate = $k[C_{11}H_{13}NO_2]^1[C_4H_3NO_2]^0[\text{iridium catalyst}]^0$
Or if we simplify:
rate = $k[C_{11}H_{13}NO_2]$

Now, to find k, we choose any experiment (here for the solution we'll choose experiment 1), plug in, and solve for k:

rate = $k[C_{11}H_{13}NO_2]$

1.70×10^{-6} mol/(L s) = $k[0.0396 \text{ mol/L}]$
$k = 4.29 \times 10^{-5}$ 1/s

Altogether, then, the rate law is:

rate = 4.29×10^{-5} 1/s[$C_{11}H_{13}NO_2$]

PROBLEM 14.4

Radioactive decay is a first-order reaction. Fluorine-18 is a radioisotope used in positron-emission tomography (PET) scans. If the rate constant for fluorine-18 to decay is 1.05×10^{-4} 1/s, how long will it take for a dose of fluorine-18 to decay by half?

$$[^{18}F]_t = [^{18}F]_0 e^{-kt}$$

For the dose to decay by half, that means that $\dfrac{[^{18}F]_t}{[^{18}F]_0} = \dfrac{1}{2}$, and we are given $k = 1.05 \times 10^{-4}$ 1/s.

$$\frac{1}{2} = e^{-1.05 \times 10^{-4} \text{ 1/s } t}$$

$$\ln\left(\frac{1}{2}\right) = -1.05 \times 10^{-4} \text{ 1/s } t$$

$$-\ln(2) = -1.05 \times 10^{-4} \text{ 1/s } t$$

$$\frac{\ln(2)}{1.05 \times 10^{-4} \text{ 1/s}} = t$$

$$t = 6.60 \times 10^3 \text{ s (or 110 minutes)}$$

PROBLEM 14.5

The hydrolysis of *tert*-butyl bromide, $(CH_3)_3CBr$, is studied in 90% aqueous acetone.[2] At 25.0 °C, the rate constant is found to be 1.30×10^{-5} 1/s, and at 50.0 °C the rate constant is found to be 1.93×10^{-4} 1/s. What is the activation energy (kJ/mol) for this reaction?

25.0 °C + 273.15 = 298.2 K

50.0 °C + 273.15 = 323.2 K

$$\ln\left(\frac{1.93 \times 10^{-4} \frac{1}{s}}{1.30 \times 10^{-5} \frac{1}{s}}\right) = -\frac{E_a}{0.008\,314 \frac{kJ}{mol\,K}}\left(\frac{1}{298.2\,K} - \frac{1}{323.2\,K}\right)$$

$$2.7 = \frac{E_a}{0.008314 \frac{kJ}{mol\,K}}\left(0.000\,259 \frac{1}{K}\right)$$

$$2.7 = \frac{E_a}{0.008\,314 \frac{kJ}{mol\,K}}\left(0.000\,259 \frac{1}{K}\right)$$

$$0.022 \frac{kJ}{mol\,K} = E_a\left(0.000\,259 \frac{1}{K}\right)$$

$$87 \frac{kJ}{mol} = E_a$$

CHAPTER 15 ANSWERS

PROBLEM 15.1
Find the enthalpy of reaction ($\Delta_r H°$) in each of the following calorimetry experiments.

a. In a calorimeter ($C = 1686.2$ J/°C) that contains 2257.0 g of water ($c_p(H_2O(l)) = 4.184 \dfrac{J}{g\,°C}$), 0.8106 g

of benzene (C_6H_6) is reacted with oxygen, which causes the temperature to rise by 3.054 °C. What is the enthalpy of reaction ($\Delta_r H°$)? The balanced equation is:

$2\,C_6H_6(l) + 15\,O_2(g) \rightarrow 12\,CO_2(g) + 6\,H_2O(g)$

$q_{bath} = mc_p(H_2O(l))\Delta T + C\Delta T$

$q_{bath} = (2257.0\text{ g})(4.184\ \dfrac{J}{g\,°C})(3.054\ °C) + (1686.2\ \dfrac{J}{°C})(3.054\ °C) = 33\,990\text{ J (or }33.99\text{ kJ)}$

$q_{reaction} = -(q_{water}) = -(33.99\text{ kJ}) = -33.99\text{ kJ}$

$0.8106\text{ g benzene}\left(\dfrac{1\text{ kg}}{1000\text{ g}}\right)\left(\dfrac{1\text{ mol benzene}}{0.078\,11\text{ kg}}\right)\left(\dfrac{1\text{ mol reaction}}{2\text{ mol benzene}}\right) = 0.005\,189\text{ mol reaction}$

$\Delta_r H° = \dfrac{-33.99\text{ kJ}}{0.005\,189\text{ mol reaction}} = -6551\text{ kJ/mol}$

b. In a coffee-cup calorimetry experiment, conducted by the author, 2.1981 g of ammonium chloride (NH_4Cl, 0.05349 kg/mol) is added to 86.6883 g water ($c_p(H_2O(l)) = 4.184\ \dfrac{J}{g\,°C}$). The

temperature goes from 16.8 °C to 15.0 °C. What is the enthalpy of reaction ($\Delta_r H°$)? The balanced equation is:

$NH_4Cl(s) \rightarrow NH_4Cl(aq)$

$q_{bath} = mc_p(H_2O(l))\Delta T$

$q_{bath} = (86.6883\text{ g})(4.184\ \dfrac{J}{g\,°C})(15.0 - 16.8\ °C) = -650\text{ J (or }-0.65\text{ kJ)}$

$q_{reaction} = -(q_{water}) = -(-0.65\text{ kJ}) = 0.65\text{ kJ}$

$2.1981\text{ g }NH_4Cl\left(\dfrac{1\text{ kg}}{1000\text{ g}}\right)\left(\dfrac{1\text{ mol ammonium chloride}}{0.053\,49\text{ kg}}\right)\left(\dfrac{1\text{ mol reaction}}{1\text{ mol ammonium chloride}}\right) = 0.041\,094$

mol reaction

$\Delta_r H° = \dfrac{0.65\text{ kJ}}{0.041\,094\text{ mol reaction}} = 16\text{ kJ/mol}$

PROBLEM 15.2
For each reaction below, calculate $\Delta_r H°$ using E_d values.

a. $2\,H_2(g) + O_2(g) \rightarrow 2\,H_2O(l)$

$\Delta_r H° = \sum nE_d(\text{reactants}) - \sum nE_d(\text{products})$

$\Delta_r H° = 2E_d(H\text{-}H) + 1E_d(O=O) - 4E_d(H\text{-}O)$

$\Delta_r H° = 2(436\text{ kJ/mol}) + 1(498\text{ kJ/mol}) - 4(497\text{ kJ/mol})$

$\Delta_r H° = -618\text{ kJ/mol}$

b. $\underset{\displaystyle H}{\overset{\displaystyle H}{H-C-H}}$ $:\ddot{C}l-\ddot{C}l:$ $\longrightarrow$ $\underset{\displaystyle H}{\overset{\displaystyle H}{H-C-\ddot{C}l:}}$ $H-\ddot{C}l:$

$\Delta_rH° = \sum nE_d(\text{reactants}) - \sum nE_d(\text{products})$

$\Delta_rH° = 1E_d(\text{C-H}) + 1E_d(\text{Cl-Cl}) - 1E_d(\text{C-Cl}) - 1E_d(\text{H-Cl})$

$\Delta_rH° = 1(439 \text{ kJ/mol}) + 1(243 \text{ kJ/mol}) - 1(350. \text{ kJ/mol}) - 1(431 \text{ kJ/mol})$

$\Delta_rH° = -99 \text{ kJ/mol}$

c. $2 \text{ Al(s)} + 3 \text{ CuCl}_2(\text{aq}) \rightarrow 2 \text{ AlCl}_3(\text{aq}) + 3 \text{ Cu(s)}$

$\Delta_rH° = \sum nE_d(\text{reactants}) - \sum nE_d(\text{products})$

$\Delta_rH° = 6E_d(\text{Cu-Cl}) - 6E_d(\text{Al-Cl})$

$\Delta_rH° = 6(378 \text{ kJ/mol}) - 6(502 \text{ kJ/mol})$

$\Delta_rH° = -744 \text{ kJ/mol}$

PROBLEM 15.3

For each equation, calculate $\Delta_rH°$ using $\Delta_fH°$ values. Notice the standard enthalpy of formation can be used not just for chemical reactions but also for phase changes and for the dissociation of ionic compounds in water.

a. $UO_2(s) + 4 \text{ HF(aq)} \rightarrow UF_4(s) + 2 \text{ H}_2O(l)$

$\Delta_rH° = [2\Delta_fH°(\text{H}_2O(l)) + 1\Delta_fH°(UF_4(s))] - [4\Delta_fH°(\text{HF(aq)}) + 1\Delta_fH°(UO_2(s))]$

$\Delta_rH° = [2(-285.8 \text{ kJ/mol}) + 1(-1914.2 \text{ kJ/mol})] - [4(-321.1 \text{ kJ/mol}) + 1(-1085.0 \text{ kJ/mol})]$

$\Delta_rH° = -116.4 \text{ kJ/mol}$

b. $HF(g) \rightarrow HF(l)$

$\Delta_rH° = [1\Delta_fH°(\text{HF(l)})] - [1\Delta_fH°(\text{HF(g)})]$

$\Delta_rH° = [1(-299.0 \text{ kJ/mol})] - [1(-273.3 \text{ kJ/mol})]$

$\Delta_rH° = -25.7 \text{ kJ/mol}$

c. $NH_4NO_3(s) \rightarrow NH_4^+(\text{aq}) + NO_3^-(\text{aq})$

$\Delta_rH° = [1\Delta_fH°(NH_4^+(\text{aq})) + 1\Delta_fH°(NO_3^-(\text{aq}))] - [1\Delta_fH°(NH_4NO_3(s))]$

$\Delta_rH° = [1(-132.5 \text{ kJ/mol}) + 1(-207.4)] - [1(-365.2 \text{ kJ/mol})]$

$\Delta_rH° = 25.3 \text{ kJ/mol}$

d. $2 \text{ Al(s)} + 3 \text{ CuCl}_2(\text{aq}) \rightarrow 2 \text{ AlCl}_3(\text{aq}) + 3 \text{ Cu(s)}$

$\Delta_rH° = [2\Delta_fH°(\text{AlCl}_3(\text{aq})) + 3\Delta_fH°(\text{Cu(s)})] - [2\Delta_fH°(\text{Al(s)}) + 3\Delta_fH°(\text{CuCl}_2(\text{aq}))]$

$\Delta_rH° = [2(-1032.6 \text{ kJ/mol}) + 3(0 \text{ kJ/mol})] - [2(0 \text{ kJ/mol}) + 3(-269.6 \text{ kJ/mol})]$

$\Delta_rH° = -1256.4 \text{ kJ/mol}$

PROBLEM 15.4

For each equation, calculate $\Delta_rH°$ using the individual reaction enthalpy values provided.

a. Overall reaction: $Ca^{2+}(\text{aq}) + 2 \text{ OH}^-(\text{aq}) + CO_2(g) \rightarrow CaCO_3(s) + H_2O(l)$

Individual reactions:
$CaCO_3(s) \rightarrow CaO(s) + CO_2(g)$ $\Delta_rH° = 178.3 \text{ kJ/mol}$
$CaO(s) + H_2O(l) \rightarrow Ca(OH)_2(s)$ $\Delta_rH° = -65.2 \text{ kJ/mol}$
$Ca(OH)_2(s) \rightarrow Ca^{2+}(\text{aq}) + 2 \text{ OH}^-(\text{aq})$ $\Delta_rH° = 16.7 \text{ kJ/mol}$

The first reaction will be flipped to put $CaCO_3$ on the right.
$CaO(s) + CO_2(g) \rightarrow CaCO_3(s)$ $\Delta_rH° = 178.3$ kJ/mol

The second reaction will be flipped to get H_2O on the right.
$Ca(OH)_2(s) \rightarrow CaO(s) + H_2O(l)$ $\Delta_rH° = 65.2$ kJ/mol

The third reaction will be flipped to get the ions on the left.
$Ca^{2+}(aq) + 2\ OH^-(aq) \rightarrow Ca(OH)_2(s)$ $\Delta_rH° = 16.7$ kJ/mol

Now we add the three reactions above and the modified enthalpy values.
$CaO(s) + CO_2\ (g) + Ca(OH)_2(s) + Ca^{2+}(aq) + 2\ OH^-(aq) \rightarrow CaCO_3(s) + CaO(s) + H_2O(l) + Ca(OH)_2(s)$ $\Delta_rH° = -96.4$ kJ/mol

Canceling terms gives us the overall equation we want so the enthalpy shown is the desired reaction enthalpy.
$CO_2(g) + Ca^{2+}(aq) + 2\ OH^-(aq) \rightarrow H_2O(l) + Ca(OH)_2(s)$ $\Delta_rH° = -96.4$ kJ/mol

b. Overall reaction: $P_4(s) + 6\ Cl_2(g) \rightarrow 4\ PCl_3(l)$

Individual reactions:
$P_4(s) + 10\ Cl_2(g) \rightarrow 4\ PCl_5(s)$ $\Delta_rH° = -1774.0$ kJ/mol
$PCl_3(l) + Cl_2(g) \rightarrow PCl_5(s)$ $\Delta_rH° = -123.8$ kJ/mol

The first reaction will be left as is because it has 1 P_4 on the left:
$P_4(s) + 10\ Cl_2(g) \rightarrow 4\ PCl_5(s)$ $\Delta_rH° = -1774.0$ kJ/mol

The second reaction will be flipped (to get PCl_3 on the right) and multiplied by 4.
$4\ PCl_5(s) \rightarrow 4\ PCl_3(l) + 4\ Cl_2(g)$ $\Delta_rH° = 495.2$ kJ/mol

Now if we add the two reactions above and the modified enthalpy values.
$P_4(s) + 10\ Cl_2(g) + 4\ PCl_5(s) \rightarrow 4\ PCl_5(s) + 4\ PCl_3(l) + 4\ Cl_2(g)$ $\Delta_rH° = -1278.8$ kJ/mol

Canceling terms gives us the overall equation we want, so the enthalpy shown is the desired reaction enthalpy.

$P_4(s) + 6\ Cl_2(g) \rightarrow 4\ PCl_3(l)$ $\Delta_rH° = -1278.8$ kJ/mol

c. Overall reaction: $C_2H_4(g) + H_2O(l) \rightarrow C_2H_5OH(l)$

Individual reactions:
$C_2H_4(g) + 3\ O_2(g) \rightarrow 2\ CO_2(g) + 2\ H_2O(l)$ $\Delta_rH° = -1411.1$ kJ/mol
$C_2H_5OH(l) + 3\ O_2(g) \rightarrow 2\ CO_2(g) + 3\ H_2O(l)$ $\Delta_rH° = -1367.5$ kJ/mol

The first reaction will be left as is because it has C_2H_4 on the left side and with the right coefficient.
$C_2H_4(g) + 3\ O_2(g) \rightarrow 2\ CO_2(g) + 2\ H_2O(l)$ $\Delta_rH° = -1411.1$ kJ/mol

The second reaction will be flipped to get C_2H_5OH on the right side. The coefficients are good.
$2\ CO_2(g) + 3\ H_2O(l) \rightarrow C_2H_5OH(l) + 3\ O_2(g)$ $\Delta_rH° = 1367.5$ kJ/mol

Now we add the two reactions above and the modified enthalpy values.
$C_2H_4(g) + 3\ O_2(g) + 2\ CO_2(g) + 3\ H_2O(l) \rightarrow 2\ CO_2(g) + 2\ H_2O(l) + C_2H_5OH(l) + 3\ O_2(g)$ $\Delta_rH° = -43.6$ kJ/mol

Canceling terms gives us the overall equation we want, so the enthalpy shown is the desired reaction enthalpy.
$C_2H_4(g) + H_2O(l) \rightarrow C_2H_5OH(l)$ $\Delta_rH° = -43.6$ kJ/mol

PROBLEM 15.5
Calculate the enthalpy of reaction for each of the following. What trend(s) do you notice? What topics do the trend(s) suggest should be investigated further to try to understand the data?

a. $2 Al(s) + 3 Cu(NO_3)_2(aq) \rightarrow 2 Al(NO_3)_3(aq) + 3 Cu(s)$

$\Delta_rH° = [3\Delta_fH°(Cu(s)) + 2\Delta_fH°(Al^{3+}(aq)) + 6\Delta_fH°(NO_3^-(aq))] - [2\Delta_fH°(Al(s)) + 3\Delta_fH°(Cu^{2+}(aq)) + 6\Delta_fH°(NO_3^-(aq))]$

$\Delta_rH° = [3\Delta_fH°(Cu(s)) + 2\Delta_fH°(Al^{3+}(aq))] - [2\Delta_fH°(Al(s)) + 3\Delta_fH°(Cu^{2+}(aq))]$

$\Delta_rH° = [3(0.0 kJ/mol) + 2(-531.0 kJ/mol)] - [2(0.0 kJ/mol) + 3(64.8 kJ/mol)]$

$\Delta_rH° = -1256.4 kJ/mol$

b. $Al(s) + 3 AgNO_3(aq) \rightarrow Al(NO_3)_3(aq) + 3 Ag(s)$

$\Delta_rH° = [3\Delta_fH°(Ag(s)) + \Delta_fH°(Al^{3+}(aq)) + 3\Delta_fH°(NO_3^-(aq))] - [\Delta_fH°(Al(s)) + 3\Delta_fH°(Ag^+(aq)) + 3\Delta_fH°(NO_3^-(aq))]$

$\Delta_rH° = [3\Delta_fH°(Ag(s)) + \Delta_fH°(Al^{3+}(aq))] - [\Delta_fH°(Al(s)) + 3\Delta_fH°(Ag^+(aq))]$

$\Delta_rH° = [3(0.0 kJ/mol) + (-531.0 kJ/mol)] - [(0.0 kJ/mol) + 3(105.6 kJ/mol)]$

$\Delta_rH° = -847.8 kJ/mol$

c. $Al(s) + Au(NO_3)_3(aq) \rightarrow Al(NO_3)_3(aq) + Au(s)$

$\Delta_rH° = [\Delta_fH°(Au(s)) + \Delta_fH°(Al^{3+}(aq)) + 3\Delta_fH°(NO_3^-(aq))] - [\Delta_fH°(Al(s)) + \Delta_fH°(Au^{3+}(aq)) + 3\Delta_fH°(NO_3^-(aq))]$

$\Delta_rH° = [\Delta_fH°(Au(s)) + \Delta_fH°(Al^{3+}(aq))] - [\Delta_fH°(Al(s)) + \Delta_fH°(Au^{3+}(aq))]$

$\Delta_rH° = [3(0.0 kJ/mol) + (-531.0 kJ/mol)] - [(0.0 kJ/mol) + (409.2 kJ/mol)]$

$\Delta_rH° = -940.2 kJ/mol$

Here as we move down a group, there is a significant decrease in the exothermicity of the reaction from Cu to Ag and then a small increase going from Ag to Au. Considering the trends at play in this electron transfer reaction, it is worth investigating the ionization energy values for Cu, Ag, and Au (Chapter 8); the trends in atomic and ionic radii; and $F_{electrostatic}$ as it relates to effective nuclear charge. Later, it shall be seen that this can and should be investigated and studied in the context of electrochemistry (Chapter 19).

CHAPTER 16 ANSWERS

PROBLEM 16.1
For each of the following, identify only whether energy and/or matter is more dispersed on the left side/before or the right side/after (do not consider enthalpy). Then identify and explain whether you think this is a product-favored or reactant-favored process.

a. $NaCl(s) \rightleftarrows NaCl(aq)$

Dispersed on the right side/after = product favored (the sodium chloride ions are being dispersed in the water)

b. $H_2O(g) \rightleftarrows H_2O(l)$

Dispersed on the left side/before = reactant favored (the water molecules are condensing into a liquid, which means they are spread over less volume and have less energy)

c. $2 C_6H_6(l) + 15 O_2(g) \rightleftarrows 12 CO_2(g) + 6 H_2O(g)$

Dispersed on the right side/after = product favored (the amount of gas is increasing; gas particles are the most dispersed phase of matter)

d. $Pb(NO_3)_2(aq) + Na_2SO_4(aq) \rightleftarrows PbSO_4(s) + 2NaNO_3(aq)$

Dispersed on the left side/before = reactant favored (we start with six moles of ions dispersed/dissolved in water and produce a solid [a highly ordered and concentrated phase] and with only four moles of dispersed/dissolved ions)

e. $2\,NaCl(s) \rightleftharpoons 2\,Na(s) + Cl_2(g)$

Dispersed on the right side/after = product favored (a gas is produced from a solid, which leads to a greater dispersal of energy and matter)

f. A container of 1.0 mol neon gas goes from 5.0 L in volume to 10.0 L in volume.

Dispersed on the right side/after = product favored (the neon molecules are now spread through a larger volume)

g. A 50. mM KBr solution is diluted to 25 mM by the addition of more water.

Dispersed on the right side/after = product favored (the potassium and bromide ions have a larger volume of water to disperse in)

h. A refrigerator gets colder on the inside than it is on the outside.

Dispersed on the left side/before = reactant favored (the energy [thermal motion] is being concentrated outside of the refrigerator)

i. A transport protein maintains more $H^+(aq)$ in the intermembrane space of a mitochondrion than there is in the matrix.

Dispersed on the left side/before = reactant favored (hydrons are being concentrated in the matrix and not dispersed throughout the mitochondrion)

PROBLEM 16.2
For each of the following, calculate $\Delta_r S°$ using Appendix 7. How do your results compare with your predictions made?

a. $NaCl(s) \rightleftharpoons NaCl(aq)$

$\Delta_r S° = \sum nS°(products) - \sum nS°(reactants)$

$\Delta_r S° = 1S°(Na^+(aq)) + 1S°(Cl^-(aq)) - 1S°(NaCl(s))$

$\Delta_r S° = 1(59.0\ J/(mol\ K)) + 1(56.5\ J/(mol\ K)) - 1(72.1\ J/(mol\ K))$

$\Delta_r S° = 43.4\ J/(mol\ K)$ (entropy is increasing as the ions dissociate and disperse, products favored)

b. $H_2O(g) \rightleftharpoons H_2O(l)$

$\Delta_r S° = \sum nS°(products) - \sum nS°(reactants)$

$\Delta_r S° = 1S°(H_2O(l)) - 1S°(H_2O(g))$

$\Delta_r S° = 1(70.0\ J/(mol\ K)) - 1(188.8\ J/(mol\ K))$

$\Delta_r S° = -118.8\ J/(mol\ K)$ (entropy is decreasing as water condenses, reactants favored)

c. $2\,C_6H_6(l) + 15\,O_2(g) \rightleftharpoons 12\,CO_2(g) + 6\,H_2O(g)$

$\Delta_r S° = \sum nS°(products) - \sum nS°(reactants)$

$\Delta_r S° = 12S°(CO_2(g)) + 6S°(H_2O(g)) - [2S°(C_6H_6(l)) + 15S°(O_2(g))]$

$\Delta_r S° = 12(213.8\ J/(mol\ K)) + 6(188.8\ J/(mol\ K)) - [2(173.4\ J/(mol\ K)) + 15(205.2\ J/(mol\ K))]$

$\Delta_r S° = 273.6\ J/(mol\ K)$ (entropy is increasing as the number of moles of gas increases, products favored)

d. $Pb(NO_3)_2(aq) + Na_2SO_4(aq) \rightleftharpoons PbSO_4(s) + 2NaNO_3(aq)$

$\Delta_r S° = \sum nS°(products) - \sum nS°(reactants)$

$\Delta_r S° = 2S°(Na^+(aq)) + 2S°(NO_3^-(aq)) + 1S°(PbSO_4(s)) - [1S°(Pb^{2+}(aq)) + 1S°(SO_4^{2-}(aq)) + 2S°(Na^+(aq)) + 2S°(NO_3^-(aq))]$

$\Delta_r S° = 1S°(PbSO_4(s)) - [1S°(Pb^{2+}(aq)) + 1S°(SO_4^{2-}(aq))]$

$\Delta_r S° = 1(148.5\ J/(mol\ K)) - [1(10.5\ J/(mol\ K)) + 1(20.1\ J/(mol\ K))]$

$\Delta_r S° = 117.9$ J/(mol K) (entropy is increasing, products favored! A surprising result, as we would assume forming a solid would decrease entropy. Suggests the story is more complicated.)

e. $2 \, NaCl(s) \rightleftarrows 2 \, Na(s) + Cl_2(g)$

$\Delta_r S° = \sum nS°(products) - \sum nS°(reactants)$

$\Delta_r S° = 2S°(Na(s)) + 1S°(Cl_2(g)) - 2S°(NaCl(s))$

$\Delta_r S° = 2(51.3 \, J/(mol \, K)) + 1(223.1 \, J/(mol \, K)) - 2(72.1 \, J/(mol \, K))$

$\Delta_r S° = 181.5$ J/(mol K) (entropy is increasing as a gas forms, products favored)

PROBLEM 16.3

When we calculate $\Delta_r S°$, we arrive at the net or total entropy change for a process. This can, however, obscure factors at the atomic level that are contributing to an increase in entropy and those that are contributing to a decrease in entropy. The model below is an idealized atom-scale view of what happens as NaCl dissolves in water.

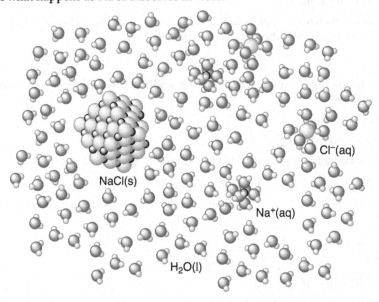

a. Referring to the model, provide an example of how entropy is increasing in the system.

The Na⁺ and Cl⁻ ions are separating from a larger solid and are dispersing in water. This dispersal of ions in water corresponds to an increase in entropy.

b. Referring to the model, provide an example of how entropy is decreasing in the system.

The water molecules solvate the sodium and chloride ions (through ion–dipole and H-bond interactions). This concentration of water molecules around the ions corresponds to a decrease in entropy.

c. When some ionic compounds dissolve, the value for $\Delta_r S° > 0$. What can you say about the two effects discussed in parts a and b?

For situations where ionic compounds dissolve and $\Delta_r S° > 0$, the entropy increase associated with the dispersal of ions must be greater than the entropy decrease associated with the solvation of ions by water molecules. (Note: This is common when ions are large and/or have a small charge).

d. When some ionic compounds dissolve, the value for $\Delta_r S° < 0$. What can you say about the two effects discussed in parts a and b?

For situations where ionic compounds dissolve and $\Delta_r S° > 0$, the entropy increase associated with the dispersal of ions must be less than the entropy decrease associated with the solvation of ions by water molecules. (Note: This is common when ions are small and/or have high charge).

CHAPTER 17 ANSWERS

PROBLEM 17.1
For each reaction, calculate $\Delta_r G°$ at 77.0 K (the temperature of liquid nitrogen), 298.15 K (room temperature), and 737 K (the surface temperature of Venus).

a. $CH_3OH(g) + H_2(g) \rightleftarrows CH_4(g) + H_2O(g)$

$\Delta_r H° = -116.0$ kJ/mol, $\Delta_r S° = 4.6$ J/(mol K)
$\Delta_r G°(77.0$ K$) = -116.0$ kJ/mol $- (77.0$ K$)(0.0046$ kJ/(mol K)$) = -116.4$ kJ/mol (product favored)
$\Delta_r G°(298.15$ K$) = -116.0$ kJ/mol $- (298.15$ K$)(0.0046$ kJ/(mol K)$) = -117.4$ kJ/mol (product favored)
$\Delta_r G°(737$ K$) = -116.0$ kJ/mol $- (737$ K$)(0.0046$ kJ/(mol K)$) = -119.4$ kJ/mol (product favored)

b. $Mg(OH)_2(s) \rightleftarrows MgO(s) + H_2O(g)$

$\Delta_r H° = 81.0$ kJ/mol, $\Delta_r S° = 152.6$ J/(mol K)
$\Delta_r G°(77.0$ K$) = 81.0$ kJ/mol $- (77.0$ K$)(0.1526$ kJ/(mol K)$) = 69.2$ kJ/mol (reactant favored)
$\Delta_r G°(298.15$ K$) = 81.0$ kJ/mol $- (298.15$ K$)(0.1526$ kJ/(mol K)$) = 35.5$ kJ/mol (reactant favored)
$\Delta_r G°(737$ K$) = 81.0$ kJ/mol $- (737$ K$)(0.1526$ kJ/(mol K)$) = -31.5$ kJ/mol (product favored)

c. $2 CH_4(g) \rightleftarrows C_2H_6(g) + H_2(g)$

$\Delta_r H° = 64.9$ kJ/mol, $\Delta_r S° = -12.9$ J/(mol K)
$\Delta_r G°(77.0$ K$) = 64.9$ kJ/mol $- (77.0$ K$)(-0.0129$ kJ/(mol K)$) = 65.9$ kJ/mol (reactant favored)
$\Delta_r G°(298.15$ K$) = 64.9$ kJ/mol $- (298.15$ K$)(-0.0129$ kJ/(mol K)$) = 68.7$ kJ/mol (reactant favored)
$\Delta_r G°(737$ K$) = 64.9$ kJ/mol $- (737$ K$)(-0.0129$ kJ/(mol K)$) = 74.4$ kJ/mol (reactant favored)

d. $CO(g) + 2 H_2(g) \rightleftarrows CH_3OH(g)$

$\Delta_r H° = -90.1$ kJ/mol, $\Delta_r S° = -219.2$ J/(mol K)
$\Delta_r G°(77.0$ K$) = -90.1$ kJ/mol $- (77.0$ K$)(-0.2192$ kJ/(mol K)$) = -73.2$ kJ/mol (product favored)
$\Delta_r G°(298.15$ K$) = -90.1$ kJ/mol $- (298.15$ K$)(-0.2192$ kJ/(mol K)$) = -24.7$ kJ/mol (product favored)
$\Delta_r G°(737$ K$) = -90.1$ kJ/mol $- (737$ K$)(-0.2192$ kJ/(mol K)$) = 71.5$ kJ/mol (reactant favored)

e. $H_2(g) + Br_2(g) \rightleftarrows 2 HBr(g)$

$\Delta_r H° = -103.71$ kJ/mol, $\Delta_r S° = 21.2$ J/(mol K)
$\Delta_r G°(77.0$ K$) = -103.7$ kJ/mol $- (77.0$ K$)(0.0212$ kJ/(mol K)$) = -105.3$ kJ/mol (product favored)
$\Delta_r G°(298.15$ K$) = -103.7$ kJ/mol $- (298.15$ K$)(0.0212$ kJ/(mol K)$) = -110.0$ kJ/mol (product favored)
$\Delta_r G°(737$ K$) = -103.7$ kJ/mol $- (737$ K$)(0.0212$ kJ/(mol K)$) = -119.3$ kJ/mol (product favored)

PROBLEM 17.2
For each reaction, determine the inversion temperature in Celsius (°C).

a. $CH_3OH(g) + H_2(g) \rightleftarrows CH_4(g) + H_2O(g)$

$\Delta_r H° = -116.0$ kJ/mol, $\Delta_r S° = 4.6$ J/(mol K)

$$T = \frac{\Delta_r H°}{\Delta_r S°} = \frac{-116.0 \text{ kJ/mol}}{0.0046 \text{ kJ/(mol K)}} = -25\,000 \text{ K (impossible); reaction will always be product}$$

favored.

b. $Mg(OH)_2(s) \rightleftarrows MgO(s) + H_2O(g)$

$\Delta_rH° = 81.0$ kJ/mol, $\Delta_rS° = 152.6$ J/(mol K)

$T = \dfrac{\Delta_rH°}{\Delta_rS°} = \dfrac{81.0 \text{ kJ/mol}}{0.1526 \text{ kJ/(mol K)}} = 531$ K $= 258$ °C; switches from reactant to product favored at 258 °C.

c. $2 CH_4(g) \rightleftarrows C_2H_6(g) + H_2(g)$

$\Delta_rH° = 64.9$ kJ/mol, $\Delta_rS° = -12.9$ J/(mol K)

$T = \dfrac{\Delta_rH°}{\Delta_rS°} = \dfrac{64.9 \text{ kJ/mol}}{-0.0129 \text{kJ/(mol K)}} = -5030$ K (impossible); reaction will always be reactant favored.

d. $CO(g) + 2 H_2(g) \rightleftarrows CH_3OH(g)$

$\Delta_rH° = -90.1$ kJ/mol, $\Delta_rS° = -219.2$ J/(mol K)

$T = \dfrac{\Delta_rH°}{\Delta_rS°} = \dfrac{-90.1 \text{ kJ/mol}}{-0.2192 \text{ kJ/(mol K)}} = 411$ K $= 138$ °C; switches from product to reactant favored at 138 °C.

e. $H_2(g) + Br_2(g) \rightleftarrows 2 HBr(g)$

$\Delta_rH° = -103.71$ kJ/mol, $\Delta_rS° = 21.243$ J/(mol K)

$T = \dfrac{\Delta_rH°}{\Delta_rS°} = \dfrac{-103.7 \text{ kJ/mol}}{0.0212 \text{ kJ/(mol K)}} = -4890$ K (impossible); reaction will always be product favored.

PROBLEM 17.3
Using the standard Gibbs energy of formation values ($\Delta_fG°$) in Appendix 7, calculate the Gibbs energy of reaction ($\Delta_rG°$) for each of the following.

a. $3 C_2H_2(g) \rightleftarrows C_6H_6(l)$

$\Delta_rG° = 1\Delta_fG°(C_6H_6(l)) - 3\Delta_fG°(C_2H_2(g))$

$\Delta_rG° = 1(124.5 \text{ kJ/mol}) - 3(209.9 \text{ kJ/mol})$

$\Delta_rG° = -505.2$ kJ/mol

b. $2 H_2O_2(l) \rightleftarrows 2 H_2O(l) + O_2(g)$

$\Delta_rG° = 1\Delta_fG°(O_2(g)) + 1\Delta_fG°(H_2O(l)) - 2\Delta_fG°(H_2O_2(l))$

$\Delta_rG° = 1(0 \text{ kJ/mol}) + 2(-237.1 \text{ kJ/mol}) - 2(-120.4 \text{ kJ/mol})$

$\Delta_rG° = -233.4$ kJ/mol

c. $Pb(s) + PbO_2(s) + 2 H_2SO_4(aq) \rightleftarrows 2 PbSO_4(s) + 2 H_2O(l)$

$\Delta_rG° = 2\Delta_fG°(PbSO_4(s)) + 2\Delta_fG°(H_2O(l)) - [1\Delta_fG°(Pb(s)) + 1\Delta_fG°(PbO_2(s)) + 2\Delta_fG°(H_2SO_4(aq))]$

$\Delta_rG° = 2(-813.0 \text{ kJ/mol}) + 2(-237.1 \text{ kJ/mol}) - [1(0 \text{ kJ/mol}) + 1(-217.3 \text{ kJ/mol}) + 2(-744.5 \text{ kJ/mol})]$

$\Delta_rG° = -393.9$ kJ/mol

d. $2 KMnO_4(aq) + 6 HCl(aq) + 5 CH_2O(g) \rightleftarrows 5 HCO_2H(aq) + 2 MnCl_2(s) + 2 KCl(aq) + 3 H_2O(l)$

$\Delta_rG° = 5\Delta_fG°(HCO_2H(aq)) + 2\Delta_fG°(MnCl_2(s)) + 2\Delta_fG°(KCl(aq)) + 3\Delta_fG°(H_2O(l)) - [2\Delta_fG°(KMnO_4(s)) + 6\Delta_fG°(HCl(aq)) + 5\Delta_fG°(CH_2O(g))]$

$\Delta_r G° = 5(-372.3 \text{ kJ/mol}) + 2(-440.5 \text{ kJ/mol}) + 2(-414.5 \text{ kJ/mol}) + 3(-237.1 \text{ kJ/mol}) - [2(-737.6 \text{ kJ/mol}) + 6(-131.2 \text{ kJ/mol}) + 5(-102.5 \text{ kJ/mol})]$

$\Delta_r G° = -1507.9 \text{ kJ/mol}$

e. $B_2O_3(s) + 3 Mg(s) \rightleftarrows 2 B(s) + 3 MgO(s)$

$\Delta_r G° = 2\Delta_f G°(B(s)) + 3\Delta_f G°(MgO(s)) - [1\Delta_f G°(B_2O_3(s)) + 3\Delta_f G°(Mg(s))]$

$\Delta_r G° = 2(0 \text{ kJ/mol}) + 3(-569.3 \text{ kJ/mol}) - [1(-1194.3 \text{ kJ/mol}) + 3(0 \text{ kJ/mol})]$

$\Delta_r G° = -513.6 \text{ kJ/mol}$

PROBLEM 17.4

Metals are often dug out of the ground as ores, compounds with oxygen, sulfur, and other elements. To isolate the pure metal, many metallurgical reactions and processes involve transformations that will produce a metal oxide. To isolate the metal, the metal oxide could be heated until it decomposes. Consider an example of iron(III) oxide – the ore is known as hematite – decomposing:

$$2 Fe_2O_3(s) \rightleftarrows 4 Fe(s) + 3 O_2(g)$$

Using the values in Appendix 7, calculate $\Delta_r G°$ for the decomposition of iron(III) oxide to pure iron. Is this a product-favored or reactant-favored process? Can you offer an explanation as to why the reaction needs to be heated?

$\Delta_r G° = 3\Delta_f G°(O_2(g)) + 4\Delta_f G°(Fe(s)) - 2\Delta_f G°(Fe_2O_3(s))$

$\Delta_r G° = 3(0 \text{ kJ/mol}) + 4(0 \text{ kJ/mol}) - 2(-742.2 \text{ kJ/mol})$

$\Delta_r G° = 1484.4 \text{ kJ/mol}$

Reactant favored ($\Delta_r G° > 0$); this is a disfavored reaction and so energy must be added to make products.

PROBLEM 17.5

Calculate the temperature iron(III) oxide must be heated to (in °C) to spontaneously decompose into iron and oxygen.

$\Delta_r H° = 3\Delta_f H°(O_2(g)) + 4\Delta_f H°(Fe(s)) - 2\Delta_f H°(Fe_2O_3(s))$

$\Delta_r H° = 3(0 \text{ kJ/mol}) + 4(0 \text{ kJ/mol}) - 2(-824.2 \text{ kJ/mol}) = 1484.4 \text{ kJ/mol}$

$\Delta_r S° = 3S°(O_2(g)) + 4S°(Fe(s)) - 2S°(Fe_2O_3(s))$

$\Delta_r S° = 3(130.7 \text{ J/(mol K)}) + 4(27.3 \text{ J/(mol K)}) - 2(87.4 \text{ J/(mol K)}) = 550.0 \text{ J/(mol K)}$

$T = \dfrac{\Delta_r H°}{\Delta_r S°} = \dfrac{148\,400 \text{ J/mol}}{550.0 \text{ J/(mol K)}} = 2997 \text{ K} = 2724 °C$

PROBLEM 17.6

Now consider the formation of carbon dioxide from solid, graphite carbon (coke):

$$C(s, \text{graphite}) + O_2(g) \rightleftarrows CO_2(g)$$

What is $\Delta_r G°$ for the oxidation of coke to carbon dioxide? Is this a product-favored or reactant-favored process?

$\Delta_r G° = 1\Delta_f G°(CO_2(g)) - [1\Delta_f G°(C(s, \text{graphite})) + 1\Delta_f G°(O_2(g))]$

$\Delta_r G° = 1(-394.4 \text{ kJ/mol}) - [1(0 \text{ kJ/mol}) + 1(0 \text{ kJ/mol})] = -394.4 \text{ kJ/mol}$

Product favored ($\Delta_r G° < 0$)

PROBLEM 17.7

Now consider the formation of water from hydrogen gas:

$$2 H_2(g) + O_2(g) \rightleftarrows 2 H_2O(l)$$

a. What is $\Delta_r G°$ for the oxidation of hydrogen to water? Is this a product-favored or reactant-favored process?

$\Delta_r G° = 2\Delta_f G°(H_2O(l)) - [2\Delta_f G°(H_2(g)) + 1\Delta_f G°(O_2(g))]$
$\Delta_r G° = 2(-237.1\ kJ/mol) - [1(0\ kJ/mol) + 1(0\ kJ/mol)] = -474.2\ kJ/mol$
Product favored ($\Delta_r G° < 0$)

b. What are the advantages of oxidizing hydrogen over oxidizing coke?

For each mole of oxygen gas consumed, more energy is produced in oxidizing hydrogen. In addition, oxidizing hydrogen produces water, a benign byproduct, whereas oxidizing coke produces carbon dioxide, a greenhouse gas.

PROBLEM 17.8

In industrial metallurgy, it is often the case that the reduction of a metal oxide to pure metal is accomplished by heating that metal oxide together with coke. Increasingly, companies are looking to switch to heating the metal oxide together with hydrogen. If you take the two equations above and add them, you get a new combined (coupled) reaction:

$$2\ Fe_2O_3(s) \rightleftarrows 4\ Fe(s) + 3\ O_2(g)$$
$$\underline{6\ H_2(g) + 3\ O_2(g) \rightleftarrows 6\ H_2O(l)}$$
$$6\ H_2(g) + 2\ Fe_2O_3(s) \rightleftarrows 4\ Fe(s) + 6\ H_2O(l),\ overall$$
$$3\ H_2(g) + Fe_2O_3(s) \rightleftarrows 2\ Fe(s) + 3\ H_2O(l),\ simplified$$

In coupling a reaction there is a different overall reaction, but one that still produces pure iron, the desired product. The new $\Delta_r G°$ for this process can be calculated as you did above by using the $\Delta_f G°$ values, but since we are adding the two reactions together, we can find $\Delta_r G°$ by adding the $\Delta_r G°$ from Problem 17.4 and Problem 17.7. What is $\Delta_r G°$ for this coupled reaction? $\Delta_r G° = 0.5(1484.4\ kJ/mol) + 1.5(-474.2\ kJ/mol) = 30.9\ kJ/mol$

PROBLEM 17.9

Considering your answer to Problem 17.8, can you explain why metallurgy couples the oxidation of hydrogen with the reduction of a metal oxide (like iron(III) oxide)?

Metallurgy couples these reactions together because it takes a reactant-favored process and combines it with a product-favored process to make an overall reaction that is significantly more product favored.

PROBLEM 17.10

Problem 17.8 states that the metal oxide is heated together with hydrogen. Given your answer to Problem 17.8 and the value of $\Delta_r G°$, can you provide an explanation as to why metal oxide and hydrogen must be heated? (There are two reasons for this.)

1. Energy is required to provide the activation energy to start the reaction (the added heat is necessary to initiate the reaction).

2. The reaction is still reactant favored and so it requires additional energy to drive the reaction to completion.

PROBLEM 17.11

Consider the hydrolysis of ATP in water:

$ATP(aq) + H_2O(l) \rightleftarrows ADP(aq) + HPO_4{}^{2-}(aq)$ $\qquad$ $\Delta_r G° = -30.5\ kJ/mol$

Looking at the $\Delta_r G°$ value, can you provide an explanation for how the body uses ATP (in a general sense)? That is, from an energetic standpoint, what is ATP doing? Consider the metallurgy example and the role of dihydrogen.

The body is using the breakdown of ATP as a reaction that can be coupled with other (reactant-favored) processes in the body. That is, a process that is reactant-favored can be coupled with the breakdown of ATP to yield a product-favored reaction.

PROBLEM 17.12

What is $\Delta_r G°$ for the reaction to make iodomethane from methanol and iodide?

$\Delta_r G° = 1\Delta_f G°(OH^-(aq)) + 1\Delta_f G°(iodomethane(l)) - [1\Delta_f G°(I^-(aq)) + 1\Delta_f G°(methanol(aq))]$
$\Delta_r G° = 1(-157.2 \text{ kJ/mol}) + 1(13.4 \text{ kJ/mol}) - [1(-51.6 \text{ kJ/mol}) + 1(-175.3 \text{ kJ/mol})]$
$\Delta_r G° = 83.1 \text{ kJ/mol}$

PROBLEM 17.13

Are we going to succeed in making iodomethane this way at 25 °C?

No, this is a reactant-favored ($\Delta_r G° > 0$) process and so it will not (on its own) make products.

PROBLEM 17.14

Given the $\Delta_f H°$ and $S°$ values for each (below), what are $\Delta_r H°$ and $\Delta_r S°$? Is this a reaction that could be made to happen by changing the temperature to be higher or lower than 25 °C?

$\Delta_r H° = 1(-15.5 \text{ kJ/mol}) + 1(-230.0 \text{ kJ/mol}) - [1(-55.2 \text{ kJ/mol}) + 1(-245.9 \text{ kJ/mol})]$
$\Delta_r H° = 55.6 \text{ kJ/mol (endothermic)}$

$\Delta_r S° = 1(163.2 \text{ J/(mol K)}) + 1(-10.8 \text{ J/(mol K)}) - [1(111.3 \text{ J/(mol K)}) + 1(133.1 \text{ J/(mol K)})]$
$\Delta_r S° = -92.6 \text{ J/(mol K) (entropy decrease)}$

This is an endothermic, entropy-decreasing reaction. It is reactant-favored at all temperatures.

PROBLEM 17.15

Calculate $\Delta_r G°$ for each reaction. Which reaction will work to make iodomethane at room temperature and which will not?

sodium iodide(aq)	methanol(aq)		iodomethane(l)	sodium hydroxide(aq)
$\Delta_f G° = -313.5 \text{ kJ/mol}$	$\Delta_f G° = -175.3 \text{ kJ/mol}$		$\Delta_f G° = 13.4 \text{ kJ/mol}$	$\Delta_f G° = -419.1 \text{ kJ/mol}$

a. $\Delta_r G° = 83.1 \text{ kJ/mol}$ (reactant favored)

hydrogen iodide(aq)	methanol(aq)	iodomethane(l)	water(l)
$\Delta_f G° = -51.6 \text{ kJ/mol}$	$\Delta_f G° = -175.3 \text{ kJ/mol}$	$\Delta_f G° = 13.4 \text{ kJ/mol}$	$\Delta_f G° = -237.1 \text{ kJ/mol}$

b. $\Delta_r G° = 3.2 \text{ kJ/mol}$ (slightly reactant favored)

iodine(s)	methanol(aq)	iodomethane(l)	
$\Delta_f G° = 0 \text{ kJ/mol}$	$\Delta_f G° = -175.3 \text{ kJ/mol}$	$\Delta_f G° = 13.4 \text{ kJ/mol}$	
phosphorus(s, red)		phosphoric acid(l)	water(l)
$\Delta_f G° = -12.2 \text{ kJ/mol}$		$\Delta_f G° = -1123.6 \text{ kJ/mol}$	$\Delta_f G° = -237.1 \text{ kJ/mol}$

c. $\Delta_r G° = -810.0 \text{ kJ/mol}$ (product favored)

PROBLEM 17.16

Consider the reactions in Problem 17.15.

a. Which reaction is (or reactions are) best in terms of Gibbs energy?

In terms of Gibbs energy, the reaction with phosphorus and iodine is best because it is the most exergonic and most product favored.

b. Which reaction is (or reactions are) best in terms of atom economy?

In terms of atom economy, the reactions with NaI and HI are best because the amount of wasted byproduct is minimized (only three atoms of waste byproduct are produced for every molecule of desired iodomethane), whereas the reaction with phosphorus and iodine produces 22 atoms of waste byproduct for every molecule of desired iodomethane.

c. Which reaction is (or reactions are) better in terms of chemical hazards (use the internet to find the safety data sheets [SDS] for each chemical other than methanol and iodomethane and identify the hazards associated with each)?

Reaction 1 uses sodium iodide, which is an irritant and toxic to aquatic life, and it produces sodium hydroxide, which is a strong base that is caustic, corrosive, and can damage living tissue.

Reaction 2 uses hydrogen iodide, which is a strong acid that is caustic, corrosive, and can damage living tissue, and it produces water, a benign byproduct.

Reaction 3 uses red phosphorus, which is a flammable solid that is harmful to aquatic life, and it uses iodine, which is an irritant and very toxic to aquatic life. It produces phosphoric acid, which is a moderate acid that can cause severe eye and skin burns and is corrosive, and it produces water, a benign byproduct.

d. Is there a single best reaction, from a green chemistry perspective, to produce iodomethane?

No, each reaction has advantages and disadvantages. The reaction with iodine and phosphorus is the best in terms of being the most product favored and uses/produces relatively benign chemicals, but it suffers in terms of atom economy. The reaction with hydrogen iodide could likely work well with only moderate heating, and although it uses a strong acid, it is good in terms of its atom economy and its benign byproduct. The reaction with sodium hydroxide is decently reactant favored and cannot be made product favored with a change in temperature, and so this is the worst single reaction of the three.

CHAPTER 18 ANSWERS

PROBLEM 18.1
Using Appendix 7, calculate $\Delta_r G°$ for this reaction (25 °C) and then convert that value to the equilibrium constant (K).

$$H_2O(l) \rightleftharpoons H^+(aq) + OH^-(aq)$$

$\Delta_r G° = 1\Delta_f G°(OH^-(aq)) + 1\Delta_f G°(H^+(aq)) - 1\Delta_f G°(H_2O(l))$

$\Delta_r G° = -157.2$ kJ/mol $+ 0$ kJ/mol $- (-237.1$ kJ/mol$)$

$\Delta_r G° = 79.9$ kJ/mol

$$K = e^{-\left(\frac{\Delta_r G°}{RT}\right)} = e^{-\left(\frac{79.9\,\text{kJ/mol}}{(0.008\,314\,\text{kJ/(mol K)})(298\,\text{K})}\right)}$$

$K = 1 \times 10^{-14}$

PROBLEM 18.2
Convert the $\Delta_r G°$ values to equilibrium constants (K) at 298 K.

a. $ATP(aq) + H_2O(l) \rightleftharpoons ADP(aq) + HPO_4^{2-}(aq)$ \hfill $\Delta_r G° = -30.5$ kJ/mol

$$K = e^{-\left(\frac{\Delta_r G°}{RT}\right)} = e^{-\left(\frac{-30.5\,\text{kJ/mol}}{(0.008\,314\,\text{kJ/(mol K)})(298\,\text{K})}\right)}$$

$K = 200\,000$

b. $H_2(g) + CO_2(g) \rightleftharpoons H_2O(g) + CO(g)$ $\qquad\qquad\qquad\qquad$ $\Delta_r G° = 28.62$ kJ/mol

$$K = e^{-\left(\frac{\Delta_r G°}{RT}\right)} = e^{-\left(\frac{28.62 \text{ kJ/mol}}{(0.008\,314 \text{ kJ/(mol K)})(298 \text{ K})}\right)}$$

$K = 9.7 \times 10^{-6}$

PROBLEM 18.3
Convert each equilibrium constant (K) to $\Delta_r G°$ at 298 K.

a. $N_2O_4(g) \rightleftharpoons 2\,NO_2(g)$ $\qquad\qquad\qquad\qquad\qquad\qquad$ $K = 0.0063$

$\quad$ $\Delta_r G° = -RT\ln K$

$\quad$ $\Delta_r G° = -((0.008\,314 \text{ kJ/(mol K)})(298 \text{ K}))\ln(0.0063)$

$\quad$ $\Delta_r G° = 12.55$ kJ/mol

b. $CH_3CO_2H(l) + C_2H_5OH(aq) \rightleftharpoons CH_3CO_2C_2H_5(l) + H_2O(l)$ $\quad$ $K = 0.4$

$\quad$ $\Delta_r G° = -RT\ln K$

$\quad$ $\Delta_r G° = -((0.008\,314 \text{ kJ/(mol K)})(298 \text{ K}))\ln(0.4)$

$\quad$ $\Delta_r G° = 2.3$ kJ/mol

c. $CuCl_2(aq) + Fe(s) \rightleftharpoons Cu(s) + FeCl_2(aq)$ $\qquad\qquad\qquad$ $K = 2.4 \times 10^{26}$

$\quad$ $\Delta_r G° = -RT\ln K$

$\quad$ $\Delta_r G° = -((0.008314 \text{ kJ/(mol K)})(298 \text{ K}))\ln(2.4 \times 10^{26})$

$\quad$ $\Delta_r G° = -150.49$ kJ/mol

PROBLEM 18.4
For each reaction, write an expression for Q_c.

a. $2\,H_2(g) + O_2(g) \rightleftharpoons 2\,H_2O(g)$ $\qquad\qquad\qquad$ $Q_c = \dfrac{\left[H_2O(g)\right]^2}{\left[H_2(g)\right]^2\left[O_2(g)\right]}$

b. $Co(H_2O)_6^{2+}(aq) + 4\,Cl^-(aq) \rightleftharpoons CoCl_4^{2-}(aq) + 6\,H_2O(l)$ $\qquad$ $Q_c = \dfrac{\left[CoCl_4^{2-}(aq)\right]}{\left[Co(H_2O)_6^{2+}(aq)\right]\left[Cl^-(aq)\right]^4}$

PROBLEM 18.5
Provided an expression for Q_c, what is the chemical reaction?

$2\,SO_2(g) + 2\,H_2O(g) \rightleftharpoons 3\,O_2(g) + 2\,H_2S(g)$ $\qquad\qquad$ $Q_c = \dfrac{[H_2S(g)]^2\left[O_2(g)\right]^3}{\left[SO_2(g)\right]^2\left[H_2O(g)\right]^2}$

PROBLEM 18.6
For the following reactions, calculate Q_c using the provided concentration values. K_c is provided for each reaction. Indicate whether the reaction is at equilibrium or not. If the reaction is not at equilibrium, will more reactants be produced (shift left) or will more products be produced (shift right)?

a. $H_2(g) + I_2(g) \rightleftharpoons 2\,HI(g)$ $\qquad$ $K_c(745 \text{ K}) = 50.0$

$\quad$ $[H_2] = 0.001\,25$ mol/L, $[I_2] = 0.001\,25$ mol/L, $[HI] = 0.0375$ mol/L

$$Q_c = \frac{\left[HI(g)\right]^2}{\left[H_2(g)\right]\left[I_2(g)\right]} = \frac{[0.0375]^2}{[0.001\,25][0.001\,25]} = 900$$

Here, $K < Q$, which means there are more products than reactants than there are at equilibrium, so the reaction will produce more reactants (shift left).

b. $O_2(g) + N_2(g) \rightleftarrows 2\,NO(g)$ $K_c(745\,K) = 1.7 \times 10^{-3}$

[O_2] = 0.25 mol/L, [N_2] = 0.25 mol/L, [NO] = 0.0103 mol/L

$Q_c = = 1.7 \times 10^{-3}$

Here, $K = Q$, so the reaction is at equilibrium.

PROBLEM 18.7

For the following reactions, the concentration of each chemical at equilibrium was measured. Calculate K_c.

a. $PCl_5(g) \rightleftarrows PCl_3(g) + Cl_2(g)$

At equilibrium at 250 °C, [PCl_5] = 4.2×10^{-5} mol/L, [PCl_3] = 1.3×10^{-2} mol/L, [Cl_2] = 3.9×10^{-3} mol/L.

$$K_c = \frac{[PCl_3(g)][Cl_2(g)]}{[PCl_5(g)]} = \frac{[0.013][0.0039]}{[0.000\,042]} = 1.2$$

b. $2\,BrF_5(g) \rightleftarrows Br_2(g) + 5\,F_2(g)$

At equilibrium at 1500 K, [BrF_5] = 0.0064 mol/L, [Br_2] = 0.0018 mol/L, [F_2] = 0.0090 mol/L.

$K_c = = 2.6 \times 10^{-9}$

PROBLEM 18.8

Consider the reaction of nitrosyl chloride, NOCl, below. Initially, the concentration of NOCl is 2.00 mol/L. When equilibrium is established at 462 °C, 0.66 mol/L NO is present. Calculate the equilibrium constant K_c for this reaction.

$$2\,NOCl(g) \rightleftarrows 2\,NO(g) + Cl_2(g) \quad Q_c = \frac{[NO(g)]^2[Cl_2(g)]}{[NOCl(g)]^2}$$

	2 NOCl(g)	⇄	2 NO(g)	Cl$_2$(g)
I	2.00 mol/L	⇄	0 mol/L	0 mol/L
C	−2x	⇄	+2x	+1x
E	2.00 − 2x mol/L	⇌	2x mol/L	x mol/L

[NO]$_{equilibrium}$ = 2x = 0.66 mol/L therefore x = 0.33 mol/L
[NOCl]$_{equilibrium}$ = 2.00 − 2x mol/L = 2.00 − 0.66 mol/L = 1.34 mol/L
[Cl$_2$]$_{equilibrium}$ = x mol/L = 0.33 mol/L

$$K_c = \frac{[0.66]^2[0.33]}{[1.34]^2} = 0.080$$

PROBLEM 18.9

Consider the equilibrium constant K_c for the reaction:

$$Br_2(g) + F_2(g) \rightleftarrows 2\,BrF(g) \qquad K_c = = \frac{[BrF(g)]^2}{[Br_2(g)][F_2(g)]} = 55.3$$

Calculate what the equilibrium concentrations (in mol/L) of all three gases will be if the initial concentration of bromine and fluorine were both 0.220 mol/L.

	Br$_2$(g)	F$_2$(g)	⇄	2 BrF(g)

I	0.220 mol/L	0.220 mol/L	$\rightleftarrows$	0 mol/L
C	$-x$ mol/L	$-x$	$\rightleftarrows$	$+2x$
E	0.220 $-x$ mol/L	0.220 $-x$ mol/L	$\rightleftarrows$	$2x$ mol/L

$$\frac{[2x]^2}{[0.220-x][0.220-x]} = 55.3$$

$$\frac{2x}{[0.220-x]} = 7.44, x = 0.173$$

$[F_2] = 0.220 - 0.173$ mol/L $= 0.047$ mol/L
$[Br_2] = 0.220 - 0.173$ mol/L $= 0.047$ mol/L
$[BrF] = 2(0.173)$ mol/L $= 0.347$ mol/L

PROBLEM 18.10

Acetic acid (vinegar) dissociates in water according to the following equation:

$$CH_3CO_2H(aq) + H_2O(l) \rightleftharpoons CH_3CO_2^-(aq) + H_3O^+(aq) \quad K_c = \frac{\left[C_2H_3O_2^-(aq))\right]\left[H_3O^+(aq)\right]}{\left[CH_3CO_2H(aq)\right]} = 1.8 \times 10^{-5}$$

If the initial concentration of CH_3CO_2H is 0.200 mol/L and the equilibrium constant $K_c = 1.8 \times 10^{-5}$, what are the concentrations of acetic acid, acetate, and oxidanium (H_3O^+) at equilibrium? What is the $-\log[H_3O^+]$ (this value is the pH of the solution)?

	$CH_3CO_2H(aq)$	$\rightleftarrows$	$CH_3CO_2^-(aq)$	$H_3O^+(aq)$
I	0.200 mol/L	$\rightleftarrows$	0 mol/L	0 mol/L
C	$-x$	$\rightleftarrows$	$+x$	$+x$
E	0.200 $-x$ mol/L	$\rightleftarrows$	x mol/L	x mol/L

$$\frac{[x][x]}{[0.20-x]} = 1.8 \times 10^{-5} \rightarrow 3.6 \times 10^{-6} - 1.8 \times 10^{-5}x - x^2 = 0 \rightarrow x = 0.0020$$

$[CH_3CO_2H(aq)] = 0.200 - 0.0020$ mol/L $= 0.198$ mol/L
$[CH_3CO_2^-(aq)] = x$ mol/L $= 0.0020$ mol/L
$[H_3O^+(aq)] = x$ mol/L $= 0.0020$ mol/L
$pH = -\log[H_3O^+] = -\log(0.0020 \frac{mol}{L}) = 2.70$

PROBLEM 18.11

To make carbonated water, gaseous CO_2 is dissolved in water according to the following equation:

$$CO_2(g) \rightleftharpoons CO_2(aq) \qquad K_c = \frac{\left[CO_2(aq)\right]}{\left[CO_2(g)\right]} = 0.837$$

At 25 °C, the equilibrium constant K_c is 0.837. If the concentration of $CO_2(aq)$ is initially 0.12 mol/L, what will be the concentration of both types of carbon dioxide at equilibrium?

	$CO_2(g)$	$\rightleftarrows$	$CO_2(aq)$
I	0 mol/L	$\rightleftarrows$	0.12 mol/L
C	$+x$	$\rightleftarrows$	$-x$
E	x mol/L	$\rightleftarrows$	0.12 $-x$ mol/L

$$\frac{[0.12-x]}{[x]} = 0.837, x = 0.0653$$

$[CO_2(aq)] = 0.12 - 0.653$ mol/L $= 0.05$ mol/L
$[CO_2(g)] = x$ mol/L $= 0.0653$ mol/L

PROBLEM 18.12
Sodium sulfate dissolves in water according to the following equation:

$$Na_2SO_4(s) \rightleftharpoons 2\,Na^+(aq) + SO_4^{2-}(aq) \qquad K_c = [Na^+(aq)]^2[SO_4^{2-}(aq)] = 0.007\,74$$

If the equilibrium constant K_c (at 25 °C) is 0.00774, what will be the concentration of sodium ions and sulfate ions at equilibrium?

	$Na_2SO_4(s)$		$2\,Na^+(aq)$	$SO_4^{2-}(aq)$
I	—	$\rightleftharpoons$	0 mol/L	0 mol/L
C	—	$\rightleftharpoons$	$+2x$	$+x$
E	—	$\rightleftharpoons$	$2x$ mol/L	x mol/L

$[2x]^2[x] = 0.007\,74$, $x = 0.125$
$[Na^+(aq)] = 2x$ mol/L $= 0.249$ mol/L
$[SO_4^{2-}(aq)] = x$ mol/L $= 0.125$ mol/L

PROB`LEM 18.13
For the reaction $2\,NOBr(g) \rightleftharpoons 2\,NO(g) + Br_2(g)$, what will happen to the equilibrium if more Br_2 is added at constant volume and temperature?

As more Br_2 is added, the value of Q will increase and $K < Q$. The reaction will shift to the left (more NOBr will be produced).

PROBLEM 18.14
For the reaction $H_2O(l) \rightleftharpoons H_2O(g)$, provide an explanation for why an open water bottle will entirely evaporate, but a sealed bottle will not.

In an open water bottle, the H_2O vapor is constantly streaming away into the atmosphere and so more liquid water is converted into water vapor. In a closed water bottle, the water vapor will stay at an equilibrium concentration as it has nowhere to go. So, a closed bottle will not evaporate away to all gas.

PROBLEM 18.15
Carbon monoxide is a deadly gas because it strongly binds to the iron in hemoglobin. For people exposed to carbon monoxide, treatment is to place them in a hyperbaric (high pressure) oxygen chamber (the chamber is at constant volume and temperature). Looking at the equilibrium below, explain why this helps (Hb = hemoglobin).

$$Hb(CO)_4(aq) + 4\,O_2(aq) \rightleftharpoons Hb(O_2)_4(aq) + 4\,CO(aq) \quad K_c < 1$$

The high-pressure oxygen chamber will increase the concentration of O_2, which will decrease the value of Q. $K > Q$, so more reactant (oxygen) will shift the reaction to the right, which will remove carbon monoxide from the hemoglobin.

PROBLEM 18.16
Plants produce sugar through photosynthesis.

$$6\,CO_2(g) + 6\,H_2O(l) \rightleftharpoons C_6H_{12}O_6(aq) + 6\,O_2(g)$$

Provide an explanation why, for a plant that wants to maximize sugar production, the following helps: the plant inhales CO_2 and exhales O_2 (assume volume is constant). The plant takes the aqueous glucose and turns it into solid starch.

Taking in CO_2 increases the concentration of a reactant (decreasing the value of Q and since $K > Q$, the reaction will shift right) and exhaling O_2 decreases the concentration of a product (decreasing the value of Q and since $K > Q$, the reaction will shift right). Both shift the reaction toward making more sugar. Also, by turning the aqueous sugar into solid starch, the sugar is removed from equilibrium (decreasing the value of Q and $K > Q$) and the reaction will also shift toward making more sugar.

PROBLEM 18.17

For the following equilibrium: $NH_4HS(s) \rightleftharpoons NH_3(g) + H_2S(g)$ $\quad\quad \Delta_r H° > 0$

a. How will the equilibrium be affected by increasing the temperature?

As an endothermic reaction, increasing temperature will shift the equilibrium to the right making more product.

b. By increasing the pressure?

Increasing pressure (by decreasing the volume) will increase the concentration of NH_3 and H_2S and increase the value of Q and $K < Q$, which will shift the equilibrium to the left (toward the side with fewer moles of gas).

c. By adding ammonium hydrosulfide ($NH_4SH(s)$) to the container, at constant volume and temperature?

Nothing. NH_4HS is a solid, so the concentration of NH_4HS does not affect the equilibrium in any way as it does not show up in the equilibrium constant or reaction quotient expressions.

d. By removing ammonia ($NH_3(g)$) from the container, at constant volume and temperature?

Removing ammonia from the container will shift the equilibrium right ($K > Q$).

PROBLEM 18.18

Consider the following equilibrium: $ClF_5(g) \rightleftharpoons ClF_3(g) + F_2(g)$. How will the equilibrium be affected if the reaction container is doubled in volume?

The equilibrium will shift toward the right (the side with more moles of gas) as the value of Q will decrease and $K > Q$.

PROBLEM 18.19

Consider the following equilibrium: $2\ HBr(g) \rightleftharpoons Br_2(g) + H_2(g)\ \Delta_r H° = 103.7$ kJ/mol.

a. How will the equilibrium be affected if the reaction container is halved in volume?

There will be no effect. There are equal numbers of moles of gas on both sides of the equilibrium, so a change in pressure or volume will not affect the equilibrium.

b. If the temperature decreases?

This is an endothermic reaction. Decreasing the temperature will shift the equilibrium to the left.

PROBLEM 18.20

Consider the reaction: cis-but-2-ene $\rightleftharpoons$ trans-but-2-ene. A measurement of K_c at various temperatures gave the following data:

Temperature (K)	Equilibrium Constant K_c
500	1.65
600	1.47
700	1.36

Is the reaction from cis-but-2-ene to trans-but-2-ene exothermic or endothermic? Explain.

As the temperature increases, the equilibrium constant decreases (becomes more reactant favored). As such, this is an exothermic reaction, where an increase in temperature shifts the equilibrium to the left.

PROBLEM 18.21
Consider the following equilibrium: $H_2O(l) \rightleftharpoons H_2O(g)$ $\Delta_rH° = 40.7$ kJ/mol.

  a. How will the equilibrium be affected if the pressure is reduced?

    The equilibrium shifts to the right, toward the side with more moles of gas.

  b. How will the equilibrium be affected if the temperature is increased?

    As an endothermic reaction, an increase in temperature shifts the equilibrium to the right.

PROBLEM 18.22
Consider the following equilibrium: $PbCl_2(s) \rightleftharpoons Pb^{2+}(aq) + 2\ Cl^-(aq)$.

  a. How will the equilibrium be affected if more water is added?

    The equilibrium shifts to the right, toward the side with more moles of ions.

  b. How will the equilibrium be affected if NaCl is added?

    The equilibrium shifts to the left. As the $[Cl^-]$ increases, this causes $K < Q$ and the equilibrium shifts to accommodate this change. This is an example of the "common ion effect," where a compound not part of the equilibrium is added but one of its ions is in common with another ion present in the equilibrium.

PROBLEM 18.23
Consider the following direct steam reforming reaction:

    $CH_4(g) + 2\ H_2O(g) \rightleftharpoons CO_2(g) + 4\ H_2(g)$ $\Delta_rH° = 165$ kJ/mol

  a. How will the equilibrium be affected if the container volume is doubled?

    The equilibrium will shift to the right (the side with more moles of gas).

  b. How will the equilibrium be affected if the temperature is decreased?

    This is an endothermic reaction and so adding heat will shift the reaction to the right.

  c. How will the equilibrium be affected if hydrogen gas is added to the reaction?

    Adding hydrogen gas increases the concentration of a product, so $K < Q$ and the reaction shifts left.

  d. How would the equilibrium be affected if pressure increased?

    The equilibrium will shift to the left (toward the side with fewer moles of gas).

  e. This reaction is, currently, the major source of hydrogen production. Hydrogen will likely serve many important roles in a decarbonized future.

    i. Why is direct steam reforming reaction problematic for a decarbonized future?

      Direct steam reforming is problematic because it uses methane (natural gas), which is typically a nonrenewable resource and a greenhouse gas. In addition, it produces carbon dioxide (a suffocating gas), a greenhouse gas.

    ii. A green chemistry production of hydrogen decomposes water into hydrogen and oxygen:

          $2\ H_2O(l) \rightleftharpoons O_2(g) + 2\ H_2(g)$

      Determine the enthalpy of reaction for water decomposition and provide an explanation as to why industry prefers direct steam reformation over water decomposition.

      The decomposition of water has an enthalpy of reaction of 571.6 kJ/mol, which is significantly endothermic. This enthalpy of reaction is more than twice as large as direct steam reforming reaction, which means that decomposing water is far more energy intensive than direct steam reforming, which couples the decomposition of water with the oxidation of carbon. As such, industry has favored direct steam reforming over water decomposition, but increasingly hydrogen is (and will be) produced by electrolysis of water.

CHAPTER 19 ANSWERS

PROBLEM 19.1
Assign the oxidation number for each atom or ion.

a.	$V(s)$	$V = 0$
b.	$Mg^{2+}(aq)$	$Mg = +2$
c.	$S_8(s)$	$S = 0$
d.	$O_2(g)$	$O = 0$
e.	$Cl^-(aq)$	$Cl = -1$

PROBLEM 19.2
Assign the oxidation number for each element in the following binary compounds.

a.	$CO(g)$	$C = +2, O = -2$
b.	$CO_2(g)$	$C = +4, O = -2$
c.	$CH_4(g)$	$C = -4, H = +1$
d.	$CuCl_2(s)$	$Cu = +2, Cl = -1$
e.	$SF_6(g)$	$S = +6, F = -1$

PROBLEM 19.3
Assign the oxidation number for each element in the following polyatomic ions.

a.	$SO_3^{2-}(aq)$	$S = +4, O = -2$
b.	$SO_4^{2-}(aq)$	$S = +6, O = -2$
c.	$Hg_2^{2+}(aq)$	$Hg = +1$
d.	$BrO_4^-(aq)$	$Br = +7, O = -2$
e.	$HF_2^-(aq)$	$H = +1, F = -1$

PROBLEM 19.4
Assign the oxidation number for each element in the following compounds.

a.	$NaNO_3(s)$	$Na = +1, N = +5, O = -2$
b.	$H_3PO_4(l)$	$H = +1, P = +5, O = -2$
c.	$Cu(CN)_2(s)$	$Cu = +2, C = +2, N = -3$
d.	$LiOH(s)$	$Li = +1, O = -2, H = +1$

PROBLEM 19.5
For each reaction, identify which element is oxidized and which element is reduced.

a. $CH_4(g) + 2 O_2(g) \rightleftarrows CO_2(g) + 2 H_2O(g)$

Reactant	Product
$C = -4$	$C = +4$
$H = +1$	$H = +1$
$O = 0$	$O = -2$

C loses electrons, C is oxidized; O gains electrons, O is reduced.

b. $2 Sb(s) + 5 Cl_2(g) \rightleftarrows 2 SbCl_5(s)$

Reactant	Product
Sb = 0	Sb = +5
Cl = 0	Cl = −1

Sb loses electrons, Sb is oxidized; Cl gains electrons, Cl is reduced.

c. $2 Al(s) + 6 HCl(aq) \rightleftarrows 2 AlCl_3(aq) + 3 H_2(g)$

Reactant	Product
Al = 0	Al = +3
H = +1	H = 0
Cl = −1	Cl = −1

Al loses electrons, Al is oxidized; H gains electrons, H is reduced.

d. $2 H_2O(l) \rightleftarrows 2 H_2(g) + O_2(g)$

Reactant	Product
H = +1	H = 0
O = −2	O = 0

O loses electrons, O is oxidized; H gains electrons, H is reduced.

PROBLEM 19.6
For each reaction in Problem 19.5, write an appropriate reduction and oxidation half-reaction.

a. Oxidation half-reaction

$CH_4(g) \rightarrow C^{4+}(g) + 4 H^+(g) + 8 e^-$

Reduction half-reaction

$2 O_2(g) + 8 e^- \rightarrow 4 O^{2-}(g)$

b. Oxidation half-reaction

$2 Sb(s) \rightarrow 2 Sb^{5+}(s) + 10 e^-$

Reduction half-reaction

$5 Cl_2(g) + 10 e^- \rightarrow 10 Cl^-(s)$

c. Oxidation half-reaction

$2 Al(s) \rightarrow 2 Al^{3+}(aq) + 6 e^-$

Reduction half-reaction

$6 HCl(aq) + 6 e^- \rightarrow 3 H_2(g) + 6 Cl^-(aq)$

d. Oxidation half-reaction

$2 H_2O(l) \rightarrow 4 H^+(g) + O_2(g) + 4 e^-$

Reduction half-reaction

$4 H^+(l) + 4 e^- \rightarrow 2 H_2(g)$

PROBLEM 19.7

For each of the following, write the reduction and oxidation half-reactions and then provide the overall balanced equation.

a. $CH_2O(g) + O_2(g) \rightleftarrows CO_2(g) + H_2O(l)$ *in acidic solution*

Oxidation balanced half-reaction: $CH_2O(g) + H_2O(l) \rightarrow 4 H^+ + CO_2(g) + 4 e^-$

Reduction balanced half-reaction: $O_2(g) + 4 e^- + 4 H^+ \rightarrow 2 H_2O(l)$

Overall balanced reaction: $CH_2O(g) + O_2(g) \rightleftharpoons CO_2(g) + H_2O(l)$

b. $H_2O_2(aq) + Co^{2+}(aq) \rightleftarrows H_2O(l) + Co^{3+}(aq)$ *in acidic solution*

Oxidation balanced half-reaction: $2 Co^{2+}(aq) \rightarrow 2 Co^{3+}(aq) + 2 e^-$

Reduction balanced half-reaction: $H_2O_2(aq) + 2 e^- + 2 H^+ \rightarrow 2 H_2O(l)$

Overall balanced reaction: $H_2O_2(aq) + 2 Co^{2+}(aq) + 2 H^+(aq) \rightleftharpoons 2 H_2O(l) + 2 Co^{3+}(aq)$

c. $Zn(s) + HgO(s) \rightleftarrows Zn(OH)_2(s) + Hg(l)$ *in acidic solution*

Oxidation balanced half-reaction: $Zn(s) + 2 OH^-(aq) \rightarrow Zn(OH)_2(s) + 2 e^-$

Reduction balanced half-reaction: $HgO(s) + 2 e^- + H_2O(l) \rightarrow Hg(l) + 2 OH^-(aq)$

Overall balanced reaction: $Zn(s) + HgO(s) + H_2O(l) \rightleftharpoons Zn(OH)_2(s) + Hg(l)$

d. $Br^-(aq) + MnO_4^-(aq) \rightleftarrows Br_2(g) + MnO_2(s)$ *in acidic solution*

Oxidation balanced half-reaction: $6 Br^-(aq) \rightarrow 3 Br_2(g) + 6 e^-$

Reduction balanced half-reaction: $2 MnO_4^-(aq) + 6 e^- + 4 H_2O(l) \rightarrow 2 MnO_2(s) + 8 OH^-(aq)$

Overall balanced reaction: $6 Br^-(aq) + 2 MnO_4^-(aq) + 4 H_2O(l) \rightleftharpoons 3 Br_2(g) + 2 MnO_2 + 8 OH^-(aq)$

e. $As_2O_3(s) + NO_3^-(aq) \rightleftarrows H_3AsO_4(aq) + NO(g)$ *in acidic solution*

Oxidation balanced half-reaction: $3 As_2O_3(s) + 15 H_2O(l) \rightarrow 6 H_3AsO_4(aq) + 12 e^- + 12 H^+(aq)$

Reduction balanced half-reaction: $16 H^+(aq) + 12 e^- + 4 NO_3^-(aq) \rightarrow 4 NO(g) + 8 H_2O(l)$

Overall balanced reaction: $3 As_2O_3(s) + 4 H^+(aq) + 4 NO_3^-(aq) + 7 H_2O(l) \rightleftharpoons 6 H_3AsO_4(aq) + 4 NO(g)$

f. $CH_3OH(aq) + Cr_2O_7^{2-}(aq) \rightleftarrows CH_2O(aq) + Cr^{3+}(aq)$ *in acidic solution*

Oxidation balanced half-reaction: $3 CH_3OH(aq) \rightarrow 3 CH_2O(aq) + 6 e^- + 6 H^+(aq)$

Reduction balanced half-reaction: $14 H^+(aq) + Cr_2O_7^{2-}(aq) + 6 e^- \rightarrow + 2 Cr^{3+}(aq) + 7 H_2O(l)$

Overall balanced reaction: $3 CH_4O(aq) + 8 H^+(aq) + Cr_2O_7^{2-}(aq) \rightleftharpoons 3 CH_2O(aq) + 2 Cr^{3+}(aq) + 7 H_2O(l)$

PROBLEM 19.8

For each of the following reactions, write the half-reaction that occurs at the cathode and the half-reaction that occurs at the anode.

a. $Zn(s) + Pb(NO_3)_2(aq) \rightleftarrows Zn(NO_3)_2(aq) + Pb(s)$

Anode (oxidation) reaction: $Zn(s) + 2 NO_3^-(aq) \rightarrow Zn(NO_3)_2(aq) + 2 e^-$

Cathode(reduction) reaction: $Pb(NO_3)_2(aq) + 2 e^- \rightarrow Pb(s) + 2 NO_3^-(aq)$

b. $2 AgNO_3(aq) + Sn(s) \rightleftarrows 2 Ag(s) + Sn(NO_3)_2(aq)$

Anode (oxidation) reaction: $Sn(s) + 2 NO_3^-(aq) \rightarrow Sn(NO_3)_2(aq) + 2 e^-$

Cathode(reduction) reaction: $2 AgNO_3(aq) + 2 e^- \rightarrow 2 Ag(s) + 2 NO_3^-(aq)$

c. $CH_2O(g) + O_2(g) \rightleftarrows CO_2(g) + H_2O(l)$

Anode (oxidation) reaction: $H_2CO(g) + H_2O(l) \rightarrow 4 H^+ + CO_2 + 4 e^-$

Cathode(reduction) reaction: $O_2(g) + 4 e^- + 4 H^+ \rightarrow 2 H_2O(l)$

PROBLEM 19.9
For each reaction in Problem 19.8, write the reaction in cell notation. For reactions where there is not a pure metal, assume that a Pt or C electrode is used.

a. $Zn(s)|Zn(NO_3)_2(aq)||Pb(NO_3)_2(aq)|Pb(s)$

b. $Sn(s)|Sn(NO_3)_2(aq)||AgNO_3(aq)|Ag(s)$

c. $Pt(s)|H_2CO(g), CO_2(g)||O_2(g)|H_2O(l)|Pt(s)$

PROBLEM 19.10
For each cell notation provided, write a balanced equation for the reaction.

a. $Fe(s)|Fe(NO_3)_2(aq)||Cu(NO_3)_2(aq)|Cu(s)$

$Cu(NO_3)_2(aq) + Fe(s) \rightleftharpoons Cu(s) + Fe(NO_3)_2(aq)$

b. $Pt(s)|Fe^{2+}(aq), Fe^{3+}(aq)||H^+, H_2O_2(aq), H_2O(l)|Pt(s)$

$2 Fe^{2+}(aq) + H_2O_2(aq) + 2 H^+(aq) \rightleftharpoons 2 Fe^{3+}(aq) + 2 H_2O(l)$

c. $Sn(s)|F^-(aq), SnF_6^{2-}(aq)||ClO_4^-(aq), ClO_3^-(aq), H_2O(l)|Pt(s)$

$Sn(s) + 6 F^-(aq) + 2 ClO_4^-(aq) + 4 H^+(aq) \rightleftharpoons 2 ClO_3^-(aq) + 2 H_2O(l) + SnF_6^{2-}(aq)$

PROBLEM 19.11
Consider these half-reactions.

Half-Reaction	$E°$ (V)	χ_P
$Au^{3+}(aq) + 3 e^- \rightarrow Au(s)$	1.52	2.54
$Pt^{2+}(aq) + 2 e^- \rightarrow Pt(s)$	1.12	2.28
$Co^{2+}(aq) + 2 e^- \rightarrow Co(s)$	−0.28	1.88
$Mn^{2+}(aq) + 2 e^- \rightarrow Mn(s)$	−1.18	1.55

a. Which is the weakest oxidizing agent?

$Mn^{2+}(aq)$

b. Which is the strongest oxidizing agent?

$Au^{3+}(aq)$

c. Which is the strongest reducing agent?

$Mn(s)$

d. Which is the weakest reducing agent?

$Au(s)$

e. Can you explain parts a through d in terms of their electronegativity values?

Yes, the electronegativity values follow the trend in $E°$ values (see the added column above). So, it makes sense that gold(3+) ions are the strongest oxidizing agent and gold is the weakest reducing agent (gold has the highest electronegativity). Similarly, manganese has the lowest electronegativity, so manganese(2+) ions are the weakest oxidizing agent and manganese is the best reducing agent.

f. Will $Co(s)$ reduce $Pt^{2+}(aq)$ to $Pt(s)$?

Yes, it has a more negative $E°$ than $Pt^{2+}(aq)$.

g. Will $Pt(s)$ reduce $Co^{2+}(aq)$ to $Co(s)$?

No, it has a higher $E°$ than $Co^{2+}(aq)$.

h. Which ions can be reduced by $Co(s)$?

$Au^{3+}(aq)$ and $Pt^{2+}(aq)$; anything with a more positive/higher $E°$ than itself.

PROBLEM 19.12

Calculate $E°_{cell}$ for each of these reactions. Indicate whether each is a product-favored reaction or a reactant-favored reaction.

a. $I_2(s) + Mg(s) \rightleftarrows Mg^{2+}(aq) + 2\ I^-(aq)$

$E°_{cell} = E°(I_2/I^-) - E°(Mg^{2+}/Mg) = 0.54\ V - (-2.37\ V) = 2.91\ V$ (product-favored reaction)

b. $Ag(s) + Fe^{3+}(aq) \rightleftarrows Ag^+(aq) + Fe^{2+}(aq)$

$E°_{cell} = E°(Fe^{3+}/Fe^{2+}) - E°(Ag^+/Ag) = 0.77\ V - 0.80\ V = -0.03\ V$ (reactant-favored reaction)

c. $Sn^{2+}(aq) + 2\ Ag^+(aq) \rightleftarrows Sn^{4+}(aq) + 2\ Ag(s)$

$E°_{cell} = E°(Ag^+/Ag) - E°(Sn^{4+}/Sn^{2+}) = 0.80\ V - 0.15\ V = 0.65\ V$ (product-favored reaction)

d. $2\ Zn(s) + O_2(g) + 2\ H_2O(l) \rightleftarrows 2\ Zn(OH)_2(s)$

$E°_{cell} = E°(O_2/OH^-) - E°(Zn(OH)_2/Zn) = 0.40\ V - (-1.25\ V) = 1.65\ V$ (product-favored reaction)

e. $Fe(s)|Fe(NO_3)_2(aq)\|Cu(NO_3)_2(aq)|Cu(s)$

$E°_{cell} = E°(Cu^{2+}/Cu) - E°(Fe^{2+}/Fe) = 0.34\ V - (-0.44\ V) = 0.78\ V$ (product-favored reaction)

f. $Pt(s)|Fe^{2+}(aq),\ Fe^{3+}(aq)\|H^+(aq),\ H_2O_2(aq),\ H_2O(l)|Pt(s)$

$E°_{cell} = E°(H_2O_2/H_2O) - E°(Fe^{3+}/Fe^{2+}) = 1.78\ V - 0.77\ V = 1.01\ V$ (product-favored reaction)

g. $Zn(s)|Zn(NO_3)_2(aq)\|Pb(NO_3)_2(aq)|Pb(s)$

$E°_{cell} = E°(Pb^{2+}/Pb) - E°(Zn^{2+}/Zn) = -0.13\ V - (-0.76)\ V = 0.63\ V$ (product-favored reaction)

h. $Ag(s)|AgNO_3(aq)\| Sn(NO_3)_2(aq)|Sn(s)$

$E°_{cell} = E°(Sn^{2+}/Sn) - E°(Ag^+/Ag) = -0.14\ V - 0.80\ V = -0.94\ V$ (reactant-favored reaction)

i. $Pt(s)|(COOH)_2(aq)|CO_2(g)\|O_2(g)|H^+(aq),\ H_2O(l)|Pt(s)$

$E°_{cell} = E°(O_2/H_2O) - E°(CO_2/(COOH)_2) = 1.23\ V - (-0.48)\ V = 1.71\ V$ (product-favored reaction)

PROBLEM 19.13

The low-voltage battery in a car (in contrast to the high-voltage, lithium-ion traction battery) is made up of six individual electrochemical cells. Each cell uses electrodes with lead: $PbSO_4(s)|SO_4^{2-}(aq)|Pb(s)$ and $PbO_2(s)|SO_4^{2-}(aq),\ H^+(aq),\ H_2O(l)|PbSO_4(s)$.

a. Using the list of $E°$ values, determine which half-cell will be the cathode and which half-cell will be the anode to produce a galvanic cell with positive voltage.

Two possible combinations:

$E°_{cell} = E°(PbSO_4/Pb) - E°(PbO_2/PbSO_4) = -0.36\ V - 1.69\ V = -2.05\ V$

$E°_{cell} = E°(PbO_2/PbSO_4) - E°(PbSO_4/Pb) = 1.69\ V - (-0.36)\ V = 2.05\ V$

A galvanic cell would have the $PbO_2(s)|SO_4^{2-}(aq),\ H^+(aq),\ H_2O(l)|PbSO_4(s)$ electrode as the cathode and the $PbSO_4(s)|SO_4^{2-}(aq)|Pb(s)$ electrode as the anode.

b. Write a half-reaction for the reaction occurring at the anode and the cathode.

Anode half-reaction: $Pb(s) + SO_4^{2-}(aq) \rightarrow PbSO_4(s) + 2\ e^-$

Cathode half-reaction: $PbO_2(s) + SO_4^{2-}(aq) + 4\ H^+(aq) + 2\ e^- \rightarrow PbSO_4(s) + 2\ H_2O(l)$

c. What is the overall, balanced reaction that is occurring in this cell?

$Pb(s) + PbO_2(s) + 2\ H_2SO_4(aq) \rightarrow 2\ PbSO_4(s) + 2\ H_2O(l)$

PROBLEM 19.14
Using Appendix 8, explain the following observations.

a. Co^{3+} is not stable in aqueous solution.

A product-favored reaction occurs with water because the reduction potential of cobalt(3+), 1.92 V, is greater than the reduction potential of oxygen gas (1.23 V).

$$4\ Co^{3+}(aq) + 2\ H_2O(l) \rightleftarrows 4\ Co^{2+}(aq) + O_2(aq) + 4\ H^+(aq) \qquad E°_{cell} = 0.69\ V$$

b. Fe^{2+} is not stable in air.

A product-favored reaction occurs because iron(3+) has a reduction potential (0.77 V) less than the reduction potential of oxygen gas (1.23 V).

$$4\ Fe^{2+}(aq) + O_2(aq) + 4\ H^+(aq) \rightleftarrows 4\ Fe^{3+}(aq) + 2\ H_2O(l) \qquad E°_{cell} = 0.46\ V$$

PROBLEM 19.15
Many metals dissolve in nitric acid (producing different nitrogen oxide gas products in the process).

a. Write a balanced equation for the reaction of iron metal with nitric acid and calculate $E°_{cell}$.

$$3\ Fe(s) + 2\ NO_3^-(aq) + 8\ H^+(aq) \rightleftarrows 3\ Fe^{2+}(aq) + 2\ NO(g) + 4\ H_2O(l)$$

$$E°_{cell} = E°(NO/NO_3^-) - E°(Fe^{2+}/Fe) = 0.96\ V - (-0.44\ V) = 1.40\ V$$

b. Gold does not dissolve in nitric acid. Provide an explanation.

$E°_{cell} = E°(NO/NO_3^-) - E°(Au^{3+}/Au) = 0.96\ V - (1.50\ V) = -0.54\ V$. This is a reactant-favored process because gold has a higher reduction potential than nitric oxide.

c. Gold will slowly dissolve in nitric acid with aqueous hydrogen chloride, a combination called aqua regia. Using the list of $E°$ values, can you explain why?

The addition of Cl^- ions lowers the oxidation potential of gold, making it easier to oxidize.

$$Au^{3+}(aq) + 3\ e^- \rightarrow Au(s) \qquad\qquad E° = 1.50\ V$$

$$AuCl_4^-(aq) + 3\ e^- \rightarrow Au(s) + 4\ Cl^-(aq) \qquad\qquad E° = 1.00\ V$$

PROBLEM 19.16
Consider the reaction of aluminium and copper(2+) ions.

$$2\ Al(s) + 3\ Cu^{2+}(aq) \rightleftarrows 2\ Al^{3+}(aq) + 3\ Cu(s)$$

a. Determine $E°_{cell}$ for this reaction.

$E°_{cell} = E°(Cu^{2+}/Cu) - E°(Al^{3+}/Al) = 0.34\ V - (-1.68\ V) = 2.02\ V$. This is product-favored.

b. Using your answer in part a, determine $\Delta_r G°$ at 25 °C.

There are six electrons transferred ($n = 6$).

$\Delta_r G° = -zFE°_{cell} = -(6)(96\ 485\ C/mol)(2.02\ V) = -1\ 170\ 000\ J/mol$ or $-1170\ kJ/mol$

This is a product-favored reaction ($\Delta_r G° < 0$).

c. What is K for this reaction?

$$K = e^{\left(zFE°_{cell}/RT\right)} = e^{\left(((6)(96\ 485\ C/mol)(2.02\ V))/((8.314\ J/(mol\ K))(298\ K))\right)} = e^{472} = 1 \times 10^{205}$$

This is *very* product favored ($K \gg 1$).

d. Is this reaction product favored or reactant favored?

According to $E°_{cell}$, $\Delta_r G°$, and K, this reaction is very product favored.

PROBLEM 19.17

Hydrazine, N_2H_4, reacts with oxygen in the air to produce nitrogen gas and water.

$$N_2H_4(l) + O_2(g) \rightleftarrows N_2(g) + 2\ H_2O(l)$$

a. Calculate $\Delta_rG°$ for this reaction ($\Delta_fG°(N_2H_4(l)) = 149.3$ kJ/mol).

$\Delta_rG° = 1\Delta_fG°(N_2(g)) + 2\Delta_fG°(H_2O(l)) - [1\Delta_fG°(N_2H_4(l)) + 1\Delta_fG°(O_2(g))]$

$\Delta_rG° = 1(0\ kJ/mol) + 2(-237.1\ kJ/mol) - [1(149.3\ kJ/mol) + 1(0\ kJ/mol)]$

$\Delta_rG° = -623.5$ kJ/mol

b. If you wanted to make a basic hydrazine–air fuel cell to power an electric vehicle, what would be the value of $E°_{cell}$?

Four electrons are transferred in this reaction.

$$E°_{cell} = -\frac{\Delta_rG°}{zF} = -\frac{-623\ 500\ J/mol}{(4)(96\ 485\ C/mol)} = 1.616\ V$$

PROBLEM 19.18

Determine the equilibrium constant for the reaction between Cd(s) and Cu^{2+}(aq) at 25 °C.

Reaction: $Cd(s) + Cu^{2+}(aq) \rightleftarrows Cd^{2+}(aq) + Cu(s)$

$E°_{cell} = E°(Cu^{2+}/Cu) - E°(Cd^{2+}/Cd) = 0.34\ V - (-0.40\ V) = 0.74\ V$

2 electrons are transferred.

$$K = e^{\left(zFE°_{cell}/RT\right)} = e^{\left(((2)(96\ 485\ C/mol)(0.74\ V))/((8.314\ J/(mol\ K))(298\ K))\right)} = e^{58} = 1 \times 10^{75}$$

PROBLEM 19.19

Calculate the equilibrium constant for the reaction between Br_2(l) and Cl^-(aq) at 25 °C.

Reaction: $Br_2(l) + 2\ Cl^-(aq) \rightleftarrows 2\ Br^-(aq) + Cl_2(g)$

$E°_{cell} = E°(Br_2/Br^-) - E°(Cl_2/Cl^-) = 1.07\ V - 1.36\ V = -0.29\ V$

2 electrons are transferred.

$$K = e^{\left(zFE°_{cell}/RT\right)} = e^{\left(((2)(96\ 485\ C/mol)(-0.29\ V))/((8.314\ J/(mol\ K))(298\ K))\right)} = e^{-22} = 2 \times 10^{-10}$$

PROBLEM 19.20

Consider the galvanic cell made of zinc and cadmium half-cells.

$$Zn(s) + Cd^{2+}(aq) \rightleftarrows Zn^{2+}(aq) + Cd(s)$$

a. Calculate $E°_{cell}$ for this cell.

$E°_{cell} = E°(Cd^{2+}/Cd) - E°(Cd^{2+}/Cd) = -0.40\ V - (-0.76\ V) = 0.36\ V$

b. If $[Cd^{2+}] = 0.068$ mol/L and $[Zn^{2+}] = 1.00$ mol/L, what is E_{cell} at 25.0 °C?

$$E_{cell} = E°_{cell} - \frac{RT}{zF}\ln(Q) = 0.36\ V - \frac{(8.314\ J/(mol\ K))(298.2\ K)}{(2)(96\ 485\ C/mol)}\ln(\frac{[1.00\ mol/L]}{[0.068\ mol/L]}) = 0.33\ V$$

c. If E_{cell} is 0.390 V and $[Cd^{2+}] = 2.00$ mol/L, what is the amount concentration (mol/L) of Zn^{2+}(aq) at 25.0 °C?

$$E_{cell} = E°_{cell} - \frac{RT}{zF}\ln(Q)$$

$$0.390\ V = 0.36\ V - \frac{(8.314\ J/(mol\ K))(298.2\ K)}{(2)(96\ 485\ C/mol)}\ln(\frac{[Zn^{2+}]}{2.00\ mol/L})$$

$$0.03 \text{ V} = (-0.0128 \text{ V})\ln(\frac{[Zn^{2+}]}{2.00 \text{ mol/L}})$$

$$-2 = \ln(\frac{[Zn^{2+}]}{2.00 \text{ mol/L}})$$

$$e^{-2} = \frac{[Zn^{2+}]}{0.068 \text{ mol/L}}$$

$$(2.00 \text{ mol/L})e^{-2} = [Zn^{2+}]$$

$$0.2 \text{ mol/L} = [Zn^{2+}]$$

PROBLEM 19.21

The standard potential of a Daniell cell, Zn(s)|Zn²⁺(aq)‖Cu²⁺(aq)|Cu(s), is 1.10 V when Zn²⁺(aq) = Cu²⁺(aq) = 1.0 mol/L. As the cell operates, the cell potential changes due to concentration changes of Zn²⁺(aq) and Cu²⁺(aq). Calculate the ratio (at 25.0 °C) of Zn²⁺/Cu²⁺ when E_{cell} = 0.05 V. What happens, over time, to the [Zn²⁺(aq)]? What happens to [Cu²⁺(aq)]?

$$0.05 \text{ V} = 1.103 \text{ V} - \frac{(8.314 \text{ J/(mol K)})(298.2 \text{ K})}{(2)(96\ 485 \text{ C/mol})} \ln(\frac{[Zn^{2+}]}{[Cu^{2+}]})$$

$$-1.05 \text{ V} = (-0.0128 \text{ V}) \ln(\frac{[Zn^{2+}]}{[Cu^{2+}]})$$

$$82.0 = \ln(\frac{[Zn^{2+}]}{[Cu^{2+}]})$$

$$e^{82.0} = \frac{[Zn^{2+}]}{[Cu^{2+}]}$$

$$4\times10^{35} = \frac{[Zn^{2+}]}{[Cu^{2+}]}, \text{ that is, the } [Zn^{2+}] \text{ increases as the } [Cu^{2+}] \text{ decreases.}$$

CHAPTER 20 ANSWERS

PROBLEM 20.1

For each of the following reactions, identify the acid and the base. For any reactions involving H⁺ transfer, also identify the conjugate acid and conjugate base.

a.

base   acid   conjugate acid   conjugate base

b.

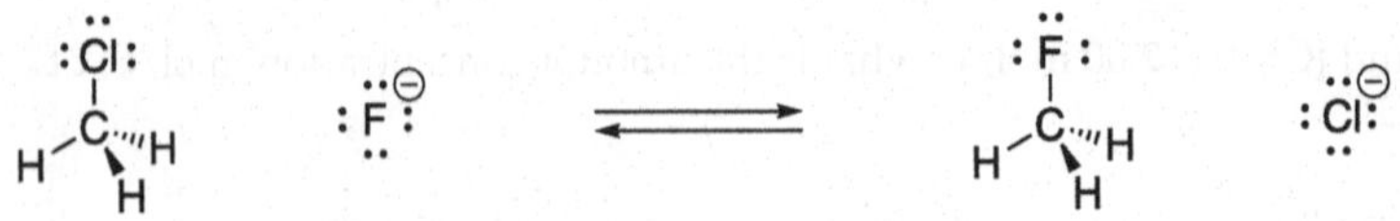

acid   base

c.

acid   base   conjugate base   conjugate acid

d.

acid   base

e.

acid   base   conjugate base   conjugate acid

f.

acid   base   conjugate base   conjugate acid

g.

base   acid   conjugate acid   conjugate base

h.

acid   base   conjugate base   conjugate acid

PROBLEM 20.2

Predict the products of each reaction. Once you have the balanced chemical equation, calculate K_c for each reaction and identify whether it is reactant favored or product favored.

a. $H_2O(l) + NH_3(aq) \rightleftarrows HO^-(aq) + NH_4^+(aq)$

$K_a(H_2O) > K_a(NH_3)$, so H_2O is the acid and NH_3 is the base.

$K_c = \dfrac{K_a(H_2O)}{K_a(NH_4^+)} = \dfrac{1.0 \times 10^{-14}}{5.6 \times 10^{-10}} = 1.8 \times 10^{-5}$ (reactant favored)

b. $HF(aq) + NaH_2PO_4(aq) \rightleftarrows Na^+(aq) + F^-(aq) + H_3PO_4(aq)$

$K_a(HF) > K_a(H_2PO_4^-)$, so HF is the acid and $H_2PO_4^-$ is the base.

$K_c = \dfrac{K_a(HF)}{K_a(H_3PO_4)} = \dfrac{6.3 \times 10^{-4}}{7.9 \times 10^{-3}} = 0.080$ (reactant favored)

c. $HCN(aq) + NaSH(aq) \rightleftarrows Na^+(aq) + CN^-(aq) + H_2S(aq)$

$K_a(HCN) > K_a(HS^-)$, so HCN is the acid and HS^- is the base.

$K_c = \dfrac{K_a(HCN)}{K_a(H_2S)} = \dfrac{3.3 \times 10^{-10}}{1.0 \times 10^{-7}} = 3 \times 10^{-3}$ (reactant favored)

d. $KHCO_3(aq) + HF(aq) \rightleftarrows K^+(aq) + F^-(aq) + H_2CO_3(aq)$

$K_a(HF) > K_a(HCO_3^-)$, so HF is the acid and HCO_3^- is the base.

$K_c = \dfrac{K_a(HF)}{K_a(H_2CO_3)} = \dfrac{6.3 \times 10^{-4}}{4.3 \times 10^{-7}} = 1.5 \times 10^3$ (product favored. Note that this is even more product

favored than the math would suggest because H_2CO_3 decomposes into CO_2 and H_2O, which decreases $[H_2CO_3]$ and leads to the creation of more product.)

e. $H_3O^+(aq) + RbF(aq) \rightleftarrows Rb^+(aq) + H_2O(l) + HF(aq)$

$K_a(H_3O^+) > K_a(HF)$, so H_3O^+ is the acid and F^- is the base.

$K_c = \dfrac{K_a(H_3O^+)}{K_a(HF)} = \dfrac{1.0}{6.3 \times 10^{-4}} = 1.6 \times 10^3$ (product favored)

f. $H_2O(l) + H_2O(l) \rightleftarrows H_3O^+(aq) + HO^-(aq)$

$K_a(H_2O) = K_a(H_2O)$, so H_2O is the acid and H_2O is the base.

$K_c = \dfrac{K_a(H2O)}{K_a(H_3O^+)} = \dfrac{1.0 \times 10^{-14}}{1.0} = 1.0 \times 10^{-14}$ (reactant favored)

PROBLEM 20.3
For each reaction, predict the products of each reaction and calculate K_c. Provide a rationalization for the position of the equilibrium (reactant favored or product favored).

a. $H_2Se(aq) + I^-(aq) \rightleftarrows HSe^-(aq) + HI(aq)$

$K_c = \dfrac{K_a(H_2Se)}{K_a(HI)} = \dfrac{1.3 \times 10^{-4}}{1 \times 10^{10}} = 1.3 \times 10^{-14}$. This reaction is reactant favored because iodine is more

electronegative than selenium. So, the iodide is less likely to share its electrons and will exist as I^-, while selenium is more likely to share its electrons with H^+.

b. $H_3O^+(aq) + HO^-(aq) \rightleftarrows H_2O(l) + H_2O(l)$

$K_c = \dfrac{K_a(H_3O^+)}{K_a(H_2O)} = \dfrac{1.0}{1.0 \times 10^{-14}} = 1.0 \times 10^{14}$. This reaction is product favored because negatively

charged hydroxide is a stronger base (because it is negative) than water and positively charged oxidanium is a stronger acid than water. So, the hydroxide is more likely to share its electrons with H^+ than water.

c. $HCl(aq) + F^-(aq) \rightleftarrows Cl^-(aq) + HF(aq)$

$K_c = \dfrac{K_a(HCl)}{K_a(HF)} = \dfrac{1 \times 10^7}{6.8 \times 10^{-4}} = 1 \times 10^{10}$. This reaction is product favored because fluoride is a smaller

anion than chloride. So, the chloride is less likely to share its electrons and will exist as Cl^-, while fluoride is more likely to share its electrons with H^+.

d. $CF_3CO_2H(aq) + CH_3CO_2^-(aq) \rightleftarrows CF_3CO_2^-(aq) + CH_3CO_2H(aq)$

$K_c = \dfrac{K_a(CF_3CO_2H)}{K_a(CH_3CO_2H)} = \dfrac{0.63}{1.8 \times 10^{-5}} = 35\,000$. This reaction is product favored because $CF_3CO_2^-$ has

three electron-withdrawing fluorine atoms ($CH_3CO_2^-$ has none). So, the $CF_3CO_2^-$ is less likely to share its electrons and exist as $CF_3CO_2^-$, while $CH_3CO_2^-$ prefers to share its electrons with H^+.

PROBLEM 20.4

For each reaction, identify the acid and the base (for reactions involving H⁺ transfer, identify the conjugate acid and the conjugate base), and then predict whether the reaction is most likely product favored or reactant-favored Explain your reasoning.

a.

Product favored. Nitrogen is more electronegative than carbon, so it is a more stable/weaker base (less likely to share its electrons than carbon).

b.

Product favored. Chlorine is a larger atom than fluorine, so it is a more stable/weaker base (it wants to share its electrons less than fluorine does).

c.

Reactant favored. There are two explanations here. First nitrate (NO_3^-) has two electron-withdrawing oxygen atoms and is stabilized by delocalization, while hydroxide (OH^-) has no electron-withdrawing groups and no delocalization. Hydroxide is therefore a stronger base and water a weaker acid.

d.

Product favored. Fluorine is more electronegative than nitrogen, so it is less likely to share its electrons than nitrogen.

PROBLEM 20.5

There are lots of chloride (Cl⁻) ions floating around in your body and there are lots of thiols (molecules that contain R–SH; note R is a common symbol used to indicate the other, unimportant to the problem, part of a larger structure). Why do we not have to worry about these combining to produce hydrogen chloride (HCl)? Estimate K_c for this reaction and provide an explanation.

$$R–SH + Cl^- \rightleftharpoons R–S^- + HCl$$

Thiols should have K_a values like H_2S ($K_a = 1.0 \times 10^{-7}$) and HCl has a K_a value of 1.0×10^7. Therefore, $K_c = \dfrac{1.0 \times 10^{-7}}{1.0 \times 10^7} = 1.0 \times 10^{-14}$. This reaction is *very* reactant favored because chlorine is much more electronegative than sulfur and therefore a more stable/weaker base and much less likely to share its electrons.

PROBLEM 20.6

Soaps are compounds that contain a large carbon-rich "tail" and a polar or ionic "head." Here are two soaps: one with a carboxylate head and one with a sulfate head.

sodium stearate (carboxylate soap)

sodium stearyl sulfate (sulfate soap)

Soap scum is the precipitation of soap with Mg^{2+} or Ca^{2+} ions. Why do carboxylate soaps cause soap scum while sulfate soaps almost never do?

Carboxylate is a stronger base (more likely to share its electrons) than sulfate because carboxylate has fewer electron-withdrawing oxygen atoms attached to it than sulfate does. With fewer electron-withdrawing groups, the carboxylate shares its electrons more readily with Mg^{2+} and Ca^{2+}, which leads to soap scum.

PROBLEM 20.7

A common structure in organic chemistry and biochemistry is the hexagonal benzene ring (with alternating double and single bonds). If you take organic chemistry in college, you will spend a lot of time with benzene and studying its chemistry, and one of the reactions you'll learn is nitration (adding $-NO_2$ groups).

$K_a = 1.1 \times 10^{-10}$	$K_a = 7.1 \times 10^{-8}$	$K_a = 1.3 \times 10^{-5}$	$K_a = 4.2 \times 10^{-1}$
phenol	4-nitrophenol	2,4-dinitrophenol	picric acid (2,4,6-trinitrophenol)

Considering the nitrated phenols (benzene with an $-OH$), what explains this trend in acidity?

This trend in acidity can be explained by the addition of the inductive electron-withdrawing nitro groups, which make the conjugate bases weaker (less likely to share their electrons). The more electron-withdrawing groups, the weaker the conjugate base; a weaker conjugate base corresponds to a stronger acid.

PROBLEM 20.8

Cysteine is one of the 20 common amino acids. Selenocysteine is a less common amino acid that is identical except that it has a selenium atom rather than a sulfur atom.

cysteine   selenocysteine

Despite their similar structure, there are some differences. Explain why selenocysteine is normally found without an H^+ attached to the selenium, whereas cysteine is normally found with an H^+ connected to the sulfur atom.

Selenium is a bigger atom than sulfur, so the selenide anion is less likely to share its electrons with H^+ and is more likely to be found as Se^-. Sulfur is a smaller atom, more likely to share its electrons, and therefore more likely to be bonded to H^+.

PROBLEM 20.9

During cellular respiration in the mitochondria, electrons are transported (through the electron transport chain). This transfer of electrons is used to create a concentration gradient (think Nernst) of H^+ ions. The cell then uses this electrochemical potential to create ATP. Here is one small step of the electron transport chain:

ubiquinol   semi-ubiquinone

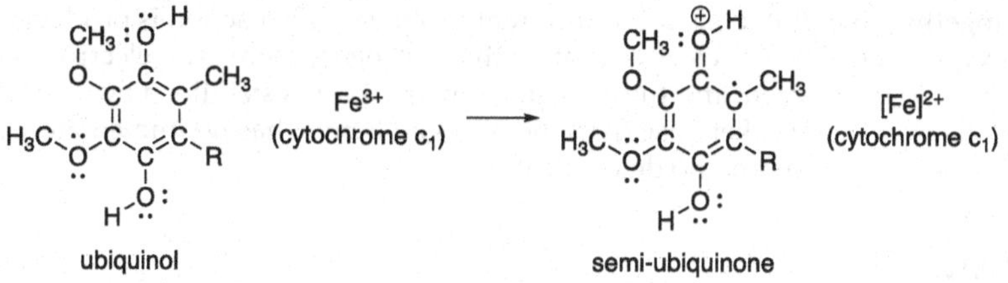

The semi-ubiquinone then donates H^+ to the inner membrane space of the mitochondria. Why is semi-ubiquinone more likely to donate H^+ than is ubiquinol?

Semi-ubiquinone is positively charged (while ubiquinol is neutral). If we consider the conjugate bases (once each gives up H^+), semi-ubiquinone's conjugate base is neutral, which is less likely to share its electrons than the negatively charged ubiquinol conjugate base.

PROBLEM 20.10

Consider the reaction between acetic acid (CH_3CO_2H) and ammonia (NH_3):

a. Use the K_a values to determine which reactant is the acid and which reactant is the base.

$K_a(CH_3CO_2H) = 1.8 \times 10^{-5}$ and $K_a(NH_3) = 1.0 \times 10^{-38}$, so acetic acid is the acid and ammonia is the base.

b. Write the products for the above reaction. See the answer to part a.

c. Calculate K_c for this reaction. Is this a product-favored or reactant-favored process?

$$K_c = \frac{K_a(CH_3CO_2H)}{K_a(NH_4^+)} = \frac{1.8 \times 10^{-5}}{5.6 \times 10^{-10}} = 32\ 000 \text{ (product favored)}$$

d. Provide a rationalization (in terms of acidity trends) for the position of the equilibrium (i.e., why is it product favored or reactant favored).

This reaction is product favored because the oxygen atom in acetate is more electronegative than nitrogen, there is an electron-withdrawing oxygen atom, and it is stabilized by delocalization, which makes acetate a weaker base than ammonia (which has no electron-withdrawing groups and no delocalization) and makes acetic acid a stronger acid than ammonium.

PROBLEM 20.11
Amino acids contain both a carboxylic acid (R-COOH) and amine (R-NH$_2$). Consider the equilibrium that exists for glycine in solution:

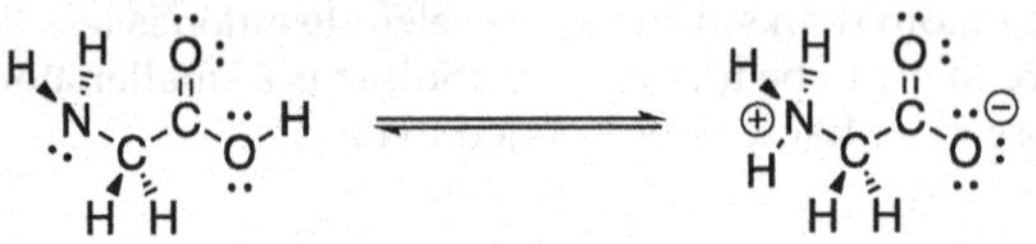

Use your understanding of acids and bases to explain whether this is a reactant-favored or product-favored equilibrium and explain your answer.

Looking at the ammonia/acetic acid example in Problem 20.10 we can see the reasons why ammonia and acetic acid proceed toward the right side to make acetate and ammonium. Here, the same processes are at work; the challenge is that the amino group and the carboxylic acid group are tethered together so that it looks like a different problem. The reaction is product favored because the oxygen atom is more electronegative than nitrogen, there is an electron-withdrawing oxygen atom next to the oxygen, and the oxygen atom lone pair is stabilized by delocalization, which makes the oxygen a weaker base than the nitrogen (which has no immediately adjacent electron-withdrawing groups and no delocalization).

PROBLEM 20.12
Aqueous hydrogen fluoride dissociates in water according to the following equation:

$$HF(aq) + H_2O(l) \rightleftharpoons F^-(aq) + H_3O^+(aq) \quad K_c = \frac{\left[F^-(aq)\right]\left[H_3O^+(aq)\right]}{\left[HF(aq)\right]} = 6.3 \times 10^{-4}$$

If the initial concentration of HF is 0.200 mol/L, what are the concentrations of aqueous hydrogen fluoride, fluoride, and oxidanium (H$_3$O$^+$) at equilibrium? What is the pH of the solution? Provide an explanation – in terms of acidity trends – for why this is a reactant-favored process.

	HF(aq)	$\rightleftharpoons$	F⁻(aq)	H$_3$O⁺(aq)
I	0.200 mol/L	$\rightleftharpoons$	0 mol/L	0 mol/L
C	−x	$\rightleftharpoons$	+x	+x
E	0.200 − x mol/L	$\rightleftharpoons$	x mol/L	x mol/L

$$\frac{[x][x]}{[0.20-x]} = 6.3 \times 10^{-4} \rightarrow 1.2 \times 10^{-4} - 6.3 \times 10^{-4}x - x^2 = 0 \rightarrow x = 0.011$$

[HF] = 0.200 − 0.011 mol/L = 0.189 mol/L
[F⁻] = x mol/L = 0.011 mol/L
[H$_3$O⁺] = x mol/L = 0.011 mol/L
pH = −log[H$_3$O⁺] = −log(0.011 mol/L) = 1.96

Explanation: Fluoride (negatively charged) is a stronger base than water (neutral) and oxidanium (positively charged) is a stronger acid than HF (neutral), so the left side is favored over the right side.

PROBLEM 20.13
The pH of a solution of unknown acid is 2.12, the concentration of HA is 0.1983 mol/L, and the concentration of A⁻ is 0.0017 mol/L. What is the pK$_a$ of the acid?

$$pH = pK_a + \log\left(\frac{[A^-]}{[HA]}\right)$$

$$pH - \log\left(\frac{[A^-]}{[HA]}\right) = pK_a$$

$$2.12 - \log\left(\frac{0.0017 \text{ mol/L}}{0.1983 \text{ mol/L}}\right) = pK_a$$

$$4.19 = pK_a$$

PROBLEM 20.14
Acetic acid is in a solution at pH 5.74. What is the ratio of [A⁻]/[HA]?

$$pH = pK_a + \log\left(\frac{[A^-]}{[HA]}\right)$$

$$pH - pK_a = \log\left(\frac{[A^-]}{[HA]}\right)$$

$$10^{(pH - pKa)} = \frac{[A^-]}{[HA]}$$

$$10^{(5.74 - 4.74)} = \frac{[A^-]}{[HA]}$$

$$10 = \frac{[A^-]}{[HA]}$$

PROBLEM 20.15
Acetic acid is in a solution at pH 3.74. What is the ratio of [A⁻]/[HA]?

$$pH = pK_a + \log\left(\frac{[A^-]}{[HA]}\right)$$

$$pH - pK_a = \log\left(\frac{[A^-]}{[HA]}\right)$$

$$10^{(pH - pKa)} = \frac{[A^-]}{[HA]}$$

$$10^{(3.74 - 4.74)} = \frac{[A^-]}{[HA]}$$

$$0.10 = \frac{[A^-]}{[HA]}$$

PROBLEM 20.16
If a researcher wanted to make a buffer with the following pH values, what acid/conjugate base pair(s) would be best?

The researcher will want to choose any acid/conjugate base pair whose pK_a is within one unit of the desired pH value.

a. pH 4.0 – citric acid/dihydrogencitrate (pK_a = 3.13), acetic acid/acetate (pK_a = 4.74), dihydrogencitrate/hydrogencitrate (pK_a = 4.76)

b. pH 6.5 – dihydrogenphosphate/hydrogenphosphate (pK_a = 7.20), hydrogencitrate/citrate (pK_a = 6.40)

c. pH 8.5 – dihydrogenphosphate/hydrogenphosphate (pK_a = 7.20), boric acid/ tetrahydroxyborate (pK_a = 9.20)

d. pH 2.5 – citric acid/dihydrogencitrate (pK_a = 3.13)

PROBLEM 20.17

A pyruvic acid ($K_a = 3.2 \times 10^{-3}$) buffer contains 0.50 mol/L pyruvic acid and 0.60 mol/L sodium pyruvate. What is the pH of the buffer?

$$pK_a = -\log K_a = -\log(3.2 \times 10^{-3}) = 2.49$$

$$pH = pK_a + \log\left(\frac{[A^-]}{[HA]}\right)$$

$$pH = 2.49 + \log\left(\frac{0.60\ \text{mol/L}}{0.50\ \text{mol/L}}\right) = 2.57$$

PROBLEM 20.18

A lactic acid ($K_a = 1.4 \times 10^{-4}$) buffer contains 0.15 mol/L each of lactic acid and sodium lactate.

a. What is the initial pH of the solution?

$$pK_a = -\log K_a = -\log(1.4 \times 10^{-4}) = 3.85$$

$$pH = pK_a + \log\left(\frac{[A^-]}{[HA]}\right)$$

$$pH = 3.85 + \log\left(\frac{0.15\ \text{mol/L}}{0.15\ \text{mol/L}}\right) = 3.85$$

b. If 0.050 mol/L HCl is added to the buffer solution, what will the new pH be?

If 0.050 mol/L HCl is added, then the concentration of lactic acid will go up by 0.050 mol/L:

[lactic acid] = 0.15 mol/L + 0.050 mol/L = 0.20 mol/L

And the concentration of lactate will decrease by 0.050 mol/L:

[lactate] = 0.15 mol/L - 0.050 mol/L = 0.10 mol/L

The new pH will be 3.54:

$$pH = 3.85 + \log\left(\frac{0.10\ \text{mol/L}}{0.20\ \text{mol/L}}\right) = 3.54$$

c. If 0.10 mol/L NaOH is added to the buffer solution, what will the new pH be?

If 0.10 mol/L NaOH is added, then the concentration of lactic acid will go down by 0.10 mol/L:

[lactic acid] = 0.15 mol/L – 0.10 mol/L = 0.05 mol/L

And the concentration of lactate will increase by 0.10 mol/L:

[lactate] = 0.15 mol/L + 0.10 mol/L = 0.25 mol/L

The new pH will be 4.55:

$$pH = 3.85 + \log\left(\frac{0.25\ \text{mol/L}}{0.05\ \text{mol/L}}\right) = 4.55$$

PROBLEM 20.19

A buffer is prepared with 0.50 mol NaH_2PO_4 and 0.30 mol Na_2HPO_4 in 0.500 L of solution.

a. Determine whether the solution can buffer the addition of 6.2 g KOH.

The initial buffer solution has a pH of 6.98.

$$pH = 7.20 + \log\left(\frac{(0.30\ \text{mol}/0.500\ \text{L})}{(0.50\ \text{mol}/0.500\ \text{L})}\right) = 6.98$$

The addition of 6.2 g of KOH corresponds to 0.11 mol KOH.

$$6.2 \text{ g KOH}\left(\frac{1 \text{ kg}}{1000 \text{ g}}\right)\left(\frac{1 \text{ mol}}{0.05611 \text{ g}}\right) = 0.11 \text{ mol KOH}$$

This is significantly fewer moles of base than there are moles of acid in the buffer. The buffer pH, then, will stay within the buffer range (1 pH unit of pK_a).

$$pH = 7.20 + \log\left(\frac{((0.30 \text{ mol} + 0.11 \text{ mol})/0.500 \text{ L})}{((0.50 \text{ mol} - 0.11 \text{ mol})/0.500 \text{ L})}\right) = 7.22$$

b. Determine whether the solution can buffer the addition of 46.0 mL of 6.0 mol/L HCl.

The initial buffer solution has a pH of 6.98.

$$pH = 7.20 + \log\left(\frac{(0.30 \text{ mol}/0.500 \text{ L})}{(0.50 \text{ mol}/0.500 \text{ L})}\right) = 6.98$$

The addition of 46.0 mL of 6.0 mol/L HCl corresponds to 0.11 mol KOH.

$$46.0 \text{ mL HCl}\left(\frac{L}{1000 \text{ mL}}\right)\left(\frac{6.0 \text{ mol}}{L}\right) = 0.28 \text{ mol HCl}$$

Given the added acid nearly matches the base, the new ratio of conjugate base to acid is 0.025, which means we have exceeded the buffer capacity and the pH will be outside the buffer range.

$$pH = 7.20 + \log\left(\frac{((0.30 \text{ mol} - 0.28 \text{ mol})/(0.500 \text{ L} + 0.046 \text{ L}))}{((0.50 \text{ mol} + 0.28 \text{ mol})/(0.500 \text{ L} + 0.046 \text{ L}))}\right) = 5.61$$

NOTES

1. Urey, H.C.; Murphy, G.M.; Brickwedde, F.G. A Name and Symbol for H^2. *J. Chem. Phys.*, **1933**, *1* (70), 512–513. DOI: 10.1063/1.1749326.

2. Bateman, L.C.; Hughes, E.D.; Ingold, C.K. 184. Mechanism of Substitution at a Saturated Carbon Atom. Part XIX. A Kinetic Demonstration of Unimolecular Solvolysis of Alkyl Halides. (Section A) Kinetics of, and Salt Effects in, the Hydrolysis of *tert.*-Butyl Bromide in Aqueous Acetone. *J. Chem. Soc. (Resumed)*. **1940**, 960–966. DOI: 10.1039/JR9400000960.

Index

Note: 'n' refers to chapter endnotes.

18-column periodic table, 19
32-column periodic table, 19

A

Absolute temperature, 120, 128
Absolute zero, 9, 162, 168, 311
Accident prevention, 2
Acid, 206
Acid–base catalysis and green chemistry, 210
Acid–base definitions, 206–208
Acidic conditions, 196–197
Acidity constants, 208
    values, 271–272
Acidity trends, 216–217
Acid speciation, 217–218
Actinoid, 18
Activation energy, 153–155
Activity, 180, 182, 216
Actual amount, 180
Actual charge, 99–100
Allen electronegativity, 83, 87
Allred–Rochow electronegativity, 83–84
α decay, 33–34
Alpha ladder, 47
Amonton's Law (Gay-Lussac's law), 124
Amount, 3, 11–12, 21, 41, 121–125, 129, 144, 146, 151
Amount concentration, 21, 146–153
Angular momentum quantum number, 64
Anion, 24, 80–81
Anode, 197–199
Antibonding, 103
Applying acidity constant values, 208–210
Areas of electron density, 101, 114–117
Argon atom, 119–120
Arrhenius, Svante, 206
Arrhenius acid-base theory, 206
Arrhenius equation, 153–154
Atom economy, 2, 154, 173, 202, 210
Atomic emission spectra, 60
Atomic number, 17, 23–25
Atomic orbital, 102–106, 115
Atomic radius, 78–80
Atomic size, 78, 97
Atomic symbol, 24–25
Atomic theory, 15–16
Atomic weapons, 51–52
Atomic weight, 16; *see also* Relative atomic mass
Atomism, 15
Atoms, 15, 23
Attraction, 76
Aufbau principle, 66, 68, 103
a value, 129, 137–138, 144
Avogadro constant, 21, 23, 124–125, 223, 277
Avogadro's law, 124–125

B

Base, 206
Basic conditions, 197
β decay, 34–39
β⁻ decay, 34–36
β⁺ decay, 34–36
Bidirectional arrows, 167–168
Big Bang, 45–46
Binding energy (E_b), 29–32, 49
Blackbody radiation, 56–58
Bohr, Niels I., 60–61
Bohr model, 60–61
Boiling point, 131–132
Bond angle, 118, 307
Bonding, 90, 138–139
    conjugation (VB theory), 107–108
    covalent bond, 94–95
        actual charge, 98–100
        bond length, 97
        contributing structures, 100–101
        formal charge, 97–99
        hypercoordinate molecules, 101
        Lewis structures, 95–97
    electron delocalization (MO theory),
        107–109
    ionic bond, 90–94
    quantum mechanical understanding, 102
        molecular orbital theory, 105–106
        valence bond theory, 102–105
    synopsis, 108–109
Bonding orbital, 103, 105, 108, 136
Bond length, 97
Born-Fajans-Haber cycle, 92–94
Bose-Einstein condensate, 141
Boyle's law, 122–123
Brønsted-Lowry acid-base theory, 206–208
Buffers, 218–220
b value, 128–129, 131–133, 135, 137–139, 144–145,
    315–316

C

Calorimeter, 159
Calorimetry, 159–160
Carbon burning, 47
Carbon-nitrogen-oxygen (CNO) cycle, 47
Catalysis, 2, 210
Catalysts, 154–155, 157
Cathode, 197–199
Cathode ray tube, 20–22
Cell potential at nonstandard conditions (*E*cell),
    202–203
Cell potentials, equilibria, and Gibbs energy values,
    199, 201–202
Chadwick, James, 24
Chain reaction, 50–51
Chalcogen, 80, 211

Charge, 76, 83, 211–212
Charged ion (anion), 90
Charged ion (cation), 90
Charles' law, 123
Chemical kinetics, 146
Chemistry, 24; *see also individual entries*
    energy sources, 12
    nomenclature reference, 256–257
    nuclear technology, 50–52
    nucleosynthesis, 45–48
Claim, evidence, and reasoning (CER), 6
Clarity, 2
Clausius–Clapeyron equation, 144
Closed system, 11
Common charge, 256
Condensation, 153
Conjugate acid, 206
Conjugate base, 206
Conjugation (VB theory), 107–108
Contributing structures, 100–101, 136
Conversion factors, 12, 223
Core electrons, 76
Coulomb constant, 76
Coulomb's law, 76–78, 90
Coupled reactions, 173–174
Covalent bond, 94–95
    actual charge, 99–100
    bond length, 97
    contributing structures, 100–101
    formal charge, 97–99
    hypercoordinate molecules, 101
    Lewis structures, 95–97
Covalent radius, 83
Critical point, 143
Crystal lattice, 92, 109
Crystallization (freezing), 40

D

Dalton, John, 15–16
Dalton's law, 125
Daniell cell, 198
Dashed bond, 113
d block, periodic table, 71–72, 287
de Broglie, Louis, 63, 75
Debye (D), 134
Debye force, 129
Decarbonize, 12, 44, 191
Decay chains, 39–41
Delocalization, 107–108, 114, 135, 212
Density, 141–142
    of gas, 119
Design for degradation, 2
Design for energy efficiency, 2
Designing safer chemicals, 2
Determining the favorability of a reaction, 170–171
Development of atomic theory, 15–16
Dipole, 129–130, 134, 315
Dipole-dipole (Keesom) interactions, 134–135
Dipole length, 134
Discovery of atomic structure, 20
    atoms, 23
    chemistry, 24

the electron, 20
the neutron, 24
the proton, 23–24
Dispersion (London) interaction, 129–135
Dissociation energy (E_d), 161–162
Dissociation values
    A-A bond dissociation energy values, 258
    A-H bond dissociation energy values, 257
    A-N bond dissociation energy values, 258
    A-O bond dissociation energy values, 258
    A-X bond dissociation energy values, 258
Diversity, 7n4–5
Donor-acceptor interaction, 107–108, 135
d orbital, 68–71

E

Effective nuclear charge, 76–77, 83
Effect of temperature on the favorability of a
        reaction, 171–172
Efficiency, 12, 154
Effusion, 125–126
Eigen cation, 206
Einstein, Albert, 51
Einstein hypothesis, 63
Electric field, 56–58
Electric potential, 9, 12
Electrochemical cells, 197–200
Electrochemistry
    and reaction control, 202
    and sustainability, 202–203
Electrode, 197, 199
Electrolytic cell, 197
Electromagnetic radiation, 56
Electromagnetic spectrum, 59
Electromagnetic wave, 56
    oscillating electric and magnetic fields, 57
Electron, 20, 22, 56, 58–61, 81
Electron affinity, 80–81, 83, 292
Electron configurations, 66–71
    and orbital energy diagrams
        electron configuration of each element,
            225–227
        ground state orbital energy diagram of
            elements hydrogen through krypton,
            228–255
    and periodic table, 71–72
Electron delocalization (MO theory), 105, 107
Electron-donating group (EDG), 211
Electronegativity, 81–87, 192, 211
Electronic stability, 72–74
Electrons and quantum numbers, 63
    electron configurations, 66–71
        and the periodic table, 71–72
    electronic stability, 72–74
    quantum mechanical energy levels, 64–66
    wave–particle duality, 63–64
Electron sharing, 206
    acid–base catalysis and green chemistry, 210
    acid–base definitions, 206–208
    acidity constants, 208
    applying acidity constant values, 208–210
    buffers, 218–220

Henderson–Hasselbalch equation and acid speciation, 217–218
insights from acidity constant values, 210–217
Electron transfer, 192
    cell potential at nonstandard conditions (Ecell), 202–203
    cell potentials, equilibria, and Gibbs energy values, 201–202
    electrochemical cells, 197–200
    electrochemistry
        and reaction control, 202
        and sustainability, 203–204
    half-reactions, 195–196
    oxidation numbers, 192–195
    redox reaction stoichiometry, 196–197
    reduction–oxidation, 195
Electronvolt (eV), 9
Electron-withdrawing group (EWG), 211
Electrostatic, 76, 138–139
Electrostatic potential, 99
    map, 133–134
Elementary charge, 22, 83
Elementary entities, 19
Endergonic, 44–45, 49
Endergonic reaction, 172–173
Endothermic, 161, 165
Energy, 9, 165
    kinetic energy, 9
    potential energy, 9
    sources, 12
    values, 161–162
English grammar, 6–7
Enrichment, 126
Ensemble, 119
Enthalpy, 21, 159
    and energy, 165
    experimental enthalpy, calorimetry, 159–160
    of formation, 162
    meaning of, 164
    of sublimation, 94, 143
    of vaporization, 144
    without direct experimentation, 160–161
        dissociation energy, 161–162
        Hess's Law, 163–164
        standard formation enthalpy and heat of formation, 162–163
Entropy, 167–168
    reaction notation: bidirectional arrows, 167–168
    second law of thermodynamics, 167
    third law of thermodynamics, 168–169
Equilibrium, 177
    constant, 180–181
    and Gibbs energy, 177–178
        importance of, 178–179
    initial, change, equilibrium (ICE), 183–185
        finding the equilibrium constant, 183–184
    measured at, 181
    perturbing equilibria, 185–190
        adding/removing chemicals, constant pressure and temperature, 187
        adding/removing chemicals, constant volume and temperature, 186–187

changing pressure/volume of a gas-phase reaction, constant temperature, 187–188
        changing temperature, 188–191
        changing volume of a solution-phase reaction at constant temperature, 188
Equilibrium constant, 177–178, 201–202
Equity, 7
Excited state, 7n4–5
Exergonic reactions, 44, 172–173
Exothermic reaction, 165, 170, 171

F

Fahrenheit vs. Celsius vs. Kelvin temperature scales, 10
Faraday constant, 201
f block, periodic table, 71–72, 79, 287, 292
Fermi, Enrico, 44, 51
First law of thermodynamics, 12–13, 167
First order (unimolecular), 150
First-order phase transition, 142
First-order reaction, 151
Fission, 48–50
Fissionable, 50
Fluorine, 132, 136
f orbital, 66, 68, 70, 286
Formal charge, 97–99
Fraction ionic character, 90, 93
Free neutrons, 46
Freezing, 153
Frequency, 56–60
Fusion, 44–45

G

Galvanic cell, 197–198
γ decay, 39
γ ray, 47–48, 59
Gas/gases, 119
    constant, 119–120, 144–145
    ideal gas law, 121–122
    mixtures, 125–126
    particles, 119, 128
    particles and energy, 119–121
    properties, 141
    specific relationships, the gas laws, 122–125
Gay-Lussac's law, 123
Geiger, Hans, 23
General Chemistry (Deming), 18
Geometry, 117
Gibbs, Josiah Willard, 174–176
Gibbs energy, 170
    and coupled reactions, 173–174
    determining the favorability of a reaction, 170–171
    effect of temperature on the favorability of a reaction, 171–172
    and reaction analysis, 172–173
    and reaction control, 174–176
    standard energy of formation, 172
Graham's law, 125

Green chemistry (↝), 1, 12, 15, 44, 59, 154, 157, 165, 173, 176, 189, 190, 202–203, 210
   12 principles, 1–2
Ground state, 39, 68, 107, 228
Ground-state electron energy
   carbon, 68
   helium, 67
   hydrogen, 67
   iron, 69
   lithium, 67
Group (periodic table), 18–19, 78–79

H

Haber-Bosch process, 189
Half-cell reactions and standard reduction potential values, 268–271
   acidic half-cell reaction, 269–270
   basic half-cell reaction, 270–271
Half life, 41
Half-reactions, 195–196
Halogen, 97, 128–129, 194, 294
Hazardous chemical syntheses, 2
Heat, 11
Heat capacity, 11, 159
Heat of formation, 162–163; *see also* Standard formation enthalpy
Heisenberg, Werner, 41n1
Heisenberg uncertainty principle, 63, 68
Helium, 67
Helium fusion, 47
Henderson–Hasselbalch equation, 217–218
Hess's Law, 163–164
High temperature, 171
Hiroshima and Nagasaki, Japan, 51–52
Hund's rule, 103
Hybridization, 103, 115–117
Hydration of ethylene, 170, 177
Hydrogen atoms, 113
Hydrogen bonds (H-bonds), 135–138
Hydron, 206–208
Hydronium, *see* Oxidanium
Hydroxide, 96
Hypercoordinate molecules, 101
Hypervalent, 101–102

I

Ideal gas law, 121–122, 128
Importance of intermolecular forces, 138
Inclusion, 1
Infrared (IR), 59
Initial change equilibrium (ICE), 183
   finding the equilibrium constant, 183–184
Initial rate, 147–150
Insights from acidity constant values, 210–217
Integrated rate law, 152–153
Intermediate, 157
Intermolecular force (IMF), 139–140
Intermolecular interactions, 138–139
Internal energy (U), 9, 13, 120, 121, 142, 165, 167, 274
International system of units (SI), 5–6
Internuclear distance, 92, 94

Iodomethane, 175–176
Ion, 80, 90, 192, 203
Ionic bond, 90–95
Ionic liquid, 94, 295
Ionization energy, 80–81, 83, 103
Isolated system, 11
Isotone, 31
Isotope, 25
IUPAC nomenclature, 256–257

J

Joule, 9

K

Kilonova, 54n23
Kinetics, 146
   energy, 9
   molecular theory, 128
Krypton, 101
   difluoride, 108

L

σ^* orbital, 103, 305–306
σ-orbital mixing
   dihydrogen, 102
   hydrogen fluoride, 104
σ orbitals (σ bond), 102–105, 108, 305–306
Lanthanoid, 18
Lattice enthalpy, 92–94
Lavoisier, 15
Law of conservation of mass, 15–16
Law of constant proportions, 16
Law of multiple proportions, 15–16
Le Chatelier, 185
Lewis acid–base theory, 206
Lewis dot symbols, 90–91
Lewis structures, 95–97, 113–114, 133–134, 210–213
Light, 12
   and electrons, 56
      atomic emission spectra, 60
      blackbody radiation, 56–58
      Bohr model, 60–61
      photoelectric effect, 58–59
      photovoltaic effect, 59–60
Linear geometry, 117
Liquids, 119–120, 141
Lithium, 67
London dispersion interactions, 129–133
Long-range order, 141
Lowest energy orbital, 108
Low temperature, 171

M

Macroscopic kinetics, 146
   catalysts, 154–155
   the integrated rate law, 152–153
   order dependence and rate law, 148–150
   rate constant and temperature dependence, 153–154

rate of reaction, 146–148
reaction order, 150–152
Magic number, 35, 71–72
Magnetic field, 56–58
Magnetic quantum number, 65
Main group elements, 18, 20
Manhattan Project, 51
Marsden, Ernest, 23
Mass defect, 29
Mass-energy equivalence, 30
Mass number, 24–25
Matter, 15
Maxwell-Boltzmann distribution, 121
Mean free path, 119
Mean kinetic energy, 120–121
Measurement, 2–3
Mendeleev, Dmitri, 17–18
Mendeleev's first periodic table (1869), 17
Mendeleev's revised periodic table (1871), 17
Methane, 113–114
Methanol, 175–176
Metric, 206
Microscopic kinetics, 146, 155–157
    catalysts, 157
Microwave, 59
Millikan, Robert, 22
Miscibility, 138
Modern chemistry, 15
Modern periodic table, 18–19
Molar mass, 20–21
Molar volume, 21
Mole (mol), 21, 159–160, 179
Molecular motion, 162
Molecular orbital theory, 105–106
Molecular shapes, 113–118
Molecules, 19
Momentum, 63–66, 286
Morse potential, 92
Mulliken electronegativity, 81, 83

N

Natural population analysis (NPA), 99
Neon burning, 48
Nernst equation, 202–203
Neutron, 24–25, 29, 31–32, 34–37, 44, 279–280
Noble gases, 18, 72–73
Noble gas notation, 67, 90
Nonbonding, 102, 108
Noncovalent interaction, 135
Noncovalent (Van der Waals) interactions, 129
    dipole–dipole (Keesom) interactions, 134–135
    (London) dispersion interactions, 129–133
Nonpolar, 134, 314, 316
Nonstandard cell potential (Ecell), 202–203
Nonstandard conditions, 202–203
Nuclear, 138–139
Nuclear chain reaction, 50
Nuclear charge, 23–24
Nuclear power, 50
Nuclear reactions, 44
    fusion, 44–45
Nuclear shell, 35–36

Nuclear technology, 50–52
Nuclear weapon, 50, 52
Nuclei, 29
Nucleon number, 25, 30, 280
Nucleosynthesis, 45–48
Nucleus, 23–24
Nuclide, 29–34
Nuclide transmutation, 39–41

O

Octahedral, 115, 307
Octet rule, 72
Open system, 11
Orbital, 65–70
Orbital images and energy values
    aluminium, 234
    argon, 239
    arsenic, 252
    beryllium, 229
    boron, 229
    bromine, 254
    calcium, 240
    carbon, 230
    chlorine, 238
    chromium, 244
    cobalt, 247
    copper, 248
    fluorine, 231
    gallium, 250
    germanium, 251
    helium, 228
    hydrogen, 228
    iron, 246
    krypton, 255
    lithium, 228
    magnesium, 233
    manganese, 245
    neon, 232
    nickel, 248
    nitrogen, 230
    oxygen, 231
    phosphorus, 236
    potassium, 239
    scandium, 241
    selenium, 253
    silicon, 235
    sodium, 232
    sulfur, 237
    titanium, 242
    vanadium, 243
    zinc, 249
Orbital mixing, 114
Order dependence, 148–150
Oxidanium, 96, 206–207
Oxidation numbers, 192–195, 206
Oxygen burning, 48

P

Partial charge, 140n6
Partial orbital energy diagram, 65–66
Partial pressure, 125

Particle-particle interaction, 119
Pauli exclusion principle, 103
Pauling, Linus, 82–83
Pauling electronegativity, 83
Pauling's scale, 83
p block, 71–72, 287
Periodic table, 16–20, 71–72
    common charge, 73
    d block, 71–72
    f block, 71–72
    p block, 72
    s block, 71–72
Periodic trends, 76
    atomic radius, 78–80
    Coulomb's law, 76–78
    electronegativity, 81–87
    ionization energy and electron affinity, 80–81
Perturbing equilibria, 185–190
pH, 185, 199, 216–220
Phase, 145
Phase changes, 141–144
Photoelectric effect, 59–60
Photon, 58–59
Photovoltaic effect, 59–60
Physical change, 145
Physical constants and conversion factors, 223
π orbital (π bond), 103, 105, 107, 305
π-orbital mixing, ethene, 105
pK_a, 209, 217–219
pK_a values, 209
Planck, Max, 56–58, 63
Planck constant, 57
Plasma, 141
Polar, 95, 134, 314–315
Polar bond, 134
Polar covalent bond, 95
Polarizability, 131–132
p orbital, 65, 68, 70, 83, 103, 105, 107, 108, 115, 117, 305
p-orbital mixing, 103
Position, 63–64
Potential energy, 9
    diagram of, rock on a cliff *vs.* valley floor, 10
Precision, 3–4, 15, 22
Pre-exponential factor, 153
Prefixes, 5–6
Pressure, 119
Prevention, 2
Principal quantum number, 64–65
Product-favored process, 167
Proposed reaction coordinate diagram, 156
Proton, 23–24
Proton-proton chain (p-p chain), 47
Prout, William, 20

Q

Quanta, 57–58
Quantum mechanical energy levels, 64–66
Quantum mechanical understanding of bonding, 102
    molecular orbital theory, 105–106
    valence bond theory, 102–105
Quantum mechanics, 35
Quantum number, 65–67

R

Radical electron, 97
Radioactive decay, 32–33
    nuclide transmutation summary and decay
        chains, 39–41
    α decay, 33–34
    β decay, 34–39
    γ decay, 39
Radioactivity, 39
Radio waves, 59
Rare earth metal, 18, 20
Rate constant, 153–154
Rate law, 148–150
Rate of reaction, 146–148, 152
Reactant-favored process, 167
Reaction analysis, 172–173
Reaction control, 174–176
Reaction control, Perturbing Equilibria,
        185–190
    adding/removing chemicals, reaction at constant
        pressure and temperature, 187
    adding/removing chemicals, reaction at constant
        volume and temperature, 186
    scenario 1, adding reactant, 186
    scenario 2, removing reactant, 186
    scenario 3, removing product, 186
    scenario 4, adding product, 187
    changing pressure/volume of a gas-phase
        reaction at constant temperature,
        187–188
    changing temperature, 188–191
    changing volume of a solution-phase reaction at
        constant temperature, 188
Reaction coordinate diagram, 155–157
Reaction notation, bidirectional arrows, 167–168
Reaction order, 150–152
Reaction quotient, 179–180
Real-time analysis, pollution prevention, 2
Redox, 90, 192
Redox reaction stoichiometry, 196–197
Reduce derivatives, 2
Reduction–oxidation, 195
Relationships of gas laws, 122–125
Relative atomic mass, 16–20, 23–24
Resonance, *see* Contributing structure;
        Delocalization
Root mean square velocity, 120–121
r-process, 48
Rutherford. Ernest, 20, 23–24
Rydberg, Johannes Robert, 60
Rydberg constant, 60

S

Safer solvents and auxiliaries, 2
Safety, 41, 176, 334
Salt bridge, 197–198
s block, periodic table, 71–72
Seaborg, Glenn T., 18
Second law of thermodynamics, 167
Second order (bimolecular), 152
Second-order reaction, 152–153

Selected thermochemical data
  aluminium, 259
  antimony, 267
  argon, 259
  arsenic, 259
  barium, 260
  beryllium, 260
  bismuth, 260
  boron, 259
  bromine, 260
  calcium, 261
  carbon, 260
  chlorine, 261
  chromium, 262
  cobalt, 262
  copper, 262
  fluorine, 262
  gold, 259
  helium, 263
  hydrogen, 263
  iodine, 263
  iron, 262
  krypton, 264
  lead, 266
  lithium, 264
  magnesium, 264
  manganese, 265
  mercury, 263
  neon, 265
  nickel, 266
  nitrogen, 265
  osmium, 266
  oxygen, 266
  phosphorus, 266
  potassium, 264
  silicon, 267
  silver, 259
  sodium, 265
  sulfur, 267
  tin, 267
  titanium, 268
  uranium, 268
  xenon, 268
  zinc, 268
Shell, 65–67, 76–78
Shielding, 79
Shielding constant, 76
Significant figures
  addition and subtraction, 5
  all leading zeros, 4
  all nonzero digits, 3–5
  all sandwiched zeros, 4
  exponential calculations, 5
  logarithmic calculations, 5
  multiplication and division, 4
  trailing zeros, 4
Silicon burning, 48
Solar system abundances, 46, 48
  of elements, 46
  nuclides, 48
Solid, 119, 141
Solution, 56, 188
s orbital, 65, 83, 103, 115

sp, 103, 116–117
sp-hybridized carbon atoms, 117
sp^2, 116–117, 306
sp^2-hybridized carbon atoms, 113, 116–117
sp^3, 115–117, 213, 306
sp^3-hybridized carbon atoms, 113, 116, 117
Spallation, 48–49
Speciation, 217–219
Specific heat capacity, 160
Spectroscopy, 56
Speed of light, 56
sp hybridized carbon atoms, 116–117
Spin quantum number, 65
Spontaneous, 170
s-process, 48–49
Stabilization, 92, 102, 136, 305
Stable, 29–36, 72, 90, 108
Stable electron configurations, 72
Standard atomic weight, 224–225
Standard enthalpy of formation, 162, 163, 165,
    168, 172
Standard formation enthalpy, 162–163
Standard Gibbs energy of formation, 172
Standard molar entropy, 168, 172
Standard reduction potential, 199, 208
Standard state, 199, 203
Standard temperature (ST), 10
State of aggregation, 141
State of matter, 141
State *vs.* phase, 141
Stelliferous Era, 46
Sublimation, 94, 143
Subshell, 65–67
Supercritical fluid (SCF), 141, 144–145
Superfluid, 141
Surroundings, 10–11
Sustainability, 1
Synthetic electrochemistry, 202
Systems, 10–11
Szilard, Leo, 50–51

T

Temperature, 9, 119–124, 160
Temperature dependence, 153–154
Tetrahedral, 115–117, 212–213, 306–307
Tetrahedral atom, 115–116
Thermal energy, 11, 170
Thermochemical data, 259
Thermodynamics, 178–179
Third law of thermodynamics, 168–169
Third-order (termolecular), 152
Thomson, Joseph John, 21–23
Three-dimensional model, 113
Three-dimensional shape, 113
Time course, 146–149
Tin–gold cell, 201
Transition metal, 18–20
Transition state, 155–156
Trigonal bipyramidal, 115, 307–308
Trigonal planar, 115–117
Triple alpha process, 47
Triple point, 143

U

Ultraviolet (UV), 59
Uncertainty principle, 63
Units, 5–6
Unstable, 29, 32, 36
Unstable electron configurations, 72
Uranium, 49–50
Uranium-235
    diagrammatic representation,
        50–51
    fission fragment, 49
Use of renewable feedstocks, 2
U.S. Strategic Bombing Survey, 52

V

Valence, 67–68
    bond theory, 102–105
    electrons, 76
Valence shell electron pair repulsion (VSEPR), 114,
    115
van der Waals (vdW), 80
    diameter, 119
    equation, 144–145
    interactions, 129, 138, 139, 142, 314, 317
    radius, 63, 78–80, 119

van der Waals equation and intermolecular forces,
        128–129
van't Hoff equation, 188
Vaporization, 143–144, 317
Visible light, 56
Voltaic cell, 202
Volume, 120–124

W

Waste, 2, 15, 334
Water, 137–138
Wavelength, 56–60
Wave–particle duality, 63–64
Wedge bond, 113
Work, 12
World War II, 50, 52
Writing, 6–7

X

X rays, 59, 62n9

Z

Zeroth order, 150–151
Zundel cation, 206

Printed in the United States
by Baker & Taylor Publisher Services